Methods for the Summation of Series

Discrete Mathematics and Its Applications
Series editors:
Miklos Bona, Donald L. Kreher, Douglas B. West

Handbook of Geometric Constraint Systems Principles
Edited by Meera Sitharam, Audrey St. John, Jessica Sidman

Introduction to Chemical Graph Theory
Stephan Wagner, Hua Wang

Extremal Finite Set Theory
Daniel Gerbner, Balazs Patkos

The Mathematics of Chip-Firing
Caroline J. Klivans

Computational Complexity of Counting and Sampling
Istvan Miklos

Volumetric Discrete Geometry
Karoly Bezdek, Zsolt Langi

The Art of Proving Binomial Identities
Michael Z. Spivey

Combinatorics and Number Theory of Counting Sequences
Istvan Mezo

Applied Mathematical Modeling
A Multidisciplinary Approach
Douglas R. Shier, K.T. Wallenius

Analytic Combinatorics
A Multidimensional Approach
Marni Mishna

50 years of Combinatorics, Graph Theory, and Computing
Edited by Fan Chung, Ron Graham, Frederick Hoffman, Ronald C. Mullin, Leslie Hogben, Douglas B. West

Fundamentals of Ramsey Theory
Aaron Robertson

Methods for the Summation of Series
Tian-Xiao He

https://www.routledge.com/Discrete-Mathematics-and-Its-Applications/book-series/CHDISMTHAPP

Methods for the Summation of Series

Tian-Xiao He

CRC Press is an imprint of the
Taylor & Francis Group, an **informa** business
A CHAPMAN & HALL BOOK

First edition published 2022
by CRC Press
6000 Broken Sound Parkway NW, Suite 300, Boca Raton, FL 33487-2742

and by CRC Press
2 Park Square, Milton Park, Abingdon, Oxon, OX14 4RN

CRC Press is an imprint of Taylor & Francis Group, LLC

ISBN: 978-0-367-50797-8 (hbk)
ISBN: 978-1-032-19500-1 (pbk)
ISBN: 978-1-003-05130-5 (ebk)

DOI: 10.1201/9781003051305

Publisher's note: This book has been prepared from camera-ready copy provided by the authors

Contents

To Yulan
To Calvin and Viola

Foreword

You are about to encounter a very special book. Summing series has been of interest for centuries, and, in an age of powerful computers, the interest has greatly intensified. Keopf's Hypergeometric Summations, The Concrete Tetrahedron by Kauers and Paule, and A=B by Petkovsek, Wilf, and Zeilberger are all impressive works devoted to this topic.

So why do we need, Methods for the Summation of Series, ostensibly devoted to the same subject? Let us begin by noting the background that the author Tian-Xiao He (Earl and Marian A. Beling Professor of Natural Sciences and Professor of Mathematics, Illinois Wesleyan University) brings to this effort. He has done important work in numerical analysis, wavelet analysis, approximation theory, and splines. These interests have led him naturally into enumerative combinatorics and the emerging field of Riordan Arrays. This diversity of interests is on full display in this book. It would be fair to say that this volume combines the charm of an ancient book like I. J. Schwatt's, An Introduction to the Operations with Series (1924), with a keen awareness of the many aspects of the most recent methods developed for the summation of series. The advantage of this mixture is that insight and context are provided for many applications.

The five chapters of this book provide a clear view of the depth of vision. The first chapter is devoted to classical methods, which, while they date back to the 19th century and before, are nonetheless effective and always timely. Symbolic methods occupy the next two chapters. This, too, is a venerable subject dating back to invariant theory; its modern combinatorial manifestations were pioneered by Gian-Carlo Rota. This is a compelling way to place the classic theory of finite differences in a modern and substantially more powerful setting.

Chapter 4 moves to the world of special functions. Of particular interest is the extensive use of Riordan Arrays, a topic in which Professor He is one of the world leaders. This is, indeed, one of the highlights of this book. The final chapter continues to build on Riordan Arrays and concludes with an account of some of the algorithms that have been so successful in doing summations via computer algebra.

This is a well-written, lucid book with many surprising gems. I am happy to recommend it to you as a valuable addition to your library.

George E. Andrews
Evan Pugh University Professor in Mathematics
Member, National Academy of Sciences (USA)
Past President, American Mathematical Society

Testimonial

In the past three months, I really enjoyed reading through the book. It is a very good monograph and text and offering an overview of several valuable techniques, and readers will find it to be a very fine reference book as well as one from which to study. I certainly give it my highest recommendation. The author presented very impressive publications and research activities.

Henry Wadsworth Gould
Professor Emeritus of Mathematics
West Virginia University, Morgantown
Fellow of the American Association for the Advancement of Science
Honorary Fellow of the Institute of Combinatorics and its Applications
July 11, 2021

Preface

This book presents methods for the summation of infinite and finite series and the related identities and inversion relations. The summation includes the column sums and row sums of lower triangular matrices. The convergence of the summation of infinite series is considered. We focus on symbolic methods and the Riordan array approach for the summation. Much of the materials in this book have never appeared before in textbook form. This book can be used as a suitable textbook for advanced courses for higher-level undergraduate and lower level graduate students. It is also an introductory self-study book for researchers interested in this field, while some materials of the book can be used as a portal for further research. In addition, this book contains hundreds of summation formulas and identities, which can be used as a handbook for people working in computer science, applied mathematics, and computational mathematics, particularly, combinatorics, computational discrete mathematics, and computational number theory. The exercises at the end of each chapter help deepen understanding.

Since the methods discussed in this book are related to the classical summation methods, we present the main classical methods in Chapter 1 with the example oriented way. This chapter provides useful supplementary materials for the people who study advanced Calculus, and training materials for the people who study applied and computational mathematics.

The *infinitesimal calculus* or differential and integral calculus is a field to treat functions of continuous independent variables. The methods to find summation of series by using infinitesimal calculus shall be surveyed Chapter 1. We will introduce one by one the following five simple and common methods: (1) Substitution method; (2) Telescoping method; (3) Method of the summation of trigonometric series; (4) Differentiation and integration method for uniformly convergent series; and (5) Abel's summation.

As is well known, the closed form representation of series has been studied extensively. It is also known that the symbolic calculus with operators Δ (difference), E (shift or displacement), and D (derivative) plays an important role in the Calculus of Finite Differences, which is often employed by statisticians and numerical analysts. The object of Chapter 2 is to make use of the classical operators Δ, E, and D to develop closed forms for the summation of power series that appear to have a certain wide scope of applications. Throughout this chapter the theory of formal power series and of differential operators will be utilized, while the convergence of the infinite series is discussed. In this chapter, we focus on the summation and identities arising from the interrelations of a number of operators in common use in combinatorics, number theory, and discrete mathematics. Various well-known results can also be found in some classical treatises in this chapter. Since all the symbolic expressions used and operated in the calculus could be formally expressed as power series in Δ (or D or E) over the real or complex number field, it is clear that the theoretical basis of the calculus may be found within the general theory of the formal power series. Worth reading is a sketch of the theory of formal series that has been given briefly in Chapter 2.

Chapter 3 presents a frame work with several source formulas, from which numerous summation formulas and identities are constructed. This frame work is due to a general substitution rule, called Mullin-Rota's substitution rule. Given a generating function or a formal power series expansion, then a certain operational formula may be obtained by using the substitution rule. Some operator summation formulas from multifold convolutions are also obtained similarly. With the aid of Mullin-Rota's substitution rule, we shall show in this chapter that the Sheffer-type differential operators together with the delta operators Δ and D could be used to construct a pair of expansion formulas that imply a wide variety of summation formulas in the discrete analysis and combinatorics. A convergence theorem is established for fruitful source formulas. Numerous new formulas are represented as illustrative examples. A kind of lifting process is used to enlarge the number of new formulas. In addition, this chapter presents further investigation on a general source formula (GSF) that has been proved capable of deducing numerous classical and new formulas for series expansions and summations besides those given in the previous parts of the book.

In the first half of Chapter 4, we shall continue the symbolic process for some special function sequences and number sequences such as the sequences of Bernoulli polynomials and numbers, Stirling numbers, Fibonacci numbers, etc. In the second half of this chapter, we construct identities and summation formulas for the function sequences and number sequences related to Riordan arrays. A Riordan array is an infinite lower triangular matrix, which columns are multiplication of certain power series g and f. The theory of Riordan arrays provides a modern method for classical umbra calculus, bringing new insights into many areas of combinatorial importance. This chapter gives an introduction to the basic and applicable materials on Riordan arrays and the Riordan group for students and researchers, who seek novel ways of working in fields such as combinatorial identities, triangles for enumerating combinatorial numbers, special polynomial sequences, orthogonal polynomials, etc.

In Chapter 5, we shall present the methods extended from the previous chapters for constructing various summation formulas, identities, and inversion relations. The formulas constructed in this chapter include the identities of high dimensions, convolution-type, Abel-type, and those related to Bernoulli polynomials and numbers, Euler polynomials and numbers, etc. Some methods represented in this chapter are related to generalized Riordan arrays and generalized Riordan groups with different bases and their Sheffer analogs, Sheffer-type polynomial sequences and the Sheffer group. Some identities and inversion relations are constructed by using dual sequences with the Riordan array representation named pseudo-Riordan involutions. Finally, an extension of W-Z algorithm and Zeilberger's creative telescoping algorithm is represented and used to construct and prove the identities for Bernoulli polynomials and numbers.

I am grateful to Professor Leetsch C. Hsu (Xu Lizhi) for guiding me into the field of enumeration combinatorics. This book is dedicated to the memory of him. The author would like to thank Professor George Andrews for his foreword and Professor Henry Gould for his testimonial and both of them for their comments and encouragements. I would like to thank all the collaborators who have published joint papers with me in the fields of symbolic methods and/or Riordan arrays over the past decades, especially Leetsch C. Hsu, Louis W. Shapiro, Renzo Sprugnoli, Henry W. Gould, and Peter J.-S. Shiue for the pleasant cooperation with them and everything I have learned from them. I would like to thank the Editors of the Discrete Mathematics and Its Applications Series, CRC Press/Taylor & Francis Group, LLC, and KGL, specially Miklos Bona, Robert Ross, Vaishali Singh, and Manisha Singh for their help and patience in the process. The author is thankful for the support given by the Earl and Marian A. Beling Professor's Fund.

Tian-Xiao He
Illinois Wesleyan University
Bloomington, IL

Biography

Tian-Xiao He received Ph.D. degrees at the Dalian University of Technology, Dalian, and Texas A&M University, College Station, respectively. Dr. He is a Professor of Mathematics and Earl and Marian A. Beling Professor of Natural Science at Illinois Wesleyan University. Dr. He has authored or co-authored over 150 research articles and 7 volumes in mathematics and is presently an editor/chief editor for several math journals.

Symbols

Symbol	Description
$\mathbb{N}$	Natural number set
$\mathbb{N}_0$	The set of natural numbers and zero
$\mathbb{Z}$	The ring of integers
$\mathbb{R}$	The field of real numbers
$\mathbb{C}$	The field of complex numbers
$\mathcal{F} = \mathbb{K}[\![t]\!]$	The ring of formal power series in variable t over a field $\mathbb{K}$.
$\mathcal{F}_r$	The set of formal power series of order r
Δ	Difference
E	Displacement
D	Derivative
(g, f)	The Riordan array generated by $g(t) \in \mathcal{F}_0$, $f(t) \in \mathcal{F}_1$
$\left[{n \atop k}\right]$	Unsigned Stirling numbers of the first kind
$\left\{{n \atop k}\right\}$	Stirling numbers of the second kind
$S(n, k)$	Generalized Stirling numbers
$C(t)$	The generating function of the Catalan numbers
$F_m(t)$	The generating function of the Fuss-Catalan numbers

1

Classical Methods from Infinitesimal Calculus

CONTENTS

Since the methods discussed in this book are related to the classical summation methods, we start this book from main classical methods for summation of series described in an example-oriented way.

The *infinitesimal calculus* or differential and integral calculus is a field to treat functions of continuous independent variables, i.e., the variables may take every possible value in a given interval. The methods to find the summation of series by using infinitesimal calculus shall be surveyed in this chapter in the example-oriented way.

The summation of a series $\sum_{n\geq 1} a_n$ is defined by

$$\sum_{n\geq 1} a_n = \lim_{n\to\infty} s_n \equiv \lim_{n\to\infty} \sum_{k=1}^{n} a_k,$$

where $s_n = \sum_{k=1}^{n} a_k$ is the partial sum of the series. On the summation of infinite series, there are following five simple and common methods.

(1) Substitution method. To get the sum of $\sum_{n\geq 0} u_n$, we may substitute $u_n = a_n k^n$ that brings the sum $f(k) = \sum_{n\geq 0} u_n$ if a known function $f(x) = \sum_{n\geq 0} a_n x^n$ can be determined.

(2) Telescoping method. In a given series $\sum_{n\geq 0} u_n$, if we have $u_n = v_n - v_{n+1}$ $(n = 0, 1, \ldots)$ and $\lim_{n\to\infty} v_n = v_\infty < \infty$, then $\sum_{n\geq 0} u_n =$

DOI: 10.1201/9781003051305-1

$v_0 - v_\infty$. In particular, if

$$u_n = \frac{1}{a_n a_{n+1} \cdots a_{n+m}},$$

where $a_k = c + kd$ $(k = 1, 2, \ldots)$, $c, d \in \mathbb{R}$, $d \neq 0$, then

$$v_n = \frac{1}{md} \frac{1}{a_n a_{n+1} \cdots a_{n+m-1}}.$$

(3) Trigonometric series summation. In order to evaluate the sums of

$$\sum_{n \geq 0} a_n \cos(nx) \text{ and } \sum_{n \geq 0} a_n \sin(nx),$$

we consider them as the real part and the imaginary part of power series

$$\sum_{n \geq 0} a_n z^n,$$

where $z = e^{-ix}$, which can be summarized. In many cases, summation

$$\sum_{n \geq 0} \frac{1}{n} z^n = \log \frac{1}{1-z} \quad (|z| < 1)$$

is useful to find the sum of the power series.

(4) Differentiation and integration method for uniformly convergent series. Namely, we may transfer a given uniform convergent series to a new series by differentiation or a suitable integration that can be summed, and the sum of original series will be found by taking an inverse transformation, i.e., integration or differentiation.

(5) Abel's summation. One may see Subsection 1.2.2 for details.

Some examples shown in this chapter are selected from or greatly influenced by [1, 44, 51, 73, 79, 84, 102, 123, 135, 139, 141, 143, 164, 165, 177, 183, 191, 200].

1.1 Use of Infinitesimal Calculus

1.1.1 Convergence of series

Example 1.1.1 *Here is an example of the first method. Since*

$$\sum_{n \geq 1} \frac{(-1)^{n-1}}{n} = \ln 2, \tag{1.1}$$

we immediately have

$$\sum_{n \geq 1} \frac{(-1)^{n-1}}{n(n+1)} = \sum_{n \geq 1} (-1)^{n-1} \left(\frac{1}{n} - \frac{1}{n+1} \right) = 2 \sum_{n \geq 1} \frac{(-1)^{n-1}}{n} - 1 = 2 \ln 2 - 1.$$

A few examples for the second method dealing with the so-called *telescoping series* are shown below.

Example 1.1.2 *Since*

$$\frac{1}{n(n+1)(n+2)} = \frac{1}{2}\left(\frac{1}{n(n+1)} - \frac{1}{(n+1)(n+2)}\right),$$

we obtain

$$\sum_{n\geq 1} \frac{1}{n(n+1)(n+2)} = \frac{1}{4}.$$

For $m \in \mathbb{N}$, we have

$$\frac{1}{n(n+m)} = \frac{1}{m}\left(\frac{1}{n} - \frac{1}{n+m}\right).$$

Thus,

$$\sum_{n\geq 1} \frac{1}{n(n+m)} = \frac{1}{m}\left(1 + \frac{1}{2} + \ldots + \frac{1}{m}\right).$$

Similarly,

$$\sum_{n\geq 1} \frac{2n+1}{n^2(n+1)^2} = \lim_{\ell\to\infty} \sum_{n=1}^{\ell} \left(\frac{1}{n^2} - \frac{1}{(n+1)^2}\right) = 1.$$

Telescoping method can also be used for the following triangular function series.

$$\sum_{n\geq 1} \tan^{-1}\left(\frac{1}{n^2+n+1}\right) = \frac{\pi}{4}.$$

In fact,

$$\begin{aligned}
\sum_{n\geq 1} \tan^{-1}\left(\frac{1}{n^2+n+1}\right) &= \lim_{\ell\to\infty} \sum_{n=1}^{\ell} [\tan^{-1}(n+1) - \tan^{-1} n] \\
= \quad & \lim_{\ell\to\infty} [\tan^{-1}(\ell+1) - \tan^{-1} 1] = \frac{\pi}{4}.
\end{aligned}$$

Similarly,

$$\sum_{n\geq 1} \tan^{-1}\left(\frac{2n+1}{n^2(n+1)^2}\right) = \frac{\pi}{4}.$$

Some series can be transferred to telescoping series as shown below.

Example 1.1.3

$$\begin{aligned}\sum_{n\geq 1}\frac{1}{n^2(n+1)^2} &= \sum_{n\geq 1}\left(\frac{1}{n}-\frac{1}{n+1}\right)^2\\ &= \sum_{n\geq 1}\left(\frac{1}{n^2}-\frac{2}{n(n+1)}+\frac{1}{(n+1)^2}\right)\\ &= \frac{\pi^2}{3}-3,\end{aligned}$$

where we use $\sum_{n\geq 1} 1/n^2 = \pi^2/6$, *and the telescoping series* $\sum_{n\geq 1} 1/n\,(n+1) = 1$.

Similarly,

$$\begin{aligned}&\sum_{n\geq 1}\frac{n}{(n+1)(n+2)(n+3)}\\ &= \sum_{n\geq 1}\frac{1}{(n+2)(n+3)}-\sum_{n\geq 1}\frac{1}{(n+1)(n+2)(n+3)}\\ &= \frac{1}{3}-\frac{1}{12}=\frac{1}{4}.\end{aligned}$$

Example 1.1.4 *Using the method of differentiation term by term to the following series in their uniform convergence intervals,* $|x|<1$, *we obtain*

$$\begin{aligned}\sum_{n\geq 1} n^2x^{n-1} &= \left(\sum_{n\geq 1} x^{n+1}\right)'' - \left(\sum_{n\geq 1} x^n\right)'\\ &= \left(\frac{1}{1-x}-1-x\right)'' - \left(\frac{1}{1-x}-1\right)' = \frac{1+x}{(1-x)^3}.\end{aligned}$$

Similarly,

$$\sum_{n\geq 1} n(n+2)x^n = \frac{x(3-x)}{(1-x)^3}\ (|x|<1) \text{ and}$$

$$\sum_{n\geq 1}\frac{2n+1}{n!}x^{2n} = (1+2x^2)e^{x^2},$$

where the Taylor's series of $1/(1-x)$ *and* e^x *are applied.*

Some techniques shown below are useful in applying differentiation method. Consider the summation of the following series

$$y(x) := \sum_{n\geq 1}\frac{a(a+d)\dots[a+(n-1)d]}{d(2d)\cdots(nd)}x^n,$$

$d > 0$. It is easy to have the differential equation

$$(1-x)y'(x) = \frac{a}{d}y(x),$$

which yields a solution $y(x) = (1-x)^{-a/d}$.

Another example for differentiation method is

$$\frac{1}{2}\sum_{n\geq 1}\frac{((n-1)!)^2}{(2n)!}(2x)^{2n} = (\sin^{-1} x)^2$$

for all $|x| < 1$. Denote $y \equiv y(x) = (\sin^{-1} x)^2$. it is easy to find y satisfies differential equation $(1-x^2)y'' - xy' - 2 = 0$ and one of its power series solution is

$$y = \frac{1}{2}\sum_{n\geq 1}\frac{((n-1)!)^2}{(2n)!}(2x)^{2n}.$$

Readers can check the correction by evaluating the first few terms of Taylor's expansion of $(\sin^{-1}(x))^2$ and compare with the series shown above. Another power series solution of the differential equation can be obtained using coefficient comparison method.

Making use of the complex function properties, one may obtain more summation formulas. Denote by $w = u + iv$ a complex number. Then $\log w = \log|w| + i(arg\ w + k\pi)$, where $arg\ w = \tan^{-1}(v/u)$ is an argument of w, and k is an arbitrary integer. Let $Re(w)$ be the real part of w. Then $Re(\log w) = \log|w|$. Denote $z = e^{ix}$. We have

$$\begin{aligned}\sum_{n\geq 1}\frac{\cos nx}{n} &= Re\left(\sum_{n\geq 1}\frac{z^n}{n}\right) = -Re(\log(1-z)) \\ &= -Re(\log(1-e^{ix})) = -\log|1-e^{ix}|.\end{aligned}$$

Using Euler formula, we obtain

$$1 - e^{ix} = 2e^{ix/2}\left(\frac{1}{2}\left(e^{-ix/2} - e^{ix/2}\right)\right) = -2e^{ix/2}\sin\frac{x}{2}.$$

Thus,

$$\sum_{n\geq 1}\frac{\cos nx}{n} = -\log\left|2\sin\frac{x}{2}\right| \tag{1.2}$$

for $0 < x < 2\pi$. More examples are given as follows.

Example 1.1.5

$$\sum_{n\geq 1}\frac{\sin nx}{n} = Im\left(\sum_{n\geq 1}\frac{z^n}{n}\right) = \frac{\pi - x}{2}, \ (0 < x < 2\pi)$$

$$\sum_{n\geq 1} \frac{\cos nx}{n!} = Re\left(\sum_{n\geq 1} \frac{z^n}{n!}\right) = Re(e^z) = e^{\cos x}\cos(\sin x), \ (|x| < \infty)$$

$$\sum_{n\geq 1} \frac{\sin nx}{n!} = Im\left(\sum_{n\geq 1} \frac{z^n}{n!}\right) = Im(e^z) = e^{\cos x}\sin(\sin x) \ (|x| < \infty).$$

Using formula $\sin na \sin nx = \frac{1}{2}[\cos n(a-x) - \cos n(a+x)]$ *and summation* (1.2) *yields*

$$\sum_{n\geq 1} \frac{\sin na \sin nx}{n} = \frac{1}{2}\log\left|\frac{\sin\frac{1}{2}(x+a)}{\sin\frac{1}{2}(x-a)}\right|.$$

Similarly, formula

$$\sin^2 na \sin nx = \frac{1}{4}\left(\sin n(2a-x) - \sin n(2a+x) + 2\sin nx\right)$$

and the first summation in Example 1.1.5 yields

$$\sum_{n\geq 1} \frac{\sin^2 na \sin nx}{n} = \frac{\pi}{4} \tag{1.3}$$

for all $0 < x < 2a < \pi$. *By taking limit* $a \to \pi/2$ *on the both sides of* (1.3), *we obtain*

$$\sum_{n\geq 1} \frac{\sin(2n-1)x}{2n-1} = \frac{\pi}{4} sgn x, \ (|x| \leq \pi). \tag{1.4}$$

From (1.4) *one can establish*

$$\sum_{n\geq 1} \frac{\cos(2n-1)x}{(2n-1)^2} = \frac{\pi^2}{8} - \frac{\pi}{4}|x|, \ (|x| \leq \pi). \tag{1.5}$$

Indeed, for $|x| = \pi$, (1.5) *is easy to be obtained from* $\sum_{n\geq 1} 1/n^2 = \pi^2/6$. *For* $|x| < \pi$, *denoting the series on the left-hand side of* (1.5) *by* $F(x)$ *and taking its derivative term by term yields*

$$\frac{d}{dx}F(x) = -\sum_{n\geq 1} \frac{\sin(2n-1)x}{2n-1} = -\frac{\pi}{4} sgn x,$$

where the last step is from (1.4). *The antiderivative of the above equation generates*

$$F(x) = -\frac{\pi}{4}\int sgn x dx = C - \frac{\pi}{4}|x|,$$

where the constant C *is determined by*

$$C = F(0) = \sum_{n\geq 1} \frac{1}{(2n-1)^2} = \frac{\pi^2}{8},$$

which implies (1.5).

We now consider a positive divergent series $\sum_{n\geq 1} 1/a_n$. Then, for $x > 0$, we have

$$\frac{a_1}{a_2+x} + \left(\frac{a_1}{a_2+x}\right)\left(\frac{a_2}{a_3+x}\right) + \left(\frac{a_1}{a_2+x}\right)\left(\frac{a_2}{a_3+x}\right) \times \left(\frac{a_3}{a_4+x}\right) + \cdots = \frac{a_1}{x}. \tag{1.6}$$

Indeed, it is easy to write

$$\begin{aligned}
\frac{a_1}{x} &= \left(\frac{a_1}{a_2+x}\right)\left(\frac{a_2+x}{x}\right) = \frac{a_1}{a_2+x} + \left(\frac{a_1}{a_2+x}\right)\left(\frac{a_2}{x}\right) \\
&= \frac{a_1}{a_2+x} + R_1, \\
\frac{a_1}{x} &= \frac{a_1}{a_2+x} + \left(\frac{a_1}{a_2+x}\right)\left(\frac{a_2}{a_3+x}\right) + \left(\frac{a_1}{a_2+x}\right)\left(\frac{a_2}{a_3+x}\right)\left(\frac{a_3}{x}\right) \\
&= \frac{a_1}{a_2+x} + \left(\frac{a_1}{a_2+x}\right)\left(\frac{a_2}{a_3+x}\right) + R_2.
\end{aligned}$$

In general, we have the expression of the remainder R_n as

$$\begin{aligned}
R_n &= \frac{a_1}{x}\left(\frac{a_2}{a_2+x}\right)\left(\frac{a_3}{a_3+x}\right)\cdots\left(\frac{a_{n+1}}{a_{n+1}+x}\right) \\
&= \frac{a_1}{x}\Pi_{k=2}^{n+1}\left(1 - \frac{x}{a_k+x}\right).
\end{aligned}$$

Hence, to prove the convergence of series of (1.6), we only need to show the infinite product tends to zero as $n \to \infty$.

Since $x > 0$, we have

$$\sum_{k\geq 2} \frac{x}{a_k+x} = \sum_{k\geq 2} \frac{1}{b_k+1},$$

where $b_k = a_k/x$ $(k = 2, 3, \ldots)$. The divergence of $\sum_{k\geq 1} 1/a_k$ implies the divergence of $\sum_{k\geq 1} 1/b_k$. If $\underline{\lim}_{k\to\infty} b_k < \infty$, then $\overline{\lim}_{k\to\infty} 1/(b_k+1) \neq 0$, which implies that $\sum_{k\geq 2} 1/(b_k+1)$ diverges to infinite. If $\underline{\lim}_{k\to\infty} b_k = \infty$, then $1/(b_k+1) \sim 1/b_k$ as $k \to \infty$. Hence, $\sum_{k\geq 1} 1/(b_k+1) = \infty$. We have shown in any case,

$$R_n = \frac{a_1}{x}\prod_{k=2}^{n+1}\left(1 - \frac{x}{a_k+x}\right) \to 0$$

as $n \to \infty$. This completes the proof of formula (1.6).

Formula (1.6) has a lot of applications. Here are two examples.

Example 1.1.6

$$\frac{1!}{x+1}+\frac{2!}{(x+1)(x+2)}+\frac{3!}{(x+1)(x+2)(x+3)}\cdots$$
$$=\frac{1}{x-1}, \quad (x>1)$$
$$\frac{x}{1-x^2}+\frac{x^2}{1-x^4}+\frac{x^4}{1-x^8}+\cdots$$
$$=\begin{cases} x/(1-x), & if\ |x|<1 \\ 1/(1-x), & if\ |x|>1. \end{cases}$$

Here, the first formula is from (1.6) *by substituting transform* $x \to x-1$ *and* $a_n = n$. *And we leave them as exercises (cf. Exercise 1.4).*

1.1.2 Limits of sequences and series

In this section, we discuss the limits of sequences and series, which will be applied in convergence of formal series and sequence approximation. The major part of this section is Toeplitz theorem on sequence transformation and its corollaries as well as the limits of sequences and series related to integrals.

First, we establish the following proposition.

Proposition 1.1.7 *Denote* $s_n = \sum_{k=1}^n v_k$, $t_n = \frac{1}{n}\sum_{k=1}^n s_k$, *and* $\tau_n = \frac{1}{n}\sum_{k=1}^n kv_k$. *Then*

(i) $s_n \to \ell$ *implies* $t_n \to \ell$ *and* $\tau_n \to 0$.
(ii) $t_n \to \ell$ *and* $\tau_n \to \tau$ *imply* $\tau = 0$ *and* $s_n \to \ell$.

Proof. (i) Write $s_n = \ell + u_n$. Thus for and $\epsilon > 0$, there exists $N \equiv N(\epsilon)$ such that $n \ge N$ implies $|u_n| < \epsilon/3$. Thus, for $n \ge N$, we have

$$t_n = \frac{1}{n}\sum_{k=1}^{N} s_k + \frac{1}{n}\sum_{k=N+1}^{n}(\ell + v_k) = \frac{1}{n}\sum_{k=1}^{N} s_k + \ell - \frac{N}{n}\ell + r_n,$$

where

$$|r_n| \equiv \left|\frac{1}{n}\sum_{k=N+1}^{n} u_k\right| \le \frac{n-N}{n}\cdot\frac{\epsilon}{3} \le \frac{\epsilon}{3}.$$

For the fixed N chosen as above, there exits $n_0 \ge N$ such that

$$\left|\frac{1}{n}\sum_{k=1}^{N} s_k\right| < \frac{\epsilon}{3}, \quad \left|\frac{N\ell}{n}\right| < \frac{\epsilon}{3}.$$

Therefore,

$$|t_n - \ell| < \frac{1}{3}\epsilon + \frac{1}{3}\epsilon + \frac{1}{3}\epsilon = \epsilon$$

when $n \ge n_0$, i.e., $t_n \to \ell$ as $n \to \infty$.

Since

$$\begin{aligned}\tau_n &= \frac{1}{n}[-(s_1 + s_2 + \cdots s_{n-1}) + ns_n] \\ &= -\frac{n-1}{n}t_{n-1} + s_n \to -\ell + \ell = 0\end{aligned}$$

when $n \to \infty$.

(ii) We have shown that

$$t_n + \tau_n = \frac{1}{n}(s_n + ns_n),$$

or equivalently,

$$s_n = \frac{n}{n+1}(t_n + \tau_n).$$

Hence, from (i), $\tau_n \to 0$ and $\lim_{n\to\infty} s_n = \lim_{n\to\infty} t_n = \ell$.

■

Proposition 1.1.8 *If $A_n \to A$ and $B_n \to B$ as $n \to \infty$, then*

$$\frac{1}{n}\sum_{k=1}^{n} A_k B_{n+1-k} \to AB, \quad (n \to \infty).$$

Proof. For any $\epsilon > 0$ ($\epsilon < 1$), there exists N such that $A_n = A + R_n$ and $B_n = B + S_n$ with $|R_n|, |S_n| < \epsilon$ whenever $n \geq N$. Hence, for all $p, q \geq N$, we have

$$\begin{aligned}&A_pB_q = AB + AS_q + BR_p + R_pS_q, \\ &|A_pB_q - AB| < \epsilon(|A| + |B| + 1).\end{aligned}$$

Let $n \geq 2N$. Then

$$\begin{aligned}&\frac{1}{n}\sum_{k=1}^{n} A_k B_{n-k+1} \\ = \ &\frac{1}{n}\sum_{k=1}^{N-1} A_k B_{n-k+1} + \frac{1}{n}\sum_{k=N}^{n-N+1} A_k B_{n-k+1} + \frac{1}{n}\sum_{k=n-N+2}^{n} A_k B_{n-k+1}.\end{aligned}$$

Since A_m, B_n are bounded, the first and the third sums on the right-hand side of above equation tend to zero as $n \to \infty$. In addition,

$$\frac{1}{n}\sum_{k=N}^{n-N+1} A_k B_{n-k+1} = AB - AB\frac{2(N-1)}{n} + H,$$

where $|H| < \epsilon(|A| + |B| + 1)$. Therefore, there exists n_0 such that $n \geq n_0(\epsilon)$ implies

$$\left|\frac{1}{n}\sum_{k=1}^{n} A_k B_{n-k+1} - AB\right| < \epsilon[1 + 1 + (|A| + |B| + 1) + 1],$$

which completes the proof of the proposition.

■

We now present the *Toeplitz's sequence transformation theorem*. Consider lower triangular matrix

$$\begin{array}{cccc} p_{00} & & & \\ p_{10} & p_{11} & & \\ p_{20} & p_{21} & p_{22} & \\ \cdots & \cdots & \cdots & \cdots \end{array} \tag{1.7}$$

where $p_{nk} \geq 0$ and the sums of all rows are one (i.e., $\sum_{k=0}^{n} p_{nk} = 1$). For a given sequence (s_n), we call sequence

$$t_n = \sum_{k=0}^{n} p_{nk} s_k, \tag{1.8}$$

$n = 0, 1, 2, \ldots$, a *transformed sequence* from (s_n) with respect to matrix $[p_{nk}]_{n \geq k \geq 0}$.

Theorem 1.1.9 *(Toeplitz's sequence transformation theorem) Assume* (t_n) *is transformed from* (s_n) *with respect to* $[p_{nk}]_{n \geq k \geq 0}$. *Then*

$$\lim_{n\to\infty} s_n = s \to \lim_{n\to\infty} t_n = s$$

holds if and only if for every fixed m

$$\lim_{n\to\infty} p_{nm} = 0.$$

Proof. Necessity. If $s_n = 0$ for $n \neq m$ and $s_m = 1$. Then $t_n = p_{nm}$ $(n \geq m)$, which yields

$$\lim_{n\to\infty} p_{nm} = \lim_{n\to\infty} t_n = \lim_{n\to\infty} s_n = 0.$$

Sufficiency. If the antecedent of theorem holds, then for any $\epsilon > 0$ there exists $N \equiv N(\epsilon)$ such that $n > N$ implies $|s_n - s| < \epsilon/2$. In addition, there exists $N' > N$ so that

$$p_{n0}, p_{n1}, \ldots, p_{nN} < \frac{\epsilon}{4(N+1)M},$$

are fulfilled for every $n > N'$, where $M = Max|s_k|$ (note that if $M = 0$, then the conclusion of the theorem is trivial). Therefore,

$$t_n - s = \sum_{k=0}^{n} p_{nk}(s_k - s)$$

satisfies

$$\begin{aligned} |t_n - s| &< (N+1)(2M)\frac{\epsilon}{4(N+1)M} \\ &\quad + \frac{\epsilon}{2} \sum_{k=N+1}^{n} p_{nk} < \epsilon \end{aligned}$$

whenever $n > N'$. This implies $t_n \to s$ as $n \to \infty$ and completes the proof for the sufficiency.

■

Example 1.1.10 *We now give some examples of applications to Theorem 1.1.9. The first one is*

$$\lim_{n\to\infty}(p_0p_1p_2\cdots p_n)^{\frac{1}{n+1}} = p, \tag{1.9}$$

where $(p_k)_{k\geq 0}$ *is a positive sequence that approaches to* p *as* $n\to\infty$. *Indeed, the left-hand side of* (1.9) *can be written as*

$$\lim_{n\to\infty} exp\left(\frac{1}{n+1}\sum_{k=0}^{n}\log p_k\right).$$

Sequence $\left\{\frac{1}{n+1}\sum_{k=0}^{n}\log p_k\right\}_{n\geq 0}$ *can be considered as the transfered matrix from* (p_k) *with respect to the transformation matrix* $[p_{nk}=1/(n+1)]$. *Since* $p_n\to p$, *from Theorem 1.1.9, we have* $\frac{1}{n+1}\sum_{k=0}^{n}\log p_k\to\log p$ *as* $n\to\infty$. *Thus,* (1.9) *holds.*

By setting

$$p_0=1, p_1=\left(\frac{2}{1}\right)^1, p_2=\left(\frac{3}{2}\right)^2, \cdots, p_n\left(\frac{n+1}{n}\right)^n, \cdots$$

in (1.9), *we obtain*

$$\lim_{n\to\infty}\left(\frac{(n+1)^{n+1}}{(n+1)!}\right)^{\frac{1}{n+1}} = \lim_{n\to\infty}\left(1+\frac{1}{n}\right)^n = e, \tag{1.10}$$

i.e., $\lim_{n\to\infty}\left(\frac{n^n}{n!}\right)^{\frac{1}{n}} = e.$

Corollary 1.1.11 *Let* (a_n) *and* (b_n) *be two sequences with* $b_n>0$, $\sum_{n\geq 0}b_n=\infty$, *and* $a_n/b_n\to s$ $(n\to\infty)$. *Then*

$$\lim_{n\to\infty}\frac{a_0+a_1+a_2+\cdots+a_n}{b_0+b_1+\cdots+b_n} = s. \tag{1.11}$$

Proof. (1.11) can be proved from Theorem 1.1.9 by setting $s_n=a_n/b_n$, $p_{nk}=b_k/(b_0+b_1+\cdots+b_n)$, and $t_n=(a_0+a_1+\cdots+a_n)/(b_0+b_1+\cdots+b_n)$.

■

Example 1.1.12 *In* (1.11), *if* $a_n=(n+1)^{\alpha-1}$ $(\alpha>0)$ *and* $b_n=(n+1)^\alpha-n^\alpha$, *then we have*

$$\lim_{n\to\infty}\frac{1^{\alpha-1}+2^{\alpha-1}+\cdots+n^{\alpha-1}}{n^\alpha} = \lim_{n\to\infty}\left(\frac{n^{\alpha-1}}{(n+1)^\alpha-n^\alpha}\right) = \frac{1}{\alpha}.$$

Similarly, if a positive sequence $(p_n)_{n\geq 0}$ *satisfies*

$$\lim_{n\to\infty}\frac{p_n}{p_0+p_1+\cdots+p_n} = 0,$$

then from an assumption $\lim_{n\to\infty} s_n = s$, *we obtain*

$$\lim_{n\to\infty} \frac{s_0 p_n + s_1 p_{n-1} + s_2 p_{n-2} + \cdots + s_n p_0}{p_0 + p_1 + \cdots + p_n} = s. \tag{1.12}$$

Let $p_n > 0$, $\sum_{k=1}^{\infty} p_k = \infty$, *and let* $\sum_{k=1}^{n} p_k = P_n$ *with* $p_n/P_n \to 0$ *as* $n \to \infty$. *Then*

$$\lim_{n\to\infty} \frac{p_1 P_1^{-1} + p_2 P_2^{-1} + \cdots + p_n P_n^{-1}}{\log P_n} = 1. \tag{1.13}$$

In fact, from (1.11) *the left-hand limit of* (1.13) *can be written as*

$$\begin{aligned} &\lim_{n\to\infty} \frac{p_1 P_1^{-1} + p_2 P_2^{-1} + \cdots + p_n P_n^{-1}}{\log P_1 + (\log P_2 - \log P_1) + (\log P_3 - \log P_2) + \cdots + (\log P_n - \log P_{n-1})} \\ = \; &\lim_{n\to\infty} \frac{p_n P_n^{-1}}{\log P_n - \log P_{n-1}} = \lim_{n\to\infty} \frac{p_n P_n^{-1}}{\log\left(1 + \frac{P_n}{P_{n-1}}\right)} = 1. \end{aligned}$$

If $p_n = 1$ $(n = 1, 2, \ldots)$, *then* (1.13) *implies*

$$1 + \frac{1}{2} + \frac{1}{3} + \cdots + \frac{1}{n} \sim \log n$$

as $n \to \infty$.

The final example is related to positive sequences (p_n) *and* (q_n) *satisfying*

$$\lim_{n\to\infty} \frac{p_1 + p_2 + \cdots p_n}{n p_n} = a, \quad \lim_{n\to\infty} \frac{q_1 + q_2 + \cdots q_n}{n q_n} = b,$$

where $0 < a, b < \infty$. *Then*

$$\lim_{n\to\infty} \frac{p_1 q_1 + 2 p_2 q_2 + 3 p_3 q_3 + \cdots + n p_n q_n}{n^2 p_n q_n} = \frac{ab}{a+b}. \tag{1.14}$$

To prove (1.14), *we first observe that* $\sum_{k=1}^{\infty} p_k = \infty$ *and* $\sum_{k=1}^{\infty} q_k = \infty$. *Otherwise* $n p_n \to 0$ *and* $n q_n \to 0$ *as* $n \to \infty$, *which imply* $a = b = \infty$ *that contradict to the assumption. Denote*

$$a_n = P_n Q_n - P_{n-1} Q_{n-1}, \; b_n = n p_n q_n,$$

where $P_n = \sum_{k=1}^{n} p_k$ *and* $Q_n = \sum_{k=1}^{n} q_k$. *Thus*

$$\frac{a_n}{b_n} = \frac{P_n}{n p_n} + \frac{Q_n}{n q_n} - \frac{1}{n} \to a + b \quad (n \to \infty).$$

Therefore, from Corollary 1.1.11 we obtain

$$\lim_{n\to\infty} \frac{a_1 + a_2 + \cdots + a_n}{b_1 + b_2 + \cdots + b_n} = \lim_{n\to\infty} \frac{P_n Q_n}{\sum_{k=1}^{n} k p_k q_k} = a + b,$$

which yields

$$\frac{\sum_{k=1}^{n} kp_kq_k}{n^2p_nq_n} = \left(\frac{\sum_{k=1}^{n} kp_kq_k}{P_nQ_n}\right)\left(\frac{P_nQ_n}{n^2p_nq_n}\right) \to \frac{ab}{a+b},$$

as $n \to \infty$.

Corollary 1.1.13 *Suppose two positive sequences* (p_n) *and* (q_n) *satisfy*

$$\lim_{n\to\infty} \frac{p_n}{p_0+p_1+\cdots+p_n} = 0, \quad \lim_{n\to\infty} \frac{q_n}{q_0+q_1+\cdots+q_n} = 0.$$

And denote $r_n = \sum_{k=0}^{n} p_kq_{n-k}$ $(n = 0, 1, \ldots)$. *Then*

$$\lim_{n\to\infty} \frac{r_n}{r_0+r_1+\cdots+r_n} = 0. \tag{1.15}$$

Proof. Denote

$$P_n = \sum_{k=0}^{n} p_k, \quad Q_n = \sum_{k=0}^{n} q_k, \quad R_n = \sum_{k=0}^{n} r_k.$$

Then

$$\begin{aligned}\frac{r_n}{R_n} &= \frac{p_0q_n+p_1q_{n-1}+\cdots+p_nq_0}{p_0Q_n+p_1Q_{n-1}+\cdots+p_nQ_0}\\ &= p_{n0}\left(\frac{q_0}{Q_0}\right)+p_{n1}\left(\frac{q_1}{Q_1}\right)+\cdots+p_{nn}\left(\frac{q_n}{Q_n}\right),\end{aligned}$$

where

$$p_{nj} = \frac{p_{n-j}Q_j}{p_0Q_n+p_1Q_{n-1}+\cdots p_nQ_0} \le \frac{p_{n-j}}{p_0+p_1+\cdots+p_{n-j}} \to 0$$

as $n \to \infty$, and $p_{n0}+p_{n1}+\cdots p_{nn} = 1$. Hence, (1.15) is simply a special case of the conclusion of Theorem 1.1.9.

■

Example 1.1.14 *Suppose that sequences* (p_n) *and* (q_n) *are defined as Corollary 1.1.13. Let* (s_n) *be any sequence. Then the existence of limits*

$$\lim_{n\to\infty} \frac{s_0p_n+s_1p_{n-1}+\cdots+s_np_0}{p_0+p_1+\cdots p_n} = \bar{p}$$

and

$$\lim_{n\to\infty} \frac{s_0q_n+s_1q_{n-1}+\cdots+s_nq_0}{q_0+q_1+\cdots q_n} = \bar{q}$$

implies $\bar{p} = \bar{q}$. *Indeed, let* (r_n) *be the sequence defined in Corollary 1.1.13, and let*

$$\tau_n := \frac{s_0r_n+s_1r_{n-1}+\cdots+s_nr_0}{r_0+r_1+\cdots r_n}.$$

Denote

$$\bar{p}_n := \frac{s_0 p_n + s_1 p_{n-1} + \cdots + s_n p_0}{p_0 + p_1 + \cdots p_n}, \text{ and } \bar{q}_n := \frac{s_0 q_n + s_1 q_{n-1} + \cdots + s_n q_0}{q_0 + q_1 + \cdots q_n}.$$

From Corollary 1.1.13 we have

$$\begin{aligned} \tau_n &= \frac{p_n Q_0 \bar{q}_0 + p_{n-1} Q_1 \bar{q}_1 + \cdots + p_0 Q_n \bar{q}_n}{p_0 Q_n + p_1 Q_{n-1} + \cdots p_n Q_0} \\ &= \frac{q_n P_0 \bar{p}_0 + q_{n-1} P_1 \bar{p}_1 + \cdots + q_0 P_n \bar{p}_n}{q_0 P_n + q_1 P_{n-1} + \cdots q_n P_0}. \end{aligned}$$

Thus, the same argument in the proof of Corollary 1.1.13 yields $\lim_{n\to\infty} \tau_n = \lim_{n\to\infty} \bar{p}_n = \lim_{n\to\infty} \bar{q}_n$. *Consequently,* $\bar{p} = \bar{q}$. *It is interesting to see that this conclusion holds even* $\lim_{n\to\infty} s_n$ *does not exist.*

Corollary 1.1.15 *If* $\sigma > 0$*, then the convergence of Dirichlet series*

$$a_1 1^{-\sigma} + a_2 2^{-\sigma} + \cdots + a_n n^{-\sigma} + \cdots$$

implies

$$(a_1 + a_2 + \cdots + a_n) n^{-\sigma} \to 0$$

as $n \to \infty$.

Proof. Denote

$$\begin{aligned} t_n &= (a_1 + a_2 + \cdots + a_n) n^{-\sigma}, \\ s_n &= a_1 1^{-\sigma} + a_2 2^{-\sigma} + \cdots + a_n n^{-\sigma}. \end{aligned}$$

Then

$$\begin{aligned} & t_n - n^{-\sigma}(n+1)^{\sigma}(s_n - s) \\ = & \frac{1}{n^{\sigma}} \left[\sum_{k=1}^{n} (k^{\sigma} - (k+1)^{\sigma})(s_k - s) + s \right]. \end{aligned}$$

Hence, from Theorem 1.1.9, we obtain the right-hand side of the above equation tends to zero as $n \to \infty$, which implies $t_n \to 0$ $(n \to \infty)$.

■

Corollary 1.1.15 can be proved by using Abel's summation by part method (see 1.2.2).

Remark 1.1.16 *We can reduce the request of the row sums of a transformation matrix to be uniformly bounded by a constant* $K > 0$. *This matrix* *is called the transformation matrix with respect to* K. *In this case, the consequence of Theorem 1.1.9 is changed to be that* $\lim_{n\to\infty} t_n = Ks$ *when* $\lim_{n\to\infty} s_n = s$. *Furthermore, if a lower triangular matrix* $[p_{nk}]_{n\geq k\geq 0}$ *has entries satisfying*

$$\sum_{k=0}^{n} p_{nk} = P_n \to 1$$

as $n \to \infty$ *and* $p_{nk} \geq 0$, *then this lower triangular matrix can be used to construct a sequence transformation. Theorem 1.1.9 still holds for those transformation matrices. We leave the proof of those claims as Exercise 1.8.*

Definition 1.1.17 *A sequence* (s_n) *is termed a null sequence if for any given* $\epsilon > 0$, *there exists an integer* $N \equiv N(\epsilon)$ *such that* $n > N$ *implies* $|s_n| < \epsilon$.

From Theorem 1.1.9 and Remark 1.1.16, we immediately have

Theorem 1.1.18 *Let* (s_n) *be a null sequence, and let* (t_n) *the transformed sequence of* (s_n) *using a transformation matrix with respect to constant* $K > 0$. *Then* (t_n) *is also a null sequence if for every fixed* m

$$\lim_{n\to\infty} p_{nm} = 0.$$

Proof. For any given $\epsilon > 0$, there exists $N \equiv N(\epsilon)$ such that for every $n > N$, $|s_n| < \epsilon/(2K)$. Then for that n,

$$|t_n| < \left|\sum_{k=0}^{N} p_{nk}s_k\right| + \frac{\epsilon}{2}.$$

By $\lim_{n\to\infty} p_{nm} = 0$ we may choose $N_0 > N$ so that for every $n > N_0$,

$$\left|\sum_{k=0}^{N} p_{nk}s_k\right| < \frac{\epsilon}{2}.$$

Therefore, we have shown that $|t_n| < \epsilon$ for these n's, which completes the proof of the theorem.

■

Theorem 1.1.9 can be extended to the following theorem.

Theorem 1.1.19 *Let positive infinite matrix* $P = [p_{nk}]_{0<n,k<\infty}$ *satisfies* $\sum_{k=0}^{\infty} p_{nk} = 1$. *And the sequence*

$$t_n = \sum_{k\geq 0} p_{nk}s_k$$

is called the transformation sequence of (s_n) *with respect to the transformation matrix* P. *Then* $\lim_{n\to\infty} s_n = s$ *implies* $\lim_{n\to\infty} t_n = s$ *if and only if* $\lim_{n\to\infty} p_{nk} = 0$ *for all* $k = 0, 1, 2, \ldots$.

The proof of Theorem 1.1.19 is similar to the proof of Theorem 1.1.9. We leave it as an exercise for the interested reader (see Exercise 1.10).

Example 1.1.20 *(i) If sequence $\sum_{k\geq 1} kc_k$ converges, then the sequence (t_n) defined by*

$$t_n = \sum_{k\geq 0}(k+1)c_{n+k}$$

converges to 0*. Actually, the limit can be proved by using Theorem 1.1.19 and the sequence (s_n) defined by $s_n = \sum_{k\geq 0}(n+k)c_{n+k} \to 0$ $(n\to\infty)$. Hence,*

$$t_n = \frac{1}{n}s_n + \left(\frac{2}{n+1} - \frac{1}{n}\right)s_{n+1} + \left(\frac{3}{n+2} - \frac{2}{n+1}\right)s_{n+2} + \cdots .$$

From Theorem 1.1.19 and noting $\lim_{n\to\infty} s_n = 0$, we have $\lim_{n\to\infty} t_n = 0$.

(ii) Let power series $f(x) = \sum_{k\geq 0} a_k x^k$ converge at $x = 1$, and let $0 < \alpha < 1$. Then the power series

$$f(\alpha) + \frac{f'(\alpha)}{1!}h + \frac{f''(\alpha)}{2!}h^2 + \cdots + \frac{f^{(n)}(\alpha)}{n!}h^n + \cdots \tag{1.16}$$

converges to $f(1)$ at $h = 1-\alpha$. To prove (1.16)*, we denote*

$$a_0 + a_1 + \cdots + a_n = s_n, \qquad \frac{f^{(n)}(\alpha)}{n!} = b_n,$$
$$b_0 + b_1(1-\alpha) + b_2(1-\alpha)^2 + \cdots + b_n(1-\alpha)^n = t_n.$$

Hence, for $|y| < 1-\alpha$,

$$\begin{aligned} & b_0 + b_1 y + b_2 y^2 + \cdots + b_n y^n + \cdots \\ = \; & a_0 + a_1(\alpha + y) + \cdots + a_n(\alpha+y)^n + \cdots . \end{aligned}$$

Consequently,

$$\begin{aligned} \sum_{k\geq 0}(1-\alpha)^{n-k}y^k \sum_{j\geq 1} b_j y^j &= \frac{(1-\alpha)^{n+1}}{1-(\alpha+y)}\sum_{j\geq 0} a_j(\alpha+y)^j \\ &= (1-\alpha)^{n+1}\sum_{j\geq 0} s_j(\alpha+y)^j. \end{aligned}$$

Comparing the coefficients of y^n on the leftmost and rightmost sides of the above equation yields

$$t_n = (1-\alpha)^{n+1}\sum_{k\geq 0}\binom{n+k}{k}\alpha^k s_{n+k}.$$

Noting that the sum of the coefficients of the series is 1*, from Theorem 1.1.19, we obtain*

$$\lim_{n\to\infty} t_n = \lim_{n\to\infty} s_n = f(1).$$

More convergence of the transformation series will be described in Subsection 1.3.1. At the end of this section, we discuss the limits of series they are equivalent to the existence of integrals. We start from the limits of some sequences and their evaluation using suitable integrals. For example, from the definition of Riemann integrals, we have

$$\lim_{n\to\infty}\sum_{k=1}^{n}\frac{1}{n+k}=\lim_{n\to\infty}\sum_{k=1}^{n}\frac{1}{1+k/n}\frac{1}{n}=\int_0^1\frac{1}{1+x}dx=\log 2.$$

Similarly,

$$\lim_{n\to\infty}\sum_{k=1}^{n}\frac{n}{n^2+k^2}=\frac{\pi}{4},$$
$$\lim_{n\to\infty}\sum_{k=1}^{n}\frac{1}{n}\sin\frac{k\pi}{n}=\frac{2}{\pi},$$
$$\lim_{n\to\infty}\sum_{k=0}^{n}\frac{1}{n}\sec^2\frac{k\pi}{4n}=\frac{4}{\pi},$$
$$\lim_{n\to\infty}\sum_{k=1}^{n}\frac{k^\alpha}{n^{\alpha+1}}=\frac{1}{\alpha+1}\quad(\alpha>-1).$$

Since

$$\frac{1}{n}\sqrt[n]{n!}=exp\left(\frac{1}{n}\sum_{k=1}^{n}\log\frac{k}{n}\right),$$

we immediately know limit

$$\lim_{n\to\infty}\frac{1}{n}\sqrt[n]{n!}=exp\left(\int_0^1\log x dx\right)=\frac{1}{e}.$$

The following limit belongs to Pólya:

$$\lim_{n\to\infty}\frac{1}{n}\sum_{k=1}^{n}\left(\left[\frac{2n}{k}\right]-2\left[\frac{n}{k}\right]\right)=2\log 2-1, \tag{1.17}$$

where $[x]$ denotes the largest integer $\le x$. Obviously,

$$0\le\left[\frac{2n}{k}\right]-2\left[\frac{n}{k}\right]=\left[\frac{2}{k/n}\right]-2\left[\frac{1}{k/n}\right]\le 1.$$

Thus,

$$\lim_{n\to\infty}\frac{1}{n}\sum_{k=1}^{n}\left(\left[\frac{2n}{k}\right]-2\left[\frac{n}{k}\right]\right)=\int_{0+}^1\left(\left[\frac{2}{x}\right]-2\left[\frac{1}{x}\right]\right)dx. \tag{1.18}$$

Denote $n = k\,[n/k] + n_k$, where $0 \le n_k < k$. Then

$$\left[\frac{2n}{k}\right] - 2\left[\frac{n}{k}\right] = \begin{cases} 0 & if\ n_k < k/2, \\ 1 & if\ n_k \ge k/2. \end{cases}$$

Therefore, for $\frac{1}{k+1} < x \le \frac{1}{k}$, we obtain

$$\left[\frac{2}{x}\right] - 2\left[\frac{1}{x}\right] = \begin{cases} 0 & if\ \frac{1}{x} - \left[\frac{1}{x}\right] < \frac{1}{2}(i.e., \frac{1}{k+1/2} < x \le \frac{1}{k}), \\ 1 & if\ \frac{1}{x} - \left[\frac{1}{x}\right] \ge \frac{1}{2}(i.e., \frac{1}{k+1} < x \le \frac{1}{k+1/2}). \end{cases}$$

Thus, the integral in (1.18) is equal to

$$\begin{aligned} & \int_{0+}^{1} \left(\left[\frac{2}{x}\right] - 2\left[\frac{1}{x}\right]\right) dx \\ = & \lim_{n\to\infty} \sum_{k=1}^{n-1} \int_{1/(k+1)}^{1/k} \left(\left[\frac{2}{x}\right] - 2\left[\frac{1}{x}\right]\right) dx \\ = & \lim_{n\to\infty} \sum_{k=1}^{n-1} \left(\frac{1}{k+1/2} - \frac{1}{k+1}\right) \\ = & 2\left(\frac{1}{3} - \frac{1}{4} + \frac{1}{5} - \frac{1}{6} + \cdots\right) = 2\log 2 - 1, \end{aligned}$$

which completes the proof of (1.17). Similarly, we give

Example 1.1.21 *Let* $n_k = n - k[n/k]$ *and* $0 < \alpha \le 1$, *and denote by* γ *the Euler constant. Then*

$$\lim_{n\to\infty} \frac{1}{n}\left(\frac{n_1}{1} + \frac{n_2}{2} + \frac{n_3}{3} + \cdots + \frac{n_n}{n}\right) = 1 - \gamma.$$

$$\lim_{n\to\infty} \frac{n_1 + n_2 + n_3 + \cdots + n_n}{n^2} = 1 - \frac{\pi^2}{12}.$$

$$\lim_{n\to\infty} \frac{1}{n} \sum_{k=1}^{n} \left(\left[\frac{n}{k}\right] - \left[\frac{n}{k} - \alpha\right]\right) = \int_0^1 \frac{1 - x^\alpha}{1 - x} dx \ (0 < \alpha \le 1).$$

The first limit can be obtained from the process

$$\begin{aligned} & \lim_{n\to\infty} \frac{1}{n} \sum_{k=1}^{n} \left(\left(\frac{n}{k}\right) - \left[\frac{n}{k}\right]\right) = \int_0^1 \left(\frac{1}{x} - \left[\frac{1}{x}\right]\right) dx \\ = & \lim_{n\to\infty} \int_{1/n}^{1} \left(\frac{1}{x} - \left[\frac{1}{x}\right]\right) dx = 1 - \lim_{n\to\infty}\left(1 + \frac{1}{2} + \frac{1}{3} + \cdots + \frac{1}{n} - \log n\right) \\ = & 1 - \gamma. \end{aligned}$$

Similarly, the second limit can be derived by

$$\lim_{n\to\infty} \frac{1}{n} \sum_{k=1}^{n} \left(1 - \left[\frac{n}{k}\right]\frac{k}{n}\right) = 1 - \int_0^1 \left[\frac{1}{x}\right] x dx$$

$$= 1 - \sum_{n\geq 1} \frac{n}{2}\left(\frac{1}{n^2} - \frac{1}{(n+1)^2}\right) = 1 - \frac{\pi^2}{12}.$$

The third limit belongs to Dirichlet and the proof is left as Exercise 1.5.

We now discuss the limit of series. First, we establish the following proposition.

Proposition 1.1.22 *Let $f(x)$ be a positive decreasing function defined on $[0, \infty)$, and let improper integral $\int_0^\infty f(x)dx$ exist. Then*

$$\lim_{h\to 0^+} h \sum_{k\geq 1} f(kh) = \int_0^\infty f(x)dx. \tag{1.19}$$

Proof. Since $f(x)$ is decreasing and approaching 0 as $x \to \infty$, we have

$$\int_h^{(n+1)h} f(x)dx \leq h[f(h) + f(2h) + \cdots + f(nh)] \leq \int_0^{nh} f(x)dx.$$

Thus, as $n \to \infty$ we obtain

$$\int_h^\infty f(x)dx \leq h \sum_{k\geq 1} f(kh) \leq \int_0^\infty f(x)dx.$$

Taking $h \to 0^+$ on all sides of the above inequalities yields (1.19). ∎

Remark 1.1.23 *The condition in Proposition 1.1.22 can be released to that $f(x)$ is defined and integrable on $[0, \infty)$ and is decreasing when $x \geq a > 0$.*

Example 1.1.24 *From Proposition 1.1.22, one can establish*

$$\lim_{t\to 1^-} \sqrt{1-t}(1 + t + t^4 + t^9 + \cdots + t^{n^2} + \cdots) = \frac{\sqrt{\pi}}{2}. \tag{1.20}$$

Indeed, let $t = e^{-h^2}$. Then, $t \to 1^-$ when $h \to 0^+$ and

$$\sqrt{1-t} = (1 - e^{-h^2})^{1/2} = (h^2 - h^4/2 + \cdots)^{1/2} = h[1 + O(h^2)]$$

as $h \to 0^+$. Thus, substituting $t = e^{-h^2}$ into the limit of (1.19) and applying Proposition 1.1.22 to the resulting limit yields

$$\lim_{h\to 0^+} h[1 + O(h^2)] \sum_{n\geq 0} e^{-(nh)^2} = \int_0^\infty e^{-x^2} dx = \frac{\sqrt{\pi}}{2}.$$

Similarly, one can find

$$\lim_{t\to 1^-} \sqrt[a]{1-t}(1 + t^{1^a} + t^{2^a} + t^{3^a} + \cdots + t^{n^a} + \cdots) = \frac{1}{a}\Gamma\left(\frac{1}{a}\right) \quad (a > 0) \tag{1.21}$$

$$\lim_{t\to 1^-}(1-t)\left(\frac{t}{1+t}+\frac{t^2}{1+t^2}+\frac{t^3}{1+t^3}+\cdots+\frac{t^n}{1+t^n}+\cdots\right)=\log 2 \tag{1.22}$$

$$\lim_{t\to 1^-}(1-t)^2\left(\frac{t}{1-t}+\frac{2t^2}{1-t^2}+\frac{3t^3}{1-t^3}+\cdots+\frac{nt^n}{1-t^n}+\cdots\right)=\frac{\pi^2}{6}. \tag{1.23}$$

1.2 Abel's Summation

At the beginning of the chapter, five methods are enlisted for summation of series using the infinitesimal calculus, and the first four methods are introduced in 1.1.1. Representation of the last method, Abel's method, on the list will be a major part of this section.

1.2.1 Abel's theorem and Tauber theorem

In this subsection, we shall present the classic (i.e., little o) Tauber theorem and Littlewood's big O Tauber type theorem. First, we establish a generalized result of sequence transformation theorems 1.1.9 and 1.1.19.

Theorem 1.2.1 *Let $(\phi_n(t))$ be a non-negative function sequence with $\sum_{n\geq 0}\phi_n(t)\equiv 1$, where $t\in(0,1)$. Then $\lim\limits_{n\to\infty} s_n = s$ implies*

$$\lim_{t\to 1^-}\sum_{k\geq 0} s_k\phi_k(t)=s$$

if and only if for every fixed $n=0,1,\ldots$

$$\lim_{t\to 1^-}\phi_n(t)=0.$$

The proof of Theorem 1.2.1 is similar as those of Theorems 1.1.9 and 1.1.19 and is left for an exercise (cf. Exercise 1.11).

Corollary 1.2.2 *Let (a_n) and (b_n) be two sequences with $b_n > 0$ and $\lim\limits_{n\to\infty} a_n/b_n = s$, where $\sum_{k\geq 0} b_k t^k$ converges for every $|t|<1$ and diverges when $t=1$. Then, (i) $\sum_{k\geq 0} a_k t^k$ converges as well for every $|t|<1$, and (ii)*

$$\lim_{t\to 1^-}\frac{\sum_{k\geq 0} a_k t^k}{\sum_{k\geq 0} b_k t^k}=s. \tag{1.24}$$

Proof. Denote $s_n = a_n/b_n$ and

$$\phi_n(t) = \frac{b_n t^n}{b_0 + b_1 t + b_2 t^2 + \cdots + b_n t^n + \cdots}.$$

Then for every fixed j and for any $\epsilon > 0$, there is a large enough n such that

$$b_0 + b_1 + b_2 + \cdots + b_n > \frac{1}{\epsilon} b_j$$

because of the divergence of series $\sum_{k\geq 0} b_k$. Hence, from

$$\phi_j(t) < \frac{b_j t^j}{b_0 + b_1 t + b_2 t^2 + \cdots + b_n t^n}$$

there holds $\lim_{t\to 1^-} \phi_j(t) < \epsilon$; i.e., $\phi_j(t) \to 0$ as $t \to 1^-$. Limit (1.24) is thus proved as a corollary of Theorem 1.2.1.

■

As another corollary of Theorem 1.2.1, one may establish the following *Abel's theorem.*

Theorem 1.2.3 *(Abel's theorem) If series $\sum_{k\geq 0} a_k = s$, then*

$$\lim_{t\to 1^-} \sum_{k\geq 0} a_k t^k = s. \tag{1.25}$$

Proof. Obviously, by using Corollary 1.2.2 and noting $(\sum_{n\geq 0} a_n)(\sum_{n\geq 0} t^n) = \sum_{n\geq 0}(a_0 + a_1 + \cdots + a_n)t^n$ we obtain

$$\begin{aligned} \lim_{t\to 1^-} \sum_{n\geq 0} a_n t^n &= \lim_{t\to 1^-} \frac{\sum_{n\geq 0}(a_0 + a_1 + \cdots a_n)t^n}{\sum_{n\geq 0} t^n} \\ &= \lim_{n\to\infty} \frac{a_0 + a_1 + \cdots a_n}{1} = s. \end{aligned}$$

■

Next theorem implies Abel's theorem.

Theorem 1.2.4 *Let $s_n = \sum_{k=0}^{n} a_k$ $(n = 0, 1, 2, \ldots)$. Then $\lim_{n\to\infty} \frac{1}{n+1} \sum_{k=0}^{n} s_k = s$ implies*

$$\lim_{t\to 1^-} \sum_{k\geq 0} a_k t^k = s. \tag{1.26}$$

Proof. Noting the absolute convergence of $\sum_{k\geq 0} s_n t^n$ and $\sum_{k\geq 0} t^n$ over $|t| < 1$, we use Corollary 1.2.2 may obtain

$$\lim_{t\to 1^-} \sum_{n\geq 0} a_n t^n = \lim_{t\to 1^-} \frac{\sum_{n\geq 0}(a_0 + a_1 + \cdots a_n)t^n}{\sum_{n\geq 0} t^n}$$

$$= \lim_{t\to 1^-} \frac{\sum_{n\geq 0}(s_0+s_1+\cdots s_n)t^n}{\sum_{n\geq 0}(n+1)t^n} = \lim_{n\to\infty} \frac{s_0+s_1+\cdots s_n}{n+1} = s,$$

completing the proof.

■

It is obvious that Theorem 1.2.4 implies Abel's theorem 1.2.3 because $s_n \to s$ implies

$$\frac{1}{n+1}(s_0+s_1+\cdots+s_n) \to s.$$

Theorem 1.2.5 *Let two sequences $(a_n)_{n=0,1,\ldots}$ and $(b_n > 0)_{n=0,1,\ldots}$ satisfy $a_n/b_n \to s$ $(n\to\infty)$, and let $\sum_{n\geq 0} b_n t^n$ be convergent for all $t\in\mathbb{R}$. Prove that $\sum_{n\geq 0} a_n t^n$ converges for all $t\in\mathbb{R}$. Furthermore,*

$$\lim_{t\to\infty} \frac{\sum_{n\geq 0} a_n t^n}{\sum_{n\geq 0} b_n t^n} = s. \tag{1.27}$$

Theorem 1.2.5 can be proved by using a similar argument of the proof of Corollary 1.2.2, which is left as an exercise (cf. Exercise 1.13).

Corollary 1.2.6 *If $\lim_{n\to\infty} s_n = s$, then*

$$\lim_{t\to\infty}\left(s_0+s_1\frac{t}{1!}+s_2\frac{t^2}{2!}+\cdots+s_n\frac{t^n}{n!}+\cdots\right)e^{-t} = s. \tag{1.28}$$

Proof. It is obvious a corollary of Theorem 1.2.5 for the case of $a_n = \frac{1}{n!}s_n$ and $b_n = \frac{1}{n!}$.

■

Example 1.2.7 *Let $g(t) = \sum_{n\geq 0} a_n \frac{t^n}{n!}$ and assume $\sum_{n\geq 0} a_n = s$. Then $\int_0^\infty e^{-t}g(t)dt = s$. Indeed, denote $s_n = a_0+a_1+\cdots+a_n$ and $s_{-1}=0$ so that*

$$\int_0^t e^{-x}g(x)dx = \sum_{k\geq 0}\left(\frac{s_k-s_{k-1}}{k!}\right)\int_0^t e^{-x}x^k dx$$

$$= \sum_{k\geq 0} s_k \int_0^t \left(\frac{x^k}{k!}-\frac{x^{k+1}}{(k+1)!}\right)e^{-x}dx = \sum_{k\geq 0} s_k\frac{t^{k+1}}{(k+1)!}e^{-t},$$

which implies the desired result from the view of Corollary 1.2.6.

We now give the inverse theorem of Abel's theorem, which is also called the *little o Tauber theorem.*

Theorem 1.2.8 *(Little O Tauber theorem) Denote $f(x) = \sum_{n\geq 0} a_n x^n$, where $a_n = o\left(\frac{1}{n}\right)$. Then $\lim_{x\to 1^-} f(x) = s$ implies $\sum_{n\geq 0} a_n = s$.*

Proof. Denote $N := \left[\frac{1}{1-x}\right]$ $(x < 1)$. Then $N \to \infty$ as $x \to 1^-$. Hence to show the theorem is true, we only need to prove

$$\lim_{x\to 1^-} \sum_{n=0}^{\infty} a_n x^n - \sum_{n=0}^{N} a_n = 0,$$

i.e.,

$$\lim_{x\to 1^-} \left(\sum_{n=0}^{N} a_n(x^n - 1) + \sum_{n=N+1}^{\infty} a_n x^n \right) = 0.$$

We denote the first and the second summations in the above equation by $\sum_1$ and $\sum_2$, respectively. Thus, we have estimates

$$\begin{aligned} \left|\sum_1\right| &\leq \sum_{n=0}^{N} \left|a_n(1-x)(1+x+x^2+\cdots+x^{n-1})\right| \\ &\leq (1-x)\sum_{n=0}^{N} na_n \leq \frac{\sum_{n=0}^{N} na_n}{\left[\frac{1}{1-x}\right]} \to 0 \end{aligned}$$

as $x \to 1^-$. Here the last step is from Proposition 1.1.8, where $N = \left[\frac{1}{1-x}\right]$, $A_\nu = \nu a_\nu$, $B_\nu = 1$, $A = 0$. In addition, for summation $\sum_2$, we have

$$\begin{aligned} \left|\sum_2\right| &= \left|\sum_{n=N+1}^{\infty} a_n x^n\right| = \left|\sum_{n=N+1}^{\infty} (na_n)\frac{x^n}{n}\right| \\ &< \frac{\epsilon}{N+1}\sum_{n=N+1}^{\infty} x^n \leq \frac{\epsilon}{(N+1)(1-x)} < \epsilon, \end{aligned}$$

where $\epsilon > 0$ is arbitrarily fixed and $|na_n| < \epsilon$ $(n > N)$, which completes the proof.

■

It is worth be mentioned that the condition $a_n = o(1/n)$ in the little o Tauber theorem, Theorem 1.2.8, can be reduced to $a_n = O(1/n)$, which elegant result established by Littlewood will be represented with the aid of the following two lemmas.

Lemma 1.2.9 *Suppose* $a_n \geq 0$ $(n = 0, 1, 2, \ldots)$ *and*

$$f(x) = \sum_{n\geq 0} a_n x^n \sim \frac{1}{1-x}$$

when $x \to 1^-$. *Then* $s_n = \sum_{\nu\geq 0} a_\nu \sim n$ *as* $n \to \infty$.

Proof. Let $g(x)$ be a continuous function defined on $[0,1]$, and let $\epsilon > 0$ be arbitrarily given. From the Weierstrass polynomial approximation theorem, there exist polynomial lower approximation $p(x)$ and upper approximation $P(x)$ such that $p(x) \geq g(x) \geq P(x)$ and

$$\int_0^1 (g(x) - p(x))dx < \epsilon, \qquad \int_0^1 (P(x) - g(x))dx < \epsilon.$$

In fact, even $g(x)$ is discontinuous at finite points and the left-hand limits and right-hand limits at those points exist (but different), the above conclusion still holds. Without loose of generality, we may assume that $g(x)$ is only discontinuous at point c with $g(c^-) < g(c^+)$. We now define $\phi(x) = g(x) + \frac{1}{2}\epsilon$ when $x < c - \delta$ and $x > c$ for a sufficient small δ, and $\phi(x) = Max\{\ell(x), g(x) + \frac{1}{4}\epsilon\}$ when $c - \delta \leq x \leq c$ for the same δ, where $\ell(x)$ is a linear function with values at end points as

$$\ell(c - \delta) = g(c - \delta) + \frac{1}{2}\epsilon, \qquad \ell(c) = g(c^+) + \frac{1}{2}\epsilon.$$

Therefore, $\phi(x)$ is continuous and $\phi(x) > g(x)$. We can use Weierstrass theorem to obtain the function $P(x)$ that approximates function $\phi(x)$ sufficiently, which also gives an upper approximation to $g(x)$. Similarly, we can construct lower approximation $p(x)$ to $g(x)$.

To prove the lemma, we first show

$$\lim_{x \to 1^-} (1 - x) \sum_{n \geq 0} a_n x^n P(x^n) = \int_0^1 P(t)dt.$$

It is sufficient to prove the above equation for $P(x) = x^k$. Indeed, the left-hand side of the above equation can be written as

$$\begin{aligned}
(1 - x) \sum_{n \geq 0} a_n x^{n+kn} &= \frac{1 - x}{1 - x^{k+1}} \left((1 - x^{k+1}) \sum_{n \geq 0} a_n (x^{k+1})^n \right) \\
&\to \frac{1}{k+1} = \int_0^1 x^k dx
\end{aligned}$$

as $x \to 1^-$. Hence, considering $P(x)$ as the upper approximation of $g(x)$, we can prove

$$\lim_{x \to 1^-} (1 - x) \sum_{n \geq 0} a_n x^n g(x^n) = \int_0^1 g(t)dt.$$

Actually, since $a_n > 0$, we have

$$\begin{aligned}\overline{\lim}_{x\to 1^-}(1-x)\sum_{n\geq 0} a_n x^n g(x^n) &\leq \overline{\lim}_{x\to 1^-}(1-x)\sum_{n\geq 0} a_n x^n P(x^n)\\ &= \int_0^1 P(t)dt < \int_0^1 g(t)dt + \epsilon.\end{aligned}$$

Therefore, $\epsilon \to 0$ yields

$$\overline{\lim}_{x\to 1^-}(1-x)\sum_{n\geq 0} a_n x^n g(x^n) \leq \int_0^1 g(t)dt.$$

Similarly, for the lower approximation $p(x)$ of $g(x)$, we obtain

$$\underline{\lim}_{x\to 1^-}(1-x)\sum_{n\geq 0} a_n x^n g(x^n) \geq \int_0^1 g(t)dt,$$

which implies the desired result.

Finally, we define

$$g(t) = \begin{cases} 0 & 0 \leq t < e^{-1}, \\ \frac{1}{t} & e^{-1} \leq t \leq 1. \end{cases}$$

Then

$$\int_0^1 g(t)dt = \int_{1/e}^1 \frac{dt}{t} = 1.$$

Let $x = e^{-1/N}$ with a positive integer N. Then

$$\sum_{n\geq 0} a_n x^n g(x^n) = \sum_{e^{-1}\leq x^n \leq 1} a_n x^n g(x^n) = \sum_{n=0}^{N} a_n = s_N.$$

It is easy to see that $s_N \sim 1/(1-x) = (1-e^{-1/N})^{-1} \sim N$, which completes the proof.

■

Lemma 1.2.10 *Assume $f(x)$ has second derivative on $[0,1)$ and satisfies $f(x) = o(1)$ and $f''(x) = O\left(\frac{1}{(1-x)^2}\right)$ when $x \to 1^-$. Then we have $f(x) = o\left(\frac{1}{1-x}\right)$ as $x \to 1^-$.*

Proof. Denote $x' = x + \delta(1-x)$, $0 < \delta < \frac{1}{2}$. Then

$$f(x') = f(x) + \delta(1-x)f'(x) + \frac{1}{2}\delta^2(1-x)^2 f''(\xi),$$

where $x < \xi < x'$. Thus,

$$\begin{aligned}(1-x)f'(x) &= \frac{f(x')-f(x)}{\delta} + \frac{1}{2}\delta(1-x)^2 f''(\xi)\\ &= \frac{f(x')-f(x)}{\delta} + O(\delta).\end{aligned}$$

The rightmost term can be as small as we wish provided δ is small enough. Let x' and x be close enough to 1. Then, we have $(1-x)f'(x) = o(1)$.

■

Theorem 1.2.11 *(Big O Tauber theorem) Let* $a_n = O\left(\frac{1}{n}\right)$, $f(x) = \sum_{n\geq 0} a_n x^n \to s$ *as* $x \to 1^-$. *Then series* $\sum_{n\geq 0} a_n$ *converges to* s.

Proof. Without loss the generality we assume $s = 0$. Hence,

$$f(x) = \sum_{n\geq 0} a_n x^n = o(1)$$

when $x \to 1^-$. Since $a_n = O\left(\frac{1}{n}\right)$,

$$\begin{aligned}f''(x) &= \sum_{n\geq 2} n(n-1)a_n x^{n-2} = O\left(\sum_{n\geq 2}(n-1)x^{n-2}\right)\\ &= O\left(\frac{1}{(1-x)^2}\right).\end{aligned}$$

From Lemma 1.2.10 we obtain

$$f'(x) = \sum_{n\geq 1} n a_n x^{n-1} = o\left(\frac{1}{1-x}\right).$$

Since $a_n = O(1/n)$ implies $|na_n| \leq c$ for some positive constant c, we have

$$\sum_{n\geq 1}\left(1 - \frac{na_n}{c}\right)x^{n-1} = \frac{1}{1-x} - \frac{f'(x)}{c} \sim \frac{1}{1-x}$$

when $x \to 1^-$. Therefore, we may apply Lemma 1.2.9 to have

$$\sum_{k=1}^{n}\left(1 - \frac{ka_k}{c}\right) \sim n.$$

Thus,

$$w_n := \sum_{k=1}^{n} ka_k = o(n), \quad w_0 := 0.$$

Furthermore,

$$f(x) - a_0 = \sum_{n\geq 1} \left(\frac{w_n - w_{n-1}}{n}\right) x^n = \sum_{n\geq 1} w_n \left(\frac{x^n}{n} - \frac{x^{n+1}}{n+1}\right)$$

$$= \sum_{n\geq 1} w_n \left(\frac{x^n - x^{n+1}}{n+1} + \frac{x^n}{n(n+1)}\right)$$

$$= (1-x)\sum_{n\geq 1} \frac{w_n}{n+1} x^n + \sum_{n\geq 1} \frac{w_n}{n(n+1)} x^n$$

$$= o(1) + \sum_{n\geq 1} \frac{w_n}{n(n+1)} x^n,$$

where we use $w_n = o(n)$ in the last step. Noting the condition $f(x) \to 0$ as $x \to 1^-$, we have

$$\sum_{n\geq 1} \frac{w_n}{n(n+1)} = -a_0.$$

Using $w_n/n(n+1) = o(1/n)$ and litle o Tauber theorem yields

$$\sum_{n\geq 1} \frac{w_n}{n(n+1)} = -a_0.$$

And the left-hand side of the above equation can be written as

$$\lim_{N\to\infty} \sum_{n=1}^{N} \frac{w_n}{n(n+1)} = \lim_{N\to\infty} \sum_{n=1}^{N} w_n \left(\frac{1}{n} - \frac{1}{n+1}\right)$$

$$= \lim_{N\to\infty} \left(\sum_{n=1}^{N} \frac{w_n - w_{n-1}}{n} - \frac{w_N}{N+1}\right) = \lim_{N\to\infty} \sum_{n=1}^{N} a_n.$$

Therefore, $\sum_{n\geq 1} a_n = -a_0$, or equivalently, $\sum_{n\geq 0} a_n = 0$, which proves the theorem.

■

1.2.2 Abel's summation method

Abel's summation method starts from the following identity called the *summation by part formula*, which can be considered as the discrete analog of the integration by parts. From the summation by parts, we shall describe Abel's lemma that can give an alternative proof of Abel's theorem (Theorem 1.2.3). Since Abel's theorem shows that function $f(x) = \sum_{n\geq 0} a_n x^n$ is left-hand continuous at point 1, it can be used to find the sum of convergent series $\sum_{n\geq 0} a_n$. This method is named *Abel's summation method.*

Proposition 1.2.12 *(Summation by parts formula) Let positive integers $m < n$. Then*

$$\sum_{k=m}^{n}(s_k - s_{k-1})b_k = s_n b_n - s_{m-1}b_m + \sum_{k=m}^{n-1} s_k(b_k - b_{k+1}). \tag{1.29}$$

In particular, if $m = 1$, $s_k = \sum_{j=1}^{k} a_j$ $(k = 1, 2, \ldots, n)$ and $s_0 = 0$, then formula (1.29) is reduced as

$$\sum_{k=1}^{n} a_k b_k = s_n b_n + \sum_{k=1}^{n-1} s_k(b_k - b_{k-1}). \tag{1.30}$$

Separating the left-hand sum of equation (1.29) and combining all terms with s_k yields the right-hand expression of (1.29). We leave the details as an exercise (cf. Exercise 1.14).

Corollary 1.2.13 *Suppose $s_n = \sum_{n=1}^{n} a_k \to s$ as $n \to \infty$. Then*

$$\sum_{k=1}^{n} a_k b_k = s b_1 + (s_n - s)b_n - \sum_{k=1}^{n-1}(s_n - s)(b_{k+1} - b_k). \tag{1.31}$$

Note that the value s can be eliminated.

In the summation by parts formula (1.29), taking $m = 1$ and substituting s_0 and s_k by $-s$ and $s_k - s$, respectively, we may derive (1.31).

Lemma 1.2.14 *(Abel's lemma) If for all $n = 1, 2, 3, \ldots$, $b_1 \geq b_2 \geq \cdots \geq b_n \geq 0$ and $m \leq a_1 + a_2 + \cdots + a_n \leq M$ for real numbers m and M, then*

$$b_1 m \leq a_1 b_1 + a_2 b_2 + \cdots + a_n b_n \leq b_1 M.$$

Proof. Applying formula (1.30) and noting $m \leq s_k \leq M$ and $b_k - b_{k+1} \geq 0$, we have

$$\sum_{k=1}^{n} a_k b_k \leq M b_n + \sum_{k=1}^{n-1} M(b_k - b_{k+1}) = M b_1,$$
$$\sum_{k=1}^{n} a_k b_k \geq m b_n + \sum_{k=1}^{n-1} m(b_k - b_{k+1}) = m b_1,$$

which completes the proof.

■

We shall see that Abel's theorem 1.2.3, represented in the last section, can be proved using Abel's lemma 1.2.14. For the sake of convenience, we re-write and re-prove Theorem 1.2.3 as follows.

Theorem 1.2.15 *(Abel's theorem) If* $\sum_{n\geq 0} a_n = s$*, then* $\lim_{x\to 1^-} \sum_{n\geq 0} a_n x^n = s$.

Proof. It is easy to see that $\sum_{n\geq 0} a_n x^n$ uniformly converges on $[0,1]$ and, thus, it defines a function, denoted by $f(x)$. Indeed, for any given $\epsilon > 0$, there exists N such that $n > N$ implies $\left|\sum_{k=n}^{n+p} a_k\right| < \epsilon$. Therefore, from Abel's lemma $\left|\sum_{k=n}^{n+p} a_k x^k\right| < x^n \epsilon \leq \epsilon$ for all $x \in [0,1]$. From the continuity theorem of the function series we have $\lim_{x\to 1^-} f(x) = f(1) = s$.

■

The following *general Abel's lemma* can be established using a similar argument of the proof of Abel's lemma 1.2.14.

Lemma 1.2.16 *Let* $a_1, a_2, \cdots, a_n$*,* $b_1, b_2, \cdots, b_n$ *be arbitrary real or complex numbers, and let*

$$A = max\ (|a_1|, |a_1 + a_2|, \cdots, |a_1 + a_2 + \cdots + a_n|).$$

Then

$$\left|\sum_{k=1}^{n} a_k b_k\right| \leq A \left(\sum_{k=1}^{n-1} |b_{k+1} - b_k| + |b_n|\right).$$

Theorem 1.2.17 *Assume decreasing sequence* $(\phi(n))$ *approaches to* 0 *as* $n \to \infty$*, and* $\sum_{n\geq 1} a_n \phi(n)$ *converges. Then*

$$\lim_{n\to\infty} (a_1 + a_2 + \cdots + a_n)\phi(n) = 0. \tag{1.32}$$

Proof. For any given $\epsilon > 0$, there exists a natural number N such that $n > N$ implies

$$-\frac{\epsilon}{2} < a_N\phi(N) + a_{N+1}\phi(N+1) + \cdots + a_n\phi(n) < \frac{\epsilon}{2}.$$

In addition, we have

$$0 < \phi(N)^{-1} \leq \phi(N+1)^{-1} \leq \cdots \leq \phi(n)^{-1}.$$

Therefore, Abel's lemma can be applied and yields

$$-\frac{\epsilon}{2}\phi(n)^{-1} < a_N + a_{N+1} + \cdots + a_n < \frac{\epsilon}{2}\phi(n)^{-1}.$$

Or equivalently,

$$|(a_N + a_{N+1} + \cdots + a_n)\phi(n)| < \frac{\epsilon}{2}.$$

Since $\phi(n) \to 0$, there exists a natural number N' such that $n > N'$ implies

$$|(a_1 + a_2 + \cdots + a_{N-1})\phi(n)| < \frac{\epsilon}{2}.$$

Hence, when $n > \max(N, N')$,

$$|(a_1 + a_2 + \cdots + a_n)\phi(n)| < \frac{\epsilon}{2} + \frac{\epsilon}{2} = \epsilon.$$

■

Theorem 1.2.18 *Let $\{\phi(n)\}_{n=1}^{\infty}$ be a positive increasing sequence that diverges to infinity, and let $\sum_{n\geq 1} a_n$ be any convergent series. Then*

$$\sum_{k=1}^{n} a_k\phi(k) = o(\phi(n)) \tag{1.33}$$

as $n \to \infty$.

Proof. This theorem can be proved using Corollary 1.2.13, which is left as an excise (cf. Exercise 1.17). We now give an alternative and much short proof using Theorem 1.2.17. Since sequence $(\phi(n)^{-1})_{n\geq 1}$ is decreasing and approaching to 0 as $n \to \infty$, and $\sum_{n\geq 1}(a_n\phi(n))\phi(n)^{-1} = \sum_{n\geq 1} a_n$ converges, we have

$$\lim_{n\to\infty} \left(\sum_{k=1}^{n} a_k\phi(k)\right) \phi(n)^{-1} = 0,$$

or equivalently, $\sum_{k=1}^{n} a_k\phi(k) = o(\phi(n))$.

■

Obviously, Theorem 1.2.17 can be derived from Theorem 1.2.18. We leave it as an exercise (cf. Exercise 1.17). We also have the following corollary of Theorem 1.2.17.

Corollary 1.2.19 *Let $\sigma > 0$. If the Dirichlet series $\sum_{n\geq 1} a_n n^{-\sigma}$ converges, then*

$$\lim_{n\to\infty} (a_1 + a_2 + \cdots + a_n)n^{-\sigma} = 0.$$

We now give some applications of Abel's theorem 1.2.15 or 1.2.3 to the convergence of series.

Theorem 1.2.20 *(Series Multiplication Theorem) Let $c_n = \sum_{k=0}^{n} a_k b_{n-k}$, and let $\sum_{n\geq 0} a_n$, $\sum_{n\geq 0} b_n$, and $\sum_{n\geq 0} c_n$ be convergent series. Then*

$$\sum_{n\geq 0} c_n = \left(\sum_{n\geq 0} a_n\right)\left(\sum_{n\geq 0} b_n\right). \tag{1.34}$$

Proof. Since multiplication can be applied to two absolutely convergent series,

$$\sum_{n\geq 0} c_n x^n = \left(\sum_{n\geq 0} a_n x^n\right)\left(\sum_{n\geq 0} b_n x^n\right) =: s_1(x)s_2(x).$$

Hence, from Abel's theorem, we have

$$\begin{aligned}\sum_{n\ge 0} c_n &= \lim_{x\to 1^-}\sum_{n\ge 0} c_n x^n = \lim_{x\to 1^-} s_1(x)s_2(x)\\ &= \lim_{x\to 1^-} s_1(x) \lim_{x\to 1^-} s_2(x) = s_1(1)s_2(1)\\ &= \left(\sum_{n\ge 0} a_n\right)\left(\sum_{n\ge 0} b_n\right).\end{aligned}$$

■

Example 1.2.21 *Use series multiplication theorem 1.2.20, we can prove*

$$\begin{aligned}&\frac{1}{2}\left(1-\frac{1}{2}+\frac{1}{3}-\frac{1}{4}+\cdots\right)^2\\ =\ &\frac{1}{2}-\frac{1}{3}\left(1+\frac{1}{2}\right)+\frac{1}{4}\left(1+\frac{1}{2}+\frac{1}{3}\right)-\frac{1}{5}\left(1+\frac{1}{2}+\frac{1}{3}+\frac{1}{4}\right)+\cdots.\end{aligned} \tag{1.35}$$

In fact, substituting $a_n = b_n = (-1)^{n+1}/n$ $(n = 1, 2, \ldots)$ and $a_0 = b_0 = 0$ in (1.34) yields

$$c_n = \sum_{k=1}^{n-1} a_k b_{n-k} = (-1)^n w_n,$$

where

$$w_n = \frac{1}{1\cdot(n-1)} + \frac{1}{2\cdot(n-2)} + \cdots + \frac{1}{(n-1)\cdot 1} \quad (n \ge 2).$$

Obviously,

$$\begin{aligned}nw_n &= \left(1+\frac{1}{n-1}\right)+\left(\frac{1}{2}+\frac{1}{n-2}\right)+\cdots+\left(\frac{1}{n-1}+1\right)\\ &= 2\left(1+\frac{1}{2}+\cdots+\frac{1}{n-1}\right) = 2\ln n + 2\gamma + o(1)\end{aligned}$$

as $n\to\infty$, where γ is the Euler constant. Hence, $w_n \to 0$ as $n\to\infty$. Since $(n+1)w_{n+1} - nw_n = 2/n$, and

$$n(w_n - w_{n+1}) = w_{n+1} - [(n+1)w_{n+1} - nw_n] = w_{n+1} - \frac{2}{n} > 0,$$

we know w_n is decreasing and tends to 0 as $n\to\infty$. Hence, $\sum_{n\ge 0} c_n = \sum_{n\ge 0}(-1)^n w_n$ converges, and (1.35) is obtained.

We now give two useful tests, *Abel's test* and *Dirichlet test*, for the convergence of the series $\sum_{n\ge 1} a_n b_n$ using the properties of series $\sum_{n\ge 1} a_n$ and $\sum_{n\ge 1} b_n$.

Theorem 1.2.22 *(Abel's test) Suppose real or complex series $\sum_{n\geq 1} a_n$ converges and $\sum_{n\geq 1}(b_n - b_{n+1})$ converges absolutely. Then series $\sum_{n\geq 1} a_n b_n$ converges.*

Proof. The theorem can be proved by using general Abel's lemma 1.2.16. Since $\sum_{n\geq 1}(b_n - b_{n+1})$ converges, limit $\lim_{n\to\infty}\sum_{k=1}^{n}(b_k - b_{k+1}) = b_1 - \lim_{n\to\infty} b_n$ exists. Hence, there exists a constant A for which

$$\left|\sum_{n=m+1}^{m+p} a_n b_n\right| \leq H_m \left\{\sum_{j=1}^{p-1} |b_{m+j} - b_{m+j+1}| + |b_{m+p}|\right\} \leq H_m A, \tag{1.36}$$

where $H_m = \max(|a_{m+1}, \cdots, |a_{m+1} + \cdots + a_{m+p}|)$. Since $\sum_{n\geq 1} a_n$ converges, for any given $\epsilon > 0$ there exists large enough m for which $H_m < \epsilon$. Therefore,

$$\left|\sum_{n=m+1}^{m+p} a_n b_n\right| < A\epsilon,$$

which implies $(s_n = \sum_{k=1}^{n} a_k b_k)$ is a Cauchy sequence and the convergence of $\sum_{n\geq 1} a_n b_n$ follows.

■

As a special case, we give the following corollary.

Corollary 1.2.23 *Let $(b_n)_{n\geq 1}$ be a bounded decreasing real sequence, and let $\sum_{n\geq 1} a_n$ be a convergent series. Then series $\sum_{n\geq 1} a_n b_n$ converges.*

Theorem 1.2.24 *(Dirichlet Test) Suppose $\sum_{k=1}^{n} a_k = O(1)$ as $n\to\infty$, and series $\sum_{n\geq 1}(b_n - b_{n+1})$ converges absolutely with $b_n \to 0$ when $n\to\infty$, where a_n and b_n can be complex numbers. Then $\sum_{n\geq 1} a_n b_n$ converges.*

Proof. From the first inequality (1.36) shown in the proof of Theorem 1.2.22, one can see the conclusion of the theorem holds because $H_m = O(1)$ and the observation $\left\{\sum_{j=1}^{p-1} |b_{m+j} - b_{m+j+1}| + |b_{m+p}|\right\} \to 0$ as $m\to\infty$.

■

Corollary 1.2.25 *If (b_n) is a decreasing sequence approaching 0 as $n\to\infty$, and $\sum_{k=1}^{n} a_k$ is bounded, then series $\sum_{n\geq 1} a_n b_n$ converges.*

Example 1.2.26 *Denote $t := \cos\theta + i\sin\theta = e^{i\theta}$. Then series $\sum_{n\geq 1}\frac{1}{n}t^n$ converges for all $\theta \neq 0$. In fact, noting*

$$\left|\sum_{n=m+1}^{m+p} t^n\right| = \left|\frac{t^{m+1}(1-t^p)}{1-t}\right| \leq \frac{2}{|1-t|}, \quad (t\neq 1),$$

and $b_n = 1/n$ implies the convergence of $\sum_{n\geq 1}|b_n - b_{n+1}|$, we immediately obtain the result.

Abel's theorem is also called Abel's power series continuity theorem since it represents the left-hand continuity of function $f(x) = \sum_{n\geq 1} a_n x^n$ at point $x = 1$. Therefore, the theorem can help us to evaluate the sums of some convergent series. We now illustrate it by using the following examples.

Example 1.2.27 *Since*

$$\sum_{n\geq 1} \frac{(-1)^{n+1}}{n} x^n = \log(1+x) \text{ and } \sum_{n\geq 0} \frac{(-1)^n}{2n+1} x^{2n+1} = \tan^{-1} x \ (0 \leq x < 1),$$

we have

$$\sum_{n\geq 1} \frac{(-1)^{n+1}}{n} = \lim_{x\to 1^-} \sum_{n\geq 1} \frac{(-1)^{n+1}}{n} x^n = \lim_{x\to 1^-} \log(1+x) = \log 2$$

and

$$\sum_{n\geq 0} \frac{(-1)^n}{2n+1} = \lim_{x\to 1^-} \sum_{n\geq 0} \frac{(-1)^n}{2n+1} x^{2n+1} = \lim_{x\to 1^-} \tan^{-1} x = \frac{\pi}{4},$$

respectively. Thus,

$$1 - \frac{1}{2} + \frac{1}{3} - \frac{1}{4} + \cdots = \log 2, \tag{1.37}$$

$$1 - \frac{1}{3} + \frac{1}{5} - \frac{1}{7} + \cdots = \frac{\pi}{4}. \tag{1.38}$$

Similarly, from

$$\sum_{n\geq 1} (-1)^{n+1} \left(\frac{n}{(n+1)(n+2)} \right) x^{n+1} = \left(1 + \frac{2}{x}\right) \log(1+x) - 2$$

for all $0 < x < 1$, *we obtain*

$$\frac{1}{2\cdot 3} - \frac{2}{3\cdot 4} + \frac{3}{4\cdot 5} - \frac{4}{5\cdot 6} + \cdots = 3\log 2 - 2.$$

The following example needs a simple calculus computation.

Example 1.2.28 *We will show that*

$$1 - \frac{1}{4} + \frac{1}{7} - \frac{1}{10} + \cdots = \frac{1}{3}\log 2 + \frac{2}{\sqrt{3}} \tan^{-1} \frac{1}{\sqrt{3}} = \frac{1}{3}\log 2 + \frac{\pi}{3\sqrt{3}}. \tag{1.39}$$

Denote

$$f(x) = x - \frac{1}{4}x^4 + \frac{1}{7}x^7 - \frac{1}{10}x^{10} + \cdots \quad (|x| < 1),$$

and $f(0) = 0$. *Since*

$$f'(x) = 1 - x^3 + x^6 - \cdots = \frac{1}{1+x^3}, \quad (|x| < 1),$$

we have

$$\begin{aligned} f(x) &= \int_0^x \frac{dt}{1+t^3} = \frac{1}{3}\log(1+x) - \frac{1}{6}\log(1-x+x^2) \\ &\quad + \frac{1}{\sqrt{3}}\left(\tan^{-1}\left(\frac{2x-1}{\sqrt{3}}\right) + \tan^{-1}\left(\frac{1}{\sqrt{3}}\right)\right), \end{aligned}$$

where the lower limit of the integer is 0 *is due to* $f(0) = 0$. *Hence, applying Abel's theorem yields* (1.39).

Proposition 1.2.29 *Suppose* $a \neq b$, $a > -1$, $b > -1$. *Then*

$$\sum_{n\geq 1} \frac{1}{(n+a)(n+b)} = \frac{1}{b-a}\int_0^{1^-} \frac{x^a - x^b}{1-x} dx. \tag{1.40}$$

Proof. It is obvious that the left-hand of (1.40) converges. Write

$$\sum_{n\geq 1} \frac{1}{(n+a)(n+b)} = \frac{1}{b-a}\sum_{n\geq 1}\left(\frac{1}{n+a} - \frac{1}{n+b}\right).$$

Denote

$$f(x) := \frac{1}{b-a}\sum_{n\geq 1}\left(\frac{x^{n+a}}{n+a} - \frac{x^{n+b}}{n+b}\right), \quad (|x| < 1).$$

Thus,

$$f'(x) = \frac{1}{b-a}\sum_{n\geq 1}(x^{n+a-1} - x^{n+b-1}) = \frac{1}{b-a}\left(\frac{x^a}{1-x} - \frac{x^b}{1-x}\right).$$

Because $f(0) = 0$,

$$f(x) = \int_0^x \frac{1}{b-a}\left(\frac{t^a - t^b}{1-t}\right) dt, \quad (0 \leq x < 1).$$

Since

$$\frac{t^a - t^b}{1-t} = \frac{(1-t^b) - (1-t^a)}{1-t} \to b - a$$

as $t \to 1^-$, we obtain the left-hand continuity of $f(x)$ at point $x = 1$. Therefore, applying Abel's theorem yields

$$\sum_{n\geq 1} \frac{1}{(n+a)(n+b)} = f(1^-),$$

and (1.40) follows. ∎

We now give an extension of Abel's theorem.

Theorem 1.2.30 *Assume positive continuous function sequence* $(v_n(x))_{n\geq 0}$ *is decreasing for all* $x \in (0,1)$ *as* n *is increasing, and* $\lim_{x\to 1^-} v_n(x) = 1$ $(n = 0,1,2,\ldots)$*. Then, the convergence of series* $\sum_{n\geq 0} a_n$ *implies*

$$\lim_{x\to 1^-} \sum_{n\geq 0} a_n v_n(x) = \sum_{n\geq 0} a_n.$$

Proof. The proof is similar as the proof of Abel's theorem. Since $v_n(x) \to 1$ as $x \to 1^-$, we may define $v_n(1) = 1$ $(n = 0,1,2,\ldots)$. Hence, there exists a positive constant A and $\delta > 0$ for which $v_0(x) < A$ for all $1-\delta \leq x \leq 1$. For an arbitrary $\epsilon > 0$, there exists $N = N_\epsilon > 0$ such that $m \geq n > N$ implies

$$\left|\sum_{k=n}^{m} a_k\right| < \epsilon, \quad (m \geq n > N).$$

Therefore,

$$\left|\sum_{k=n}^{m} a_k v_k(x)\right| \leq v_0(x) \max_{m\geq n}\left|\sum_{k=n}^{m} a_k\right| < \epsilon A, \quad (1-\delta \leq x \leq 1).$$

Hence, $\sum_{n\geq 0} a_n v_n(x)$ is uniformly convergent on $[1-\delta, 1]$, which implies the left-hand continuity of $\sum_{n\geq 0} a_n v_n(x)$ at point $x = 1$.

■

Corollary 1.2.31 *Let series* $\sum_{n\geq 1} n a_n$ *converge. Denote* $f(x) = \sum_{n\geq 1} a_n x^n$ $(|x| < 1)$*. Then*

$$\sum_{n\geq 1} n a_n = \lim_{x\to 1^-} \frac{f(1) - f(x)}{1-x}.$$

Proof. Noting $\sum_{n\geq 1} a_n = \sum_{n\geq 1}(n a_n)\frac{1}{n}$ and using Corollary 1.2.25, we obtain the convergence of $\sum_{n\geq 1} a_n$ from the convergence of $\sum_{n\geq 1} n a_n$. Since $v_n(x) = (1-x^n)/(n(1-x))$ generates a decreasing sequence with $\lim_{x\to 1^-} v_n(x) = 1$ $(n = 1,2,\ldots)$, from the convergence of $\sum_{n\geq 1}(n a_n)v_n(x)$ and Theorem 1.2.30 we obtain the desired result.

■

1.3 Series Method

In the first section, we describe the summation of series using the infinitesimal calculus. However, for the functions of the independent variable taking only the given discrete values x_i $(i = 0, 1, \ldots, n)$, the methods of infinitesimal calculus are not applicable. The *calculus of finite difference* deals especially with those functions while it may be applied to the functions with continuous independent variables as well. In the second part of this section, we will present the methods by applying the Euler-Maclaurin formula and the Bernoulli polynomials.

1.3.1 Use of the calculus of finite difference

The calculus of finite difference was originally introduced by Brook Taylor in his Methodus Incrementorum (Londo, 1717). And the solid foundation was established by Jacob Stirling in his Methodus Differentialis (London, 1730), which applied useful methods to solve very advanced problems including Stirling numbers and their applications. As what was pointed in Jordan's book [141], "(Jacob Stirling's work) formed the backbone of the Calculus of Finite Differences". In this subsection, we shall present some symbols used in this section and the later contents. The first symbol is Δ, the *difference*, which is first introduced in Leonhardo Eulero, Institutiones Calculi Differentialis (Academiae Imperialis Scientiarum Petropolitanae, 1755).

A function $f(x)$ is defined at $x = x_0, x_1, \ldots, x_n$, where the values of variable x are equidistant for the sake of the real advantages of the theory of finite differences. Hence we denote

$$h = x_{i+1} - x_i, \quad (i = 0, 1, \ldots, n-1),$$

i.e., h is independent of i. The first difference is defined by

$$\Delta_h f(x) := f(x+h) - f(x). \tag{1.41}$$

In particular, we denote $\Delta \equiv \Delta_1$. The difference of the first difference is called the second difference denoting by Δ_h^2 and $\Delta^2 \equiv \Delta_1^2$ for $h = 1$. Hence,

$$\begin{aligned} \Delta_h^2 f(x) &= \Delta_h(\Delta_h f(x)) = \Delta_h f(x+h) - \Delta_h f(x) \\ &= f(x+2h) - 2f(x+h) + f(x). \end{aligned}$$

In general, we define the n–th difference of $f(x)$ by

$$\Delta_h^n f(x) = \Delta_h(\Delta_h^{n-1} f(x)) = \Delta_h^{n-1} f(x+h) - \Delta_h^{n-1} f(x). \tag{1.42}$$

Obviously, $\Delta_h^n f(x) = \Delta_h^{n-1}(\Delta_h f(x))$. By means of mathematical induction, one can prove the following expression of Δ_h^n (cf. Exercise 1.28).

$$\Delta_h^n f(x) = \sum_{j=0}^{n} \binom{n}{j} (-1)^{n-j} f(x+jh). \tag{1.43}$$

We may use the following table to evaluate $\Delta^n(x)$ successively. In general, we write the values of function f in the first column of the table, the first differences in the second column, the second differences in the third column, and so on. More precisely, in the table below, we begin the first column with $f(x)$, then we shall write the first difference $\Delta_h f(x)$ in the second column and along the row between $f(x)$ and $f(x+h)$. The second difference $\Delta_h^2 f(x)$ will be put into the third column and along the row between $\Delta_h f(x)$ and $\Delta_h f(x+h)$ and so on.

$$
\begin{array}{lllll}
f(x) & & & & \\
 & \Delta_h f(x) & & & \\
f(x+h) & & \Delta_h^2 f(x) & & \\
 & \Delta_h f(x+h) & & \Delta_h^3 f(x) & \\
f(x+2h) & & \Delta_h^2 f(x+h) & & \Delta_h^4 f(x) \\
 & \Delta_h f(x+2h) & & \Delta_h^3 f(x+h) & \\
f(x+3h) & & \Delta_h^2 f(x+2h) & & \\
 & \Delta_h f(x+3h) & & & \\
f(x+4h) & & & &
\end{array} \tag{1.44}
$$

A polynomial in terms of operator Δ_h,

$$P(\Delta_h) = p_0 + p_1\Delta_h + p_2\Delta_h^2 + \cdots + p_n\Delta_h^n,$$

is also an operator, called operator polynomial, where coefficients p_i $(i = 0, 1, \ldots, n)$ are real or complex numbers. Applying $P(\Delta_h)$ to function $f(x)$ yields

$$P(\Delta_h)f(x) = p_0 f(x) + p_1\Delta_h f(x) + \cdots + p_n\Delta_h^n f(x).$$

we also denote by I(or 1) and 0 the unit and zero operators, respectively, which are defined by

$$I = 1 = \Delta_h^0, \; If(x) = 1f(x) = \Delta_h^0 f(x) = f(x), \; 0f(x) = 0.$$

We also define $\Delta_h^s + 0 = 0 + \Delta_h^s$.

It is easy to show that difference of a sum is

$$\Delta_h(f(x) + g(x)) = \Delta_h f(x) + \Delta_h g(x).$$

and if c is a constant that

$$\Delta_h cf(x) = c\Delta_h f(x).$$

Hence, all operator polynomials in the real or complex field with respect to polynomial addition and regular scale multiplication form an Abelian ring. If $Q(\Delta_h)$ is another operator polynomial, then

$$P(\Delta_h)(Q(\Delta_h)f(x)) = (P(\Delta_h)Q(\Delta_h))f(x) = Q(\Delta_h)(P(\Delta_h)f(x)).$$

In infinitesimal calculus the first derivative of a function $f(x)$, generally denoted by $Df(x)$ or $df(x)/dx$, is given by

$$Df(x) = \lim_{h\to 0} \frac{f(x+h)-f(x)}{h} = \lim_{h\to 0} \frac{\Delta_h f(x)}{h}.$$

Moreover, it is shown that the n-th derivative of $f(x)$ is

$$D^n f(x) = \lim_{h\to 0} \frac{\Delta_h^n f(x)}{h^n}.$$

In Boole's Calculus of Finite Differences, an important operator, called *displacement* or *shift*, was introduced. This consists for a given $f(x)$ in increasing the variable x by h. Denoting this shift operator by E_h, we have

$$E_h f(x) = f(x+h).$$

In particular, $E_1 \equiv E$. The operation E_h^2 is defined by

$$E_h^2 f(x) = E_h(E_h f(x)) = E_h f(x+h) = f(x+2h).$$

In the same way,

$$E_h^n f(x) = E_h(E_h^{n-1} f(X)) = f(x+nh),$$

and $E_h^0 = I = 1$. Obviously, $E_h^n f(x) = E_h(E_h^{n-1} f(X))$. It is easy to extend this operation to negative indices of E_h so that

$$\frac{1}{E_h} f(x) = E_h^{-1} f(x) = f(x-h), \ \frac{1}{E_h^n} f(x) = E_h^{-n} f(x) = f(x-nh).$$

It is easy to see the symbol E_h is also distributive and commutative:

$$E_h^n(f(x)+g(x)) = E_h^n f(x) + E_h^n g(x),$$
$$E_h^n E_h^m f(x) = E_h^m E_h^n f(x) = E_h^{n+m} f(x).$$

In addition, if c is a constant, then we have $E_h c f(x) = cE_h f(x)$. The operation E_h^{-n} behaves exactly like E_h^m. For instance,

$$E_h^{-n} E_h^m = E_h^m E_h^{-n} = E_h^{m-n}.$$

Moreover,

$$E_h(f(x)g(x)) = f(x+h)g(x+h) = E_h f(x) E_h g(x).$$

These are not true for the other symbols introduced.

In this book, we mainly discuss the symbols Δ_h, D_h, and E_h, particularly, Δ, D, and E. Some other operations such as operation of mean (M_h) defined by $M_h f(x) = (f(x) + f(x+h))/2$, central difference operator (δ_h) defined by

$\delta_h f(x) = f(x+h/2) - f(x-h/2)$, etc. will be introduced later when they are used.

From the definition of the operations Δ_h, D_h, and E_h, it is easy to see that their symbols are connected, for instance, by the following relations

$$E_h = 1 + \Delta_h = e^{hD_h}, \ \Delta_h = E_h - 1 = e^{hD_h} - 1, \ hD_h = \log(1+\Delta_h). \quad (1.45)$$

To prove the first of (1.45) we write

$$(1+\Delta_h)f(x) = f(x) + \Delta_h f(x) = f(x+h) = E_h f(x)$$

and

$$E_h f(x) = f(x+h) = f(x) + hD_h f(x) + \frac{h^2}{2!}D_h^2 f(x) + \cdots = e^{hD_h} f(x).$$

Others of (1.45) can be proved similarly.

Using (1.45), we have

$$E_h^n = (1+\Delta_h)^n = \sum_{k=0}^{n} \binom{n}{k} \Delta_h^k.$$

Applying the above operation on $f(x)$ yields expression

$$f(x+nh) = f(x) + \binom{n}{1}\Delta_h f(x) + \binom{n}{2}\Delta_h^2 f(x) + \cdots + \binom{n}{n}\Delta_h^n f(x).$$

Similarly, relation

$$\Delta_h^n = (E_h - 1)^n = \sum_{k=0}^{n} (-1)^k \binom{n}{k} E_h^{n-k}$$

generating

$$\Delta_h^n f(x) = f(x+nh) - \binom{n}{1} f(x+(n-1)h) + \cdots + (-1)^n \binom{n}{n} f(x).$$

For any real number x, we define $E^x f(0) = f(x)$. Thus from *Newton's formula*

$$f(x) = f(0) + \binom{x}{1}\Delta f(0) + \binom{x}{2}\Delta^2 f(0) + \cdots, \quad (1.46)$$

we obtain the series expression of the operator E^x,

$$E^x = 1 + \binom{x}{1}\Delta + \binom{x}{2}\Delta^2 + \cdots,$$

if the series in (1.46) converges. If certain conditions are satisfied, the infinite series expansion of operators are permitted.

Lemma 1.3.1 *Let $f(x)$ be a real-valued function, and let x be an integer satisfying $0 \le x \le n$. Then*

$$f(x) = f(0) + \binom{x}{1}\Delta f(0) + \binom{x}{2}\Delta^2 f(0) + \cdots + \binom{x}{n}\Delta^n f(0). \tag{1.47}$$

Proof. The expansion (1.47) can be proved directly using symbolic method.

$$\begin{aligned} f(x) &= E^x f(0) = (1+\Delta)^x f(0) \\ &= \left(\sum_{k=0}^{x} \binom{x}{k}\Delta^k\right) f(0) = \sum_{k=0}^{x}\binom{x}{k}\Delta^k f(0). \end{aligned}$$

■

Let $f(x)$ be a function of real variable. We write Newton's formula (1.46) with remainder $R_n(x)$ as follows.

Theorem 1.3.2 *Let $f(x)$ be a real-valued function. Then*

$$f(x) = f(0) + \binom{x}{1}\Delta f(0) + \cdots + \binom{x}{n}\Delta^n f(0) + R_n(x), \tag{1.48}$$

where

$$R_n(x) = \begin{vmatrix} 1 & 0 & \cdots & 0 & f(0) \\ 1 & 1^1 & \cdots & 1^n & f(1) \\ \vdots & \vdots & \cdots & \vdots & \vdots \\ 1 & n^1 & \cdots & n^n & f(n) \\ 1 & x^2 & \cdots & x^n & f(x) \end{vmatrix} \Bigg/ \begin{vmatrix} 1^1 & 1^2 & \cdots & 1^n \\ 2^1 & 2^2 & \cdots & 2^n \\ \vdots & \vdots & \cdots & \vdots \\ n^1 & n^2 & \cdots & n^n \end{vmatrix}. \tag{1.49}$$

In particular, if $f(x)$ is a polynomial of degree n then $\Delta^{n+1} f(x) = 0$ and the corresponding series is finite:

$$f(x) = f(0) + \binom{x}{1}\Delta f(0) + \cdots + \binom{x}{n}\Delta^n f(0).$$

Proof. After expanding the numerator determinant on the right-hand side of (1.49) in terms of the last row of the determinant, we may see $R_n(x) = f(x) - p_n(x)$, where $p_n(x)$ is a polynomial of degree $\le n$. It is obvious that $R(x) = 0$ at $x = 0, 1, \ldots, n$. Thus, $p_n(x) = f(x)$ at $x = 0, 1, \ldots, n$. On the other hand, $q_n(x) = \sum_{k=0}^{n}\binom{x}{k}\Delta^k f(0)$ is also a polynomial of degree $\le n$. From Lemma 1.3.1, $q_n(x) = f(x)$ at $x = 0, 1, \ldots, n$. Since two polynomials of degree $\le n$ satisfy $p_n(x) = q_n(x)$ at $n+1$ distinct points $x = 0, 1, \ldots, n$, we have $p_n(x) = q_n(x)$, which implies (1.48).

■

We now present an inverse formula of Lemma 1.3.1.

Theorem 1.3.3 *Let $f(x)$ be a real-valued function. Then*

$$\Delta^n f(0) = f(n) - \binom{n}{1} f(n-1) + \binom{n}{2} f(n-2) - \cdots + (-1)^n \binom{n}{n} f(0), \quad (1.50)$$

which is an inverse relation of (1.47) at $x = n$; i.e., one can be derived from another one.

Proof. Applying symbol relation $\Delta^n = (E-1)^n = \sum_{k=0}^{n} \binom{n}{k}(-1)^k E^{n-k}$ to $f(0)$, we obtain (1.50) immediately. Since $\Delta^n = (E-1)^n$ and $E^n = (\Delta+1)^n$, when $x = n$ (1.47) and (1.50) form a pair of inverse relation. ■

From the inverse relations (1.47) and (1.50), we can establish a pair of more general inverse formulas.

Theorem 1.3.4 *Let $(g(k))_{k\geq 0}$ be the real-valued sequence of function $g(x)$ and sequence $(f(n))_{n\geq 0}$ is defined by*

$$f(n) = \sum_{k=0}^{n} \binom{n}{k} (-1)^k g(k), \quad n = 0, 1, 2, \ldots. \quad (1.51)$$

Then

$$g(n) = \sum_{k=0}^{n} \binom{n}{k} (-1)^k f(k), \quad n = 0, 1, 2, \ldots, \quad (1.52)$$

where $f(0) = g(0)$. Conversely, (1.52) implies (1.51).

Proof. For given $\{g(k)\}_{k=0}^{\infty}$, we can define a function such that $\Delta^k f(0) = (-1)^k g(k)$ $(k = 0, 1, \ldots)$; i.e., $f(0) = g(0)$ and $f(k) = (-1)^k g(k) - \sum_{j=0}^{k-1} \binom{k}{j}(-1)^{k-j} f(j)$ $(k \geq 1)$ successively using (1.43). It can be seen that (1.51) and (1.47) are equivalently under the transformation $\Delta^k f(0) = (-1)^k g(k)$ $(k = 0, 1, \ldots)$. Since (1.47) implies (1.50), we obtain (1.52) from (1.51). Similarly, we can show (1.52) implies (1.51).

The inverse relationship of (1.51) and (1.52) can also be proved by substitution. More precisely, to prove (1.51) $\Rightarrow$ (1.52), i.e., given (1.51) $g(n)$ can be found from (1.52), we substitute (1.51) into (1.52) and obtain

$$\begin{aligned} &\sum_{k=0}^{n} \binom{n}{k} (-1)^k \left(\sum_{\ell=0}^{k} \binom{k}{\ell} (-1)^\ell g(\ell) \right) \\ = &\sum_{\ell=0}^{k} (-1)^\ell g(\ell) \left(\sum_{k=0}^{n} (-1)^k \binom{n}{k} \binom{k}{\ell} \right). \end{aligned} \quad (1.53)$$

Noting

$$\sum_{k=\ell}^{n} (-1)^k \binom{n}{k} \binom{k}{\ell} = (-1)^n \delta_{n,\ell},$$

where Kronecker delta $\delta_{n,\ell} = 0$ when $\ell \neq n$ and $\delta_{n,n} = 1$, one may know the right-hand side of (1.53) yields $g(n)$.

We will see in Section 4.3, the lower triangular matrix $((-1)^k \binom{n}{k})$ is a Riordan array involution. Thus, the inverse relationship between (1.51) and (1.52) is a consequence of the Riordan involutions.

■

Theorem 1.3.5 *Let n be a non-negative integer. Then we have inverse formulas*

$$f(n) = \sum_{k=0}^{m+n} \binom{m+n}{k} g(k) \leftrightarrow \Delta^n f(0) = \sum_{k=0}^{m} \binom{m}{k} g(n+k) \tag{1.54}$$

for $n = 0, 1, 2, \ldots$.

Proof. Obviously, the left equation of (1.54) can be written as

$$f(n) = \left(\sum_{k=0}^{m+n} \binom{m+n}{k} E^k \right) g(0) = (1+E)^{m+n} g(0) \tag{1.55}$$

for $n = 0, 1, 2, \ldots$. And the right equation of (1.54) is equivalently

$$\begin{aligned} \Delta^n f(0) &= (1+E)^m E^n g(0) \\ &= (-1)^n (1+E)^m (1-(1+E))^n g(0) \\ &= \sum_{k=0}^{n} (-1)^{n-k} \binom{n}{k} \{(1+E)^{m+k} g(0)\} \end{aligned} \tag{1.56}$$

for $n = 0, 1, 2, \ldots$. Applying Theorem 1.3.4 to (1.56), we obtain the inverse expression

$$(1+E)^{m+n} g(0) = \sum_{k=0}^{n} \binom{n}{k} (-1)^{k-k} \Delta^k f(0) = f(n)$$

for $n = 0, 1, 2, \ldots$. This shows that (1.56) implies (1.55). Conversely, substituting (1.55) into the rightmost of (1.56) and noting (1.50), we obtain

$$\sum_{k=0}^{n} (-1)^{n-k} \binom{n}{k} f(k) = \Delta^n f(0)$$

for $n = 0, 1, 2, \ldots$, which means that (1.55) can be derived from (1.56).

■

Another example of symbol expansion is the *Newton's backward formula.* Since $E_h - \Delta_h = 1$,

$$E_h^x = \left(\frac{E_h}{E_h - \Delta_h} \right)^x = 1 / \left(1 - \frac{\Delta_h}{E_h} \right)^x .$$

For $h = 1$, we expand the denominator of the rightmost term into infinite series

$$E^x = 1/\left(1-\frac{\Delta}{E}\right)^x = \sum_{k\geq 0}\binom{x+k-1}{k}\frac{\Delta^k}{E^k}, \tag{1.57}$$

which is applied to function f at point 0 and generates

$$f(x) = f(0) + \binom{x}{1}\Delta f(-1) + \binom{x+1}{2}\Delta^2 f(-2) + \cdots \tag{1.58}$$

if f is defined on $\mathbb{Z}$ and the series on the right-hand side converges.

An analog of (1.57) with $x = 1$ and general h is written as

$$E_h = \sum_{k\geq 0}\left(\frac{\Delta_h}{E_h}\right)^k,$$

and it can be performed on $f(x)$ and yields

$$f(x+h) = \sum_{k\geq 0}\Delta_h^k f(x-kh).$$

We may obtain a modified formula of (1.57) with general h with starting form

$$E_h^{x+1} = 1/\left(1-\frac{\Delta_h}{E_h}\right)^{x+1}.$$

Thus, expanding the right-hand term yields

$$E_h^x = \sum_{k\geq 0}\binom{x+k}{k}\frac{\Delta_h^k}{E_h^{k+1}}.$$

In the same manner,

$$E_h^{-n} = \frac{1}{(1+\Delta_h)^n} = \sum_{k\geq 0}(-1)^k\binom{n+k-1}{k}\Delta_h^k,$$

which give

$$f(x-nh) = \sum_{k\geq 0}(-1)^k\binom{n+k-1}{k}\Delta_h^k f(x).$$

Theorem 1.3.6 *Let $f(x)$ be a polynomial of degree k. Then*

$$\begin{aligned} f(n) &= f(-1) + \binom{n+1}{1}\Delta f(-2) + \binom{n+2}{2}\Delta^2 f(-3) \\ &\quad + \cdots + \binom{n+k}{k}\Delta^k f(-k-1). \end{aligned} \tag{1.59}$$

Proof. The proof is similar as the way in deriving (1.58). Since $f(x)$ is a polynomial of degree k, $\Delta^n f(x) = 0$ $(n = k+1, k+2, \ldots)$. Hence, from (1.57)

$$\begin{aligned} & (1 - E^{-1}\Delta)^{n+1}\{1 + \binom{n+1}{1} E^{-1}\Delta + \binom{n+2}{2}(E^{-1}\Delta)^2 + \cdots \\ & + \binom{n+k}{k}(E^{-1}\Delta)^k\} f(x) = (1 - E^{-1}\Delta)^{n+1} \frac{1}{(1 - E^{-1}\Delta)^{n+1}} f(x) \\ = \; & f(x) \end{aligned}$$

■

Theorem 1.3.7 *(Bernoulli Summation Formula) Let $f(x)$ be any function defined at $x = 0, 1, 2, \ldots$. Then for any $n \in \mathbb{N}$*

$$\sum_{k=1}^{n} f(k) = \sum_{k=1}^{n} \binom{n}{k} \Delta^{k-1} f(1). \tag{1.60}$$

Proof. (1.60) can be proved using Lemma 1.3.1. Here, we give a direct proof using symbol relation

$$\frac{E^{n+1} - E}{E - 1} = \sum_{k=1}^{n} E^k.$$

Hence,

$$\begin{aligned} \sum_{k=1}^{n} E^k &= \frac{E}{E-1}[(E - 1 + 1)^n - 1] \\ = \; & \frac{E}{E-1} \sum_{k=1}^{n} \binom{n}{k} (E-1)^k = \sum_{k=1}^{n} \binom{n}{k} \Delta^{k-1} E. \end{aligned}$$

Applying the leftmost side and the rightmost sides of the above operator equation to $f(0)$ yields (1.60).

■

Example 1.3.8 *In Bernoulli summation formula, substituting $f(x) = t^x$ and noting*

$$\Delta^k t^x \big|_{x=1} = \sum_{j=0}^{k} \binom{k}{j} (-1)^{k-j} t^{x+j} \big|_{x=1} = t(t-1)^k,$$

we obtain Identity

$$t + t^2 + \cdots + t^n = nt + \binom{n}{2} t(t-1) + \binom{n}{3} t(t-1)^2 + \cdots + t(t-1)^{n-1}.$$

Example 1.3.9 *Using Bernoulli summation formula, we can establish*

$$\sum_{k=1}^{n} k^4 = \binom{n}{1} + 15\binom{n}{2} + 50\binom{n}{3} + 60\binom{n}{4} + 24\binom{n}{5}. \tag{1.61}$$

In fact, by setting $f(k) = k^4$ *and using* (1.44) *to find* $\Delta f(1) = 15$, $\Delta^2 f(1) = 50$, $\Delta^3 f(1) = 60$, *and* $\Delta^4 f(1) = 24$, (1.61) *follows from* (1.60).

In next chapter, we shall give more construction of summation formulas as well as series transformation formulas using symbolic method. A series transformation formula represents a relation between different series in order to evaluate sums of (infinite) series or accelerate the convergence rate of convergent (infinite) series. Here is the first formula derived by Montmort.

Theorem 1.3.10 *(Montmort Series Transformation Formula) Suppose* $|x| < 1$, $|x/(1-x)| < 1$. *Then*

$$\sum_{n\geq 1} f(n)x^n = \sum_{n\geq 1}\left(\frac{x}{1-x}\right)^n \Delta^{n-1} f(1), \tag{1.62}$$

where the left-hand series is assumed to be convergent for all $|x| < 1$.

Proof. Denote $y = x/(1-x)$. From the given conditions

$$|y| = \left|\frac{x}{1-x}\right| < 1, \qquad \left|\frac{y}{1+y}\right| = |x| < 1.$$

Since

$$\begin{aligned}(1+y)^{-n} &= \sum_{k\geq 0}(-1)^k\binom{n+k-1}{k}y^k \\ &= 1 - \binom{n}{1}y + \binom{n+1}{2}y^2 - \binom{n+2}{3}y^3 + \cdots,\end{aligned}$$

we have

$$\begin{aligned}S : &= \sum_{n\geq 1} f(n)x^n = \sum_{n\geq 1} f(n)\left(\frac{y}{1+y}\right)^n \\ &= \sum_{n\geq 1}\sum_{k\geq 0}\binom{n+k-1}{k} f(n)y^{n+k}.\end{aligned}$$

Because $|y| < 1$ and $|y/(1+y)| < 1$, the above double series is convergent absolutely, and it can be rearranged as

$$S = \sum_{j\geq 1}\sum_{n=1}^{j}(-1)^{j-n}\binom{j-1}{j-n} f(n)y^j$$

$$= \sum_{j\geq 1}\left(\sum_{n=0}^{j-1}(-1)^{j-1-n}\binom{j-1}{j-1-n}f(n+1)\right)y^j$$
$$= \sum_{j\geq 1}\Delta^{j-1}f(1)y^j = f(1)y+\Delta f(1)y^2+\Delta^2 f(1)y^3+\Delta^3 f(1)y^4+\cdots.$$

Here, we have used the expansion of Δ^n shown in (1.43) into the step before the last one. After substituting $y = x/(1-x)$ into the rightmost series, we obtain (1.62).

■

We now give a special case of Montmort formula–*Euler series transformation formula.* First, we establish the following lemma.

Lemma 1.3.11 *If sequence $(a_n)_{n\geq 0}$ converges to zero, then sequence*

$$b_n = \frac{1}{2^n}\sum_{k=0}^{n}\binom{n}{k}a_k$$

$(n = 0, 1, 2, \ldots)$ *is also convergent to zero.*

Proof. For any $m \geq n$, denote $\epsilon_m = \sup\{a_m, a_{m+1}, \ldots\}$. Then

$$|b_n| \leq \frac{\epsilon_0}{2^n}\sum_{k=0}^{m-1}\binom{n}{k}+\frac{\epsilon_m}{2^n}\sum_{k=m}^{n}\binom{n}{k}$$
$$\leq \epsilon_0\frac{mn^{m-1}}{2^n}+\epsilon_m,$$

which implies $\overline{\lim}_{n\to\infty}|b_n| \leq \epsilon_m$. Since $a_n \to 0$ as $n \to \infty$, we have $\epsilon_m \to 0$ as $m \to \infty$. Therefore, $\overline{\lim}_{n\to\infty}|b_n| = 0$, i.e., $\lim_{n\to\infty} b_n = 0$.

■

Corollary 1.3.12 *(Euler series transformation formula) Let sequence $(f(n))_{n\geq 1}$ be decreasing and convergent to 0 as $n \to \infty$. Then*

$$\sum_{n\geq 0}(-1)^n f(n+1) = \sum_{k\geq 0}\frac{1}{2^{k+1}}(-1)^k\Delta^k f(1) \tag{1.63}$$

$$= \sum_{k\geq 0}^{n}\frac{1}{2^{k+1}}(-1)^k\Delta^k f(1)+\frac{(-1)^{n+1}}{2^{n+1}}\sum_{j\geq 0}(-1)^j\Delta^{n+1}f(j+1) \tag{1.64}$$

Proof. From Theorem 1.3.10, by replacing x to $-x$, we have

$$\sum_{n\geq 1}(-1)^n f(n)x^n = \sum_{n\geq 1}(-1)^n\Delta^{n-1}f(1)\left(\frac{x}{1+x}\right)^n.$$

Hence, applying Abel's continuity theorem 1.2.3 or 1.2.15 to both sides of the above equation yields

$$\sum_{n\geq 1}(-1)^n f(n) = \sum_{n\geq 1}(-1)^n \Delta^{n-1} f(1)\left(\frac{1}{2}\right)^n.$$

To prove (1.64), we only need to show that the second series tends to zero when $n \to \infty$. Indeed, after substituting the expansion of Δ^n shown in (1.43) into the series, we obtain

$$\begin{aligned} &\frac{(-1)^{n+1}}{2^{n+1}}\sum_{j\geq 0}(-1)^j \Delta^{n+1} f(j+1) \\ =\ & \frac{1}{2^{n+1}}\sum_{k=0}^{n+1}\binom{n+1}{k}\sum_{j\geq 0}(-1)^{j-k} f(j+1+k) \\ =\ & \frac{1}{2^{n+1}}\sum_{k=0}^{n+1}\binom{n+1}{k}\sum_{j\geq k}(-1)^j f(j+1), \end{aligned}$$

where $\sum_{j\geq k}(-1)^j f(j+1)$ is the remainder of convergent series $\sum_{j\geq 0}(-1)^j f(j+1)$, which approaches to zero as $k \to \infty$. Thus, from Lemma 1.3.11, the rightmost term of the above equation tends to zero when $n \to \infty$.

■

Example 1.3.13 *We now apply the Euler series transformation formula to accelerate the convergent rate of the following given series.*

$$1 - \frac{1}{3} + \frac{1}{5} - \frac{1}{7} + \cdots = \frac{\pi}{4}, \quad 1 - \frac{1}{2} + \frac{1}{3} - \frac{1}{4} + \cdots = \log 2.$$

For the first series, let $f(n) = 1/(2n-1)$, *thus,*

$$\Delta^k f(1) = (-1)^k 2^k \frac{k!}{(2k+1)!!}, \quad (k = 1, 2, \ldots).$$

Thus, from (1.63),

$$\frac{\pi}{2} = 1 + \frac{1}{3} + \frac{1\cdot 2}{3\cdot 5} + \frac{1\cdot 2\cdot 3}{3\cdot 5\cdot 7} + \cdots.$$

Similarly,

$$\log 2 = \frac{1}{2} + \frac{1}{2\cdot 2^2} + \frac{1}{3\cdot 2^3} + \frac{1}{4\cdot 2^4} + \cdots.$$

Exercise 1.35 shows that the Euler series transformation formula does not always change alternating series to the series with higher convergent rates. To guarantee the Euler series transformation formula works, Exercise 1.37

presents a sufficient condition (See also [135]). More series transformation formulas will be given in Chapter 2.

The connection among other symbols, for example,

$$\delta_h = \frac{\Delta_h}{E_h^{1/2}}, \ \delta_h^{2n} = \frac{\Delta_h^{2n}}{E_h^n},$$

will also be described in Chapter 2.

1.3.2 Application of Euler-Maclaurin formula and the Bernoulli polynomials

In 1736, by comparing summation of $f(k)$ and integer of $f(x)$ for a positive and strictly decreasing function, Euler obtained what came to known as Euler's summation (*cf.* [60] and [61]). The formula is a powerful tool for estimating sums by integers and also for evaluating integers in terms of sums. Colin Maclaurin discovered the formula independently and used it in his *Treatise of Fluxions*, published in 1742. The formula was referred to as the Euler-Maclaurin summation formula and was widely used in numerical analysis, analytic number theory, and asymptotic expansions. The formula contains *Bernoulli numbers, periodic Bernoulli functions*, and, in particular, *Bernoulli polynomials*, a kind of *Sheffer type polynomials*. Hence, we start from the following concept.

Definition 1.3.14 *Let $f(t)$ and $g(t)$ be two formal power series defined in the real number field $\mathbb{R}$ or the complex number field $\mathbb{C}$ with $f(0) = 1$, $g(0) = 0$ and $g'(0) \neq 0$. Then the polynomials $p_n(x)$ $(n = 0, 1, 2, \cdots)$ defined by the generating function (GF)*

$$f(t)e^{xg(t)} = \sum_{n\geq 0} \frac{p_n(x)}{n!} t^n \tag{1.65}$$

are called Sheffer-type polynomials, where $p_0(x) = 1$. In particular, if $g(t) = t$, the corresponding Sheffer-type polynomials are called Appell type polynomials. Accordingly, $p_n(D)$ with $D = d/dt$ is called Sheffer-type differential operator of degree n associated with $f(t)$ and $g(t)$. In particular, $p_0(D) = I$ is the identity operator.

Furthermore, numbers P_n defined by

$$f(t) = \sum_{n\geq 0} \frac{P_n}{n!} t^n \tag{1.66}$$

are called the Sheffer-type numbers. Clearly, $P_n = p_n(0)$, i.e., P_n is the value of Sheffer-type polynomial at point $x = 0$.

For $g(t) = t$, if $f(t) = t/(e^t - 1)$ and $f(t) = 2/(e^t + 1)$, then corresponding Appell (Sheffer) type polynomials are denoted by $B_n(x)$ and $E_n(x)$ and called the Bernoulli polynomials and Euler polynomials, respectively; namely,

$$\frac{t}{e^t - 1} e^{xt} = \sum_{n \geq 0} \frac{B_n(x)}{n!} t^n \tag{1.67}$$

$$\frac{2}{e^t + 1} e^{xt} = \sum_{n \geq 0} \frac{E_n(x)}{n!} t^n. \tag{1.68}$$

Bernoulli polynomials and Euler polynomials play fundamental roles in various branches of mathematics including combinatorics, number theory, special functions and analysis, see for example [34, 57, 142, 163, 167, 207, 205, 206, 219]. Since the function $f(t)$ has value 1 at $x = 0$, its power series expansion shown in (1.66) has a positive convergence radius. Hence, P_n is well defined by (1.66). When $f(t) = t/(e^t - 1)$, the corresponding Sheffer-type numbers are called Bernoulli numbers and denoted by B_n. Hence,

$$\sum_{n \geq 0} \frac{B_n}{n!} t^n = \frac{t}{e^t - 1}. \tag{1.69}$$

Theorem 1.3.15 *Bernoulli polynomials $B_n(x)$ defined by (1.65) with $f(t) = t/(e^t - 1)$ and $g(t) = t$ can be written in terms of the Bernoulli numbers as*

$$B_n(x) = \sum_{k=0}^{n} \binom{n}{k} B_k x^{n-k}, \tag{1.70}$$

where the Bernoulli numbers B_n satisfy

$$\begin{aligned} B_0 = 1,\ B_n &= -\sum_{k=0}^{n-1} \frac{n!}{k!(n-k+1)!} B_k \\ &= -\sum_{k=0}^{n-1} \binom{n}{k} \frac{B_k}{n-k+1},\ (n = 1, 2, \ldots) \end{aligned} \tag{1.71}$$

$$B_1 = -\frac{1}{2},\ B_{2n+1} = 0,\ (n = 1, 2, \ldots). \tag{1.72}$$

Proof. From (1.69), we have

$$\begin{aligned} \frac{t}{e^t - 1} e^{xt} &= \left(\sum_{k \geq 0} \frac{B_k}{k!} t^k \right) \left(\sum_{n \geq 0} \frac{(xt)^n}{n!} \right) \\ &= \sum_{n \geq 0} \sum_{k \geq 0} \frac{B_k}{n!k!} x^n t^{n+k} \end{aligned}$$

$$= \sum_{n\geq 0}\sum_{k=0}^{n} \frac{B_k}{(n-k)!k!} x^{n-k} t^n$$

$$= \sum_{n\geq 0}\left(\sum_{k=0}^{n}\binom{n}{k} B_k x^{n-k}\right)\frac{t^n}{n!}.$$

Comparing the above expansion of $te^{xt}/(e^t-1)$ with (1.67), we obtain (1.70).

To prove (1.71) and (1.72), we rewrite (1.69) as

$$\begin{aligned} 1 &= \frac{e^t-1}{t}\sum_{k\geq 0}\frac{B_k}{k!}t^k = \sum_{n\geq 0}\frac{t^n}{(n+1)!}\sum_{k\geq 0}\frac{B_k}{k!}t^k \\ &= \sum_{n\geq 0}\sum_{k\geq 0}\frac{B_k}{k!(n+1)!}t^{n+k} = \sum_{n\geq 0}\sum_{k=0}^{n}\frac{B_k}{k!(n-k+1)!}t^n. \end{aligned}$$

Comparing the coefficients, we obtain (1.71). In addition,

$$\frac{-t}{e^{-t}-1} = \frac{-te^t}{1-e^t} = t + \frac{t}{e^t-1}$$

implies

$$\sum_{n\geq 0}\frac{B_n}{n!}(-t)^n = t + \sum_{n\geq 0}\frac{B_n}{n!}t^n.$$

Thus, we obtain (1.72). ■

From Theorem 1.3.15, we can define Bernoulli numbers, b_n, as follows (see also in Bourbaki):

$$\frac{t}{e^t-1} = 1 - \frac{1}{2}t + \sum_{n\geq 1}(-1)^{n+1}\frac{b_n}{(2n)!}t^{2n},$$

where b_n is positive and equals to $(-1)^{n+1}B_{2n}$.

Theorem 1.3.16 *Define Euler numbers E_n by*

$$\frac{2e^t}{e^{2t}+1} = \sum_{n\geq 0}\frac{E_n}{n!}t^n.$$

Then

$$E_n(x) = \sum_{k=0}^{n}\binom{n}{k}\frac{E_k}{2^k}\left(x-\frac{1}{2}\right)^{n-k}.$$

In addition, for E_n we have $E_n = 2^n E_n(1/2)$ and

$$E_0 = 1,\ E_n = -\sum_{k=0}^{n-1}\binom{n}{k}\frac{1+(-1)^{n-k}}{2}E_k,\ (n=1,2,\ldots)$$

$$E_{2k-1} = 0,\ (k=1,2,\ldots),$$

or equivalently,

$$E_0 = 1, \ E_{2k} = -\sum_{j=0}^{k-1} \binom{2k}{2j} E_{2j}, \ E_{2k-1} = 0, \ (k = 1, 2, \ldots).$$

The proof is similar as the proof of Theorem 1.3.15. We leave it as an exercise (cf. Exercise 1.38).

In 1713, Jacob Bernoulli published under the title Summae Potestatum the following expression of the sum of the k powers of the n first integers as a $(k+1)$th-degree polynomial function of n, with coefficients involving Bernoulli numbers B_n, now called Bernoulli numbers. The formula is known as Bernoulli's formula or Faulhaber's formula (cf. [45, 107]).

Proposition 1.3.17 *Let $n, k \in \mathbb{N}$. Then*

$$1^k + 2^k + \cdots + n^k = \frac{1}{k+1} \sum_{j=0}^{k} (-1)^j \binom{k+1}{j} B_j n^{k-j+1}$$

$$= \frac{n^{k+1}}{k+1} + \frac{n^k}{2} + \sum_{j=2}^{k} (-1)^j \frac{B_j}{j!} k(k-1) \cdots (k-j+2) n^{k-j+1}$$

$$= \frac{n^{k+1}}{k+1} + \frac{n^k}{2} + \sum_{j=1}^{[k/2]} \frac{B_{2j}}{(2j)!} k(k-1) \cdots (k-2j+2) n^{k-2j+1}. \quad (1.73)$$

Proof. From Bernoulli summation formula (1.60) and the fact that jth difference of a kth ($k < j$) degree polynomial is zero, we immediately know that $P(n) := 1^k + 2^k + \cdots + n^k$ is a $k+1$st degree polynomial in terms of n. Hence, we can write

$$P(n) = C_0 n^{k+1} + C_1 n^k + \cdots + C_k n + C_{k+1}.$$

Comparing the coefficients of the identity $P(n+1) - P(n) = (n+1)^k$, we obtain the first equation of (1.73). Using the Bernoulli numbers $B_0 = 1$ and $B_1 = -1/2$, we obtain the second equation of (1.73) from its first equation. Noting $B_{2n+1} = 0$, we get the third equation of (1.73) from its second equation.

■

From Proposition 1.3.17, we obtain

$$B_2 = \frac{1}{6}, \ B_4 = -\frac{1}{30}, \ B_6 = \frac{1}{42}, \ B_8 = -\frac{1}{30}, \ldots.$$

We now present some properties of Bernoulli polynomials and Euler polynomials.

Theorem 1.3.18 *Bernoulli polynomials satisfy*

$$B_0(x) = 1, \ B_n'(x) = nB_{n-1}(x), \ \int_0^1 B_n(x)dx = 0, \ (n = 1, 2, \ldots). \quad (1.74)$$

Conversely, the above conditions determine Bernoulli polynomials uniquely.

Proof. From (1.70), we immediately have $B_0(x) = 1$ and

$$
\begin{aligned}
B_n'(x) &= \sum_{k=0}^{n} \binom{n}{k} B_k (n-k) x^{n-k-1} \\
&= n \sum_{k=0}^{n-1} \binom{n-1}{k} B_k x^{n-k-1} = nB_{n-1}(x).
\end{aligned}
$$

Similarly,

$$
\begin{aligned}
\int_0^1 B_n(x)dx &= \sum_{k=0}^{n} \binom{n}{k} B_n \int_0^1 x^{n-k} dx \\
&= \sum_{k=0}^{n} \binom{n}{k} \frac{B_n}{n-k+1} = 0,
\end{aligned}
$$

where the last step is due to $B_n = -\sum_{k=0}^{n-1} \binom{n}{k} \frac{B_k}{n-k+1}$ shown in (1.71).

Conversely, starting from $B_0(x) = 1$ and taking integral successively to evaluate $B_n(x)$ $(n = 1, 2, \ldots)$, where the integral constants are determined by $\int_0^1 B_n(x)dx = 0$, we obtain the sequence of Bernoulli polynomials.

■

Corollary 1.3.19 *Bernoulli polynomials have properties*

$$B_n^{(n-1)}(1) - B_n^{(n-1)}(0) = n!, \quad (n = 1, 2, \ldots), \tag{1.75}$$

$$B_n^{(k)}(1) = B_n^{(k)}(0) = \frac{n!}{(n-k)!} B_{n-k}, \quad (n = 2, 3, \ldots; k = 0, 1, \ldots, n-1, n), \tag{1.76}$$

$$B_n(x+1) - B_n(x) = nx^{n-1} \quad (n = 1, 2, \ldots). \tag{1.77}$$

Proof. Formulas (1.75) and (1.76) can be proved from Theorem 1.3.18, which is left as Exercise 1.39. To prove (1.77), we write the Taylor's expansion of $B_n(x+1)$ and use (1.75) and (1.76) to obtain

$$
\begin{aligned}
B_n(x+1) &= \sum_{k=0}^{n} \frac{B_n^{(k)}(1)}{k!} x^k \\
&= \sum_{k=0}^{n} \frac{B_n^{(k)}(0)}{k!} x^k + \sum_{k=0}^{n} \frac{B_n^{(k)}(1) - B_n^{(k)}(0)}{k!} x^k \\
&= B_n(x) + nx^{n-1}.
\end{aligned}
$$

■

Use (1.77) we immediately have

Corollary 1.3.20 *Let $n, k \in \mathbb{N}$. Then*

$$\sum_{j=1}^{n} j^k = \frac{1}{k+1}(B_{k+1}(n+1) - B_{k+1}(1)). \tag{1.78}$$

It can be seen the equivalence between formulas (1.78) and (1.73). Since the former uses the Bernoulli polynomial values, its expression of $\sum_{j=1}^{n} j^k$ looks more compact.

Similar to Theorem 1.3.18, we have

Theorem 1.3.21 *Euler polynomials satisfy $E_0(x) = 1$ and*

$$E'_n(x) = nE_{n-1}(x), \tag{1.79}$$
$$E_n(x+1) + E_n(x) = 2x^n. \tag{1.80}$$

We now give *symmetry formulas* of Bernoulli polynomials and Euler polynomials.

Theorem 1.3.22 *Bernoulli polynomials and Euler polynomials have properties*

$$B_n(1-x) = (-1)^n B_n(x), \quad E_n(1-x) = (-1)^n E_n(x),$$
$$(-1)^n B_n(-x) = B_n(x) + nx^{n-1}, \quad (-1)^{n+1} E_n(-x) = E_n(x) - 2x^n.$$

The Darboux formula, first given in 1876, is an expression formula for an analytic function. The Taylor formula is one of its many special cases.

Let $f(z)$ be normal analytic on the line connecting points a and z, and $\phi(t)$ a polynomial of degree n. The derivative of $\phi(t)$ with respect to t $(0 \le t \le 1)$ can be written as

$$\begin{aligned} & \tfrac{d}{dt} \textstyle\sum_{v=1}^{n} (-1)^v (z-a)^v \phi^{(n-v)}(t) f^{(v)}(a + t(z-a)) \\ = \; & -(z-a)\phi^{(n)}(t) f'(a + t(z-a)) \\ & +(-1)^n (z-a)^{n+1} \phi(t) f^{(n+1)}(a + t(z-a)) \end{aligned}$$

By taking the integral in terms of t from 0 to 1 and noting that $\phi^{(n)}(t) = \phi^{(n)}(0)$, we obtain

$$\begin{aligned} & \phi^{(n)}(0)[f(z) - f(a)] \\ = \; & \sum_{v=1}^{n} (-1)^{v-1} (z-a)^v \left[\phi^{(n-v)}(1) f^{(v)}(z) - \phi^{(n-v)}(0) f^{(v)}(a)\right] \\ & +(-1)^n (z-a)^{n+1} \int_0^1 \phi(t) f^{n+1}(a + t(z-a)) dt. \end{aligned} \tag{1.81}$$

Equation (1.81) is called the Darboux formula. Now we will consider several of its special cases.

Let $\phi(t) = (t-1)^n$ in equation (1.81). Then $\phi^{(n)}(0) = n!$, $\phi^{(n-v)}(1) = 0$, and $\phi^{(n-v)}(0) = (-1)^v n!/v!$, for $1 \le v \le n$. This is of course the Taylor formula with a Cauchy integral form remainder.

In equation (1.81), changing n to $2n$, setting $\phi(t) = t^n(t-1)^n$, and taking $n \to \infty$ yields

$$f(z) - f(a) = \sum_{v=1}^{\infty} \frac{(-1)^{v-1}(z-a)^v}{2^v \cdot v!}[f^{(v)}(z) + (-1)^{v-1} f^{(v)}(a)].$$

Here the series is assumed to be convergent.

Without a loss of generality, let the highest power term of $\phi(t)$ be t^n, and let $F(t)$ be an antiderivative $f(t)$, where $F(t)$ is assumed to be n order continuously differentiable. Then the following integral quadrature formula will be obtained by setting $a = 0$ and $z = 1$ in equation (1.81)

$$\begin{aligned}\int_0^1 F(t)dt &= \sum_{v=1}^{n} \frac{(-1)^{v-1}}{n!}\left[\phi^{(n-v)}(t)F^{(v-1)}(t)\right]_{t=0}^{t=1} \\ &+ \frac{(-1)^n}{n!}\int_0^1 \phi(t)F^{(n)}(t)dt \end{aligned} \tag{1.82}$$

In fact, equation (1.82) can easily be verified by applying n times integral by parts to its integral form remainder. One may rewrite the above formula into a more general form (a suitable transformation of variable is needed).

$$\int_a^b F(t)dt = \sum_{v=1}^{n} \frac{(-1)^{v-1}}{n!}\left[\phi^{(n-v)}(t)F^{(v-1)}(t)\right]_{t=a}^{t=b} + R_n \tag{1.83}$$

where remainder R_n is

$$R_n = \frac{(-1)^n}{n!}\int_a^b \phi(t)F^{(n)}(t)dt. \tag{1.84}$$

Equations (1.83) and (1.84) can be understood as the integral form of the Darboux formula. However, the applications of the Darboux formula to numerical integration were only given attention to after 1940 (see [172]).

The formula shown in equation (1.83) has two features: (1) Except for the remainder, the right-hand side of the equation consists of the values of the integrand and its derivatives at endpoints of the integral interval. Hence, the formula is a boundary-type quadrature formula (BTQF). (2) By choosing a suitable weight function $\phi(t)$ in the remainder (1.84), we can make $R_n \equiv R_n(F)$ has the smallest possible estimate in various norms.

In addition, setting $\phi(t)$ as the nth order Bernoulli polynomial in equation (1.83), we obtain the *Euler-Maclaurin formula.* More precisely, defferenting the identity (1.76) $n-k$ times gives

$$B_n^{(n-k)}(t+1) - B_n^{(n-k)}(t) = n(n-1)\cdots kt^{k-1}.$$

Plugging in $t = 0$ yields $B_n^{(n-k)}(1) = B_n^{(n-k)}(0)$ for $k = 2, 3, \ldots$. From the Maclaurin series (i.e., the Taylor series at 0) of $B_n(t)$ with $k > 0$, we have $B_n^{(n-2k-1)}(0) = 0$, $B_n^{(n-2k)}(0) = n!B_{2k}/(2k)!$, $B_n^{(n-1)}(0) = n!/2$, and $B_n^{(n)}(0) = n!$. Substituting these values of $B_n^{(n-k)}(1)$ and $B_n^{(n-k)}(0)$ into Darboux's formula (1.83) gives

$$\begin{aligned}(z-a)f'(a) &= f(z) - f(a) - \frac{z-a}{2}[f'(z) - f'(a)] \\ &+ \sum_{k=1}^{n-1} \frac{B_{2k}(z-a)^{2k}}{(2k)!}\left[f^{(2k)}(z) - f^{(2k)}(a)\right] \\ &- \frac{(z-a)^{2n+1}}{(2n)!}\int_0^1 B_{2n}(t) f^{(2n+1)}(a - (z-a)t)dt, \end{aligned} \tag{1.85}$$

which is called the *Euler-Maclaurin integration formula.* It holds when the function $f(z)$ is analytic in the integration region. In certain cases, the last term of (1.85) tends to 0 as n tends to infinity, and an infinite series can then be obtained for $f(z) - f(a)$. Thus, the sums may be converted to integrals by inverting the formula to obtain the following *Euler-Maclaurin summation formula* (cf. [79, p.471])

$$\begin{aligned}&\sum_{k=1}^{n-1} f(k) \\ =& \int_0^n f(t)dt - \frac{1}{2}[f(n) + f(0)] + \sum_{k\geq 1} \frac{B_{2k}}{(2k)!}\left[f^{(2k-1)}(n) - f^{(2k-1)}(0)\right] \\ =& \int_0^n f(t)dt + \sum_{k\geq 1} \frac{B_k}{k!}\left[f^{(k-1)}(n) - f^{(k-1)}(0)\right]. \end{aligned} \tag{1.86}$$

Also, if $n = 2m$ and $\phi(t) = (t-a)^m(t-b)^m$, then equation (1.83) will give the following Petr formula when the Leibniz higher order differentiation rule is applied.

$$\begin{aligned}\int_a^b F(t)dt &= \sum_{v=1}^{m} \frac{\binom{m}{v}}{\binom{2m}{v}} \frac{(b-a)^v}{v!}\left[F^{(v-1)}(a)\right. \\ &+ \left.(-1)^{v-1}F^{(v-1)}(b)\right] + R_m, \end{aligned} \tag{1.87}$$

where remainder R_m is given by equation (1.84). Using the mean value theorem of integration, we can write it as

$$R_m = \frac{(-1)^m}{2^{m+1}}\left[\frac{m!}{(2m)!}\right]^2 F^{(2m)}(\xi)(b-a)^{2m+1}, \qquad a \leq \xi \leq b.$$

Similarly, setting $n = m+k$ and $\phi(t) = (a-t)^k(b-t)^m$ in equation (1.83), we obtain the so-called *Obreschkoff formula*

$$\begin{aligned}\int_a^b F(t)dt &= \frac{m!k!}{(m+k)!}\sum_{v=1}^{m}\begin{pmatrix} m+k-v \\ k \end{pmatrix}\frac{(b-a)^v}{v!}f^{(v)}(a) \\ &- \frac{m!k!}{(m+k)!}\sum_{v=1}^{k}(-1)^k\begin{pmatrix} m+k-v \\ k \end{pmatrix}\frac{(b-a)^v}{v!}f^{(v)}(b) \\ &- \frac{m!k!}{(m+k)!}R_{m+k}, \qquad (1.88)\end{aligned}$$

where

$$R_{m+k} = -\frac{1}{k!m!}\int_a^b (b-t)^m(a-t)^k f^{(m+k+1}(t)dt. \qquad (1.89)$$

All of the above formulas are useful in approximating definite integrals.

At the end of this section, we will give one more example of the Darboux formula. In equation (1.83), we set $a = -1$, $b = 1$, and $\phi(t) = T_n(t) + 1/2^{n-1}$, where $T_n(t) = 2^{1-n}\cos(n\cos^{-1}t)$ is the Chebyshev polynomial of degree n. Then,

$$\phi(1) = \frac{1}{2^{n-2}} \qquad \phi(-1) = \frac{1+(-1)^n}{2^{n-1}}.$$

Since $T_n(t)$ satisfies differential equation $(1-t^2)T_n''(t) - tT_n'(t) + n^2T_n(t) = 0$, by taking derivatives consecutively, we obtain

$$\begin{aligned}\phi^{(n-v)}(-1) &= (-1)^v\phi^{(n-v)}(1) \\ 2^{n-1}\phi^{(n-v)}(1) &= \tfrac{n^2}{1}\cdot\tfrac{n^2-1^2}{3}\cdots\tfrac{n^2-(n-v-1)^2}{2n-2v-1},\end{aligned}$$

where $v = 1, 2, \cdots, n-1$. Therefore, when n is an even integer, we can derive the following BTQF from equation (1.83).

$$\begin{aligned}&\int_{-1}^{1} F(t)dt = F(1) + F(-1) \\ &+ \sum_{v=2}^{n-1}\frac{(-1)^{v-1}}{2^v\cdot v!}\frac{\begin{pmatrix} 2n-v-1 \\ v \end{pmatrix}}{\begin{pmatrix} n-1 \\ v \end{pmatrix}}\left[F^{(v-1)}(1) + (-1)^{v-1}F^{(v-1)}(-1)\right] \\ &- \frac{1}{2^{n-2}\cdot n!}\left[F^{(n-1)}(1) - F^{(n-1)}(-1)\right] \\ &+ \begin{pmatrix} n^2-2 \\ n^2-1 \end{pmatrix}\frac{F^{(n)}(\xi)}{2^{n-2}\cdot n!}, \qquad (1.90)\end{aligned}$$

where $-1 \le \xi \le 1$.

It is obvious that the BTQF derived from equation (1.83) by deleting the remainder R_n possesses algebraic precision degree $n-1$. Here, the degree of algebraic precision of a quadrature formula is the largest positive integer r such that the formula is exact for x^i, when $i = 0, 1, \cdots, r$. Hence, choosing a suitable $\phi(t)$, we can construct various BTQFs for different purposes and make the corresponding remainders the smallest possible under different norms. More details can be seen in [83].

Exercises

1.1 Use the method shown in Examples 1.1.1–1.1.3 to prove formula

$$\sum_{n\geq 1} \frac{1}{n^2(n+1)^2(n+2)^2} = \frac{\pi^2}{4} - \frac{39}{16}.$$

1.2 Make use of the technique of Example 1.1.5 to show

$$(a) \sum_{n\geq 0} a^n \cos nx = \frac{1 - a\cos x}{1 - 2a\cos x + a^2}$$

$$(b) \sum_{n\geq 0} a^n \sin nx = \frac{a \sin x}{1 - 2a\cos x + a^2}$$

for all $|a| < 1$.

1.3 Use (1.2) or similar techniques shown in Example 1.1.5 to prove each of the summation formulas.

$$(a) \quad \sum_{n\geq 2} (-1)^n \frac{\cos nx}{n^2 - 1} = \frac{1}{2}\left(1 - \frac{1}{2}\cos x\right) - \frac{x}{2}\sin x, \ (|x| < \pi).$$

$$(b) \quad \sum_{n\geq 1} (-1)^{n-1} \frac{\cos(2n+1)x}{(2n-1)(2n+1)(2n+3)} = \frac{\pi}{8}\cos^2 x - \frac{1}{3}\cos x, \ (|x| < \pi).$$

$$(c) \quad \sum_{n\geq 1} \frac{\cos 2nx}{(2n-1)(2n+1)} = \frac{1}{2} - \frac{\pi}{4}\sin x, \ (0 < x < \pi).$$

$$(d) \quad \sum_{n\geq 1} \frac{\sin(2n-1)x}{(2n-1)^3} = \frac{1}{8}\pi x(\pi - x), \ (0 < x < \pi).$$

Hints: (a) Use (1.2) and the following factoring:

$$\sum_{n\geq 2} (-1)^n \frac{\cos nx}{n^2 - 1} = \frac{1}{2}\left(\sum_{n\geq 2} (-1)^n \frac{\cos nx}{n-1} - \sum_{n\geq 2} (-1)^n \frac{\cos nx}{n+1}\right).$$

(d) Denote the left-hand series by $F(x)$ and study $F''(x)$ and make use of (1.4).

1.4 Prove two summation formulas given in Example 1.1.6.

1.5 Use the technique shown in Example (1.17) to prove

$$\lim_{n\to\infty} \frac{1}{n} \sum_{k=1}^{n} \left(\left[\frac{n}{k}\right] - \left[\frac{n}{k} - \alpha\right]\right) = \int_0^1 \frac{1 - x^\alpha}{1 - x} dx,$$

where $\alpha \in \mathbb{R}$ and $0 < \alpha \leq 1$.

1.6 Prove limits (1.21)–(1.23).

1.7 Suppose $(P_1, P_2, \cdots, P_n, \cdots)$ is a positive sequence, and denote $Q_n^{-1} = P_1^{-1} + P_2^{-1} + \cdots + P_n^{-1}$. Prove that the positive series $\sum_{n=1}^{n} a_n$ converges if

$$\overline{\lim_{n\to\infty}} Q_n(\log(P_n a_n)) < 0.$$

Hint: Let the upper limit in the inequality $\le -2k$ for some positive k. Denote $u_n = P_n^{-1}$, $s_n = u_1 + \cdots + u_n$. Hence, from the assumption, there exists n_0 such that when $n > n_0$, we have

$$Q_n \log(P_n a_n) < -k,$$

i.e.,

$$a_n < u_n e^{-kQ_n^{-1}} = u_n e^{-ks_n} \le u_n.$$

Hence, (i) $\sum_{n\ge1} u_n < \infty$ implies $\sum_{n\ge1} a_n < \infty$, and (ii) when $\sum_{n\ge1} u_n = +\infty$, from $\int_1^\infty f(x)dx < \infty$ we have $\sum_{n\ge1} f(s_n)u_n < \infty$. Taking $f(x) = e^{-kx}$ yields $\sum_{n\ge1} u_n e^{-ks_n} < \infty$, which implies $\sum_{n\ge1} a_n < \infty$.

1.8 Prove the claim given in Remark 1.1.16.

Hint: Note that $t_n = P_n s + \sum_{k=0}^{n} p_{nk}(s_k - s)$.

1.9 Let $\lim\limits_{n\to\infty} a_n = a > 0$. Prove

$$\lim_{n\to\infty} \frac{\sqrt{a_1a_2a_3} + \sqrt{a_2a_3a_4} + \cdots + \sqrt{a_na_{n+1}a_{n+2}}}{a_1a_2 + a_2a_4 + \cdots a_na_{n+2}} = \frac{1}{\sqrt{a}}.$$

Hint: Using Corollary 1.1.11, the left-hand limit of the above equation is

$$\lim_{n\to\infty} \frac{\sqrt{a_na_{n+1}a_{n+2}}}{a_na_{n+2}} = \frac{1}{\sqrt{a}}.$$

1.10 Prove Theorem 1.1.19.

Hint: The argument is similar to the proof of Theorem 1.1.9.

1.11 Verify Theorem 1.2.1.

Hint: See the proof of Theorem 1.1.9.

1.12 Suppose $b_n > 0$, $\sum_{n\ge0} b_n = \infty$, and $\lim\limits_{n\to\infty} \sum\limits_{k=0}^{n} a_k / \sum\limits_{k=0}^{n} b_k = s$. Show that

$$\lim_{t\to1^-} \frac{\sum_{n\ge0} a_n t^n}{\sum_{n\ge0} b_n t^n} = s.$$

Hint: Multiplying $\sum_{n\ge0} t^n$ to both numerator and denominator of the left-hand side of the last expression yields

$$\lim_{t\to1^-} \frac{\sum_{n\ge0} a_n t^n}{\sum_{n\ge0} b_n t^n} = \lim_{t\to1^-} \frac{\sum_{n\ge0}(a_0 + a_1 + \cdots + a_n)t^n}{\sum_{n\ge0}(b_0 + b_1 + \cdots + b_n)t^n} = s.$$

1.13 Prove Theorem 1.2.5.

1.14 Prove Proposition 1.4.1.

1.15 Use Abel's lemma 1.2.14 to prove the following statement: Suppose for all positive integers n, $f_n \geq f_{n+1} \geq 0$ and denote

$$A = max\ (|a_1|, |a_1| + |a_2|, \cdots, |a_1| + |a_2| + \cdots + |a_m|).$$

Show

$$\left|\sum_{n=1}^{m} a_n f_n\right| \leq A f_1.$$

1.16 Prove general Abel's lemma, Lemma 1.2.16.

1.17 (a) Use Theorem 1.2.18 to prove Theorem 1.2.17.

(b) Use Corollary 1.2.13 to prove Theorem 1.2.18

(c) Show that Theorem 1.2.17 and Theorem 1.2.18 implies each other.

Hint: (a) Using the transformations $a_n\phi(n) \mapsto a_n$ and $\phi(n) \mapsto \phi(n)^{-1}$ in Theorem 1.2.18, we immediately obtain

$$\sum_{k=1}^{n} a_k = \sum_{k=1}^{n} (a_k\phi(k))\phi(k)^{-1} = o\left(\phi(n)^{-1}\right)$$

as $n \to \infty$, which implies (1.32).

(b) Let $s = \sum_{n\geq 0} a_n$, and let $1 < m < n$. Then

$$\begin{aligned}
&\sum_{k=1}^{n} a_k\phi(k) \\
= \;& s\phi(1) + (s_n - s)\phi(n) - \sum_{k=1}^{n-1} (s_k - s)(\phi(k+1) - \phi(k)) \\
= \;& O(1) + o(\phi(n)) - \sum_{k=1}^{m-1} (s_k - s)(\phi(k+1) - \phi(k)) \\
& - \sum_{k=m}^{n-1} (s_k - s)(\phi(k+1) - \phi(k)).
\end{aligned}$$

Hence, for an arbitrarily fixed m

$$\left|\sum_{k=1}^{n} a_k\phi(k)\right| \leq o(\phi(n)) + \epsilon_m \sum_{k=m}^{n-1} (\phi(k+1) - \phi(k)),$$

where $\epsilon_m = \max_{k\geq m} |s_k - s|$. Noting $\phi(n)$ is increasing and diverges to ∞, and $\epsilon_m \to 0$ as $m \to \infty$, for any arbitrarily given positive number ϵ, we can find a large enough m such that

$$\left|\sum_{k=1}^{n} a_k\phi(k)\right| \leq o(\phi(n)) + \epsilon_m\phi(n) \leq o(\phi(n)) + \epsilon\phi(n).$$

Since the left-hand side of the above expression is independent of ϵ, we obtain (1.33) by taking $\epsilon \to 0$ in the rightmost side of the above inequality.

1.18 Assume that $(z_n)_{n\geq 1}$ is a complex sequence with $\sum_{n\geq 1} \left|z_{n+1}^{-1} - z_n^{-1}\right| = \infty$, and series $\sum_{n\geq 1} a_n z_n$ converges. Prove

$$\lim_{N\to\infty} \left(\sum_{n=1}^{N} a_n\right) \left(\sum_{n=1}^{N} \left|z_{n+1}^{-1} - z_n^{-1}\right|\right)^{-1} = 0.$$

Hint: Apply Exercise 1.16 and a similar argument of the proof of Theorem 1.2.17.

1.19 Suppose that $\sum_{n\geq 0} a_n x^n$, $\sum_{n\geq 0} b_n x^n$, and $\sum_{n\geq 0} c_n x^n$ are absolutely convergent, and $d_n = \sum_{i+j+k=n} a_i b_j c_k$. Prove

$$\left(\sum_{n\geq 0} a_n x^n\right)\left(\sum_{n\geq 0} b_n x^n\right)\left(\sum_{n\geq 0} c_n x^n\right) = \sum_{n\geq 0} d_n x^n.$$

1.20 Assume that at least one of series $\sum_{n\geq 0} a_n$ and $\sum_{n\geq 0} b_n$ is absolutely convergent, and $c_n = \sum_{k=0}^{n} a_k b_{n-k}$. Show $\sum_{n\geq 0} c_n$ converges and

$$\left(\sum_{n\geq 0} a_n\right)\left(\sum_{n\geq 0} b_n\right) = \left(\sum_{n\geq 0} c_n\right).$$

Hint: Assume that $\sum_{n\geq 0} b_n$ converges absolutely. Hence, from the given conditions,

$$s_n = \sum_{k=0}^{n} a_k \to s, \quad s_n' = \sum_{k=0}^{n} b_k \to s', \quad \sum_{k=0}^{n} |b_k| < \infty.$$

Denote $\sum_n := |b_0| + |b_1| + \cdots + |b_n|$, and $\sigma_n := c_0 + c_1 + \cdots + c_n$. Then,

$$\begin{aligned} \sigma_n &= (a_0 + a_1 + \cdots + a_n)(b_0 + b_1 + \cdots + b_n) \\ &\quad - b_1 a_n - b_2(a_n + a_{n-1}) - b_3(a_n + a_{n-1} + a_{n-2}) - \cdots \\ &\quad - b_n(a_n + a_{n-1} + \cdots + a_1) \\ &= s_n s_n' - b_1(s_n - s_{n-1}) - b_2(s_n - s_{n-2}) - \cdots - b_n(s_n - s_0). \end{aligned}$$

Therefore,

$$\begin{aligned} |\sigma_n - ss'| &\leq |s_n s_n' - ss'| + |b_1||s_n - s_{n-1}| \\ &\quad + |b_2||s_n - s_{n-2}| + \cdots + |b_n||s_n - s_0| \\ &= |s_n s_n' - ss'| + \sum_{k=1}^{m} |b_k||s_n - s_{n-k}| \\ &\quad + \sum_{j=m+1}^{n} |b_j||s_n - s_{n-j}|, \qquad (m < n). \end{aligned}$$

Since

$$|s_n - s_{n-k}| \to 0 \ (n-k \to \infty), \qquad \sum_{j=m+1}^{n} |b_j| \to 0 \ (n > m \to \infty),$$

for any given $\epsilon > 0$ there exist n, m with large enough $n - m$ such that

$$|\sigma_n - ss'| < |s_n s_n' - ss'| + \epsilon \sum_{k=1}^{m} |b_k| + \epsilon A,$$

where $A = \max\{|s_n - s_{n-j}| : m+1 \le j \le n\}$.

1.21 Discuss the convergence or divergence of the square of series $\sum_{n\ge 1}(-1)^{n-1}\frac{1}{n^\alpha}$, $(0 < \alpha < 1)$.

1.22 Prove Corollary 1.2.23.

1.23 Prove Corollary 1.2.25.

1.24 Extend the result shown in Example 1.2.26 to the convergence of series $\sum_{n\ge1} a_n t^n$, where $t = e^{i\theta}$ $(0 < \theta < 2\pi)$, and (a_n) is a decreasing sequence with $\lim_{n\to\infty} a_n = 0$. Furthermore,

$$\left|\sum_{n=m}^{m+p} a_n t^n\right| \le \frac{2a_m}{1-t}.$$

1.25 Use Corollary 1.2.25 to prove if (b_n) is a decreasing sequence approaching to 0 as $n \to \infty$, then

$$\sum_{n\ge1} b_n \sin n\theta \cos n^2\theta, \qquad \sum_{n\ge1} b_n \sin n\theta \sin n^2\theta$$

converge.

Hint: Discuss the convergence of series $\sum_{n\ge1} b_n u_n$ and $\sum_{n\ge1} b_n v_n$, where

$$\sum_{k=1}^{n} u_k = \sin\left(n + \frac{1}{2}\right)^2 \theta, \qquad \sum_{k=1}^{n} v_k = \cos\left(n+\frac{1}{2}\right)^2 \theta.$$

1.26 Use the technique shown in Examples 1.2.27 and 1.2.28 to establish

$$\sum_{n\ge1} (-1)^{\binom{n}{2}} \frac{1}{n} = \frac{\pi}{4} - \frac{1}{2}\log 2.$$

1.27 Prove

$$\lim_{x\to1^-} \sum_{n\ge1} (-1)^{n-1} \frac{x^n}{n(1+x^n)} = \frac{1}{2}\log 2 = \lim_{x\to1^-} \sum_{n\ge1} (-1)^{n-1} \frac{x^n(-x)}{1-x^{2n}}.$$

Hint: Write

$$\sum_{n\ge1} (-1)^{n-1} \frac{x^n}{n(1+x^n)} = \frac{1}{2} \sum_{n\ge1} \frac{(-1)^{n-1}}{n} \left(\frac{2x^n}{1+x^n}\right),$$

denote $a_n = (-1)^{n-1}/n$ and $v_n(x) = 2x^n/(1+x^n)$ and apply Theorem 1.2.30.

1.28 Prove the expansion (1.43) of $\Delta_h^n f(x)$ is true.

1.29 Let $f(x)$ be a polynomial of degree k. Prove $\Delta^s f(x) = 0$ for all $s \geq k+1$.

1.30 Let $\phi(x) = a_0 + a_1 x + a_2 x(x-1) + a_3 x(x-1)(x-2) + \cdots$, where x are non-negative integers. Prove

$$\begin{aligned}&\binom{n}{0}\phi(0) - \binom{n}{1}\phi(1) + \binom{n}{2}\phi(2) - \cdots + (-1)^n \binom{n}{n}\phi(n) \\ = \; &(-1)^n a_n n!.\end{aligned}$$

Hint: Use difference and shift operators.

1.31 Use Lemma 1.3.1 to prove

$$\sum_{k=1}^{n} (-1)^{n-k} \binom{n}{k} k^n = n!.$$

Hint: It can also be proved using Exercise 1.30 with $\phi(x) = x^n$ and hence, $a_n = 1$.

1.32 Define $\Delta^{-1} f(x) = \sum_{k=0}^{x-1} f(k)$, $\Delta^{-n} = \Delta^{-1}\Delta^{-(n-1)}$. Prove

$$\Delta^n \binom{x}{m} = \Delta\Delta^{n-1}\binom{x}{m} = \binom{x}{m-n}$$

for all integers $n \leq m$.

1.33 Use a similar argument of Example 1.3.9 to find summation formula of $\sum_{k=1}^{n} k^5$.

1.34 Prove the following formula holds if its both sides series converge.

$$\begin{aligned}\sum_{k=1}^{n} f(k)x^k \;=\; &[f(1) - x^n f(n+1)]\left(\frac{x}{1-x}\right) \\ &+[\Delta f(1) - x^n \Delta f(n+1)]\left(\frac{x}{1-x}\right)^2 \\ &+[\Delta^2 f(1) - x^n \Delta^2 f(n+1)]\left(\frac{x}{1-x}\right)^3 + \cdots.\end{aligned}$$

1.35 Suppose $|x| < 1$ and $|x| < |1-x|$. Prove

$$\sum_{n\geq 0} (-1)^n \cos(n\theta) x^{n+1} = -\sum_{n\geq 0} 2^n \cos^n\left(\frac{\theta}{2}\right)\cos\left(\frac{n\theta}{2}\right)\left(\frac{x}{x-1}\right)^{n+1},$$

$$\sum_{n\geq 0} 2^n \cos^n\left(\frac{\theta}{2}\right)\cos\left(\frac{n\theta}{2}\right) x^{n+1} = -\sum_{n\geq 0} (-1)^n \cos(n\theta)\left(\frac{x}{x-1}\right)^{n+1}.$$

Hint: Use Euler series transformation formula.

1.36 Apply Euler series transformation formula to series $\sum_{n\geq 0}\frac{(-1)^n}{4^n}$, and explain the resulting series has slower convergent rate.

Hint: The resulting series is $\frac{1}{2}\sum_{n\geq 0}\left(\frac{3}{8}\right)^n$.

1.37 (*cf* [135]) If convergent series $\sum_{n\geq 1}(-1)^{n-1}f(n)$ satisfies (i) $(-1)^k\Delta^k f(n) > 0$ ($k = 0, 1, \ldots$; $n = 1, 2, \ldots$) and (ii) there exists $a > 1/2$ such that for all $n \in \mathbb{N}$, $f(n+1)/f(n) \geq a$, prove that the series changed from the given series by using Euler series transformation formula has a higher convergent rate.

Hint: Denote

$$R_n = \sum_{k\geq n+1}(-1)^{k-1}f(k), \quad r_n = \sum_{k\geq n+1}\frac{(-1)^k}{2^{k+1}}\Delta^k f(1).$$

We need to show $\lim_{n\to\infty} r_n/R_n = 0$. In fact, from (i)

$$\begin{aligned}(-1)^k\Delta^k f(n) &= (-1)^{k-1}\Delta^{k-1}f(n) - (-1)^{k-1}\Delta^{k-1}f(n+1)\\ &< (-1)^{k-1}\Delta^{k-1}f(n) < \cdots < f(n).\end{aligned}$$

Therefore,

$$r_n = \sum_{k\geq n+1}\frac{(-1)^k}{2^{k+1}}\Delta^k f(1) \leq \sum_{k\geq n+1}\frac{1}{2^{k+1}}f(1) = \frac{f(1)}{2^{n+1}}.$$

On the other hand,

$$\begin{aligned}R_n &= \sum_{k\geq n+1}(-1)^{k-1}f(k)\\ &= \frac{(-1)^n}{2}\left(f(n+1) + \sum_{k\geq 0}(-1)^{k+1}\Delta f(n+1+k)\right).\end{aligned}$$

Since $-\Delta f(n) > 0$, the series on the rightmost side is an alternating series. Because $\Delta^2 f(n) > 0$, the absolute value of the general term of the series tends to zero monotonically. Hence, from condition (ii)

$$|R_n| > \frac{1}{2}f(n+1) \geq \frac{f(1)}{2}a^n.$$

Therefore,

$$\left|\frac{r_n}{R_n}\right| \leq \frac{1}{(2a)^n},$$

which implies $\lim_{n\to\infty}|r_n/R_n| = 0$.

1.38 Prove the power series expansions

$$x\cot x = 1 + \sum_{n\geq 1}(-1)^n \frac{2^{2n}B_{2n}}{(2n)!}x^{2n},$$

$$\tan x = \sum_{n\geq 1}(-1)^{n-1}\frac{2^{2n}(2^{2n}-1)B_{2n}}{(2n)!}x^{2n-1}.$$

Hint: Use

$$x\cot x = x\frac{\cos x}{\sin x} = ix\frac{e^{ix}+e^{-ix}}{e^{ix}-e^{-ix}} = ix + \frac{2ix}{e^{2ix}-1}$$

and $\tan x = \cot x - 2\cot(2x)$.

1.39 Prove Theorem 1.3.16.

Hint: Using transformation $t \mapsto 2t$ and substituting $x = 1/2$ into (1.68), the definition of $E_n(x)$, we obtain

$$\frac{2e^t}{e^{2t}+1} = \sum_{n\geq 0}\frac{2^n E_n(1/2)}{n!}t^n,$$

which is compared with the definition of Euler numbers E_n, and $E_n = 2^n E_n(1/2)$ is followed. $E_n(x) = \sum_{k=0}^{n}\binom{n}{k}\frac{E_k}{2^k}\left(x-\frac{1}{2}\right)^{n-k}$ can be proved similarly as the proof of (1.70). In addition, comparing the coefficients of the leftmost term and the rightmost term of the following equation

$$\begin{aligned}
1 &= \frac{e^{2t}+1}{2e^t}\sum_{k\geq 0}\frac{E_k}{k!}t^k \\
&= \frac{1}{2}(e^t+e^{-t})\sum_{k\geq 0}\frac{E_k}{k!}t^k \\
&= \sum_{n\geq 0}\sum_{k\geq 0}\frac{1+(-1)^n}{2\cdot n!k!}E_k t^{n+k} \\
&= \sum_{n\geq 0}\sum_{k=0}^{n}\frac{1+(-1)^{n-k}}{2\cdot(n-k)!k!}E_k t^n,
\end{aligned}$$

yields $E_0 = 1$ and $\sum_{k=0}^{n}\frac{1+(-1)^{n-k}}{2\cdot(n-k)!k!}E_k = 0$, which implies

$$E_n = -n!\sum_{k=0}^{n-1}\frac{1+(-1)^{n-k}}{2\cdot(n-k)!k!}E_k.$$

1.40 Prove Corollaries 1.3.19 and 1.3.20.

1.41 Prove Theorem 1.3.21.

2

Symbolic Methods

CONTENTS

As is well known, the closed form representation of series has been studied extensively. See, for examples, Comtet [44], Jordan [141], Egorechev [58], Roman-Rota [187], Sofo [200], Wilf [216], Graham-Knuth-Partashnik [79], Petkovšek-Wilf-Zeilberger [177], He, Hsu, and Shiue [104], and He, Hsu, Shiue, and Torney [99], He, Hsu, and Shiue [104, 106], He, Hsu, and Ma [97], and He, Hsu, and Yin [107]. The object of this chapter is to make use of the classical operators Δ (difference), E (shift), and D (derivative) to develop a method for the summation of power series that appears to have a certain wide scope of applications.

It is known that the symbolic operations Δ (difference), E (displacement) and D (derivative) play an important role in the Calculus of Finite Differences as well as in many topics of computational methods. For various classical results, we may see Jordan [141], Milne-Thomson [164], etc. Certainly, the theoretical basis of the symbolic methods could be found within the theory of formal power series, inasmuch as all the symbolic expressions treated are

DOI: 10.1201/9781003051305-2

expressible as power series in Δ, E, or D, and all the operations employed are just the same as those applied to formal power series. For some easily accessible references on formal series, we may recommend Ch. 4.5 of Bourbaki [23], Comtet [44], Gould [73], Graham-Knuth-Partashnik [79], Petkov sek-Wilf-Zeilberger [177], and Wilf [216].

Throughout this chapter the theory of formal power series and of differential operators will be utilized. $\mathbb{R}$, $\mathbb{C}$, $\mathbb{N}$, $\mathbb{N}_0 = \mathbb{N} \cup \{0\}$, and $\mathbb{Z}$ denote, respectively, the sets of real numbers, complex numbers, natural numbers, natural numbers including 0, and integers. Generally, we will use $A(t), g(t), f(t), \varphi(t)$, etc. to denote either the formal power series (fps) in $\mathbb{K}[[t]]$, the ring of formal power series in the real field when $\mathbb{K} = \mathbb{R}$ or the complex field when $\mathbb{K} = \mathbb{C}$, or the infinitely differentiable functions (members of C^∞) defined in $\mathbb{R}$ or $\mathbb{C}$.

We will make use of the ordinary operators Δ (difference), D (differentiation) and E (shift operator) which are defined by the relations, respectively,

$$Ef(t) = f(t+1), \quad \Delta f(t) = f(t+1) - f(t), \quad Df(t) = \frac{d}{dt} f(t).$$

Powers of these operators are defined in the usual way. In particular for any real numbers x, one may define $E^x f(t) = f(t+x)$. Also, the number 1 is defined as an identity operator, viz. $1f(t) = f(t)$. Let $f \in \mathbb{C}[n]$. By Taylor theorem,

$$f(n+x) = \sum_{k\geq 0} D^k f(n) \frac{x^k}{k!}.$$

Putting $x = 1$ yields

$$f(n+1) = \left(\sum_{k\geq 0} \frac{D^k}{k!} \right) f(n) = e^D f(n),$$

which implies $\Delta f(n) = (e^D - 1) f(n)$. Thus, we have Symbolically

$$E = 1 + \Delta = e^D, \quad \Delta = E - 1 = e^D - 1, \quad \text{and} \quad D = \log(1+\Delta) = \log E.$$

In this chapter, we focus on summations and identities arising from the interrelations of a number of operators in common use in combinatorics, number theory, and discrete mathematics.

2.1 Symbolic Approach to Summation Formulas of Power Series

Note that $E^k f(0) = \left[E^k f(t)\right]_{t=0} = f(k)$ so that any power series of the form $\sum_{k=0}^\infty f(k) x^k$ can be written symbolically as

$$\sum_{k\geq 0} f(k) x^k = \sum_{k\geq 0} x^k E^k f(0) = \sum_{k\geq 0} (xE)^k f(0) = (1 - xE)^{-1} f(0).$$

This shows that the symbolic operator $(1-xE)^{-1}$ with parameter x can be applied to $f(t)$ (at $t=0$) to yield a power series or a generating function for $\{f(k)\}$.

We shall show in §2.1.2 that $(1-xE)^{-1}$ can be expanded into series in various ways to derive numerous symbolic operational formulas as well as summation formulas for $\sum_{k\geq 0} f(k)x^k$. Note that the closed form representation of series has been studied extensively. For example, Sofo [200] who presents a unified treatment of summation of series using function theoretic method. Some consequences of the summation formulas as well as the examples will be shown, which are useful for computational purpose and for accelerating the series convergence. In §2.1.3, we shall give the remainders of the summation formulas.

2.1.1 Wellknown symbolic expressions

We shall need several definitions as follows.

Definition 2.1.1 *The expression $f(t) \in C^m_{[a,b)}$ $(m \geq 1)$ means that $f(t)$ is a real function continuous together with its mth derivative on $[a,b)$.*

Definition 2.1.2 *$\langle x, x_0, x_1, \cdots, x_n\rangle$ represents a least interval containing x and the numbers $x, x_0, x_2, \cdots, x_n$.*

Definition 2.1.3 *$\alpha_k(x)$ is called an Eulerian fraction and may be expressed in the form (cf. Comtet [44])*

$$\alpha_k(x) = \frac{A_k(x)}{(1-x)^{k+1}}, \quad (x \neq 1),$$

where $A_k(x)$ is the kth degree Eulerian polynomial having the expression

$$A_k(x) = \sum_{j=1}^{k} A(k,j)x^j, \quad A_0(x) \equiv 1,$$

and $A(k,j)$ are referred to as Eulerian numbers, which are expressible as

$$A(k,j) = \sum_{i=0}^{j} (-1)^i \binom{k+1}{i} (j-i)^k, \quad (1 \leq j \leq k).$$

Definition 2.1.4 *δ is Sheppard central difference operator defined by the relation $\delta f(t) = f\left(t+\frac{1}{2}\right) - f\left(t-\frac{1}{2}\right)$ so that (cf. Jordan [141])*

$$\delta = \Delta E^{-1/2} = \Delta / E^{1/2}, \quad \delta^{2k} = \Delta^{2k} E^{-k}.$$

Moreover, in this section, we shall make use of several simple and well-known propositions which may be stated as lemmas as follows.

Lemma 2.1.5 *There is a simple binomial identity*

$$\sum_{m=k}^{\infty} \binom{m}{k} x^m = \frac{x^k}{(1-x)^{k+1}}, \quad (|x| < 1).$$

The proof of Lemma 2.1.5 is left as Exercise 2.1.

Lemma 2.1.6 *Newton's symbolic expression for E^x is given by*

$$E^x = (1+\Delta)^x = \sum_{k=0}^{\infty} \binom{x}{k} \Delta^k.$$

For $f \in C^{n+1}_{[0,\infty)}$ we have Newton's interpolation formula

$$f(x) = E^x f(0) = \sum_{k=0}^{n} \binom{x}{k} \Delta^k f(0) + \binom{x}{n+1} f^{(n+1)}(\xi),$$

where $x \in (0, \infty)$ and $\xi \in \langle x, 0, 1, \cdots, n\rangle$.

Proof. The first equation is straightforward from the definitions of operators E and Δ. We apply E^x to function f at 0 to obtain

$$f(x) = E^x f(0) = (1+\Delta)^x f(0) = \sum_{k=0}^{\infty} \binom{x}{k} \Delta^k.$$

If the function $f(x)$ is a polynomial of degree n, then

$$\Delta^{n+k} f(x) = 0$$

for all $k > 0$. Therefore the expression of $E^x f(0)$ will be finite:

$$f(x) = f(0) + \binom{x}{1} \Delta f(0) + \cdots + \binom{x}{n} \Delta^n f(0).$$

The right-hand side of the above equation is clearly the interpolation of $f(x)$ at $x = 0, 1, \ldots, n$. To use it for an approximation to an $n+1$ differentiable function $f(x)$, we may add the remainder of interpolation:

$$R_{n+1} = \frac{(x-0)(x-1)\cdots(x-n)}{(n+1)!} D^{n+1} f(\xi),$$

where ξ is included in the smallest interval containing $0, 1, 2, \ldots, n$, and x. Hence,

$$f(x) = f(0) + \binom{x}{1} \Delta f(0) + \cdots + \binom{x}{n} \Delta^n f(0) + \binom{x}{n+1} f^{(n+1)}(\xi).$$

■

Lemma 2.1.7 *Euler's summation formula for the arithmetic-geometric series is given by*

$$\sum_{j=0}^{\infty} j^k x^j = \frac{A_k(x)}{(1-x)^{k+1}} = \alpha_k(x), \quad (|x|<1),$$

where k is a positive integer, and $\alpha_k(x)$ is the Eulerian fraction.

Proof. For $|x|<1$, it easily follows that

$$\begin{aligned}
A_k(x) &= \sum_{j=0}^{\infty} A(k,j)x^j = \sum_{j=0}^{\infty}\sum_{i=0}^{j}(-1)^i\binom{k+1}{i}(j-i)^k x^j \\
&= \sum_{i=0}^{\infty}\sum_{j=i}^{\infty}(-1)^i\binom{k+1}{i}(j-i)^k x^j = \sum_{i=0}^{\infty}\sum_{j=0}^{\infty}(-1)^i\binom{k+1}{i}j^k x^{j+i} \\
&= \sum_{i=0}^{\infty}(-1)^i\binom{k+1}{i}x^i\sum_{j=0}^{\infty}j^k x^j = (1-x)^{k+1}\sum_{j=0}^{\infty}j^k x^j.
\end{aligned}$$

■

Lemma 2.1.8 *Gauss's first symbolic expression for E^x is given by*

$$E^x = \sum_{k=0}^{\infty}\left(\binom{x+k-1}{2k}\frac{\Delta^{2k}}{E^k} + \binom{x+k}{2k+1}\frac{\Delta^{2k+1}}{E^k}\right).$$

Gauss's second symbolic expression for E^x is given by

$$E^x = \sum_{k=0}^{\infty}\left(\binom{x+k}{2k}\frac{\Delta^{2k}}{E^k} + \binom{x+k}{2k+1}\frac{\Delta^{2k+1}}{E^{k+1}}\right).$$

For $f \in C^{2m}_{(-\infty,\infty)}$ we have the first Gauss interpolation formula

$$\begin{aligned}
f(x) &= \sum_{k=0}^{m-1}\left(\binom{x+k-1}{2k}\Delta^{2k}f(-k) + \binom{x+k}{2k+1}\Delta^{2k+1}f(-k)\right) \\
&\quad + \binom{x+m-1}{2m}f^{(2m)}(\xi),
\end{aligned}$$

and the second Gauss interpolation formula

$$\begin{aligned}
f(x) &= \sum_{k=0}^{m-1}\left(\binom{x+k}{2k}\Delta^{2k}f(-k) + \binom{x+k}{2k+1}\Delta^{2k+1}f(-k-1)\right) \\
&\quad + \binom{x+m}{2m}f^{(2m)}(\xi),
\end{aligned}$$

where $x \in (-\infty,\infty)$ and $\xi \in (-m, m-1)$.

Proof. Let us multiply the right-hand side of the summation in Newton's symbolic expression for E^x from the term Δ^2 up by $(\Delta+1)/E = 1$. Then E^x expression will be changed to

$$E^x = 1 + \binom{x}{1}\Delta + \binom{x}{2}\frac{\Delta^2}{E} + \sum_{k=2}^{\infty} \frac{\Delta^{k+1}}{E}\left[\binom{x}{k} + \binom{x}{k+1}\right]$$

$$= 1 + \binom{x}{1}\Delta + \binom{x}{2}\frac{\Delta^2}{E} + \binom{x+1}{3}\frac{\Delta^3}{E} + \sum_{k=3}^{\infty}\binom{x+1}{k+1}\frac{\Delta^{k+1}}{E}.$$

Repeating the operation on the last series, from the term Δ^4 up, we obtain

$$E^x = 1 + \binom{x}{1}\Delta + \binom{x}{2}\frac{\Delta^2}{E} + \binom{x+1}{3}\frac{\Delta^3}{E} + \binom{x+1}{4}\frac{\Delta^4}{E^2} + \sum_{k=4}^{\infty}\binom{x+2}{k+1}\frac{\Delta^{k+1}}{E^2}.$$

Then the above operation is repeated from Δ^6 up, and so on. Finally, we have Gauss's first symbolic expression for E^x. By applying E^x to $f \in C^{2m}_{(-\infty,\infty)}$ at 0, we may obtain the first Gauss interpolation formula. The second Gauss symbolic expression and the second Gauss interpolation formula can be proved similarly. The only difference is multiplying the terms of Newton's symbolic expression for E^x from Δ up by $(\Delta+1)/E = 1$.

■

Lemma 2.1.9 *For $n \ge 1$ we have Everett's symbolic expression* (cf. *Jordan [141], §129).*

$$E^x = \sum_{k=0}^{\infty}\left(\binom{x+k}{2k+1}\frac{\Delta^{2k}}{E^{k-1}} - \binom{x+k-1}{2k+1}\frac{\Delta^{2k}}{E^k}\right).$$

For $f \in C^{2m}_{(-\infty,\infty)}$ we have Everett's interpolation formula

$$f(x) = \sum_{k=0}^{m-1}\left(\binom{x+k}{2k+1}\delta^{2k}f(1) - \binom{x+k-1}{2k+1}\delta^{2k}f(0)\right) + \binom{x+m-1}{2m}f^{(2m)}(\xi),$$

where $x \in (-\infty,\infty)$ and $\xi \in \langle x, 0, \pm 1, \cdots, \pm m, m+1\rangle$.

Proof. From the first and the second Gauss symbolic expressions, we have

$$E^x = 1 + \sum_{k=1}^{\infty}\left[\binom{x+k-1}{2k-1}\frac{\Delta^{2k-1}}{E^k}\left(\frac{E+1}{2}\right) + \frac{\Delta^{2k}}{2E^k}\left[\binom{x+k}{2k} + \binom{x+m-1}{2k}\right]\right].$$

After simplification and denoting $(E+1)/2 = M$, this gives the symbolic expression of *Stirling's formula*

$$E^x = 1 + \sum_{k=1}^{\infty} \binom{x+k-1}{2k-1} \left[\frac{M\Delta^{2k-1}}{E^k} + \frac{x}{2k} \frac{\Delta^{2k}}{E^k} \right].$$

Stirling's formula is, in reality, a formula of central difference

$$E^x = 1 + \sum_{k=1}^{\infty} \left[\mu\delta^{2k-1} + \frac{x}{2k}\delta^{2k} \right] \binom{x+k-1}{2k-1}$$

by using

$$\frac{\Delta^{2k}}{E^k} = \delta^{2k} \quad \text{and} \quad \frac{M\Delta^{2k-1}}{E^k} = \mu\delta^{2k-1},$$

where $\mu\delta = E - 1 - \delta^2/2$. Thus

$$E^x = 1 + \sum_{k=1}^{\infty} \left[E\delta^{2k-2} - \delta^{2k-2} - \frac{1}{2}\delta^{2k} + \frac{x}{2k}\delta^{2k} \right] \binom{x+k-1}{2k-1}$$

If we write in the first two terms in the square brackets of the preceding sum $k+1$ instead of k, then k varies in these terms from 0 to ∞. Consequently, they can be written as

$$\sum_{k=0}^{\infty} \binom{x+k}{2k+1} [E\delta^{2k} - \delta^{2k}].$$

The third and the fourth terms in the square brackets give

$$\sum_{k=0}^{\infty} \binom{x+k-1}{2k-1} \frac{x-k}{2k}\delta^{2k} = \sum_{k=1}^{\infty} \binom{x+k-1}{2k} \delta^{2k}.$$

Therefore the above formula about E^x becomes

$$E^x = \sum_{k=0}^{\infty} \binom{x+k}{2k+1} E\delta^{2k} + \sum_{k=0}^{\infty} \left[\binom{x+k-1}{2k} - \binom{x+k}{2k+1} \right] \delta^{2k},$$

which implies Everett's symbolic expression.

■

Theorem 2.1.10 *(Mean Value Theorem) Let $\sum_{n=0}^{\infty} a_n x^n$ with $a_n \geq 0$ be a convergent series for $x \in (0,1)$. Suppose that $\phi(t)$ is a bounded continuous function of t on $(-\infty, \infty)$, and (t_n) is a sequence of real numbers. Then there is a number $\xi \in (-\infty, \infty)$ such that*

$$\sum_{n=0}^{\infty} a_n \phi(t_n) x^n = \phi(\xi) \sum_{n=0}^{\infty} a_n x^n.$$

2.1.2 Summation formulas related to the operator $(1-xE)^{-1}$

We now state and prove the following proposition of various expansions of $(1-xE)^{-1}$.

Proposition 2.1.11 *The operator $(1-xE)^{-1}$ has four symbolic expansions, as follows.*

$$(1-xE)^{-1} = \sum_{k=0}^{\infty} \frac{x^k}{(1-x)^{k+1}} \Delta^k, \tag{2.1}$$

$$(1-xE)^{-1} = \sum_{k=0}^{\infty} \frac{\alpha_k(x)}{k!} D^k, \tag{2.2}$$

$$(1-xE)^{-1} = 1 + \sum_{k=0}^{\infty} \left(\frac{x}{(1-x)^2}\right)^{k+1} \left(\frac{\Delta^{2k}}{E^{k-1}} - x\frac{\Delta^{2k}}{E^k}\right), \tag{2.3}$$

$$(1-xE)^{-1} = 1 + \sum_{k=0}^{\infty} \left(\frac{x}{(1-x)^2}\right)^{k+1} \left(x^{-1}\frac{\Delta^{2k}}{E^k} - \frac{\Delta^{2k}}{E^{k+1}}\right), \tag{2.4}$$

where the condition $x \neq 1$ is assumed, and moreover, $x \neq 0$ for (2.4).

Proof. Here we present a proof in the sense of symbolic calculus, viz., every series expansion is considered as a formal series.

Clearly, (2.1) may be derived as follows.

$$\begin{aligned}(1-xE)^{-1} &= (1-x(1+\Delta))^{-1} = (1-x-x\Delta)^{-1} \\ &= (1-x)^{-1}(1-x\Delta/(1-x))^{-1} = \sum_{k=0}^{\infty} \frac{x^k \Delta^k}{(1-x)^{k+1}}.\end{aligned}$$

For proving (2.2) it suffices to make use of $E = e^D$ and Lemma 2.1.7. Indeed we have

$$\begin{aligned}&(1-xE)^{-1} = (1-xe^D)^{-1} = \sum_{k=0}^{\infty} x^k e^{kD} \\ &= \sum_{k=0}^{\infty} x^k \sum_{j=0}^{\infty} \frac{(kD)^j}{j!} = \sum_{j=0}^{\infty} \left(\sum_{k=0}^{\infty} x^k k^j\right) \frac{D^j}{j!} = \sum_{j=0}^{\infty} \alpha_j(x) \frac{D^j}{j!}.\end{aligned}$$

Expansions (2.3) and (2.4) can be justified in an entirely similar manner by using Lemma 2.1.5, Lemma 2.1.9, and Lemma 2.1.8, respectively. Indeed, (2.4) may be derived as follows.

$$(1-xE)^{-1} - 1 = \sum_{j=1}^{\infty} (xE)^j$$

$$
\begin{aligned}
&= \sum_{k=0}^{\infty}\left\{\left(\sum_{j=1}^{\infty}\binom{j+k}{2k}x^j\right)\frac{\Delta^{2k}}{E^k}+\left(\sum_{j=1}^{\infty}\binom{j+k}{2k+1}x^j\right)\frac{\Delta^{2k+1}}{E^{k+1}}\right\}\\
&= \sum_{k=0}^{\infty}\left\{\frac{x^k}{(1-x)^{2k+1}}\frac{\Delta^{2k}}{E^k}+\frac{x^{k+1}}{(1-x)^{2k+2}}\frac{\Delta^{2k+1}}{E^{k+1}}\right\}\\
&= \sum_{k=0}^{\infty}\left(\frac{x}{(1-x)^2}\right)^{k+1}\left(\frac{1-x}{x}\frac{\Delta^{2k}}{E^k}+\frac{\Delta^{2k+1}}{E^{k+1}}\right)\\
&= \sum_{k=0}^{\infty}\left(\frac{x}{(1-x)^2}\right)^{k+1}\left(x^{-1}\frac{\Delta^{2k}}{E^k}-\frac{\Delta^{2k}}{E^{k+1}}\right).
\end{aligned}
$$

Once (2.3) is derived by the aid of Lemma 2.1.5 and Lemma 2.1.9, it can also be verified by symbolic computations. In fact we have

$$
\begin{aligned}
&RHS\ of\ (2.3) = 1+\frac{x}{(1-x)^2}\sum_{k=0}^{\infty}\left(\frac{x}{(1-x)^2}\right)^k\left(\frac{\Delta^2}{E}\right)^k(E-x)\\
&= 1+\frac{x}{(1-x)^2}\frac{E-x}{1-\frac{x}{(1-x)^2}\frac{\Delta^2}{E}} = 1+\frac{x(E-x)}{(1-x)^2-x\frac{\Delta^2}{E}}\\
&= 1+\frac{Ex(E-x)}{(1-x)^2E-x(E-1)^2} = 1+\frac{Ex}{1-xE}\\
&= (1-xE)^{-1} = LHS\ of\ (2.3).
\end{aligned}
$$

Certainly (2.4) could also be verified in the like manner as above.

■

Remark 2.1.12 *Note that all the operators displayed on the right-hand sides of (2.1)–(2.4) involve Δ or D so that they will yield finite expressions when they are applied to any polynomial $f(t)$ at $t = 0$. In particular, we see that for the pth degree polynomial $f(t)$, (2.1) gives a generating function (GF) in the form*

$$
\sum_{k=0}^{\infty} f(k)x^k = \sum_{k=0}^{p}\frac{x^k}{(1-x)^{k+1}}\Delta^k f(0). \tag{2.5}
$$

Actually, this is a well-known formula and was mentioned in Jordan [141], §11. Moreover, an exact formula parallel to (2.5) may be obtained from (2.2), namely

$$
\sum_{k=0}^{\infty} f(k)x^k = \sum_{k=0}^{p}\frac{\alpha_k(x)}{k!}D^k f(0). \tag{2.6}
$$

Certainly, both (2.5)) and (2.6) may be used either as summation formulas for the power series $\sum_{k=0}^{\infty} f(k)x^k$ with $|x| < 1$, or as a tool for getting GF's for the sequence $(f(k))$.

Remark 2.1.13 *Observe that Euler's formula as given by Lemma 2.1.7 is a particular case of (2.6) with $f(t) = t^p$ $(p \geq 1)$. Obviously, Euler's formula may also be deduced from (2.5) by recalling the fact that* (cf. *Hsu and Shiue [133])*

$$\alpha_k(x) = \sum_{j=0}^{k} j! \left\{ {k \atop j} \right\} \frac{x^j}{(1-x)^{j+1}},$$

where $\left\{ {k \atop j} \right\}$ is a Stirling number of the second kind, which counts the number of ways to partition a set of k objects into j non-empty subsets. Some literatures denote Stirling numbers of the second kind $\left\{ {k \atop j} \right\}$ as $S(k, j)$.

Proposition 2.1.14 *Let $(f(k))$ be a given sequence of numbers (real or complex), and let $h(t)$ be infinitely differentiable at $t = 0$. Then we have formally*

$$\sum_{k=0}^{\infty} f(k)x^k = \sum_{k=0}^{\infty} \frac{x^k}{(1-x)^{k+1}} \Delta^k f(0), \tag{2.7}$$

$$\sum_{k=0}^{\infty} h(k)x^k = \sum_{k=0}^{\infty} \frac{\alpha_k(x)}{k!} D^k h(0), \tag{2.8}$$

$$\sum_{k=1}^{\infty} f(k)x^k = \sum_{k=0}^{\infty} \left(\frac{x}{(1-x)^2} \right)^{k+1} \left(\delta^{2k} f(1) - x\delta^{2k} f(0) \right), \tag{2.9}$$

$$\sum_{k=1}^{\infty} f(k)x^k = \sum_{k=0}^{\infty} \left(\frac{x}{(1-x)^2} \right)^{k+1} \left(x^{-1}\delta^{2k} f(0) - \delta^{2k} f(-1) \right), \tag{2.10}$$

where we always assume that $x \neq 0$ and $x \neq 1$.

Proof. Clearly, (2.7)–(2.10) are merely consequences of (2.1)–(2.4) by applying the operators to $f(t)$ or $h(t)$ at $t = 0$. ■

As in the case of (2.5) and (2.6), we have a corollary from (2.9) and (2.10).

Corollary 2.1.15 *If $f(t)$ is a polynomial in t of degree p, then*

$$\sum_{k=1}^{\infty} f(k)x^k = \sum_{k=0}^{[p/2]} \left(\frac{x}{(1-x)^2} \right)^{k+1} \left(\delta^{2k} f(1) - x\delta^{2k} f(0) \right), \tag{2.11}$$

$$\sum_{k=1}^{\infty} f(k)x^k = \sum_{k=0}^{[p/2]} \left(\frac{x}{(1-x)^2} \right)^{k+1} \left(x^{-1}\delta^{2k} f(0) - \delta^{2k} f(-1) \right). \tag{2.12}$$

Certainly, (2.11) and (2.12) may also be used as rules for obtaining GF's of $\{f(k)\}$.

2.1.3 Consequences and examples

As observed in §2.1.2, any of the formulas (2.5), (2.6), (2.11) and (2.12) solves generally the summation problem of power series $\sum_{k=0}^{\infty} f(k)x^k$ in the case $f(t)$ is a polynomial. Thus for instance, a few summation formulas of the forms

$$\sum_{k=0}^{\infty}(k+\lambda|\theta)_p x^k = \sum_{k=0}^{p}\frac{k!S(p,k,\lambda|\theta)x^k}{(1-x)^{k+1}} \quad \text{and} \tag{2.13}$$

$$\sum_{k=0}^{\infty} D_p(k,\alpha)x^k = \sum_{k=0}^{p}\frac{x^\alpha k!S(p,k,\alpha|\theta)x^k}{(1-x)^{k+1}} \tag{2.14}$$

as given in Hsu and Shiue [133] are just particular cases of (2.5) in which $f(t) = (t+\lambda|\theta)_p$ and $f(t) = D_p(t,\alpha)$ are known as the generalized falling factorial and the Dickson polynomial, respectively, or more precisely

$$(t+\lambda|\theta)_p = \Pi_{j=0}^{p-1}(t+\lambda-j\theta),\ (p\geq 1), \quad (t+\lambda|\theta)_0 = 1, \quad \text{and}$$

$$D_p(t,\alpha) = \sum_{j=0}^{[p/2]}\frac{p}{p-j}\binom{p-j}{j}(-\alpha)^j t^{p-2j}, \quad D_0(t,\alpha) = 2.$$

Moreover, $S(p,k,\lambda|\theta)$ denotes *Howard's degenerate weighted Stirling numbers.* (For more details, *cf.* [133] loc. cit.)

Another important consequence of Proposition 2.1.14 is that (2.7), (2.9) and (2.10) with $x=-1$ yield three series transforms, respectively

$$\sum_{k=0}^{\infty}(-1)^k f(k) = \sum_{k=0}^{\infty}\frac{(-1)^k}{2^{k+1}}\Delta^k f(0), \tag{2.15}$$

$$\sum_{k=1}^{\infty}(-1)^{k-1} f(k) = \sum_{k=0}^{\infty}\frac{(-1)^k}{4^{k+1}}\left(\delta^{2k}f(1)+\delta^{2k}f(0)\right), \tag{2.16}$$

$$\sum_{k=1}^{\infty}(-1)^k f(k) = \sum_{k=0}^{\infty}\frac{(-1)^k}{4^{k+1}}\left(\delta^{2k}f(0)+\delta^{2k}f(-1)\right). \tag{2.17}$$

Note that (2.15) is the well-known Euler series transformation formula shown in Corollary 1.3.12. As what we have seen, this formula can be used to convert a slowly convergent alternating series $\sum_{k=0}^{\infty}(-1)^k f(k)$ with $f(k)\downarrow 0$ (as $k\to\infty$) into a rapidly convergent series. A few examples are given in Example 1.3.13. For instance, the series

$$\ln 2 = 1 - \frac{1}{2} + \frac{1}{3} - \frac{1}{4} + \frac{1}{5} - \cdots \tag{2.18}$$

can be converted using (2.15) with $f(k) = \frac{1}{k+1}$ $(k=0,1,2,\ldots)$ into a quickly convergent series of the form

$$\ln 2 = \frac{1}{2} + \frac{1}{2^2\cdot 2} + \frac{1}{2^3\cdot 3} + \frac{1}{2^4\cdot 4} - \cdots \tag{2.19}$$

Actually, the above expression can be derived by substituting

$$\Delta^k f(0) = \sum_{j=0}^{k} \binom{k}{j} (-1)^{k-j} (j+1)^{-1} \tag{2.20}$$

into (2.20). Thus,

$$\begin{aligned} \ln 2 &= \sum_{k=0}^{\infty} (-1)^k f(k) = \sum_{k=0}^{\infty} \frac{(-1)^k}{2^{k+1}} \sum_{j=0}^{k} \binom{k}{j} (-1)^{k-j} (j+1)^{-1} \\ &= \sum_{k=0}^{\infty} \frac{1}{(k+1)2^{k+1}} \sum_{j=0}^{k} (-1)^j \binom{k+1}{j+1} \\ &= \sum_{k=0}^{\infty} \frac{1}{(k+1)2^{k+1}} = \sum_{k=1}^{\infty} \frac{1}{k2^k}. \end{aligned} \tag{2.21}$$

Remark 2.1.16 *Obviously, the convergence of the series shown in (2.19) with a rate of* $O(1/2^n)$ *is much faster than the convergence of the series in (2.18), which has the rate of* $O(1/n)$. *For instance, to arrive the accuracy of the five digits of* $\ln 2 = 0.69315$, *we only need to sum the first* 15 *terms of the series in (2.19), while the partial sum of the first* 40,000 *terms of the series in (2.18) is* 0.69313. *(2.16) and (2.17) appear to be novel, and they could also be used to convert slowly convergent alternating series* $\sum_{k=1}^{\infty}(-1)^k f(k)$ *into quickly convergent ones if a definition for* $f(k) = 0$ $(k = 0, -1, -2, \ldots)$ *is introduced. A general discussion on the comparison of the convergence rate of the given alternating series and its Euler transformation can be found in Exercise 1.37 and its hint. And a later work will give the comparison on the rate of the convergence of series (2.15)–(2.17) for the positive decreasing functions. Exercise 1.36 shows that for some alternating series, their Euler transformations may have slower convergent rates.*

We now give some examples of the summations shown in Proposition 2.1.14. Our first example is for function $f(x) = 1/(x+1)^2$. Similar to expression (2.20) we obtain

$$\Delta^k f(0) = \sum_{j=0}^{k} \binom{k}{j} (-1)^{k-j} (j+1)^{-2}.$$

Substituting the above expression into (2.2) and noting the well-known identity

$$\sum_{j=1}^{k} (-1)^{j-1} \binom{k}{j} \frac{1}{j} = \sum_{\ell=1}^{k} \frac{1}{\ell},$$

(see [139]), we then use the process similar to that in (2.21) to obtain

$$\sum_{k=0}^{\infty} (-1)^k f(k) = \sum_{k=0}^{\infty} \frac{1}{2^{k+1}} \sum_{j=0}^{k} \binom{k}{j} \frac{(-1)^j}{(j+1)^2}$$

$$= \sum_{k=0}^{\infty} \frac{1}{(k+1)2^{k+1}} \sum_{j=0}^{k} \binom{k+1}{j+1} \frac{(-1)^j}{j+1} = \sum_{k=1}^{\infty} \frac{1}{k2^k} \sum_{j=1}^{k} \binom{k}{j} \frac{(-1)^{j-1}}{j}$$

$$= \sum_{k=1}^{\infty} \frac{1}{k2^k} \sum_{\ell=1}^{k} \frac{1}{\ell} = \sum_{j=0}^{\infty} \sum_{\ell=1}^{\infty} \frac{1}{\ell(j+\ell)2^{j+\ell}} = \sum_{\ell=1}^{\infty} \frac{1}{\ell^2 2^\ell} + \sigma,$$

where summation 360(c) in Jolley [139] gives the first sum as $\frac{\pi^2}{12} - \frac{1}{2}\ln^2 2$, and

$$\sigma = \sum_{j=1}^{\infty} \sum_{\ell=1}^{\infty} \frac{1}{\ell(j+\ell)2^{j+\ell}}$$

is easily to be seen to equal

$$\frac{1}{2} \sum_{j=1}^{\infty} \sum_{\ell=1}^{\infty} \frac{1}{j\ell 2^{j+\ell}} = \frac{1}{2} \ln^2 \frac{1}{2}.$$

Hence,

$$\sum_{k=0}^{\infty} \frac{(-1)^k}{(k+1)^2} = \frac{\pi^2}{12}. \tag{2.22}$$

Remark 2.1.17 *Although formula (2.22) can be easily derived by using Fourier cosine expansion of x^2, we give a different approach here by using formula (2.15) because it converts the series in (2.22) into the following quickly convergent series:*

$$\frac{\pi^2}{12} = \sum_{k=0}^{\infty} \frac{(-1)^k}{(k+1)^2} = \sum_{k=1}^{\infty} \frac{1}{k2^k} \sum_{\ell=1}^{k} \frac{1}{\ell}.$$

Hence, we can use the last series to evaluate $\zeta(2)$ as

$$\zeta(2) = \frac{\pi^2}{6} = \sum_{k=1}^{\infty} \frac{1}{k2^{k-1}} \sum_{\ell=1}^{k} \frac{1}{\ell}.$$

The sum of the first 13 *terms of the last series gives,* 1.6449, *the first* 5 *digits of $\pi^2/6$, while the sum of the first* 5,000 *terms of the series in (2.22) is only* 1.6447. *This example shows that formula (2.15) is indeed to convert an alternating series into a faster convergent series .*

We now consider another example generated by function $f(x) = (g(t))^x$, where $g : \mathbb{R} \mapsto \mathbb{R}$ and f is defined on $\mathbb{N}_0$. Obviously, we have

$$\Delta^k f(0) = \sum_{j=0}^{k} \binom{k}{j} (g(t))^j (-1)^{k-j} = (g(t)-1)^k \tag{2.23}$$

and for $i = 0, 1$

$$\begin{aligned} \delta^{2k} f(i) &= \Delta^{2k} E^{-k} f(i) = \Delta^{2k} (g(t))^{i-k} \\ &= \sum_{j=0}^{2k} \binom{2k}{j} (g(t))^{i-k+j} (-1)^{2k-j} = (g(t)-1)^{2k} (g(t))^{i-k}. \end{aligned} \tag{2.24}$$

Hence, substituting (2.23) into (2.7) yields

$$\begin{aligned} \sum_{k=0}^{\infty} (g(t))^k x^k &= \sum_{k=0}^{\infty} \frac{x^k}{(1-x)^{k+1}} (g(t)-1)^k \\ &= \frac{1}{1-x} \frac{1}{1 - \frac{x(g(t)-1)}{1-x}} = \frac{1}{1 - xg(t)}. \end{aligned} \tag{2.25}$$

Similarly, substituting (2.24) into (2.9), we obtain the following summation formula

$$\begin{aligned} &\sum_{k=0}^{\infty} \left(\frac{x}{(1-x)^2} \right)^{k+1} (g(t)-1)^{2k} \left\{ (g(t))^{1-k} - x(g(t))^{-k} \right\} \\ &= \frac{g(t)(g(t)-x)}{(g(t)-1)^2} \sum_{k=0}^{\infty} \left(\frac{x(g(t)-1)^2}{g(t)(1-x)^2} \right)^{k+1} \\ &= \frac{g(t)(g(t)-x)}{(g(t)-1)^2} \frac{x(g(t)-1)^2}{g(t)(1-x)^2 - x(g(t)-1)^2} \\ &= \frac{xg(t)(g(t)-x)}{g(t)(1+x^2) - x((g(t))^2+1)} = \frac{xg(t)}{1-xg(t)}. \end{aligned} \tag{2.26}$$

As examples, we take $g(t) = e^{it}$, with $i = \sqrt{-1}$, and $g(t) = t$. Thus, from (2.25) we have, respectively

$$\sum_{k=0}^{\infty} e^{itk} x^k = \sum_{k=0}^{\infty} \frac{x^k}{(1-x)^{k+1}} (e^{it}-1)^k = \frac{1}{1-xe^{it}} \tag{2.27}$$

and

$$\sum_{k=0}^{\infty} (tx)^k = \sum_{k=0}^{\infty} \frac{x^k}{(1-x)^{k+1}} (g(t)-1)^k = \frac{1}{1-xt}. \tag{2.28}$$

By applying (2.26) for $g(t) = e^{it}$ and t we obtain

$$\sum_{k=0}^{\infty} \left(\frac{x}{(1-x)^2} \right)^{k+1} (e^{it}-1)^{2k} \left\{ e^{-i(k-1)t} - xe^{-ikt} \right\} = \frac{xe^{it}}{1-xe^{it}} \tag{2.29}$$

and

$$\sum_{k=0}^{\infty}\left(\frac{x}{(1-x)^2}\right)^{k+1}(t-1)^{2k}\left\{t^{1-k}-xt^{-k}\right\}=\frac{tx}{1-tx}, \tag{2.30}$$

respectively.

We now illustrate (2.8) with $h(x)=(g(t))^x$ with $g:\mathbb{R}\mapsto\mathbb{R}$ and $g(t)>0$. Hence, $D^k h(0)=(\ln g(t))^k$ and from (2.8) and Definition 2.1.3,

$$\sum_{k=0}^{\infty}(g(t))^k x^k = \frac{1}{1-xg(t)} = \sum_{k=0}^{\infty}\frac{\alpha_k(x)}{k!}(\ln g(t))^k$$
$$= \sum_{k=0}^{\infty}\frac{A_k(x)}{k!(1-x)^{k+1}}(\ln g(t))^k. \tag{2.31}$$

Replacing $\ln g(t)$ by t and $t(1-x)$, respectively, then Equation (2.31) yields *GF's* (*cf.* Section 6.5 in [44] and [210])

$$\sum_{k=0}^{\infty}\alpha_k(x)\frac{t^k}{k!}=\frac{1}{1-xe^t} \tag{2.32}$$

and

$$\sum_{k=0}^{\infty}A_k(x)\frac{t^k}{k!}=\frac{1-x}{1-xe^{t(1-x)}}, \tag{2.33}$$

respectively. Some other *GF's* such as $(5i)-(5k)$ shown in Section 6.5 in [44] can be derived from Equation (2.21). In addition, from Equation (2.22), we can establish the recurrence relation for $\alpha_k(x)$ by multiplying both sides of the equation by $(1-xe^t)$. The details can be found in [210].

Finally, we consider a special case of (2.31) by letting $g(t)=e^{it}$ with $t\in\mathbb{R}$, and $x=-1$ in (2.31), we obtain

$$\sum_{k=0}^{\infty}e^{ikt}(-1)^k=\frac{1}{1+e^{it}}=\frac{1}{2}\left\{1-i\tan\frac{t}{2}\right\}=\sum_{k=0}^{\infty}\frac{A_k(-1)}{k!2^{k+1}}(it)^k.$$

Therefore, direct verification of the rightmost equality would be effected by the identity

$$\sum_{k=1}^{\infty}\frac{A_k(-1)z^k}{k!}=-\tanh z, \tag{2.34}$$

implying $A_k(-1)$ equals, modulo a sign, the respective tangent coefficient [141]. Note that

$$A_k(-1)=\sum_{j=1}^{k}A(k,j)(-1)^j=-\sum_{j=1}^{k}j!\left\{ {k \atop j} \right\}(-2)^{k-j}, \tag{2.35}$$

with $\left\{ {k \atop j} \right\}$ denoting the Stirling number of the second kind (*cf.* Formula [51] in 6.5 of [44] and [23, 133]). Therefore, implementing exponential GFs, in z, on both sides of (2.35), (2.34) follows from $j! \sum_{k=j}^{\infty} \left\{ {k \atop j} \right\} z^k/k! = (e^z - 1)^j$ (*cf.* (21.1.4B) [1]).

2.1.4 Remainders of summation formulas

In this subsection, we will establish four summation formulas with remainders whose forms are suggested by Lemmas 2.1.6, 2.1.8, and 2.1.9.

Theorem 2.1.18 *Let $f(t) \in C^m_{[0,\infty)}$ $(m \geq 1)$, with bounded derivative $f^{(m)}(t)$ in $[0,\infty)$, and let $\sum_{k=0}^{\infty} f(k)x^k$ be convergent for $|x| < 1$. Then for $x \in (0,1)$ we have*

$$\sum_{k=M}^{N-1} f(k)x^k = \sum_{k=0}^{m-1} \left(x^M \Delta^k f(M) - x^N \Delta^k f(N)\right) \frac{x^k}{(1-x)^{k+1}} + \rho_m, \tag{2.36}$$

where the remainder ρ_m has a form with $\xi \in [0,\infty)$ as follows

$$\rho_m = \left(x^M f^{(m)}(M+\xi) - x^N f^{(m)}(N+\xi)\right) \frac{x^m}{(1-x)^{m+1}}. \tag{2.37}$$

Proof. Let $\phi(t) = \phi(t,x) = x^M f(t+M) - x^N f(t+N)$ so that $\phi(t) \in C^m_{[0,\infty)}$. Then by Lemma 2.1.6 and using the Mean-Value Theorem (cf. Theorem 2.1.10) with $a_n = \binom{n}{m}$, we obtain

$$\begin{aligned}
\sum_{k=M}^{N-1} f(k)x^k &= \sum_{n=0}^{\infty} \phi(n)x^n \\
&= \sum_{n=0}^{\infty} \left\{ \sum_{k=0}^{m-1} \Delta^k \phi(0) \binom{n}{k} \right\} x^n + \sum_{n=0}^{\infty} \phi^{(m)}(\xi_n) \binom{n}{m} x^n \\
&= \sum_{k=0}^{m-1} \Delta^k \phi(0) \left(\sum_{n=0}^{\infty} \binom{n}{k} x^n \right) + \phi^{(m)}(\xi) \sum_{n=0}^{\infty} \binom{n}{m} x^n \\
&= \sum_{k=0}^{m-1} \Delta^k \phi(0) \frac{x^k}{(1-x)^{k+1}} + \phi^{(m)}(\xi) \frac{x^m}{(1-x)^{m+1}} \\
&= RHS \ of \ (2.36)
\end{aligned}$$

with ρ_m being given by (2.37), where $\xi_n \in \langle n, 0, 1, 2, \cdots, m-1 \rangle$.

■

Note that the RHS of (2.36) without ρ_m may be regarded as a rational approximation to the series on the LHS. In particular, if $x^N\Delta^k f(N) \to 0$ $(N \to \infty, 0 \le k \le m-1)$, then (2.36) reduces to

$$\sum_{k=M}^{\infty} f(k)x^k = \sum_{k=0}^{m-1} x^M \Delta^k f(M) \frac{x^k}{(1-x)^{k+1}} + x^M f^{(m)}(M+\xi) \frac{x^m}{(1-x)^{m+1}}. \tag{2.38}$$

Theorem 2.1.19 *Under the same condition of Theorem 2.1.18, we have*

$$\sum_{k=M}^{N-1} f(k)x^k = \sum_{k=0}^{m-1} \left(x^M f^{(k)}(M) - x^N f^{(k)}(N) \right) \frac{\alpha_k(x)}{k!} + \rho_m, \tag{2.39}$$

where the remainder is given by

$$\rho_m = \left(x^M f^{(m)}(M+\xi) - x^N f^{(m)}(N+\xi) \right) \frac{\alpha_m(x)}{m!}. \tag{2.40}$$

Proof. Denote $\phi(t) = x^M f(t+M) - x^N f(t+N)$ so that $\phi(t) \in C^m_{[0,\infty)}$. Clearly, by using Taylor's expansion with Lagrange's remainder, we have

$$\begin{aligned} & \sum_{k=M}^{N-1} f(k)x^k = \sum_{n=0}^{\infty} \phi(n)x^n \\ = & \sum_{n=0}^{\infty} \left(\sum_{k=0}^{m-1} \frac{1}{k!}\phi^{(k)}(0)n^k \right) x^n + \sum_{n=0}^{\infty} \frac{1}{m!}\phi^{(m)}(\xi_n)n^m x^n, \ (0 < \xi_n < n) \\ = & \sum_{k=0}^{m-1} \frac{1}{k!}\phi^{(k)}(0) \left(\sum_{n=0}^{\infty} n^k x^n \right) + S_2. \end{aligned}$$

Here we can apply Theorem 2.1.10 to the series S_2 and obtain

$$\begin{aligned} S_2 &= \frac{1}{m!}\phi^{(m)}(\xi) \left(\sum_{n=0}^{\infty} n^m x^n \right) \quad (0 < \xi < \infty) \\ &= \frac{1}{m!}\phi^{(m)}(\xi)\alpha_m(x) = \rho_m. \end{aligned}$$

Hence, in accordance with Lemma 2.1.7, we get (2.39) and (2.40).

■

Remark 2.1.20 *Theorem 2.1.19 with expressions (2.39) and (2.40) is of similar nature as that of Theorems 1 and 2 in Wang and Hsu [210]. However, Theorem 2.1.19 appears to be a little more restrictive since we have assumed here the condition* $0 < x < 1$ *and the convergence of* $\sum_{k=0}^{\infty} f(k)x^k$ *for* $|x| < 1$. *In what follows we shall give formulas using the central difference operators* $\delta^{2k} = \Delta^{2k}/E^k$ *which appear to be more available for numerical computations.*

Theorem 2.1.21 *Let $f(t) \in C^{2m}_{(-\infty,\infty)}$ with bounded derivative $f^{(2m)}(t)$ in $(-\infty,\infty)$ and let $\sum_{k=0}^{\infty} f(k)x^k$ be convergent for $|x| < 1$. Then for $x \in (0,1)$, we have*

$$\sum_{k=M}^{N-1} f(k)x^k = \sum_{k=0}^{m-1} \left(\delta^{2k}\phi(1) - x\delta^{2k}\phi(0)\right)\left(\frac{x}{(1-x)^2}\right)^{k+1} + \rho_m, \tag{2.41}$$

where $\delta^{2k}\phi(t) = x^M\delta^{2k}f(t+M) - x^N\delta^{2k}f(t+N)$ and ρ_m is given by the following expression with $\xi \in [-m,\infty)$

$$\rho_m = \left(x^M f^{(2m)}(M+\xi) - x^N f^{(2m)}(N+\xi)\right)\frac{x^{m+1}}{(1-x)^{2m+1}}. \tag{2.42}$$

Proof. Denote $\phi(t) = x^M f(t+M) - x^N f(t+N)$ so that $\phi(t) \in C^{2m}_{(-\infty,\infty)}$. Let us now make use of Everett's formula in Lemma 2.1.9 for $\phi(t)$ at $t = n$,

$$\begin{aligned}
\phi(n) &= \sum_{k=0}^{m-1}\left[\binom{n+k}{2k+1}\delta^{2k}\phi(1) - \binom{n+k-1}{2k+1}\delta^{2k}\phi(0)\right] \\
&\quad + \binom{n+m-1}{2m}\phi^{(2m)}(\xi_n),
\end{aligned}$$

where $\xi_n \in \langle n, 0, \pm 1, \pm 2, \cdots, \pm m, m+1\rangle$. Clearly, we have

$$\begin{aligned}
&\sum_{k=M}^{N-1} f(k)x^k = \sum_{n=0}^{\infty}\phi(n)x^n, \ (0 < x < 1) \\
= &\sum_{n=0}^{\infty}\sum_{k=0}^{m-1}\left[\binom{n+k}{2k+1}\delta^{2k}\phi(1) - \binom{n+k-1}{2k+1}\delta^{2k}\phi(0)\right]x^n \\
&+\sum_{n=0}^{\infty}\binom{n+m-1}{2m}\phi^{(2m)}(\xi_n)x^n \\
= &\sum_{k=0}^{m-1}\delta^{2k}\phi(1)\left(\sum_{n=0}^{\infty}\binom{n+k}{2k+1}x^n\right) \\
&-\sum_{k=0}^{m-1}\delta^{2k}\phi(0)\left(\sum_{n=0}^{\infty}\binom{n+k-1}{2k+1}x^n\right) + \rho_m \\
= &\sum_{k=0}^{m-1}\delta^{2k}\phi(1)\frac{x^{k+1}}{(1-x)^{2k+2}} - \sum_{k=0}^{m-1}\delta^{2k}\phi(0)\frac{x^{k+2}}{(1-x)^{2k+2}} + \rho_m \\
= &\sum_{k=0}^{m-1}\left(\delta^{2k}\phi(1) - x\delta^{2k}\phi(0)\right)\left(\frac{x}{(1-x)^2}\right)^{k+1} + \rho_m.
\end{aligned}$$

Here an application of Theorem 2.1.10 to the series representation of ρ_m yields

$$\rho_m = \phi^{(2m)}(\xi)\left(\sum_{n=0}^{\infty}\binom{n+m-1}{2m}x^n\right) = \phi^{(2m)}(\xi)\frac{x^{m+1}}{(1-x)^{2m+1}},$$

where $\xi \in [-m, \infty)$. Hence the theorem is proved.

■

Theorem 2.1.22 *Under the same condition of Theorem 2.1.21, we have*

$$\sum_{k=M}^{N-1} f(k)x^k = \sum_{k=0}^{m-1}\left(x^{-1}\delta^{2k}\phi(0) - \delta^{2k}\phi(-1)\right)\left(\frac{x}{(1-x)^2}\right)^{k+1} + \rho_m, \quad (2.43)$$

where $x \neq 0$ *and* $\delta^{2k}\phi(t) = x^M\delta^{2k}f(t+M) - x^N\delta^{2k}f(t+N)$, *and*

$$\rho_m = \left(x^M f^{(2m)}(M+\xi) - x^N f^{(2m)}(N+\xi)\right)\frac{x^m}{(1-x)^{2m+1}}. \quad (2.44)$$

Proof. As before, denote $\phi(t) = x^M f(t+M) - x^N f(t+N)$ and let $x \in (0,1)$. Using Gauss interpolation formula with remainder in Lemma 2.1.8 for $\phi(t)$ at $t = n$, we get as in the case of proving Theorem 2.1.21 the following expressions

$$\begin{aligned}
&\sum_{k=M}^{N-1} f(k)x^k = \sum_{n=0}^{\infty}\phi(n)x^n \\
= &\sum_{k=0}^{m-1}\Delta^{2k}\phi(-k)\left(\sum_{n=0}^{\infty}\binom{n+k}{2k}x^n\right) \\
&+\sum_{k=0}^{m-1}\Delta^{2k+1}\phi(-k-1)\left(\sum_{n=0}^{\infty}\binom{n+k}{2k+1}x^n\right) \\
&+\sum_{n=0}^{\infty}\binom{n+m}{2m}\phi^{(2m)}(\xi_n)x^n \\
= &\sum_{k=0}^{m-1}\Delta^{2k}\phi(-k)\frac{x^k}{(1-x)^{2k+1}} + \sum_{k=0}^{m-1}\Delta^{2k+1}\phi(-k-1)\frac{x^{k+1}}{(1-x)^{2k+2}} + \rho_m \\
= &\sum_{k=0}^{m-1}\left(\frac{1-x}{x}\Delta^{2k}\phi(-k) + \left[\Delta^{2k}\phi(-k) - \Delta^{2k}\phi(-k-1)\right]\right) \\
&\times\left(\frac{x}{(1-x)^2}\right)^{k+1} + \rho_m \\
= &\sum_{k=0}^{m-1}\left(x^{-1}\Delta^{2k}\phi(-k) - \Delta^{2k}\phi(-k-1)\right)\left(\frac{x}{(1-x)^2}\right)^{k+1} + \rho_m.
\end{aligned}$$

Finally, an application of Theorem 2.1.10 to the series expression of ρ_m gives

$$\rho_m = \phi^{(2m)}(\xi) \sum_{n=0}^{\infty} \binom{n+m}{2m} x^n = \phi^{(2m)}(\xi) \frac{x^m}{(1-x)^{2m+1}},$$

where $\xi \in (-\infty, \infty)$. Hence Theorem 2.1.22 is proved.

■

Remark 2.1.23 *The uniform boundedness conditions for $f^{(m)}(t)$ in $[0,\infty)$ as well as for $f^{(2m)}(t)$ in $(-\infty,\infty)$ imply that $x^N f^{(m)}(N+\xi) \to 0$ and $x^N f^{(2m)}(N+\xi) \to 0$ as $N \to \infty$ and $0 < x < 1$. Thus, if in addition, $x^N f^{(k)}(N) = o(N)$ $(N \to \infty,\ 0 \le k \le m-1)$, then (2.39) and (2.40) yield*

$$\sum_{k+M}^{\infty} f(k)x^k = \sum_{k=0}^{m-1} x^M f^{(k)}(M) \frac{\alpha_k(x)}{k!} + x^M f^{(m)}(M+\xi) \frac{\alpha_m(x)}{m!}. \tag{2.45}$$

Similar consequences from Theorems 2.1.21 and 2.1.22 may also be deduced by providing additional conditions such as $x^N \delta^{2k} f(N) \to 0$ $(N \to \infty,\ 0 < x < 1)$.

2.1.5 Q-analog of symbolic operators

In this subsection, we shall present q-extensions of several linear operators including a novel q-*analog* of the derivative operator D shown in Dancs and the author [48]. Here, a q-analog of an operator is referred to as a q-*operator.* As sample applications, we shall show how the q-*substitution rules* can be used to construct q-analog of the symbolic summation and series transformation formulas shown in the previous subsections, which include, particularly, the q-analog of the ordinary Euler transformation for accelerating the convergence of alternating series.

We first give definitions and basic identities. Unless otherwise stated, we consider all operators to act on formal power series in the single variable t, with coefficients possibly depending on q. We assume $0 < |q| < 1$. We will use 1 to denote the identity operator and define the following operators:

1. $E_q f(t) = f(tq)$, (forward multiplicative shift),
2. $\Delta_q f(t) = f(tq) - f(t)$, (forward q-difference),
3. $L_q f(t) = t(\log q) f'(t)$, (forward logarithmic shift).

The first two of these can be regarded as q-analog of the ordinary (additive) shift and forward difference operators, respectively. L_q will play a role similar to that of the derivative D.

The operator inverse of E_q (which we denote as E_q^{-1}) clearly exists and is equal to $E_{q^{-1}}$. We define the central q-difference operator δ_q by

$$\delta_q = f(tq^{1/2}) - f(tq^{-1/2}) \tag{2.46}$$

and note that $\delta_q = \Delta_q E_q^{-1/2} = \Delta_q / E_q^{1/2}, \quad \delta_q^{2k} = \Delta_q^{2k} E_q^{-k}$.

The q-operators above are linear and satisfy some familiar identities, for example, $E_q = 1 + \Delta_q$. The binomial identity

$$\Delta_q^n = \sum_{k=0}^{n} (-1)^{n-k} \binom{n}{k} E_q^k \tag{2.47}$$

can be established by induction, or by considering the operator expansion of $(E_q - 1)^n$.

Treating these operators formally, we need only consider their effect on non-negative integer powers of t. E_q, Δ_q, and L_q are "diagonal" in the sense that each maps $t^k \mapsto M(q,k)t^k$, with the function M depending on the particular operator. For example, $\Delta_q[t^k] = (q^k - 1)t^k$ for $k > 0$, and $\Delta_q[1] = 0$. Similarly, $L_q[t^k] = t^k \log(q^k)$.

With this observation, it is easy to verify many additional identities. For example, considering the alternating geometric series $\sum_{n=0}^{\infty} (-1)^n \Delta_q^n$ applied to t^k, we have

$$\begin{aligned} \sum_{n=0}^{\infty} (-1)^n \Delta_q^n [t^k] &= t^k \sum_{n=0}^{\infty} (-1)^n (q^k - 1)^n \\ &= t^k \tfrac{1}{1-(1-q^k)} = t^k q^{-k}. \end{aligned}$$

In other words, this formal power series gives the operator $E_{q^{-1}}$. Stated differently,

$$(1 + \Delta_q)^{-1} = (E_q)^{-1} = E_{q^{-1}} = \sum_{n=0}^{\infty} (-1)^n \Delta_q^n, \tag{2.48}$$

which is exactly the result we should expect. We may establish the following identities in similar fashion:

$$(1 - \Delta_q)^{-1} = \sum_{n=0}^{\infty} \Delta_q^n, \tag{2.49}$$

$$\log(1 + \Delta_q) = \sum_{n=1}^{\infty} \frac{(-1)^{n+1}}{n} \Delta_q^n = L_q, \tag{2.50}$$

$$e^{L_q} = \sum_{n=0}^{\infty} \frac{1}{n!} L_q^n = E_q. \tag{2.51}$$

In addition to these last two identities, L_q obeys the product rule

$$L_q \left[f(t)g(t) \right] = L_q[f(t)]g(t) + f(t)L_q[g(t)], \tag{2.52}$$

so that L_q is a q-analog of the ordinary derivative operator D.

We begin with some q-analog of the symbolic substitution rules in the previous subsections (specifically, equations (2.3) and (2.4)).

Proposition 2.1.24 *Let $F(t)$ have the formal power series expansion $F(t) = \sum_{k\geq 0} f_k t^k$, with coefficients possibly dependent on q. We may obtain operational formulas according to the following rules:*

1. *The substitution $t \mapsto E_q$ leads to the symbolic formula*

$$F(E_q) = \sum_{k=0}^{\infty} f_k E_q^k. \tag{2.53}$$

2. *If $F(t) = G(t, e^t)$, the substitution $t \mapsto L_q$ leads to*

$$G(L_q, E_q) = \sum_{k=0}^{\infty} f_k L_q^k. \tag{2.54}$$

3. *If $F(t) = G(t, \log(1+t))$, the substitution $t \mapsto \Delta_q$ leads to*

$$G(\Delta_q, L_q) = \sum_{k=0}^{\infty} f_k \Delta_q^k. \tag{2.55}$$

Note that each of the identities in equations (2.49)–(2.51) can be obtained from elementary Maclaurin series by applying one of these substitution rules. We now present a less trivial example.

For k a positive integer, let $\alpha_k(x)$ denote the Eulerian fraction (cf. [44] P. 245). We recall Lemma 2.1.7 and have

$$\sum_{j=0}^{\infty} j^k x^j = \frac{A_k(x)}{(1-x)^{k+1}} = \alpha_k(x), \quad (|x| < 1), \tag{2.56}$$

where $A_k(x)$ is the kth Eulerian polynomial. Additionally, (2.32) gives the formula (cf. also [210], P. 24)

$$(1 - xe^t)^{-1} = \sum_{k=0}^{\infty} \alpha_k(x) \frac{t^k}{k!}. \tag{2.57}$$

Substituting $t \mapsto L_q$ leads to the formal identity

$$(1 - xE_q)^{-1} = \sum_{k=0}^{\infty} \frac{\alpha_k(x)}{k!} L_q^k, \tag{2.58}$$

which is a q-analog of (2.2).

We can obtain additional identities in this fashion from other expansions of $(1 - xe^t)^{-1}$ as the following theorem.

Theorem 2.1.25 *Let E_q, Δ_q, and L_q be the operators defined before, and let $\alpha_k(x)$ denote the Eulerian fraction. Then we have the q-analog of (2.2) shown in (2.58). If $x \neq 0$ and $x \neq 1$, we have the following q-analog of (2.1), (2.3), and (2.4):*

$$(1-xE_q)^{-1} = \sum_{k=0}^{\infty} \frac{x^k}{(1-x)^{k+1}} \Delta_q^k, \tag{2.59}$$

$$(1-xE_q)^{-1} = \sum_{k=0}^{\infty} \left(\frac{x}{(1-x)^2}\right)^{k+1} \left(x^{-1}\frac{\Delta_q^{2k}}{E_q^k} - \frac{\Delta_q^{2k}}{E_q^{k+1}}\right), \tag{2.60}$$

$$(1-xE_q)^{-1} = 1 + \sum_{k=0}^{\infty} \left(\frac{x}{(1-x)^2}\right)^{k+1} \left(\frac{\Delta_q^{2k}}{E_q^{k-1}} - x\frac{\Delta_q^{2k}}{E_q^k}\right). \tag{2.61}$$

The proofs of (2.58)–(2.61) will be given after establishing the following lemma.

Lemma 2.1.26 *Let $\beta = 1+\alpha$ with $0<\alpha<1$, and let x be any real number. We have symbolic identities involving the first Gauss series:*

$$\beta^x = \sum_{k=0}^{\infty} \left[\binom{x+k}{2k}\frac{\alpha^{2k}}{\beta^k} + \binom{x+k}{2k+1}\frac{\alpha^{2k+1}}{\beta^{k+1}}\right] \tag{2.62}$$

and a modified q-form of Gauss's first symbolic expression (cf §127 of [141]):

$$\begin{aligned} E_q^x &= \sum_{k=0}^{\infty} \left[\binom{x+k}{2k}\frac{\Delta_q^{2k}}{E_q^k} + \binom{x+k}{2k+1}\frac{\Delta_q^{2k+1}}{E_q^{k+1}}\right] \\ &= \sum_{k=0}^{\infty} \left[\binom{x+k}{2k}\delta_q^{2k} + \binom{x+k}{2k+1}\Delta_q\delta_q^{2k}E_{q^{-1}}\right]. \end{aligned} \tag{2.63}$$

Proof. Starting from Newton's formula:

$$\beta^x = \sum_{k=0}^{\infty} \binom{x}{k}\alpha^k.$$

We multiply $\frac{\alpha+1}{\beta}=1$ to the summation from the term α up and obtain

$$\begin{aligned} \beta^x &= 1 + \binom{x}{1}\frac{\alpha}{\beta} + \binom{x}{1}\frac{\alpha^2}{\beta} + \sum_{k=2}^{\infty}\binom{x}{k}\frac{\alpha^k(1+\alpha)}{\beta} \\ &= 1 + \binom{x}{1}\frac{\alpha}{\beta} + \binom{x+1}{2}\frac{\alpha^2}{\beta} + \sum_{k=2}^{\infty}\frac{\alpha^{k+1}}{\beta}\left[\binom{x}{k} + \binom{x}{k+1}\right] \\ &= 1 + \binom{x}{1}\frac{\alpha}{\beta} + \binom{x+1}{2}\frac{\alpha^2}{\beta} + \sum_{k=2}^{\infty}\binom{x+1}{k+1}\frac{\alpha^{k+1}}{\beta} \end{aligned}$$

$$= \quad 1+\binom{x}{1}\frac{\alpha}{\beta}+\binom{x+1}{2}\frac{\alpha^2}{\beta}+\binom{x+1}{3}\frac{\alpha^3}{\beta}+\sum_{k=3}^{\infty}\binom{x+1}{k+1}\frac{\alpha^{k+1}}{\beta}.$$

Repeating the operation on the series from the term α^3 up yields

$$\begin{aligned}\beta^x &= 1+\binom{x}{1}\frac{\alpha}{\beta}+\binom{x+1}{2}\frac{\alpha^2}{\beta}+\binom{x+1}{3}\frac{\alpha^3}{\beta^2}+\binom{x+1}{3}\frac{\alpha^4}{\beta^2}\\ &\quad+\sum_{k=3}^{\infty}\binom{x+1}{k+1}\frac{\alpha^{k+1}(1+\alpha)}{\beta^2}\\ &= 1+\binom{x}{1}\frac{\alpha}{\beta}+\binom{x+1}{2}\frac{\alpha^2}{\beta}+\binom{x+1}{3}\frac{\alpha^3}{\beta^2}+\binom{x+2}{4}\frac{\alpha^4}{\beta^2}\\ &\quad+\sum_{k=3}^{\infty}\binom{x+2}{k+2}\frac{\alpha^{k+2}}{\beta^2}.\end{aligned}$$

The above operation is repeated from α^5 up, and so on. We obtain

$$\beta^x=\sum_{k=0}^{\infty}\left[\binom{x+k}{2k}\frac{\alpha^{2k}}{\beta^k}+\binom{x+k}{2k+1}\frac{\alpha^{2k+1}}{\beta^{k+1}}\right]. \tag{2.64}$$

Substituting $\beta=E_q$ and $\alpha=\Delta_q$ into the above identity, we obtain the desired result.

Here we present the proofs of (2.58)–(2.61) in the sense of symbolic calculus, viz., every series expansion is considered as a formal series.

Proof. For proving (2.58), it suffices to make use of $E_q = e^{L_q}$ and (2.56). Indeed we have

$$\begin{aligned}(1-xE_q)^{-1}&=(1-xe^{L_q})^{-1}=\sum_{k=0}^{\infty}x^k e^{kL_q}\\ &=\sum_{k=0}^{\infty}x^k\sum_{j=0}^{\infty}\frac{(kL_q)^j}{j!}=\sum_{j=0}^{\infty}\left(\sum_{k=0}^{\infty}x^k k^j\right)\frac{L_q^j}{j!}=\sum_{j=0}^{\infty}\alpha_j(x)\frac{L_q^j}{j!}.\end{aligned}$$

q-operator expression (2.59) may be derived as follows:

$$\begin{aligned}(1-xE_q)^{-1}&=(1-x(1+\Delta_q))^{-1}\\ &=(1-x)^{-1}\left(1-\frac{x\Delta_q}{1-x}\right)^{-1}=\sum_{k=0}^{\infty}\frac{x^k\Delta_q^k}{(1-x)^{k+1}}.\end{aligned}$$

q-operator expressions (2.60) and (2.61) can be proved using Lemma 2.1.26, the first Gauss symbolic expression (2.63), and the following q-form of the Everett's symbolic expression (cf [141], §129), respectively.

$$E_q^x \quad = \quad \sum_{k=0}^{\infty}\left[\binom{x+k}{2k+1}\frac{\Delta_q^{2k}}{E_q^{k-1}}-\binom{x+k-1}{2k+1}\frac{\Delta_q^{2k}}{E_q^k}\right]$$

$$= \sum_{k=0}^{\infty}\left[\binom{x+k}{2k+1}E_q\delta_q^{2k} - \binom{x+k-1}{2k+1}\delta_q^{2k}\right]. \qquad (2.65)$$

Indeed, using (2.63) and noting the identity

$$\sum_{m=k}^{\infty}\binom{m}{k}x^m = \frac{x^k}{(1-x)^{k+1}}, \quad (|x|<1).$$

one may derive (2.60) as follows:

$$\begin{aligned}
(1-xE_q)^{-1} &= \sum_{j=0}^{\infty}(xE_q)^j \\
&= \sum_{k=0}^{\infty}\left\{\left(\sum_{j=0}^{\infty}\binom{j+k}{2k}x^j\right)\frac{\Delta_q^{2k}}{E_q^k} + \left(\sum_{j=0}^{\infty}\binom{j+k}{2k+1}x^j\right)\frac{\Delta_q^{2k+1}}{E_q^{k+1}}\right\} \\
&= \sum_{k=0}^{\infty}\left\{\frac{x^k}{(1-x)^{2k+1}}\frac{\Delta_q^{2k}}{E_q^k} + \frac{x^{k+1}}{(1-x)^{2k+2}}\frac{\Delta_q^{2k+1}}{E_q^{k+1}}\right\} \\
&= \sum_{k=0}^{\infty}\left(\frac{x}{(1-x)^2}\right)^{k+1}\left(\frac{1-x}{x}\frac{\Delta_q^{2k}}{E_q^k} + \frac{\Delta_q^{2k+1}}{E_q^{k+1}}\right) \\
&= \sum_{k=0}^{\infty}\left(\frac{x}{(1-x)^2}\right)^{k+1}\left(x^{-1}\frac{\Delta_q^{2k}}{E_q^k} - \frac{\Delta_q^{2k}}{E_q^{k+1}}\right).
\end{aligned}$$

q-operator expression (2.61) can be proved similarly using (2.65). However, it can also be verified by a direct symbolic computations. In fact, we have

$$\begin{aligned}
RHS\ of\ (2.61) &= 1 + \frac{x}{(1-x)^2}\sum_{k=0}^{\infty}\left(\frac{x}{(1-x)^2}\right)^k\left(\frac{\Delta_q^2}{E_q}\right)^k(E_q - x) \\
&= 1 + \frac{x}{(1-x)^2}\frac{E_q - x}{1-\frac{x}{(1-x)^2}\frac{\Delta_q^2}{E_q}} = 1 + \frac{x(E_q-x)}{(1-x)^2 - x\frac{\Delta_q^2}{E_q}} \\
&= 1 + \frac{E_q x(E_q-x)}{(1-x)^2E_q - x(E_q-1)^2} = 1 + \frac{E_q x}{1-xE_q} \\
&= (1-xE_q)^{-1} = LHS\ of\ (2.61).
\end{aligned}$$

This complete the proofs of (2.58)–(2.61).

Proposition 2.1.27 *For a given analytic function $f(t)$, define $F_q(x) = \sum_{k\geq 0} f(q^k)x^k$. If $x \neq 0$ and $x \neq 1$,*

$$F_q(x) = \sum_{k=0}^{\infty}\frac{\alpha_k(x)}{k!}L_q^k f(1), \qquad (2.66)$$

$$F_q(x) = \sum_{k=0}^{\infty} \frac{x^k}{(1-x)^{k+1}} \Delta_q^k f(1), \tag{2.67}$$

$$F_q(x) = \sum_{k=0}^{\infty} \left(\frac{x}{(1-x)^2}\right)^{k+1} (x^{-1}\delta_q^{2k} f(1) - \delta_q^{2k} f(q^{-1})), \tag{2.68}$$

$$F_q(x) = 1 + \sum_{k=0}^{\infty} \left(\frac{x}{(1-x)^2}\right)^{k+1} (\delta_q^{2k} f(q) - x\delta_q^{2k} f(1)). \tag{2.69}$$

Proof. Clearly, these follow by applying the q-operators expressions shown in (2.58)–(2.61) to the function $f(t)$ and then evaluating at $t = 1$.

As an application, taking $f(t) = \frac{1}{\log_q(t)+1}$ and $x = -1$ in (2.67) leads to

$$\begin{aligned}
\sum_{k\geq 0}(-1)^k \frac{1}{k+1} &= \sum_{k\geq 0} \frac{(-1)^k}{2^{k+1}} \sum_{j=0}^{k} \binom{k}{j} (-1)^{k-j} \frac{1}{j+1} \\
&= \sum_{k=0}^{\infty} \frac{1}{(k+1)2^{k+1}} \sum_{j=0}^{k} (-1)^j \binom{k+1}{j+1} \\
&= \sum_{k=0}^{\infty} \frac{1}{(k+1)2^{k+1}} = \sum_{k=1}^{\infty} \frac{1}{k2^k},
\end{aligned}$$

which gives again

$$\ln 2 = \frac{1}{2} + \frac{1}{2\cdot 2^2} + \frac{1}{3\cdot 2^3} + \frac{1}{4\cdot 2^4} + \cdots.$$

The rate of convergence of this series is $O(1/2^n)$, much faster than

$$\ln 2 = \sum_{k\geq 0}(-1)^k \frac{1}{k+1},$$

whose convergence rate is $O(1/n)$.

As for a second application, we may substitute $x = -1$ in Proposition 2.1.27, obtaining the following series transformation formulas:

$$\sum_{k\geq 0}(-1)^k f(q^k) = \sum_{k=0}^{\infty} \frac{\alpha_k(-1)}{k!} L_q^k f(1), \tag{2.70}$$

$$\sum_{k\geq 0}(-1)^k f(q^k) = \sum_{k=0}^{\infty} \frac{(-1)^k}{2^{k+1}} \Delta_q^k f(1), \tag{2.71}$$

$$\sum_{k\geq 0}(-1)^k f(q^k) = \sum_{k=0}^{\infty} \frac{(-1)^k}{4^{k+1}} (\delta_q^{2k} f(1) + \delta_q^{2k} f(q^{-1})), \tag{2.72}$$

$$\sum_{k\geq 1}(-1)^k f(q^k) = 1 + \sum_{k=0}^{\infty} \left(\frac{-1}{4}\right)^{k+1} (\delta_q^{2k} f(q) + \delta_q^{2k} f(1)). \tag{2.73}$$

These four identities appear to be novel and could be used to accelerate slowly convergent alternating series $\sum_{k=1}^{\infty}(-1)^k f(q^k)$. We consider them as q-analog of the ordinary Euler transformations.

All operational formulas represented in Proposition 2.1.27 can be extended, and the corresponding symbolic q-substitution formulas can be established accordingly. For example, we may consider a generating function of the form

$$\sum_{k\geq 0} f_k t^k = F(t, e^t, e^{\alpha t}).$$

Letting $t \mapsto L_q$ gives

$$\sum_{k\geq 0} f_k L_q^k = F(L_q, E_q, E_q^{\alpha}).$$

Applying this to the well-known identity

$$\sum_{k\geq 0} \frac{4^k}{(2k)!} B_{2k} t^{2k} = t \coth t = t\frac{e^t + e^{-t}}{e^t - e^{-t}},$$

with B_n being the nth Bernoulli number, we obtain

$$\sum_{k\geq 0} \frac{4^k}{(2k)!} B_{2k} L_q^{2k} = L_q \frac{E_q + E_q^{-1}}{E_q - E_q^{-1}}.$$

Hence, we obtain a symbolic formula

$$\sum_{k\geq 0} \frac{4^k}{(2k)!} B_{2k} L_q^{2k-1}(E_q - E_q^{-1}) = E_q + E_q^{-1}. \tag{2.74}$$

Applying this to an infinitely differentiable function $f(t)$ at $t = 1$ yields

$$\sum_{k\geq 0} \frac{4^k}{(2k)!} B_{2k} L_q^{2k-1}(E_q - E_q^{-1}) f(1) = (E_q + E_q^{-1}) f(1). \tag{2.75}$$

Similarly, using the symbolic relation

$$L_q \frac{E_q + E_q^{-1}}{E_q - E_q^{-1}} = L_q \left(1 + \Delta_q^{-1} - (E_q + 1)^{-1}\right),$$

we obtain another operational formula

$$-1 + \sum_{k\geq 0} \frac{4^k}{(2k)!} B_{2k} L_q^{2k-1} + (E_q + 1)^{-1} = \Delta_q^{-1},$$

from which one may construct a series transformation formula.

Another extension is a q-analog of the symbolic formulas represented in [63], which is actually a Newton series type extension of the symbolic expansions given in §2.1.1. Consider

$$(1+E_q)^x f(1) = \sum_{k\geq 0} \binom{x}{k} E_q^k f(1) = \sum_{k\geq 0} f(q^k) \frac{(x)_k}{k!},$$

where $(x)_k = x(x-1)\cdots(x-k+1)$. We have the following results.

Theorem 2.1.28 *Let E_q, Δ_q, and δ_q be the operators defined before. Then we have*

$$(1+E_q)^x = 2^x \sum_{k=0}^{\infty} \frac{(x)_k}{2^k k!} \Delta_q^k, \tag{2.76}$$

$$\begin{aligned}(1+E_q)^x = \sum_{k=0}^{\infty} \binom{x}{k} \Big[& {}_2F_1(k-x, 2k+1; k+1; -1) \\ & + \frac{x-k}{k+1}\, {}_2F_1(k+1-x, 2k+2; k+2; -1)\, E_q^{-1}\Delta_q \Big] \delta_q^{2k},\end{aligned} \tag{2.77}$$

$$\begin{aligned}(1+E_q)^x = 1 + \sum_{k=0}^{\infty} \binom{x}{k+1} \Big[& {}_2F_1(k+1-x, 2k+2; k+2; -1)\, E_q \\ & - \frac{x-k-1}{k+2}\, {}_2F_1(k+2-x, 2k+2; k+3; -1) \Big] \delta_q^{2k}.\end{aligned} \tag{2.78}$$

In addition, we may present an extension of (2.58) by using Bell polynomials (see, for example, pg. 134 in [44]) as follows.

$$(1+E_q)^x = 2^x \sum_{k=0}^{\infty} P_k^{(x)} \left(\frac{1}{2}, \frac{1}{2}, \cdots, \right) \frac{L_q^k}{k!}, \tag{2.79}$$

where the values of potential Bell polynomials at $(1/2, 1/2, \ldots)$ are defined by

$$2^x P_k^{(x)} \left(\frac{1}{2}, \frac{1}{2}, \cdots, \right) = \sum_{\ell=0}^{\infty} \ell^k \frac{(x)_\ell}{\ell!}. \tag{2.80}$$

Proof.
The proof of (2.76) is straightforward:

$$\begin{aligned}(1+E_q)^x &= (2+\Delta_q)^x = 2^x \left(1 + \frac{\Delta_q}{2}\right)^x \\ &= 2^x \sum_{k\geq 0} \frac{(x)_k}{k! 2^k} \Delta_q^k.\end{aligned}$$

To prove (2.77), we use (2.63) to find

$$
\begin{aligned}
(1+E_q)^x &= 1+\sum_{j\ge 1}\binom{x}{j}E_q^j \\
&= 1+\sum_{j\ge 1}\binom{x}{j}\sum_{k\ge 0}\left[\binom{k+j}{2k}\frac{\Delta_q^{2k}}{E_q^k}+\binom{k+j}{2k+1}\frac{\Delta_q^{2k+1}}{E_q^{k+1}}\right] \\
&= \sum_{k\ge 0}\left[\frac{\Delta_q^{2k}}{E_q^k}\sum_{j\ge k}\binom{x}{j}\binom{k+j}{2k}+\frac{\Delta_q^{2k+1}}{E_q^{k+1}}\sum_{j\ge k+1}\binom{x}{j}\binom{k+j}{2k+1}\right] \\
&= \sum_{k\ge 0}\left[\binom{x}{k}\ {}_2F_1(k-x,2k+1;k+1;-1)\frac{\Delta_q^{2k}}{E_q^k}\right. \\
&\quad \left.+\binom{x}{k+1}\ {}_2F_1(k+1-x,2k+2;k+2;-1)\frac{\Delta_q^{2k}}{E_q^{k+1}}\right],
\end{aligned}
$$

which implies (2.77). q-operator expression (2.78) can be proved similarly by using Everett's symbolic expression (2.65).

For (2.79), we first have

$$
\begin{aligned}
(1+E_q)^x &= \left(1+e^{L_q}\right)^x=\sum_{j\ge 0}\binom{x}{j}e^{jL_q} \\
&= \sum_{j\ge 0}\binom{x}{j}\sum_{k\ge 0}\frac{(jL_q)^k}{k!}=\sum_{k\ge 0}\left(\sum_{j\ge 0}\frac{(x)_j j^k}{j!}\right)\frac{L_q^k}{k!}.
\end{aligned}
$$

Using (2.80), we may write the part in the parenthesis of the rightmost term as $2^xP_k^{(x)}(1/2,1/2,\ldots)$ to finish the proof.

For a given analytic function $f(t)$, define $F_q(x)=\sum_{k\ge 0}f(q^k)(x)_k/k!$, from (2.76)–(2.79) we obtain series transformation formulas by simply applying (2.76)–(2.79) to f.

$$
F_q(x) = 2^x\sum_{k=0}^{\infty}\frac{(x)_k}{2^kk!}\Delta_q^kf(1), \tag{2.81}
$$

$$
\begin{aligned}
F_q(x) &= \sum_{k=0}^{\infty}\binom{x}{k}\Big[\ {}_2F_1\left(k-x,2k+1;k+1;-1\right) \\
&\quad \left.+\frac{x-k}{k+1}\ {}_2F_1\left(k+1-x,2k+2;k+2;-1\right)E_q^{-1}\Delta_q\right]\delta_q^{2k}f(1),
\end{aligned} \tag{2.82}
$$

$$\begin{aligned} F_q(x) &= 1+\sum_{k=0}^{\infty}\binom{x}{k+1}\Big[\,{}_2F_1\left(k+1-x,2k+2;k+2;-1\right)E_q \\ &\quad -\frac{x-k-1}{k+2}\,{}_2F_1\left(k+2-x,2k+2;k+3;-1\right)\Big]\,\delta_q^{2k}f(1), \quad (2.83)\\ F_q(x) &= 2^x\sum_{k=0}^{\infty}P_k^{(x)}\left(\frac{1}{2},\frac{1}{2},\cdots,\right)\frac{L_q^k}{k!}f(1). \quad (2.84)\end{aligned}$$

As an example, substituting $f(t)=t^n$ into (2.84) and noting

$$L_qf(1)=\sum_{j=1}^{\infty}\frac{(-1)^{j+1}}{j}\Delta_q^jf(1)=\sum_{j=1}^{\infty}\frac{(-1)^{j+1}}{j}(q^n-1)^j=\log(1+(q^n-1))=n\log q,$$

from (2.50) and (2.84), we obtain the series transformation formula

$$\sum_{k=0}^{\infty}q^k\frac{(x)_k}{k!}=2^x\sum_{j=0}^{\infty}P_j^{(x)}\left(\frac{1}{2},\frac{1}{2},\cdots\right)\frac{(n\log q)^j}{j!}.$$

2.2 Series Transformation

Now suppose that $\Phi(t)$ is an analytic function of t or a formal power series in t, say

$$\Phi(t)=\sum_{k=0}^{\infty}c_kt^k,\qquad c_k=[t^k]\Phi(t), \quad (2.85)$$

where c_k can be either real or complex numbers. Then, formally we have a sum of general form

$$\Phi(xE)f(0)=\sum_{k=0}^{\infty}c_kf(k)x^k. \quad (2.86)$$

The operator $\Phi(xE)=\Phi(x+x\Delta)=\Phi(xe^D)$ can be expressed as some power series involving operators Δ^k or D^k's. Then it may be possible to compute the right-hand side of (2.86) by means of operator-series in Δ^k or D^k's. This idea could be readily applied to various elementary functions $\Phi(t)$. Indeed, if we take $\Phi(t)$ to be any of the following functions

$$\begin{array}{lll} (i)\ (1+t)^{\alpha}, & (ii)\ (1-t)^{-\alpha-1}, & (iii)\ e^t \\ (iv)\ -\log(1-t), & (v)\ \sin t, & (vi)\ \cosh t, \end{array} \quad (2.87)$$

etc., thus, using suitable expressions of $\Phi(xE)=\Phi(x+x\Delta)=\Phi(xe^D)$ in terms of Δ^k or D^k, we can obtain various transformation formulas as well as summation formulas for the series of the form (2.86).

The results represented in this section are a significant improvement of the previous work shown in the Section 2.1, in which Proposition 2.1.14 is a special case of Theorem 2.2.2 in this section.

Remark 2.2.1 *Obviously, $\Phi(t)$ is not limited to the functions shown in (2.87). For instance, we may choose (cf. for example Hsu and Shiue [134] and [111])*

$$\Phi(t) = (1 - mzt + yt^m)^{-\lambda} = \sum_{k=0}^{\infty} P_k(m, z, y, \lambda)t^k, \tag{2.88}$$

the generating functions (GFs) of the so-called Gegenbauer-Humbert-type polynomials. As special cases of (2.88), we consider $P_k(m, z, y, \lambda)$ as follows

$P_k(2, z, 1, 1) = U_k(z)$, *Chebyshev 2nd kind polynomial,*
$P_k(2, z, 1, 1/2) = \psi_k(z)$, *Legendre polynomial,*
$P_k(2, z, -1, 1) = P_{k+1}(z)$, *Pell polynomial,*
$P_k(2, z/2, -1, 1) = F_{k+1}(z)$, *Fibonacci polynomial,*
$P_k(2, z/2, 2, 1) = \Phi_{k+1}(z)$, *Fermat 1st kind polynomial,*

where $F_{k+1} = F_{k+1}(1)$ is the Fibonacci number.

The expansion (2.88) is a special case of the generalized Humbert polynomials studied by Gould in [72], in which a generalized Humbert polynomial $P_n(m, x, y, p, C)$ is defined by means of

$$(C - mxt + yt^m)^p = \sum_{n=0}^{\infty} t^n P_n(m, x, y, p, C),$$

where m is an integer ≥ 1 and the other parameters are unrestricted. In that paper, Gould first obtained some recurrences satisfied by the P_n and then gives a formula for ${D_x}^k P_{n+k}$ that generalizes a formula of Catalan for the kth derivative of the Legendre polynomial. He also showed that if the function $f(x, t)$ satisfies $(tD_t)f(x, t) = (x - yt^{m-1})D_x f(x, t)$, then

$$(tD_t)^r f(x, t) = \sum_{j=1}^{r} {Q_j}^r(m, x, y, t){D_x}^j f(x, t) \quad (r \geq 1),$$

where

$$p!(-mt)^p {Q_p}^r(m, x, y, t) = \sum_{n=0}^{mp} n^r t^n P_n(m, x, y, p, mxt - yt^m) \quad (1 \leq p \leq r).$$

Some notations and an extension of Eulerian fractions will be given in next subsection. Two lists of transformation and summation formulas will be displayed in Subsections §2.2.3. Many illustrative examples will be given in Subseection §2.2.4. Finally, we will discuss the convergence of the transformation series in Subsection §2.2.5.

2.2.1 An extension of Eulerian fractions

It is well-known that the Eulerian fraction defined in Lemma 2.1.7 is a powerful tool to study the Eulerian polynomial, Euler function and its generalization, Jordan function, etc. (cf. Comtet [44]).

The ordinary Eulerian fraction, $\alpha_m(x)$, can be expressed in the form

$$\alpha_m(x) = \frac{A_m(x)}{(1-x)^{m+1}} \quad (x \neq -1), \tag{2.89}$$

where $A_m(x)$ is the mth degree Eulerian polynomial of the form

$$A_m(x) = \sum_{j=0}^{m} j! \left\{ {m \atop j} \right\} x^j (1-x)^{m-j}, \tag{2.90}$$

and $\left\{ {m \atop j} \right\}$ are Stirling numbers of the second kind, i.e., $j!\left\{ {m \atop j} \right\} = \left[\Delta^j t^m\right]_{t=0}$. $\left\{ {m \atop j} \right\}$ is also denoted by $S(m,j)$. Evidently $\alpha_m(x)$ can be written in the form (see [99])

$$\alpha_m(x) = \sum_{j=0}^{m} \frac{j!\left\{ {m \atop j} \right\} x^j}{(1-x)^{j+1}}.$$

In order to express some new formulas for certain general types of power series, we need to introduce an extension of Euler fraction associated with an infinitely differentiable function $g(x)$ defined as

$$A_m(x, g(x)) := \sum_{j=0}^{m} \left\{ {m \atop j} \right\} g^{(j)}(x) x^j, \tag{2.91}$$

where $g^{(j)}(x)$ is the j-th derivative of $g(x)$. Obviously, $\alpha_m(x)$ defined by (2.89) can be represented as

$$\alpha_m(x) = A_m(x, (1-x)^{-1}).$$

From (2.91), two kinds of generalized Eulerian fractions in terms of $g(x) = (1+x)^\alpha$ and $g(x) = (1-x)^{-\alpha-1}$, with real number α as a parameter, can be introduced respectively, namely

$$\begin{aligned} A_m(x,\alpha) &\equiv A_m\left(x,(1+x)^\alpha\right) = \frac{A_m(x)_\alpha}{(1+x)^{m-\alpha}} \\ &= \sum_{j=0}^{m} \binom{\alpha}{j} \frac{j!\left\{ {m \atop j} \right\} x^j}{(1+x)^{j-\alpha}} \quad (x \neq -1), \end{aligned} \tag{2.92}$$

$$\begin{aligned}\tilde{A}_m(x,\alpha) &\equiv A_m\left(x,(1-x)^{-\alpha-1}\right)=\frac{\tilde{A}_m(x)_\alpha}{(1-x)^{\alpha+m+1}}\\ &=\sum_{j=0}^{m}\binom{\alpha+j}{j}\frac{j!\left\{ {m \atop j} \right\}x^j}{(1-x)^{\alpha+j+1}}\quad (x\neq 1).\end{aligned}\tag{2.93}$$

These may be called, respectively, the 1st kind and 2nd kind of generalized Eulerian fractions. Correspondingly $A_m(x,\alpha)$ and $\tilde{A}_m(x,\alpha)$ are called the mth degree generalized Eulerian polynomials, having explicit expressions as follows:

$$A_m(x,\alpha) = \sum_{j=0}^{m}\binom{\alpha}{j}j!\left\{ {m \atop j} \right\}x^j(1+x)^{m-j},\tag{2.94}$$

$$\tilde{A}_m(x,\alpha) = \sum_{j=0}^{m}\binom{\alpha+j}{j}j!\left\{ {m \atop j} \right\}x^j(1-x)^{m-j}.\tag{2.95}$$

As easily seen, $\tilde{A}_m(x,0)=\alpha_m(x)=A_m(-x,-1)$.

2.2.2 Series-transformation formulas

All formulas represented in this subsection are formal identities in which we always assume that $x\neq 1$ or $x\neq -1$ according as $(1-x)^{-1}$ or $(1+x)^{-1}$ appears in the formulas.

Theorem 2.2.2 *Let $(f(k))$ be a given sequence of numbers (real or complex), and let $g(t)$ and $h(t)$ be infinitely differentiable on $[0,\infty)$. Then we have formally*

$$\sum_{k=0}^{\infty}f(k)g^{(k)}(0)\frac{x^k}{k!}=\sum_{k=0}^{\infty}\Delta^k f(0)g^{(k)}(x)\frac{x^k}{k!},\tag{2.96}$$

$$\sum_{k=0}^{\infty}h(k)g^{(k)}(0)\frac{x^k}{k!}=\sum_{k=0}^{\infty}\frac{1}{k!}h^{(k)}(0)A_k(x,g(x)),\tag{2.97}$$

where $A_m(x,g(x))$ is an extension of Euler fraction in terms of $g(x)$ defined as in (2.91).

Proof. To prove (2.96), we apply the operator $g(xE)$ to $f(t)$ at $t=0$, where E is the shift operator. Therefore,

$$g(xE)f(t)|_{t=0}=\sum_{k=0}^{\infty}\frac{1}{k!}g^{(k)}(0)\,(xE)^k f(t)\Big|_{t=0}=\sum_{k=0}^{\infty}f(k)g^{(k)}(0)\frac{x^k}{k!}.$$

On the other hand, we have

$$\begin{aligned} g(xE)f(t)|_{t=0} &= g(x+x\Delta)f(t)|_{t=0} \\ &= \sum_{k=0}^{\infty} \frac{1}{k!} g^{(k)}(x)\,(x\Delta)^k f(t)\big|_{t=0} = \sum_{k=0}^{\infty} \Delta^k f(0) g^{(k)}(x) \frac{x^k}{k!}. \end{aligned}$$

Similarly, for the infinitely differentiable function $h(t)$, we can present

$$g(xE)h(t)|_{t=0} = g(xe^D)h(t)\big|_{t=0} = \sum_{j=0}^{\infty} \frac{1}{j!} g^{(j)}(0)\left(xe^D\right)^j h(t)|_{t=0}$$

$$= \sum_{j=0}^{\infty} \frac{x^j}{j!} g^{(j)}(0) \sum_{k=0}^{\infty} \frac{j^k}{k!} h^{(k)}(0) = \sum_{k=0}^{\infty} \left(\sum_{j=0}^{\infty} g^{(j)}(0) j^k \frac{x^j}{j!} \right) \frac{1}{k!} h^{(k)}(0).$$

By applying (2.96) to the inner sum of the rightmost side of the above equation for $f(t) = t^k$ and noting $\left\{ {k \atop j} \right\} = \left(\Delta^j t^k\right)_{t=0}/j!$, we obtain

$$\begin{aligned} g(xE)h(t)|_{t=0} &= \sum_{k=0}^{\infty} \left(\sum_{j=0}^{k} \left(\Delta^j t^k\right)_{t=0} g^{(j)}(x) \frac{x^j}{j!} \right) \frac{1}{k!} h^{(k)}(0) \\ &= \sum_{k=0}^{\infty} \left(\sum_{j=0}^{k} \left\{ {k \atop j} \right\} g^{(j)}(x) x^j \right) \frac{1}{k!} h^{(k)}(0) \\ &= \sum_{k=0}^{\infty} \frac{1}{k!} h^{(k)}(0) A_k(x, g(x)). \end{aligned}$$

This completes the proof of the theorem.

■

Remark 2.2.3 *The series transformation formulas (2.96) and (2.97) could have numerous applications by setting different infinitely differentiable functions for $g(x)$. For examples, we have*

$$\sum_{k=0}^{\infty} f(k)x^k = \sum_{k=0}^{\infty} \frac{x^k}{(1-x)^{k+1}} \Delta^k f(0) \quad (g(x) = (1-x)^{-1}), \tag{2.98}$$

$$\sum_{k=0}^{\infty} h(k)x^k = \sum_{k=0}^{\infty} \frac{\alpha_k(x)}{k!} D^k h(0) \quad (g(x) = (1-x)^{-1}), \tag{2.99}$$

$$\sum_{k=0}^{\infty} \binom{\alpha}{k} f(k)x^k = \sum_{k=0}^{\infty} \binom{\alpha}{k} \frac{x^k}{(1+x)^{k-\alpha}} \Delta^k f(0)$$
$$(g(x) = (1+x)^{\alpha}), \tag{2.100}$$

$$\sum_{k=0}^{\infty}\binom{\alpha+k}{k}f(k)x^k=\sum_{k=0}^{\infty}\binom{\alpha+k}{k}\frac{x^k}{(1-x)^{\alpha+k+1}}\Delta^k f(0)$$
$$(g(x)=(1-x)^{-\alpha-1}),$$

$$\sum_{k=0}^{\infty}\frac{f(k)x^k}{k!}=e^x\sum_{k=0}^{\infty}\frac{x^k}{k!}\Delta^k f(0)\quad (g(x)=e^x), \tag{2.101}$$

$$\sum_{k=1}^{\infty}\frac{f(k)x^k}{k}=-f(0)\ln(1-x)+\sum_{k=1}^{\infty}\frac{1}{k}\left(\frac{x}{1-x}\right)^k\Delta^k f(0)$$
$$(g(x)=-\ln(1-x)), \tag{2.102}$$

$$\sum_{k=0}^{\infty}\binom{\alpha}{k}h(k)x^k=\sum_{k=0}^{\infty}\frac{A_k(x,\alpha)}{k!}D^k h(0)\quad (g(x)=(1+x)^\alpha), \tag{2.103}$$

$$\sum_{k=0}^{\infty}\binom{\alpha+k}{k}h(k)x^k=\sum_{k=0}^{\infty}\frac{\tilde{A}_k(x,\alpha)}{k!}D^k h(0)$$
$$(g(x)=(1-x)^{-\alpha-1}), \tag{2.104}$$

$$\sum_{k=m}^{\infty}\binom{k}{m}f(k)x^k=\sum_{k=0}^{\infty}\binom{k+m}{m}\frac{x^{k+m}}{(1-x)^{k+m+1}}\Delta^k f(m)$$
$$(g(x)=(1-x)^{-m-1}), \tag{2.105}$$

$$\sum_{k=m}^{\infty}\binom{k}{m}h(k)x^k=\sum_{k=0}^{\infty}\frac{\tilde{A}_k(x,m)x^m}{k!}D^k h(m)$$
$$(g(x)=(1-x)^{-m-1}), \tag{2.106}$$

$$\sum_{k=0}^{\infty}\frac{(-1)^k f(2k+1)x^{2k+1}}{(2k+1)!}=\sin x\sum_{k=0}^{\infty}\frac{(-1)^k x^{2k}\Delta^{2k}f(0)}{(2k)!}$$
$$+\cos x\sum_{k=0}^{\infty}\frac{(-1)^k x^{2k+1}\Delta^{2k+1}f(0)}{(2k+1)!}\quad (g(x)=\sin x), \tag{2.107}$$

$$\sum_{k=0}^{\infty}\frac{(-1)^k f(2k)x^{2k}}{(2k)!}=\cos x\sum_{k=0}^{\infty}\frac{(-1)^k x^{2k}\Delta^{2k}f(0)}{(2k)!}$$
$$+\sin x\sum_{k=0}^{\infty}\frac{(-1)^{k+1} x^{2k+1}\Delta^{2k+1}f(0)}{(2k+1)!}\quad (g(x)=\cos x), \tag{2.108}$$

$$\sum_{k=0}^{\infty}\frac{f(2k)x^{2k}}{(2k)!}=\frac{e^x}{2}\sum_{k=0}^{\infty}\frac{x^k\Delta^k f(0)}{k!}+\frac{e^{-x}}{2}\sum_{k=0}^{\infty}\frac{(-x)^k\Delta^k f(0)}{k!}, \tag{2.109}$$

$$\sum_{k=0}^{\infty}\frac{f(2k+1)x^{2k+1}}{(2k+1)!}=\frac{e^x}{2}\sum_{k=0}^{\infty}\frac{x^k\Delta^k f(0)}{k!}-\frac{e^{-x}}{2}\sum_{k=0}^{\infty}\frac{(-x)^k\Delta^k f(0)}{k!}, \tag{2.110}$$

where (2.109) and (2.110) are obtained by replacing $g(x)$ *by* e^x *and* e^{-x} *and adding and subtracting the resulting formulas, respectively.*

Note that (2.98) and (2.99) are well-known and have been utilized to construct summation formulas with estimable remainders. See, e.g., He, Hsu, Shiue, and Torney [99]. The particular cases of (2.100) with $\alpha = m$ $(m \in \mathbb{N})$ *and of (2.101) with* $f(x)$ *denoting a* r*th degree polynomial of* x *have been expounded in Problems* (1109) *and* (1110) *of Jolley's book [139]. The rest of the above list appears to be not easily found in literatures, and the formulas (2.103)–(2.106) are believed to be first introduced in [106].*

Apparently, (2.98) is implied by (2.100) (with $\alpha = -1$, $x \mapsto -x$*) and (2.101) (with* $\alpha = 0$*). Also, (2.99) is a particular case of (2.103) (with* $\alpha = -1$, $x \mapsto -x$*) and a particular case of (2.104) (with* $\alpha = 0$*). Moreover, it is easily observed that (2.105) and (2.106) can be derived from (2.101) and (2.104), respectively, by substituting* $\alpha = m$*, applying operator* E^m*, and multiplying* x^m *on the both sides of the former two formulas, respectively.*

The transformation formulas given in the list is useful for accelerating convergence of power series because $\Delta^k f(0)$ *and* $D^k f(0)$ *decreases to zero rapidly as* $k \to \infty$.

Remark 2.2.4 *From (2.88) we can derive Gegenbauer type series transformation formulas. For examples, we consider*

$$\Phi(t) = (1 - 2zt + t^2)^{-\lambda} = \sum_{k=0}^{\infty} C_k^{(\lambda)}(z) t^k,$$

the GF of $C_k^{(\lambda)}(z) \equiv P_k(2, z, 1, \lambda)$*, where* $P_k(2, z, 1, \lambda)$ *was shown in (2.88), and* $C_k^{(1)}(z) = U_k(z)$ *and* $C_k^{(1/2)}(z) = \psi_k(z)$ *are respectively the* 2*nd kind Chebyshev and Legengre polynomials. Using the same argument to derive (2.101) we obtain the following Gegenbauer type series transformation formula*

$$\begin{aligned}
\Phi(xE)f(0) &= \sum_{k=0}^{\infty} C_k^{(\lambda)}(z) x^k f(k) \\
&= \sum_{i=0}^{\infty}\sum_{j=0}^{\infty} \binom{\lambda+i-1}{i}\binom{\lambda+j-1}{j} (z+\delta)^i (z-\delta)^j x^{i+j} f(i+j) \\
&= (1 - 2zx + x^2)^{-\lambda} \sum_{i=0}^{\infty}\sum_{j=0}^{\infty} \binom{\lambda+i-1}{i}\binom{\lambda+j-1}{j} \\
&\quad \times \frac{(z+\delta)^i (z-\delta)^j x^{i+j}}{(1-z+\delta)x)^i (1-(z-\delta)x)^j} \Delta^{i+j} f(0).
\end{aligned} \tag{2.111}$$

Formula (2.111) can also be verified directly as follows. By denoting $\delta = \sqrt{z^2 - 1}$ *we can expand* $\Phi(xE)$ *formal power series in terms of operator* Δ *as*

$$\Phi(xE) = (1 - (z+\delta)xE)^{-\lambda}(1 - (z-\delta)xE)^{-\lambda}$$

$$
\begin{aligned}
&= (1-(z+\delta)x-(z+\delta)x\Delta)^{-\lambda}(1-(z-\delta)x-(z-\delta)x\Delta)^{-\lambda} \\
&= [1-(z+\delta)x]^{-\lambda}\left[1-\frac{(z+\delta)x\Delta}{1-(z+\delta)x}\right]^{-\lambda}[1-(z-\delta)x]^{-\lambda} \\
&\quad \left[1-\frac{(z-\delta)x\Delta}{1-(z-\delta)x}\right]^{-\lambda} \\
&= (1-2zx+x^2)^{-\lambda}\sum_{i=0}^{\infty}\sum_{j=0}^{\infty}\binom{\lambda+i-1}{i}\binom{\lambda+j-1}{j} \\
&\quad \times\frac{(z+\delta)^i(z-\delta)^j x^{i+j}\Delta^{i+j}}{(1-z+\delta)x)^i(1-(z-\delta)x)^j}.
\end{aligned}
$$

Thus, (2.111) is obtained.

In series transformation formula (2.111), we assume $f(t)$ to be a rth degree polynomial, denoted by $\phi(t)$, and obtain the generating function

$$
\begin{aligned}
GF\left\{\phi(k)C_k^{(\lambda)}(z)\right\} &= \sum_{k=0}^{\infty}\left(C_k^{(\lambda)}(z)\phi(k)\right)x^k \\
&= (1-2zx+x^2)^{-\lambda}\sum_{i=0}^{r}\sum_{j=0}^{r}\binom{\lambda+i-1}{i}\binom{\lambda+j-1}{j} \\
&\quad \times\frac{(z+\delta)^i(z-\delta)^j x^{i+j}\Delta^{i+j}\phi(0)}{(1-z+\delta)x)^i(1-(z-\delta)x)^j}.
\end{aligned}
\tag{2.112}
$$

In particular, for $\lambda = 1$ and $1/2$ we have generating functions

$$
\begin{aligned}
GF\{\phi(k)U_k(z)\} &= (1-2zx+x^2)^{-1}\sum_{i=0}^{r}\sum_{j=0}^{r}\binom{\lambda+i-1}{i}\binom{\lambda+j-1}{j} \\
&\quad \times\frac{(z+\delta)^i(z-\delta)^j x^{i+j}\Delta^{i+j}\phi(0)}{(1-z+\delta)x)^i(1-(z-\delta)x)^j}, \\
GF\{\phi(k)\psi_k(z)\} &= (1-2zx+x^2)^{-1/2}\sum_{i=0}^{r}\sum_{j=0}^{r}\binom{\lambda+i-1}{i}\binom{\lambda+j-1}{j} \\
&\quad \times\frac{(z+\delta)^i(z-\delta)^j x^{i+j}\Delta^{i+j}\phi(0)}{(1-z+\delta)x)^i(1-(z-\delta)x)^j}.
\end{aligned}
$$

Remark 2.2.5 *Evidently, when $f(t)$ is a polynomial, all the formulas in this subsection become closed summation formulas with a finite number of terms. Moreover, the Right-hand side of each formula may also be viewed as a GF for the sequence of coefficeients contained in the power series on the left-hand side. Thus, for the rth degree polynomial $\phi(t)$, from (2.96) and (2.97) we obtain two type GF's of $\{\phi(k)g^{(k)}(0)\}$:*

$$
\sum_{k=0}^{\infty}\phi(k)g^{(k)}(0)\frac{x^k}{k!} = \sum_{k=0}^{r}\Delta^k\phi(0)g^{(k)}(x)\frac{x^k}{k!} \tag{2.113}
$$

$$\sum_{k=0}^{\infty}\phi(k)g^{(k)}(0)\frac{x^k}{k!}=\sum_{k=0}^{r}\frac{1}{k!}\phi^{(k)}(0)A_k(x,g(x)). \tag{2.114}$$

Replacing f and h in (2.98)–(2.110) by polynomial ϕ, we obtain the special cases of (2.113) and (2.114). For instance,

$$\sum_{k=0}^{\infty}\phi(k)x^k=\sum_{k=0}^{r}\frac{x^k}{(1-x)^{k+1}}\Delta^k\phi(0)\quad (g(x)=(1-x)^{-1}), \tag{2.115}$$

$$\sum_{k=0}^{\infty}\phi(k)x^k=\sum_{k=0}^{r}\frac{\alpha_k(x)}{k!}D^k\phi(0)\quad (g(x)=(1-x)^{-1}), \tag{2.116}$$

$$\sum_{k=0}^{\infty}\binom{\alpha}{k}\phi(k)x^k=\sum_{k=0}^{r}\binom{\alpha}{k}\frac{x^k}{(1+x)^{k-\alpha}}\Delta^k\phi(0)$$
$$(g(x)=(1+x)^{\alpha}), \tag{2.117}$$

$$\sum_{k=0}^{\infty}\binom{\alpha+k}{k}\phi(k)x^k=\sum_{k=0}^{r}\binom{\alpha+k}{k}\frac{x^k}{(1-x)^{\alpha+k+1}}\Delta^k\phi(0)$$
$$(g(x)=(1-x)^{-\alpha-1}), \tag{2.118}$$

$$\sum_{k=0}^{\infty}\frac{\phi(k)x^k}{k!}=e^x\sum_{k=0}^{r}\frac{x^k}{k!}\Delta^k\phi(0)\quad (g(x)=e^x), \tag{2.119}$$

$$\sum_{k=1}^{\infty}\frac{\phi(k)x^k}{k}=-f(0)\ln(1-x)+\sum_{k=1}^{r}\frac{1}{k}\left(\frac{x}{1-x}\right)^k\Delta^k\phi(0)$$
$$(g(x)=-\ln(1-x)), \tag{2.120}$$

$$\sum_{k=0}^{\infty}\binom{\alpha}{k}\phi(k)x^k=\sum_{k=0}^{r}\frac{A_k(x,\alpha)}{k!}D^k\phi(0)\quad (g(x)=(1+x)^{\alpha}), \tag{2.121}$$

$$\sum_{k=0}^{\infty}\binom{\alpha+k}{k}\phi(k)x^k=\sum_{k=0}^{r}\frac{\tilde{A}_k(x,\alpha)}{k!}D^k\phi(0)$$
$$(g(x)=(1-x)^{-\alpha-1}), \tag{2.122}$$

$$\sum_{k=m}^{\infty}\binom{k}{m}\phi(k)x^k=\sum_{k=0}^{r}\binom{k+m}{m}\frac{x^{k+m}}{(1-x)^{k+m+1}}\Delta^k\phi(m)$$
$$(g(x)=(1-x)^{-m-1}), \tag{2.123}$$

$$\sum_{k=m}^{\infty}\binom{k}{m}\phi(k)x^k=\sum_{k=0}^{r}\frac{\tilde{A}_k(x,m)x^m}{k!}D^k\phi(m)$$
$$(g(x)=(1-x)^{-m-1}). \tag{2.124}$$

2.2.3 Illustrative examples

Certainly a great variety of special examples could be given via applications of the formulas displayed in the previous subsection. In what follows we merely present some selective examples for references.

Example 2.2.6 *Taking $\alpha = -1$ and $x \mapsto -x$ in (2.100), we get (2.98), which is a well-known formula utilized in the construction of a summation formula with a remainder in the previous section (cf. also [99, 102]). Putting $x = -1$ in (2.98) and (2.102) we get*

$$\sum_{k=0}^{\infty}(-1)^k f(k) = \sum_{k=0}^{\infty}\frac{(-1)^k\Delta^k f(0)}{2^{k+1}}$$
$$= \sum_{k=1}^{\infty}(-1)^k\frac{f(k)}{k} = -f(0)\log 2 + \sum_{k=1}^{\infty}\frac{(-1)^k\Delta^k f(0)}{k2^k}$$

These are known as Euler's series transformation formula and its analog, which may sometimes be used to convert slowly convergent series into rapidly convergent ones.

Example 2.2.7 *From (2.100) and noting (2.89) and (2.90), we obtain the sum of the Euler's arithmetic-geometric series*

$$\sum_{k=0}^{\infty}k^p x^k = \sum_{k=0}^{p}\frac{x^k\left[\Delta^k t^p\right]_{t=0}}{(1-x)^{k+1}} = \sum_{k=0}^{p}\frac{k!\left\{{p \atop k}\right\}x^k}{(1-x)^{k+1}} = \alpha_p(x),$$

where $\alpha_p(x)$ is known as Eulerian fraction (see Wang and Hsu [210]).

Example 2.2.8 *In (2.100) taking $\alpha = n$, $f(t) = \binom{t}{j}$, a jth degree polynomial so that $f(k) = \binom{k}{j}$, we get*

$$\sum_{k=0}^{n}\binom{n}{k}\binom{k}{j}x^k = \sum_{\nu=0}^{j}\binom{n}{\nu}\frac{x^\nu}{(1+x)^{\nu-n}}\Delta^\nu\binom{t}{j}_{t=0}$$
$$= \sum_{\nu=0}^{j}\binom{n}{\nu}\frac{x^\nu}{(1+x)^{\nu-n}}\binom{t}{j-\nu}_{t=0} = \sum_{\nu=0}^{j}\binom{n}{\nu}\frac{x^\nu}{(1+x)^{\nu-n}}\delta_{j\nu}$$
$$= \binom{n}{j}x^j(1+x)^{n-j},$$

where we use $\Delta^k\binom{t}{r}_{t=0} = \binom{t}{r-k}_{t=0} = \binom{0}{r-k} = \delta_{rk}$, the Kronecker symbol. This is (3.118) of the Gould's book [73].

Example 2.2.9 *The series transformation formulas can be applied to construct a set of identities by substituting certain functions.*

Similar to Example 2.2.8, taking $f(t) = \binom{t}{r}$ so that $f(k) = \binom{k}{r}$ in (2.101) yields (for $|x| < 1$)

$$\begin{aligned}\sum_{k=r}^{\infty}\binom{\alpha+k}{k}\binom{k}{r}x^k &= \sum_{k=0}^{r}\binom{\alpha+k}{k}\frac{x^k}{(1-x)^{\alpha+k+1}}\Delta^k\binom{t}{r}_{t=0}\\ &= \binom{\alpha+r}{r}\frac{x^r}{(1-x)^{\alpha+r+1}}.\end{aligned}$$

Consequently,

$$\sum_{k=r}^{\infty}\binom{\alpha+k}{k}\binom{k}{r}\frac{1}{2^k} = \binom{\alpha+r}{r}2^{\alpha+1}.$$

Similarly, for $f(t) = \binom{t}{r}$, ($r \in \mathbb{N}_0$), from (2.101) and (2.102) and (2.109) and (2.110) we obtain, respectively,

$$\sum_{k=0}^{\infty}\binom{k}{r}\frac{x^k}{k!} = e^x\frac{x^r}{r!},$$

$$\sum_{k=1}^{\infty}\binom{k}{r}\frac{x^k}{k} = -\log(1-x) + \frac{1}{r}\left(\frac{x}{1-x}\right)^r \quad (r \geq 1),$$

$$\sum_{k=0}^{\infty}\binom{2k}{r}\frac{x^{2k}}{(2k)!} = \frac{e^x}{2}\frac{x^r}{r!} + \frac{e^{-x}}{2}\frac{(-x)^r}{r!},$$

$$\sum_{k=0}^{\infty}\binom{2k+1}{r}\frac{x^{2k+1}}{(2k+1)!} = \frac{e^x}{2}\frac{x^r}{r!} - \frac{e^{-x}}{2}\frac{(-x)^r}{r!}.$$

Example 2.2.10 *In (2.101) taking $f(t) = t^r$ so that $f(k) = k^r$, we get*

$$\sum_{k=0}^{\infty}\binom{\alpha+k}{k}k^r x^k = \sum_{k=0}^{r}\binom{\alpha+k}{k}\frac{x^k}{(1-x)^{\alpha+k+1}}k!\left\{{r \atop k}\right\}.$$

Formula (1.126) in [73] can be written as

$$\begin{aligned}\sum_{k=0}^{n}\binom{n}{k}k^r x^k &= (1+x)^n\sum_{j=0}^{r}\binom{n}{j}\frac{x^j}{(1+x)^j}\sum_{k=0}^{j}(-1)^{j-k}\binom{j}{k}k^r\\ &= \sum_{j=0}^{r}\binom{n}{j}\frac{x^j}{(1+x)^{j-n}}\left[\Delta^j t^r\right]_{t=0}.\end{aligned}$$

This is obviously a particular case of formula (2.100) with $\alpha = n$ and $f(t) = t^r$.

Example 2.2.11 *Series transformation (2.105) can be used to extend Gould and Wetweerapong's comparable finite sum formula (see Gould and Wetweerapong [78]) to the infinite sum setting, namely,*

$$\sum_{k=0}^{\infty}\binom{k}{j}k^p x^k = \sum_{k=0}^{\infty}\binom{k+j}{j}\frac{x^{k+j}}{(1-x)^{k+j+1}}\sum_{i=0}^{k}(-1)^{k-i}\binom{k}{i}(i+j)^p.$$

Example 2.2.12 *It is known that there is a GF for Bell numbers*

$$\sum_{k=0}^{\infty}W(k)\frac{x^k}{k!} = e^{e^x-1}.$$

Also, for $g(x) = e^x$ and $f(k) = W(k+1)$, formula (2.103) implies

$$\begin{aligned}
&\sum_{k=0}^{\infty}\frac{1}{k!}\Delta^k W(1)x^k = e^{-x}\sum_{k=0}^{\infty}\frac{1}{k!}W(k+1)x^k \\
=\ & e^{-x}\frac{d}{dx}\left(\sum_{k=0}^{\infty}\frac{1}{(k+1)!}W(k+1)x^{k+1}\right) = e^{-x}\frac{d}{dx}\left(e^{e^x-1}-1\right) \\
=\ & e^{e^x-1} = \sum_{k=0}^{\infty}\frac{1}{k!}W(k)x^k.
\end{aligned}$$

Comparing the coefficients of x^k in the leftmost and the rightmost expressions, we get $W(k) = \Delta^k W(1)$, which is called the Aitken identity (cf. Theorem B in §5.4 of Comtet [44]).

Example 2.2.13 *Our series transformation can be used to reconstruct the Dobinski's formula (see [4a] in §5.4 of [44]). If $g(t) = e^t$ and $f(t) = t^r$ $(r \in \mathbb{N})$. Then (2.98) or (2.103) implies*

$$\sum_{k=0}^{\infty}k^r\frac{x^k}{k!} = e^x\sum_{k=0}^{r}\left\{ {r \atop k} \right\}x^k.$$

This leads to

$$e^{-1}\sum_{k=0}^{\infty}\frac{k^r}{k!} = \sum_{k=0}^{r}\left\{ {r \atop k} \right\} = W(r).$$

This is the well-known formula of Dobinski for the Bell number $W(r)$, the number of all possible partition of a set with r distinct elements.

Example 2.2.14 *In (2.101)–(2.104) and (2.109)–(2.110), we substitute* $f(t) = t^r$ *and* $h(t) = t^r$ *and obtain, respectively,*

$$\sum_{k=0}^{\infty} \frac{k^r x^k}{k!} = e^x \sum_{k=0}^{r} \left\{ {r \atop k} \right\} x^k,$$

$$\sum_{k=1}^{\infty} \frac{k^r x^k}{k} = -f(0)\log(1-x) + \sum_{k=1}^{r} (k-1)! \left\{ {r \atop k} \right\} \left(\frac{x}{1-x} \right)^k,$$

$$\sum_{k=0}^{\infty} \binom{\alpha}{k} k^r x^k = A_r(x,\alpha), (r \in \mathbb{N}_0),$$

$$\sum_{k=0}^{\infty} \binom{\alpha+k}{k} k^r x^k = \tilde{A}_r(x,\alpha), (r \in \mathbb{N}_0),$$

$$\sum_{k=0}^{\infty} \frac{(2k)^r x^{2k}}{(2k)!} = \frac{e^x}{2} \sum_{k=0}^{r} \left\{ {r \atop k} \right\} x^k + \frac{e^{-x}}{2} \sum_{k=0}^{r} \left\{ {r \atop k} \right\} (-x)^k,$$

$$\sum_{k=0}^{\infty} \frac{(2k+1)^r x^{2k+1}}{(2k+1)!} = \frac{e^x}{2} \sum_{k=0}^{r} \left\{ {r \atop k} \right\} x^k - \frac{e^{-x}}{2} \sum_{k=0}^{r} \left\{ {r \atop k} \right\} (-x)^k.$$

Example 2.2.15 *In (2.100) taking* $f(t) = r^t$, $(r > 0,\ r \neq 1)$ *so that* $f(k) = r^k$ *and* $\Delta^k f(0) = \sum_{j=0}^{k} \binom{k}{j} (-1)^{k-j} r^j = (r-1)^k$, *we get*

$$\sum_{k=0}^{\infty} \binom{\alpha}{k} (rx)^k = \sum_{k=0}^{\infty} \binom{\alpha}{k} \frac{((r-1)x)^k}{(1+x)^{k-\alpha}}.$$

Similarly, from (2.101)–(2.104) and (2.107)–(2.110) we have, respectively,

$$\sum_{k=0}^{\infty} \binom{\alpha+k}{k} (rx)^k = \sum_{k=0}^{\infty} \binom{\alpha+k}{k} \frac{((r-1)x)^k}{(1-x)^{\alpha+k+1}},$$

$$\sum_{k=0}^{\infty} \frac{(rx)^k}{k!} = e^x \sum_{k=0}^{\infty} \frac{((r-1)x)^k}{k!},$$

$$\sum_{k=1}^{\infty} \frac{(rx)^k}{k} = -f(0)\log(1-x) + \sum_{k=1}^{\infty} \frac{1}{k} \left(\frac{(r-1)x}{1-x} \right)^k,$$

$$\sum_{k=0}^{\infty} \binom{\alpha}{k} (rx)^k = \sum_{k=0}^{\infty} \frac{A_k(x,\alpha)}{k!} (\ln r)^k.$$

Example 2.2.16 *Recall that Bernoulli polynomials* $B_n(t)$*'s are generated by the expression*

$$e^{tx} \frac{x}{e^x - 1} = \sum_{n=0}^{\infty} \frac{B_n(t)}{n!} x^n$$

and enjoy the properties

$$\frac{d}{dt}B_n(t) = nB_{n-1}(t), \quad (n = 1, 2, \cdots)$$

with $B_0(t) = 1$ *and* $B_n(0) = B_n$*, the Bernoulli numbers. From Theorem 1.3.18,* $D^k B_n(t) = (n)_k B_{n-k}(t)$ *so that* $D^k B_n(0) = (n)_k B_{n-k}(0) = (n)_k B_{n-k}$*, where* $(n)_k$ *are* k*th falling factorial of* n *with step length* 1*. Now let* $g(x) = B_n(x)$ *and* $f(k) = k^r$ *(*n *and* r *are integers with* $0 \le r \le n$*) Then,* $g^{(k)}(0) = B_n^{(k)}(0) = (n)_k B_{n-k}$ *and* $f^{(k)}(x) = (n)_k B_{n-k}(x)$ *so that Theorem 2.2.2 implies*

$$\sum_{k=0}^{n} \binom{n}{k} B_{n-k} k^r x^k = \sum_{k=0}^{r} \binom{n}{k} B_{n-k}(x) k! \left\{ {r \atop k} \right\} x^k. \tag{2.125}$$

Since $S(0,0) = 1$ *and* $S(0,k) = 0$ *for all* $k \ge 1$*, the particular case* $r = 0$ *gives the well-known expression*

$$B_n(x) = \sum_{k=0}^{n} \binom{n}{k} B_{n-k} x^k.$$

Of course the right-hand side of (2.125) can be regarded as a generating function of $\{\binom{n}{k} B_{n-k} k^r\}_{k=0}^n$*.*

In addition, taking $f(t) = B_n(t)$ *in (2.103) and (2.104), respectively, we easily obtain*

$$\sum_{k=0}^{\infty} \binom{\alpha}{k} B_n(k) x^k = \sum_{k=0}^{n} \binom{n}{k} A_k(x, \alpha) B_{n-k}, \tag{2.126}$$

$$\sum_{k=0}^{\infty} \binom{\alpha + k}{k} B_n(k) x^k = \sum_{k=0}^{n} \binom{n}{k} \tilde{A}_k(x, \alpha) B_{n-k}. \tag{2.127}$$

Recalling that $\tilde{A}_k(x, 0) = \alpha_k(x)$ *(the ordinary Eulerian fraction), we can find that the last identity implies (with* $\alpha = 0$*)*

$$\sum_{k=0}^{\infty} B_n(k) x^k = \sum_{k=0}^{n} \binom{n}{k} \alpha_k(x) B_{n-k}. \tag{2.128}$$

Surely similar identities of some interest may be found for other classical special polynomials.

Example 2.2.17 *Let* λ *and* θ *be any real numbers. The generalized falling factorial* $(t + \lambda|\theta)_p$ *is usually defined by*

$$(t + \lambda|\theta)_p = \Pi_{j=0}^{p-1} (t + \lambda - j\theta) \quad \text{for} \quad p \ge 1 \quad \text{and} \quad (t + \lambda|\theta)_0 = 1.$$

It is known that Howard's degenerate weighted Stirling numbers (cf. Howard [119]) may be defined by the finite differences of $(t+\lambda|\theta)_p$ at $t=0$:

$$S(p,k,\lambda|\theta) := \frac{1}{k!}\left[\Delta^k(t+\lambda|\theta)_p\right]_{t=0}.$$

Then, using (2.117) and (2.118) with $\phi(t)=(t+\lambda|\theta)_p$, we get

$$\sum_{k=0}^{\infty}\binom{\alpha}{k}(k+\lambda|\theta)_p x^k = \sum_{k=0}^{p}\binom{\alpha}{k}\frac{k!S(p,k,\lambda|\theta)x^k}{(1+x)^{k-\alpha}}, \tag{2.129}$$

$$\sum_{k=0}^{\infty}\binom{\alpha+k}{k}(k+\lambda|\theta)_p x^k = \sum_{k=0}^{p}\binom{\alpha+k}{k}\frac{k!S(p,k,\lambda|\theta)x^k}{(1-x)^{\alpha+k+1}}. \tag{2.130}$$

The particular case of (2.130) with $\alpha=0$ was considered in [134]. It is also obvious that the ordinary Euler's summation formula for the arithmetic-geometric series (cf, for example, Lemma 2.1.7 or Lemma 2.7 in [78]) is implied by (2.129) with $\lambda=\theta=0$, $\alpha=-1$, $x\mapsto -x$, or by (2.130) with $\lambda=\theta=0$ and $\alpha=0$.

Example 2.2.18 *For any given positive integer m denote $\phi(t)=\binom{t}{m}$. It is easy to find that $\Delta^k\phi(m)=\binom{t}{m-k}_{t=m}=\binom{m}{k}$. Thus an application of (2.123) to $\binom{t}{m}$ gives*

$$\sum_{k=m}^{\infty}\binom{k}{m}^2 x^k = \sum_{k=0}^{m}\binom{k+m}{k}\binom{m}{k}\frac{x^{k+m}}{(1-x)^{k+m+1}}. \tag{2.131}$$

This shows that the generating function of the number sequence $\left(\binom{k}{m}^2\right)$ is given by

$$GF\left\{\binom{k}{m}^2\right\} = \sum_{k=0}^{m}\binom{k+m}{k}\binom{m}{k}\frac{x^{k+m}}{(1-x)^{k+m+1}}. \tag{2.132}$$

Naturally one may ask to find $GF\left\{\binom{k}{m}^3\right\}$. Actually, this can be worked out as follows.

Takes the left-hand side of (2.131) as $\Phi(x)$. Then using (2.131) we find

$$\Phi(xE)\phi(0) = \sum_{k=m}^{\infty}\binom{k}{m}^2(xE)^k\phi(0) = \sum_{k=m}^{\infty}\binom{k}{m}^3 x^k$$

$$= \sum_{k=0}^{m}\binom{k+m}{k}\binom{m}{k}\frac{(xE)^{k+m}}{(1-xE)^{k+m+1}}\phi(0)$$

$$= \sum_{k=0}^{m}\binom{k+m}{k}\binom{m}{k}\frac{x^{k+m}}{(1-x)^{k+m+1}}\left(1-\frac{x\Delta}{1-x}\right)^{-k-m-1}E^{k+m}\phi(0)$$

$$= \sum_{k=0}^{m} \binom{k+m}{k}\binom{m}{k}\frac{x^{k+m}}{(1-x)^{k+m+1}}\sum_{j=0}^{m}\binom{k+m}{j}\left(\frac{x}{1-x}\right)^j \Delta^j\phi(k+m)$$

$$= \sum_{k=0}^{m}\sum_{j=0}^{m}\binom{k+m}{k}\binom{k+m}{j}\binom{m}{k}\binom{k+m}{k+j}\frac{x^{k+m+j}}{(1-x)^{k+m+j+1}}.$$

Thus we obtain

$$GF\left\{\binom{k}{m}^3\right\} =$$

$$\sum_{k=0}^{m}\sum_{j=0}^{m}\binom{k+m}{k}\binom{k+m}{j}\binom{m}{k}\binom{k+m}{k+j}\frac{x^{k+m+j}}{(1-x)^{k+m+j+1}}. \quad (2.133)$$

A similar process can be applied to find $\mathrm{GF}\left\{\binom{k}{m}^n\right\}$ for $n = 4, 5\cdots$. However, we have not yet known the closed form of $\sum_{k=0}^{\infty}\binom{k}{m}^{\ell} x^k$ for general ℓ.

Example 2.2.19 *Suppose that $\phi(t)$ is an integral polynomial, namely, all its coefficients (including the constant term) are integers. It is easily seen that $\Delta^k\phi(0)/k!$ $(k = 0, 1, 2, \cdots)$ are integers as well. In fact, each term $a_m t^m$ $(m \geq 0)$ of $\phi(t)$ has a difference at zero: $\left[\Delta^k a_m t^m\right]_{t=0} = a_m k! \left\{ {m \atop k} \right\}$ with $\left\{ {0 \atop 0} \right\} = 1$ and $\left\{ {m \atop k} \right\} = 0$ $(k > m)$. So $\Delta^k\phi(0)/k!$ is a linear combination of Stirling numbers of the second kind with integer coefficients. Thus formula (2.119) implies that $\sum_{k=0}^{\infty}\phi(k)x^k/k!$ is equal to e^x multiplying by an integral polynomial. In particular, for $x = 1$, this implies that*

$$\frac{\phi(0)}{0!} + \frac{\phi(1)}{1!} + \frac{\phi(2)}{2!} + \cdots + \frac{\phi(k)}{k!} + \cdots$$

is an integral multiple of e.

Example 2.2.20 *Every formula in Subsection §2.2.2 may be used to yield a pair of related formulas involving the trigonometric functions* $\cos k\theta$ *and* $\sin k\theta$. *For instance, setting $x = \rho e^{i\theta} = \rho(\cos\theta + \sin\theta)$ with $\rho = |x| > 0$ and $i^2 = -1$, we can obtain a pair of formulas from (2.118) as follows*

$$\sum_{k=0}^{\infty}\binom{\alpha+k}{k}\phi(k)\rho^k\cos k\theta$$

$$= \sum_{k=0}^{r}\binom{\alpha+k}{k}\Delta^k\phi(0)Re\left(\frac{(\rho e^{i\theta})^k}{(1-\rho e^{i\theta})^{\alpha+k+1}}\right) \quad (2.134)$$

$$\sum_{k=0}^{\infty}\binom{\alpha+k}{k}\phi(k)\rho^k\sin k\theta$$

$$= \sum_{k=0}^{r}\binom{\alpha+k}{k}\Delta^k\phi(0)Im\left(\frac{(\rho e^{i\theta})^k}{(1-\rho e^{i\theta})^{\alpha+k+1}}\right), \quad (2.135)$$

where $Re(z)$ and $Im(z)$ denote, respectively, the real part and imaginary part of the complex number z. Obviously (2.134) and (2.135) could be specialized in various ways.

Remark 2.2.21 *In this subsection we have mostly considered the operator method for the cases when $\phi(t)$ takes various elementary functions. From Remarks 2.2.1, 2.2.4, and 2.2.5, we can see that the method also apply to the cases where $\phi(t)$ may take various suitable special functions. However, it still remains much to be investigated for the interested readers.*

2.3 Summation of Operators

It is known that the problem for the computation of sums of convolved powers of this type

$$S(m; i, j) = \sum_{0 \le k \le m} k^i (m-k)^j \tag{2.136}$$

was first investigated by Glaisher [67] and [68] in 1911-1912, and he found a summation formula by using Bernoulli numbers. Some further investigations and extensions were given, during the years 1977-1978, by Neumann-Schonbach [169], Carlitz [30] and Gould [74], respectively, in which Eulerian numbers as well as Stirling numbers of the second kind had been utilized. Various numerical examples were also represented by Gould[74].

Actually, the most general formulation of the computational problem for convolved polynomial sums was given in Hsu [120], where a kind of general summation formula was found via several lemmas. However, [120] contains some notational errors, and all related formulas were given in quite complicated forms. This may be the reason why the general result of [120] could not be used much in practice.

Having done some practical computations, Hsu [121] eventually got realized that a kind of symbolic operator approach adopted should be the most effective way for dealing with general convolved polynomial sums. The object of this section is to show the operator method conceived previously. We will present certain general operator summation formulas that could be specialized and applied in various ways (cf. Subsections §2.3.4 and §2.3.5).

2.3.1 Summation formulas involving operators

Throughout the section $f(x)$ and $f_i(x) (i = 1, \ldots, n)$ are assumed to be arbitrary polynomials over the real or complex number field, with degrees denoted by ∂f and ∂f_i, respectively. We are concerned with the problem for the computation of convolved polynomial sums of the type

$$S(m,[f_1]\cdots[f_n]) := \sum_{(m;n;0)} f_1(x_1)\cdots f_n(x_n) \tag{2.137}$$

where m is any given positive integer, and the sum on the RHS of (2.137) is taken over all the n-compositions of m with non-negative integer components, namely, over the set $(m;n;0)$ of all integer solutions of $x_1+\cdots+x_n = m$ with each $x_i \geq 0\,(i=1,\cdots,n)$, i.e.,

$$(m;n;0) = \{(x_1,x_2,\ldots,x_n) : \sum_{i=1}^{n} x_i = m, x_i \geq 0\}.$$

If $f_1 = f_2 = \cdots = f_n$, we denote the corresponding $S(m,[f_1]\cdots[f_n])$ by $S_m^n(f)$ and call it the n-fold convolution, namely,

$$S_m^n(f) := \sum_{(m;n;0)} f_1(x_1)\cdots f_n(x_n). \tag{2.138}$$

Obviously, $S(m;i,j)$ of (2.136) just corresponds to the special case of (2.137) with $n=2$, $f_1(x)=x^i$ and $f_2(x)=x^j$.

Powers of difference operator Δ and the shift operator E are defined in the usual way with $\Delta^0 = E^0 = 1$ denoting the identity operator so that $\Delta^0 f(x) = E^0 f(x) = 1f(x) = f(x)$.

Definition 2.3.1 *For any given polynomial $f(x)$ with degree $\partial f \geq 0$, there are two operator polynomials constructed from $f(x)$ as follows:*

$$\Lambda(\Delta,f) := \sum_{k=0}^{\partial f} \Delta^k f(0)\Delta^k, \tag{2.139}$$

$$\Lambda^*(E,f) := \sum_{k=0}^{\partial f} \Delta^k f(-k-1)E^k. \tag{2.140}$$

These are called Λ-operators associated with f. In particular, $\Lambda(\Delta,f) = \Lambda^*(E,f) = f(0)\cdot 1$ for the case of $\partial f = 0$.

Note that computations of backward differences $\Delta^k f(-v-1)$ are as easy as that of $\Delta^k f(0)$. Thus Λ^* and Λ are equally useful for practical computations. In fact, we have the following result.

Proposition 2.3.2 *Let Λ and Λ^* be the operators defined by (2.139) and (2.140), respectively. Then they are the same operator, namely we have the operator identity $\Lambda(\Delta,f) = \Lambda^*(E,f)$.*

Proof. The equivalence of operators Λ and Λ^* can be verified by starting with $E = 1 + \Delta$. From

$$\Delta^k = (E-1)^k = (-1)^k \sum_{j=0}^{k} \binom{k}{j} (-E)^j,$$

we have

$$\sum_{k=0}^{\partial f} \Delta^k f(0) \Delta^k = \left\{ \sum_{k=0}^{\partial f} \sum_{j=0}^{k} \binom{k}{j} (-\Delta)^k f(0) \right\} (-E)^j$$

$$= \sum_{j=0}^{\partial f} \left(\sum_{k=j}^{\partial f} \binom{k}{j} (-\Delta)^k f(0) \right) (-E)^j,$$

where the inner terms of the rightmost summation can be simplified as follows because f is a polynomial of k-th degree:

$$\sum_{k=j}^{\infty} \binom{k}{j} (-\Delta)^k f(0) = (-\Delta^j) \sum_{k=0}^{\infty} \binom{k+j}{j} (-\Delta)^k f(0)$$

$$= \frac{(-\Delta)^j}{(1+\Delta)^{j+1}} f(0) = (-\Delta)^j E^{-j-1} f(0).$$

Hence by substitution we get

$$\sum_{k=0}^{\partial f} \Delta^k f(0) \Delta^k = \sum_{j=0}^{\partial f} (-\Delta)^j E^{-j-1} f(0) (-E)^j,$$

which implies the equivalence between (2.139) and (2.140).

■

A main proposition to be studied and given applications in this section is the following one.

Theorem 2.3.3 *Let $f_1(x), \ldots, f_n(x)$ be any given polynomials. Then there hold a pair of summation formulas as follows*

$$S(m, [f_1] \cdots [f_n]) = \left(\prod_{i=1}^{n} \Lambda(\Delta, f_i) \right) \binom{x}{m}_{x=m+n-1}, \tag{2.141}$$

$$S(m, [f_1] \cdots [f_n]) = \left(\prod_{i=1}^{n} \Lambda^*(E, f_i) \right) \binom{x}{m}_{x=m+n-1}. \tag{2.142}$$

Proof. It suffices to verify (2.141) since (2.142) $\Leftrightarrow$ (2.141). Let us recall that there is a well-known identity in Combinatorics, namely

$$\sum_{(m;n;0)} \binom{x_1}{v_1} \cdots \binom{x_n}{v_n} = \binom{m+n-1}{m-(v_1+\cdots+v_n)} \tag{2.143}$$

where $v_i \geq 0\,(1 \leq i \leq n)$, and $m \geq (v_1 + \cdots + v_n)$. Also there are simple relations

$$\Delta^v \binom{x}{m} = \binom{x}{m-v} \quad \text{for} \quad 0 \leq v \leq m \quad \text{and} \quad \Delta^v \binom{x}{m} = 0 \quad \text{for} \quad v > m.$$

Thus the RHS of (2.143) may be expressed in the form

$$\binom{m+n-1}{m-(v_1+\cdots+v_n)} = \Delta^{v_1}\cdots\Delta^{v_n}\binom{x}{m}_{x=m+n-1}. \tag{2.144}$$

Consequently, employing Newton's formula for $f(x)$ and making use of (2.143)-(2.144), one may compute the LHS of (2.141) as follows

$$\begin{aligned}
S(m,[f_1]\cdots[f_n]) &= \sum_{(m;n;0)} \prod_{i=1}^{n} \left(\sum_{v_i=0}^{\partial f_i} \Delta^{v_i} f_i(0) \binom{x_i}{v_i} \right) \\
&= \sum_{v_i=0}^{\partial f_i} \cdots \sum_{v_n=0}^{\partial f_n} \Delta^{v_i} f_i(0) \cdots \Delta^{v_n} f_n(0) \sum_{(m;n;0)} \prod_{i=1}^{n} \binom{x_i}{v_i} \\
&= \sum_{v_i=0}^{\partial f_i} \cdots \sum_{v_n=0}^{\partial f_n} (\Delta^{v_1} f_i(0)\Delta^{v_1}) \cdots (\Delta^{v_n} f_n(0)\Delta^{v_n}) \binom{x}{m}_{x=m+n-1} \\
&= \prod_{i=1}^{n} \left(\sum_{v_i=0}^{\partial f_i} \Delta^{v_i} f_i(0)\Delta^{v_1} \right) \binom{x}{m}_{x=m+n-1} \\
&= \left(\prod_{i=1}^{n} \Lambda(\Delta, f_i) \right) \binom{x}{m}_{x=m+n-1}
\end{aligned}$$

Hence (2.141) is proved.

■

Corollary 2.3.4 *For the case $f_1(x) = \cdots = f_n(x) = f(x)$ there are summation formulas for $S(m,[f]^n)$:*

$$\sum_{(m;n;0)} f(x_1)\cdots f(x_n) = (\Lambda(\Delta, f))^n \binom{x}{m}_{x=m+n-1}, \tag{2.145}$$

$$\sum_{(m;n;0)} f(x_1)\cdots f(x_n) = (\Lambda^*(E, f))^n \binom{x}{m}_{x=m+n-1}. \tag{2.146}$$

Corollary 2.3.5 *For the monomials $f_i(x) = x^{p_i}$ with $p_i \geq 0\,(i = 1,\ldots,n)$, there is a summation formula of the form*

$$\sum_{(m;n;0)} x_1^{p_1}\cdots x_n^{p_n} = \left(\prod_{i=1}^{n} \left(\sum_{v=0}^{p_i} v! \left\{ {p_i \atop v} \right\} \Delta^v \right) \right) \binom{x}{m}_{x=m+n-1}, \tag{2.147}$$

where $\left\{ {p_i \atop v} \right\}$ *are Stirling numbers of the second kind with* $\left\{ {0 \atop 0} \right\} = 1$. *In particular, for the case* $p_i \geq 1 (i = 1, \ldots, n)$, (2.147) *can be replaced by the form*

$$\sum_{(m;n;1)} x_1^{p_1} \cdots x_n^{p_n} = \left(\prod_{i=1}^{n} \left(\sum_{v=1}^{p_i} v! \left\{ {p_i \atop v} \right\} \Delta^v \right) \right) \binom{x}{m}_{x=m+n-1}, \tag{2.148}$$

where $(m;n;1)$ *denotes the set of* n*-compositions of* m *with each component* $x_i \geq 1$ $(i = 1, \ldots, n)$.

Observe that (2.147) with $n = 2$ is an old result given by Gould [74], in which several numerical instances have been displaced. Also, note that (2.145) and (2.146) are even much older results that had been derived and employed in [121] and [122], respectively.

Obviously, the set $(m;n;0)$ under the summation of (2.137) may be replaced by $(m;n;1)$ in case $f_1(0) = \cdots = f_n(0) = 0$. In what follows we will present a few examples, requiring a bit of algebraic computations.

Example 2.3.6 *Suppose we want to find a formula for the summation of the form*

$$S(m, [x^2][x^3][x^4]) \equiv \sum_{(m;n;1)} x_1^2 \cdot x_2^3 \cdot x_3^4. \tag{2.149}$$

It requires that the summation formula should be consisting of a least number of terms.

In accordance with (2.141) we have to do computations

$$\Lambda(\Delta, x^2) = \Delta + 2\Delta^2,$$
$$\Lambda(\Delta, x^3) = \Delta + 6\Delta^2 + 6\Delta^3,$$
$$\Lambda(\Delta, x^4) = \Delta + 14\Delta^2 + 36\Delta^3 + 24\Delta^4.$$

Clearly, we may rewrite $\Lambda(\Delta, x^3) = \Delta(1 + 6\Delta E)$. *Moreover, using a simple factorization technique, we find*

$$\Lambda(\Delta, x^4) = \Delta(1 + 2\Delta)(1 + 12\Delta E). \tag{2.150}$$

Consequently, we obtain

$$\begin{aligned} &\Lambda(\Delta, x^2)\Lambda(\Delta, x^3)\Lambda(\Delta, x^4) = \Delta^3(1 + 2\Delta)^2(1 + 6\Delta E)(1 + 12\Delta E) \\ &= \Delta^3(1 + 4\Delta E)(1 + 6\Delta E)(1 + 12\Delta E) \\ &= \Delta^3(1 + 22\Delta E + 144(\Delta E)^2 + 288(\Delta E)^3). \end{aligned}$$

Hence an application of (2.141) (with $n = 3$*) to the sum (2.149) gives a formula as follows*

$$\sum_{(m;n;1)} x_1^2 \cdot x_2^3 \cdot x_3^4 = \binom{m+2}{5} + 22\binom{m+3}{7} + 144\binom{m+4}{9} + 288\binom{m+5}{11}. \tag{2.151}$$

Similarly, noting that $\Lambda(\Delta, x) = \Delta$ *and using (2.141) with* $n = 4$, *we may obtain*

$$\sum_{(m;n;1)} x_1^1 \cdot x_2^2 \cdot x_3^3 \cdot x_4^4 = \binom{m+3}{7} + 22\binom{m+4}{9} + 144\binom{m+5}{11} + 288\binom{m+6}{13}. \tag{2.152}$$

Surely, (2.151) and (2.152) are the shortest formulas for the sums in question. Obviously (2.151) and (2.152) involve the following asymptotic estimates.

$$\sum_{(m;n;1)} x_1^2 \cdot x_2^3 \cdot x_3^4 \cdot x_4^4 = 288\binom{m+5}{11}(1 + O(\frac{1}{m^2})), m \to \infty,$$

$$\sum_{(m;n;1)} x_1^1 \cdot x_2^2 \cdot x_3^3 \cdot x_4^4 = 288\binom{m+6}{13}(1 + O(\frac{1}{m^2})), m \to \infty.$$

From the above example we have

$$\begin{aligned}(\Lambda(\Delta, x^2))^n &= \Delta^n(1 + 2\Delta)^n, \\ (\Lambda(\Delta, x^3))^n &= \Delta^n(1 + 6\Delta E)^n, \\ (\Lambda(\Delta, x^4))^n &= \Delta^n(1 + 2\Delta)^n(1 + 12\Delta E)^n.\end{aligned}$$

Accordingly, as particular consequences of (2.145) of Corollary 2.3, we may state the following.

Example 2.3.7 *There are* 3 *formulas as follows*

$$\sum_{(m;n;1)} (x_1 \cdot x_2 \cdots x_n)^2 = \sum_{v=0}^{n} 2^v \binom{n}{v}\binom{m+n-1}{2n+v-1}, \tag{2.153}$$

$$\sum_{(m;n;1)} (x_1 \cdot x_2 \cdots x_n)^3 = \sum_{v=0}^{n} 6^v \binom{n}{v}\binom{m+n+v-1}{2n+2v-1}, \tag{2.154}$$

$$\sum_{(m;n;1)} (x_1 \cdot x_2 \cdots x_n)^4 = \sum_{v=0}^{n}\sum_{\mu=0}^{n} 2^v \cdot 12^\mu \binom{n}{v}\binom{n}{\mu}\binom{m+n+\mu-1}{m-n-v-\mu}. \tag{2.155}$$

Certainly, these formulas are especially useful when m *is much bigger than* n. *Note that (2.153) and (2.154) have appeared previously [168, 150] and that (2.155) could be replaced by a formula of similar nature via the operator* $\Lambda(\Delta, x^4) = \Delta^n(\Delta + E)^n(1 + 12\Delta E)^n$. *However it appears to be impossible to get a simpler formula consisting of* $(n + 1)$ *terms for the sum of (2.155).*

Remark 2.3.8 *Having correctly defined $S(m,[f_1]\cdots[f_n])$ as that of (2.137), one may verify that the following formulas given by Lemmas 3 and 4 of [150] (viz. (4)' and (6) of [150])*

$$S(m,[f]^n) = n! \sum_{(m;n;0)} \binom{m+n-1}{1\cdot p_1+\cdots+k\cdot p_k+n-1} \frac{\beta_0^{p_0}\beta_1^{p_1}\cdots\beta_k^{p_k}}{p_0!p_1!\cdots p_k!} \tag{2.156}$$

$$S(m,[f_1]\cdots[f_n]) = \frac{1}{n!}\sum_{(v_1\cdots v_k)\in(1\cdots n)} (-1)^{n-k} S(m,[f_{v_1}+\cdots+f_{v_k}]^n) \tag{2.157}$$

are logically equivalent to (2.145) and (2.141), respectively, wherein $f(x)$ has the degree $\partial f = k, \beta_i$ may be rewritten as $\Delta^i f(0)(i=0,1,\ldots,k)$, and the summation on the RHS of (3.1) is taken over all the $(k+1)$-compositions of n, viz. *$p_0+\cdots+p_k = n$ with each $p_i \geq 0$; and the RHS summation of (3.2) is over all the different combinations (sub-sets) $v_1,\cdots,v_k$ out of the set $\{1,\ldots,n\}\,(k=1,\ldots,n)$.*

Indeed the RHS of (2.156) may be rewritten as

$$\begin{aligned}
&\sum_{(m;n;0)} \frac{n!}{p_0!p_1!\cdots p_k!}\binom{m+n-1}{m-(p_1+2p_2+\cdots+kp_k)} \\
&\quad (f(0))^{p_0}(\Delta f(0))^{p_1}\cdots(\Delta^k f(0))^{p_k} \\
=& \sum_{(m;n;0)} \frac{n!}{p_0!\cdots p_k!}(f(0)\Delta^0)^{p_0}\cdots(\Delta^k f(k)\Delta^k)^{pk}\binom{x}{m}_{x=m+n-1} \\
=& \left(\sum_{v=0}^{k}\Delta^v f(0)\Delta^v\right)^n \binom{x}{m}_{x=m+n-1} = RHS \text{ of } (2.145).
\end{aligned}$$

Also, for every given set $\{v_1,\ldots,v_k\}\subset\{1,\ldots,n\}$, it is obvious that the summand within the summation of (2.157) may be expressed in the following forms

$$\begin{aligned}
&(-1)^{n-k}S(m,[f_{v_1}+\cdots+f_{v_k}]^n) \\
=& (-1)^{n-k}(\Lambda(\Delta,f_{v_1}+\cdots+f_{v_k}))^n\binom{x}{m}_{x=m+n-1} \\
=& (-1)^{n-k}[\Lambda(\Delta,f_{v_1})+\cdots+\Lambda(\Delta,f_{v_k})]^n\binom{x}{m}_{x=m+n-1} \\
=& (-1)^{n-k}\sum_{q_1+\cdots+q_t=n}\frac{n!}{q_1!\cdots q_t!}(\Lambda(\Delta,f_{p_1}))^{q_1}\cdots(\Lambda(\Delta,f_{p_t}))^{q_t}\binom{x}{m}_{x=m+n-1},
\end{aligned}$$

where the last summation is taken over all the subsets of $\{p_1,\ldots,p_t\}$ of $\{v_1,\ldots,v_k\}$ $(t=1,\ldots,k,)$ and all the compositions $(n;1;q)$. Certainly the *general term (summand) of the last summation is contained in all such terms of the RHS summation of (2.157) that* $\{p_1,\dots,p_t\} \subset \{v_1,\dots,v_k\} \subset \{1,\dots,n\}$*. The number of occurrences is obviously* $\binom{n-t}{k-t}$*. Thus the total number of occurrences in the RHS summation of (2.157) is given by*

$$\sum_{k=t}^{n}(-1)^{n-k}\binom{n-t}{k-t} = (1-1)^{n-t} = \begin{cases} 0.\ t < n, \\ 1,\ t = n. \end{cases}$$

This means that the general term vanishes except that $t = n$ *so that* $k = n, \{p_1,\dots,p_t\} = \{v_1,\dots,v_k\} = \{1,\dots,n\},$*, and* $q_1 = \cdots = q_n = 1$*. Consequently, we get*

$$RHS\ of\ (2.157) = \Lambda(\Delta, f_1)\dots\Lambda(\Delta, f_n)\binom{x}{m}_{x=m+n-1.}$$

Hence, as a conclusion we may say that Lemmas 3 and 4 of [120] are implicitly involving (2.141), and the deduction of (2.141) from (2.157) plus (2.156) may be regarded as a different proof for(2.141).

Remark 2.3.9 *Observe that for* $n \geq 3$ *the summation on the LHS of (2.147) may be rewritten as*

$$S(m, [x^{p_1}]\dots[x^{p_n}]) = \sum x_1^{p_1}\cdots x_{n-1}^{p_{n-1}}(m - x_1 - \cdots - x_{n-1})^{p_n}$$

where the RHS summation extends over all the non-negative integers $x_1,\dots,x_{n-1}$ *such that* $x_1 + \cdots + x_{n-1} \leq m$*. Apparently, such a sum may be viewed as a discrete analog of the Dirichlet multiple integral*

$$\begin{aligned} &\int\cdots\int_S t_1^{\alpha_1}\cdots t_{n-1}^{\alpha_{n-1}}(1 - t_1 - \cdots - t_{n-1})^{\alpha_n}\mathrm{d}t_1\cdots\mathrm{d}t_{n-1} \\ = \ &\frac{\Gamma(\alpha_1+1)\cdots\Gamma(\alpha_n+1)}{\Gamma(\alpha_1+\cdots+\alpha_n+n)}, \end{aligned}$$

where the domain of integration is defined by the $(n-1)$*-dimensional set*

$$S : t_1 \geq 0,\dots,t_{n-1} \geq 0,\ t_1 + \cdots + t_{n-1} \leq 1,$$

and $\alpha_i (i = 1,\dots,n-1)$ *are real numbers such that* $\alpha_i + 1 > 0$*. Certainly, the formula (2.147) is much more complicated than the integration formula displayed above. However, it should be possible to verify that (2.147) implies the following limit*

$$\lim_{m\to\infty} S(m, [x^{p_1}]\cdots[x^{p_n}])/m^{p_1+\cdots+p_n+n-1} = \frac{p_1!\cdots p_n!}{(p_1 + \cdots + p_n + n - 1)!}$$

which is consistent with the integral formula when taking $a_i = p_i$.

2.3.2 Some special convolved polynomial sums

Here we will present several examples showing how to make use of the summation formulas (2.141) and (2.145) to evaluate some convolution sums that consist of certain classical polynomials. Evidently, in order to get explicit results for $S(m,[f]^n)$ and $S(m,[f_1]\cdots[f_n])$ by using (2.145) and (2.141), respectively, it requires firstly to find explicit expressions for $\Delta^v f(0)$ and $\Delta^v f_i(0)\,(v=1,2,\ldots;i=1,\ldots,n)$. In particular, related computations could be greatly shortened, in cases $f(x)$ and $f_i(x)$ are known to have explicit expressions in Newton interpolation series.

Example 2.3.10 *We wish to evaluate the convolution*

$$S(m,[B_k]^n)=\sum_{(m;n;0)} B_k(x_1)\cdots B_k(x_n),\quad k\geq 1,$$

where $B_k(x)$ is the $k-th$ degree Bernoulli polynomial defined by the expansion

$$\frac{te^{xt}}{e^t-1}=\sum_{k=0}^{\infty} B_k(x)\frac{t^k}{k!},\quad |t|<2\pi, \tag{2.158}$$

and $B_k(0)=B_k$ $(k=0,1,2,\ldots)$ are known as Bernoulli numbers.

It is known that $B_k(x)$ can be expanded in terms of Newton's polynomials with Stirling numbers of the second kind as coefficient (cf. §78 of [141]), namely

$$B_k(x)=B_k+k\sum_{v=1}^{k}(v-1)!\left\{ {k-1 \atop v-1} \right\}\binom{x}{v}. \tag{2.159}$$

This implies that $\Delta^v B_k(0)=k(v-1)!\left\{ {k-1 \atop v-1} \right\}$. Consequently, using (2.141), we get

$$S(m,[B_k]^n)=\left(B_k+k\sum_{v=1}^{k}(v-1)!\left\{ {k-1 \atop v-1} \right\}\Delta^v\right)^n\binom{x}{m}_{x=m+n-1}. \tag{2.160}$$

Certainly, this is a useful formula when m is much bigger than k and n. Also, an application of (2.145) with $n=2$ yields the formula

$$\sum_{(m;n;0)} B_p(x_1)B_q(x_2)=\left(B_p+p\sum_{v=1}^{p}(v-1)!\left\{ {p-1 \atop v-1} \right\}\Delta^v\right)$$

$$\left(B_q+q\sum_{\mu=1}^{q}(\mu-1)!\left\{ {q-1 \atop \mu-1} \right\}\Delta^\mu\right)\binom{x}{m}_{x=m+1} \tag{2.161}$$

where p and q are given positive integers.

Example 2.3.11 *It is known that the Bernoulli polynomial of the second kind of degree may be written as (see §89 of [141]).*

$$\Psi_k(x) = \int_0^1 \binom{x+t}{k} dt. \tag{2.162}$$

Accordingly, $b_k = \Psi_k(0)$ *may be called Bernoulli numbers of the second kind, viz.*

$$b_k = \int_0^1 \binom{t}{k} \mathrm{dt}, \quad k = 0, 1, 2, \ldots \tag{2.163}$$

where $b_0 = 1, b_1 = \frac{1}{2}, b_2 = -\frac{1}{12}$, *etc. A table of* b_k*'s for* $k \le 10$ *can be found in §89 of [141]. Note that* $\Delta^v f_k(0) = \Psi_{k-v}(0) = B_{k-v}, 0 \le v \le k$ *so that* $\Lambda(\Delta, \Psi_k) = \sum_0^k b_{k-v}\Delta^v = \sum_0^k b_v \Delta^{k-v}$. *Consequently, (2.141) and (2.145) imply the following special formulas*

$$S(m, [\Psi_k]^n) = \left(\sum_{v=0}^{k} b_v \Delta^{k-v}\right)^n \binom{x}{m}_{x=m+n-1} \tag{2.164}$$

and

$$\sum_{(m;n;0)} \Psi_p(x_1)\Psi_q(x_2) = \left(\sum_{v=0}^{p} b_v \Delta^{p-v}\right)\left(\sum_{\mu=0}^{q} b_\mu \Delta^{q-\mu}\right)\binom{x}{m}_{x=m+1}, \tag{2.165}$$

respectively.

Example 2.3.12 *As is known, Boole's polynomial of degree* k *can be expressed in form Section* 113 *of [141].*

$$\xi_k(x) = \sum_{v=0}^{k} \left(-\frac{1}{2}\right)^{k-v} \binom{x}{v}. \tag{2.166}$$

This implies that $\Delta^v \xi_k(0) = (\frac{1}{2})^{k-v}$. *Consequently, (2.141) and (2.145) with* $n = 2$ *yield the following special formulas*

$$S(m, [\xi_k]^n) = \left(\sum_{v=0}^{k} \left(-\frac{1}{2}\right)^v \Delta^{k-v}\right)^n \binom{x}{m}_{x=m+n-1} \tag{2.167}$$

and

$$\sum_{(m;n;0)} \xi_p(x_1)\xi_q(x_2) = \left(\sum_{v=0}^{p} \left(-\frac{1}{2}\right)^v \Delta^{p-v}\right)\left(\sum_{\mu=0}^{q} (-\frac{1}{2})^\mu \Delta^{q-\mu}\right)\binom{x}{m}_{x=m+1}, \tag{2.168}$$

respectively.

Example 2.3.13 *Let us consider the Mittag-Leffler polynomials defined by the power-type generating function [21]*

$$\left(\frac{1+t}{1-t}\right)^x = \left(1+\sum_{n=1}^{\infty} 2t^n\right)^x = \sum_{k=0}^{\infty}(ML)_k(x)\cdot t^k \tag{2.169}$$

where $(ML)_0(x) = 1$ *and* $(ML)_k(x)$ *is of degree of* k..

Instead of $(m;n;0)$, *we shall use the set* $(m;n;\mathbf{v};0)$ $(\mathbf{v} = (v_1, v_2, \ldots, v_n))$ *consisting of all the non-negative integer solutions of the equation* $v_1 + \cdots + v_n = m$. *It is easily seen that for any given set of real numbers* $\{x_1, \ldots, x_n\}$, *there holds the rather simple convolution sum*

$$\sum_{(m;n;\mathbf{v};0)} (ML)_{v_1}(x_1)\cdots(ML)_{v_n}(x_n) = (ML)_m(x_1+\cdots+x_n). \tag{2.170}$$

Actually this follows from the expansion of $((1+t)/(1-t))^{x_1+\cdots+x_n}$ *in terms of* t^m, *and may be called the convolution "in degrees".*

On the other hand, the summation (with fixed $k \geq 1$*)*

$$S(m,[(ML)_k]^n) = \sum_{(m;n;0)} (ML)_k(x_1)\cdots(ML)_k(x_n) \tag{2.171}$$

should be properly called the convolution "in arguments". Let us now evaluate (2.171) and the following sum

$$S(m,[(ML)_p][(ML)_q]) = \sum_{(m;n;0)} (ML)_p(x_1)\cdot(ML)_q(x_2) \tag{2.172}$$

by means of (2.141) and (2.145) with $n = 2$. *First, using the extracting-coefficient operator* $[t^k]$, *we find*

$$(ML)_k(x) = [t^k]\left(\frac{1+t}{1-t}\right)^x = [t^k]\left(1+\frac{2t}{1-t}\right)^x = \sum_{v\geq 0}\binom{x}{v}[t^k]\left(\frac{2t}{1-t}\right)^v$$

$$= \sum_{v\geq 0} 2^v\binom{x}{v}[t^{k-v}]\sum_{j\geq 0}\binom{v+j-1}{j}t^j = \sum_{v=1}^{k} 2^v\binom{k-1}{v-1}\binom{x}{v}.$$

Consequently, we have $\Delta^v(ML)_k(0) = 2^v\binom{k-1}{v-1}$ *and we get*

$$\textit{RHS of } (2.171) = \left(\sum_{v=1}^{k} 2^v\binom{k-1}{v-1}\Delta^v\right)^n\binom{x}{m}_{x=m+n-1} \tag{2.173}$$

and

$$\textit{RHS of } (2.172) = \left(\sum_{v=1}^{p} 2^v\binom{p-1}{v-1}\Delta^v\right)\left(\sum_{\mu=1}^{q} 2\mu\binom{q-1}{\mu-1}\Delta^\mu\right)\binom{x}{m}_{x=m+1}. \tag{2.174}$$

2.3.3 Convolution of polynomials and two types of summations

Generally, a sequence $(f_k(x))_{k\geq 0}$ of polynomials with $f_0(x) = 1$ and $\partial f_k(x) = k$ $(k = 0, 1, 2, \ldots)$, is called a convolution polynomial sequence, if there holds the convolution identity

$$\sum_{k=0}^{n} f_k(x) f_{n-k}(y) = f_n(x+y), \quad n = 0, 1, 2, \ldots.$$

Of course, this identity implies the multifold convolution in degrees

$$\sum_{(m;n;\mathbf{v};0)} f_{v_1}(x_1) \cdots f_{v_n}(x_n) = f_m(x_1 + \cdots + x_n), \tag{2.175}$$

Apparently, the sequence $((ML)_k(x))$ gives a special example.

It is known that there are various noticeable properties enjoyed by convolution polynomials. For details the reader is referred to Knuth's fundamental paper [144]. In what follows we will show that, for any given convolution polynomial sequence, there exist summation formulas for multifold convolutions in arguments.

Note that convolution polynomials can always be generated by power-type generating functions. Let $\phi(t) = 1 + a_1 t + a_2 t^2 + \cdots$ be a formal power series over the real or complex number field. Then the formal series expansion

$$(\phi(t))^x = \sum_{k=0}^{\infty} f_k(x) t^k =: \sum_{k=0}^{\infty} \begin{bmatrix} x \\ k \end{bmatrix}_\phi t^k \tag{2.176}$$

yields the convolution polynomials $f_k(x) = \left[\begin{smallmatrix} x \\ k \end{smallmatrix}\right]_\phi (k = 0, 1, 2, \ldots)$. Here we adopt the notation $\left[\begin{smallmatrix} x \\ k \end{smallmatrix}\right]_\phi$ just for expressiveness. Thus, for instances we have the special convolution polynomials:

$$\begin{bmatrix} x \\ k \end{bmatrix}_{1+t} = \binom{x}{k}, \quad \begin{bmatrix} x \\ k \end{bmatrix}_{(1-t)^{-1}} = \binom{x+k-1}{k}, \quad \begin{bmatrix} x \\ k \end{bmatrix}_{e^t} = \frac{x^k}{k!},$$
$$\begin{bmatrix} x \\ k \end{bmatrix}_{(1+t)/(1-t)} = (ML)_k(x), \quad \begin{bmatrix} x \\ k \end{bmatrix}_{exp(e^t-1)} = T_k(x),$$

where $T_k(x)$ are known as Touchard's polynomials.

Certainly (2.175) may be rewritten in the form

$$\sum_{(m;n;\mathbf{v};0)} \begin{bmatrix} x_1 \\ v_1 \end{bmatrix}_\phi \cdots \begin{bmatrix} x_n \\ v_n \end{bmatrix}_\phi = \begin{bmatrix} x_1 + \cdots + x_n \\ m \end{bmatrix}_\phi. \tag{2.177}$$

For given non-negative integers $k_1, \ldots, k_n$, we want to evaluate the multifold convolution in arguments:

$$S\left(m, \begin{bmatrix} x \\ k_1 \end{bmatrix}_\phi \cdots \begin{bmatrix} x \\ k_n \end{bmatrix}_\phi\right) = \sum_{(m;n;0)} \begin{bmatrix} x_1 \\ k_1 \end{bmatrix}_\phi \cdots \begin{bmatrix} x_n \\ k_n \end{bmatrix}_\phi.$$

Denote $\phi_1(t) = \phi(t) - 1 = \sum_{i \geq 1} a_i t^i$ with $a_1 \neq 0$. Then we have

$$\begin{aligned} \begin{bmatrix} x \\ k \end{bmatrix}_\phi &= [t^k](\phi(t_1))^x = [t^k](1 + \phi_1(t))^x \\ &= \sum_{j=0}^{k} \binom{x}{j} [t^k](\phi_1(t))^j =: \sum_{j=0}^{k} s(k, j, \phi_1) \binom{x}{j}. \end{aligned} \tag{2.178}$$

Here the numbers $s(k, j, \phi_1)$ defined by

$$s(k, j, \phi_1) = [t^k](\phi_1(t))^j, \quad 0 \leq j \leq k \tag{2.179}$$

form a simple special Riordan matrix (cf. Definition 3.2.3 and Section 4.3) whose elements may be called modified Stirling-type numbers, since $(k!/j!)s(k, j, \phi_1)$ just give the two kinds of ordinary Stirling numbers by taking $\phi_1(t) = log(1 + t)$ and $\phi_1(t) = e^t - 1$, viz.

$$\frac{k!}{j!} s(k, j, \log(1 + t)) = S_1(k, j), \quad \frac{k!}{j!} s(k, j, e^t - 1) = S_2(k, j) = \left\{ \begin{matrix} k \\ j \end{matrix} \right\}.$$

We may now state the following theorem.

Theorem 2.3.14 *There holds a summation formula for the convolution in arguments of the form*

$$S\left(m, \begin{bmatrix} x \\ k_1 \end{bmatrix}_\phi \cdots \begin{bmatrix} x \\ k_n \end{bmatrix}_\phi\right) = \left(\prod_{i=1}^{n} \left(\sum_{j=0}^{k_i} s(k_i, j, \phi_1) \Delta^j\right)\right) \binom{x}{m}_{x=m+n-1} \tag{2.180}$$

where $s(k, j, \phi_1)$ are given by (2.179).

Proof. From (2.178) we see that

$$\left(\Delta^j \begin{bmatrix} x \\ k \end{bmatrix}_\phi\right)_{x=0} = s(k, j, \phi_1).$$

Thus (2.180) follows from (2.141) as a consequence.

Example 2.3.15 *As is known, Touchard's polynomials are given by generating function [21]*

$$e^{x(e^t - 1)} = \sum_{k=0}^{\infty} T_k(x) t^k. \tag{2.181}$$

Moreover, $T_k(x)$ has an explicit expression

$$T_k(x) = \frac{1}{k!}\sum_{j=0}^{k}\left\{ {k \atop j} \right\} x^j. \tag{2.182}$$

Accordingly, we get the values of differences at zero

$$\Delta^v T_k(0) = (\Delta^v T_k(x))_{x=0} = \frac{v!}{k!}\sum_{j=v}^{k}\left\{ {k \atop j} \right\}\left\{ {j \atop v} \right\}. \tag{2.183}$$

. Let us denote $\delta(k,v) := \Delta^v T_k(0)$. Then using (2.180) we obtain

$$S(m,[T_{k_1}]\cdots[T_{k_n}]) = \left(\prod_{i=1}^{n}\left(\sum_{v=0}^{k_i}\delta(k_i,v)\Delta^v\right)\right)\binom{x}{m}_{x=m+n-1}. \tag{2.184}$$

In particular we have

$$S(m),[T_k]^n) = \left(\sum_{v=0}^{k}\delta(k,v)\Delta^v\right)^n \binom{x}{m}_{x=m+n-1}. \tag{2.185}$$

Certainly, (2.184) and (2.185) could be used to get the exact numerical results whenever $k_1,\ldots,k_n,k$, and m are given concretely.

Remark 2.3.16 *Convolutions of polynomials "in degrees" and "in arguments" may be called two types of convolutions. Note that the convolution in degrees can only be obtained from convolution polynomials (by definition). Thus one may infer from Theorem 2.3.14 that only the class of convolution polynomials could lead to the two types of convolutions which are both computable with really available summation formulas. Also, one may guess that both (2.141) and (2.180) could be extended to the cases of $q-$polynomials.*

2.3.4 Multifold Convolutions

What is worth commenting on is that summation formulas using operators (such as (2.141), (2.145), (2.148) etc.) should be the most available formulas for the practical computation of polynomial convolutions.

Usually, a good summation formula is such a formula that consists of a least number of easily computed terms. From this view-point, one may find that (2.151) and 2.152) could be regarded as good formulas, each consisting of only 4 terms. As a matter of fact, if the summation formula (2.148) is replaced by the equivalent formula

$$\sum_{(m;n;1)} x_1^{p_1}\cdots x_n^{p_n} = \sum_{\substack{1\le v_i\le p_i\\ 1\le i\le n}} v_1!\cdots v_n!\left\{ {p_1 \atop v_1} \right\}\cdots\left\{ {p_n \atop v_n} \right\}\binom{m+n-1}{m-v_1-\cdots-v_n}$$

without using $\Delta-$operators, and if it is applied to the sums of (2.151) and 2.152), one will get particular formulas, each consisting of 2 x 3 x 4 = 24 terms (since $p_1 = 1, p_2 = 2, p_3 = 3, p_4 = 4$). Thus, (2.151) and (2.152) just provide examples showing that the operator summation formulas such as (2.141) etc., may sometimes lead to much briefer formulas. Here the real reason is that the products of $\Lambda(\Delta, f_1)\cdots\Lambda(\Delta, f_n)$ or the like may sometimes be reduced to rather simple forms via the algebraic manipulations.

It is known that there is a general formula for expressing a multifold convolution of arbitrary real-valued functions defined on the set on non-negative integers. The relationship between multifold convolutions and partition sums can be derived and used for generating a type combinatorial series and identities.

Denote by $\sigma(n)$ the set of partition of $n \in \mathbb{N}$, usually written as $1^{k_1}2^{k_2}\cdots n^{k_n}$ referring to $k_1 + 2k_2 + \cdots + nk_n = n$, $k_i \in \mathbb{N}$. A subset of $\sigma(n)$, denoted by $\sigma(n,k)$, consists the partitions of n with k parts, i.e., partitions $1^{k_1}2^{k_2}\cdots n^{k_n}$ subject to $k_1 + k_2 + \cdots + k_n = k$. Thus

$$\sigma(n,k) := \left\{(x_1, x_2, \ldots, x_k) : \sum_{i=1}^{n} ix_i = n, \sum_{i=1}^{n} x_i = k, x_i \geq 0\right\}.$$

We now define fold convolutions from (2.137) for the same functions $f_i(x)$.

Definition 2.3.17 *Let $f(x)$ be a real-valued or complex-valued function defined on $\mathbb{N}_0$ with $f(0) = 1$. Then the k-fold convolution and the n/k-partition sum associated with $f(x)$ are respectively defined by the following summations:*

$$S_n^k(f) = \sum_{(n;k;0)} f(x_1)f(x_2)\cdots f(x_k), \tag{2.186}$$

$$T_n^k(f) = \sum_{\sigma(n,k)} \frac{f^{k_1}(1)f^{k_2}(2)\cdots f^{k_n}(n)}{k_1!k_2!\cdots k_n!}, \tag{2.187}$$

where the sums on the right-hand sides of (2.186) *and* (2.187) *range over the sets $(n;k;0)$ and $\sigma(n,k)$, respectively.*

Note that $(n;k;0)$ and $\sigma(n,k)$ have no meaning for $k = 0$. For convenience, we define $S_n^0(f) = T_n^0(f) = 0$, and thereby have sequences $(S_n^k(f))_{n,k\geq 0}$ and $(T_n^k(f))_{n,k\geq 0}$. Also it is obvious that $S_n^1(f) = f(n)$, $S_1^k(f) = kf(1)$ and $T_n^1(f) = f(n)$ for $n \geq 1$ and $T_1^k(f) = 0$ for $k > 1$. Moreover, it is easy to see that the RHS of (2.187) with replacement $f(n) \to t_n (n \in \mathbb{N})$ is in agreement with the incomplete Bell polynomial in t_i $(1 \leq i \leq n)$, up to a constant factor $n!$ (*cf.* Comtet [44]).

In what follows we assume that $G = \sum_{k\geq 0} g(k)t^k \in \mathbb{C}[[t]]$, the ring of formal power series in the complex field. As usual, $G^{(k)}(t) = D^kG(t)$ denotes the kth formal derivative of $G(t)$.

Theorem 2.3.18 *For $m, n \in \mathbb{N}$, we have the following identities:*

$$S_n^m(f) = \sum_{k=1}^{m} (m)_k T_n^k(f), \tag{2.188}$$

$$T_n^k(f) = \frac{1}{k!} \Delta^k S_n^t(f)\big|_{t=0}, \tag{2.189}$$

$$\sum_{k=1}^{\infty} g(k) S_n^k(f) t^k = \sum_{k=1}^{n} G^{(k)}(t) T_n^k(f) t^k, \tag{2.190}$$

where the left-hand side of (2.190) *is a formal power series.*

Proof. To justify (2.188), according to Definition 2.3.17, we only need to compute the LHS of 2.188 in this way: for $1 \le k \le m$, consider first the finite sum

$$\sum_{(n;m;0)_k} f(x_1) f(x_2) \cdots f(x_m), \tag{2.191}$$

where $(n, m, 0)_k$ denotes the subset of $(n, m, 0)$, being composed of all compositions $(x_1, x_2, \ldots, x_m)$ of n with just k components $x_i \ge 1$. In other words, there are $m - k$ components $x_i = 0$ in $(x_1, x_2, \cdots, x_m)$. Recall that $f(0) = 1$ and such factors will take $m - k$ ordered places in $\binom{m}{m-k} = \binom{m}{k}$ different ways. Meanwhile, the number of all possible permutations of the factors in the product $f^{k_1}(1) f^{k_2}(2) \cdots f^{k_n}(n)$ over the set $\sigma(n, k)$ is enumerated by $k!/(k_1! k_2! \cdots k_n!)$. Thus the sum (2.191) boils down to

$$\binom{m}{k} \sum_{\sigma(n,k)} \frac{k!}{k_1! k_2! \ldots k_n!} f^{k_1}(1) f^{k_2}(2) \cdots f^{k_n}(n) = (m)_k T_n^k(f).$$

Summing on k, $1 \le k \le m$, we therefore obtain (2.188).

With (2.188) we are able to reformulate $S_n^t(f)$ in the form

$$S_n^t(f) = \sum_{j \ge 1} j! \binom{t}{j} T_n^j(f), \quad S_n^0(f) = 0.$$

Then it follows that

$$\Delta^k S_n^t(f)\big|_{t=0} = \sum_{j \ge 1} j!\, \Delta^k \binom{t}{j}\bigg|_{t=0} T_n^j(f)$$

$$= \sum_{j \ge 1} j! \binom{0}{j-k} T_n^j(f) = k! T_n^k(f).$$

Thus (2.189) is proved.

Once again, by using (2.188) we may compute the LHS of (2.190) formally as follows:

$$\sum_{m=1}^{\infty} g(m) S_n^m(f) t^m = \sum_{m=1}^{\infty} g(m) t^m \sum_{k=1}^{m} (m)_k T_n^k(f)$$

$$= \sum_{k=1}^{\infty}\left(\sum_{m=k}^{\infty}(m)_k g(m)t^m\right)T_n^k(f)$$

$$= \sum_{k=1}^{\infty}D^k\left(\sum_{m=k}^{\infty}g(m)t^m\right)T_n^k(f)t^k = \sum_{k=1}^{n}G^{(k)}(t)T_n^k(f)t^k.$$

The last equality follows from the fact that $T_n^k(f) = 0$ for $k > n$. This completes the proof of (2.190).

Put $G(t) = 1/(1-t)$ for $|t| < 1$ and e^t for $t \in \mathbb{R}$ in (2.190) in succession. Then we may get the following result.

Corollary 2.3.19 *For $n \in \mathbb{N}$, the sequence $(S_n^k(f))_{k\geq 0}$ has the ordinary and exponential generating functions, respectively, as follows:*

$$\sum_{k=0}^{\infty}S_n^k(f)t^k = \sum_{k=1}^{n}\frac{k!}{1-t}\left(\frac{t}{1-t}\right)^k T_n^k(f), \tag{2.192}$$

$$\sum_{k=0}^{\infty}S_n^k(f)\frac{t^k}{k!} = e^t\sum_{k=1}^{n}T_n^k(f)t^k. \tag{2.193}$$

Remark 2.3.20 *In the series-transformation formula* (2.96),

$$\sum_{k=0}^{\infty}f(k)g^{(k)}(0)\frac{x^k}{k!} = \sum_{k=0}^{\infty}\Delta^k f(0)g^{(k)}(x)\frac{x^k}{k!}$$

by substituting $f(t) = S_n^t(f)$ and $g(t) = G(t)$ and simplifying the resulting identity, we may also obtain (2.190).

We now consider the convergence of (2.190).

Theorem 2.3.21 *If $G(t) = \sum_{k\geq 0}g(k)t^k$ is absolutely convergent for $|t| < r$, then so is the infinite series on the LHS of* (2.190), *thereby* (2.190) *is an exact formula for $|t| < r$.*

Proof. Comparing the LHS of (2.190) with $G(t)$ and using Cauchy's root test for the convergence of infinite series, we only need to show that for $n \in \mathbb{N}$,

$$\overline{\lim}_{k\to\infty}|S_n^k(f)|^{1/k} \leq 1.$$

To this end, assume $\max\limits_{0\leq x\leq n}|f(x)| = \rho$. By the definition of $S_n^k(f)$, it is easily found that every product $f(x_1)f(x_2)\cdots f(x_k)$ restricted by

$$x_1 + x_2 + \cdots + x_k = n \tag{2.194}$$

contains at most n factors $f(x_i)$ with $x_i \geq 1$. Owing to the facts that $f(0) = 1$ and $|f(x)| < \rho$ for $x \neq 0$, it follows directly that $|f(x_1)f(x_2)\cdots f(x_k)| \leq \rho^n$.

Meanwhile, the total number of such terms is $\binom{n+k-1}{n}$, a fact coming from the number of solutions in non-negative integers for the Diophantine equation (2.194). Thus we have

$$|S_n^k(f)| \le \binom{n+k-1}{n} \rho^n,$$

leading to

$$\overline{\lim_{k\to\infty}} |S_n^k(f)|^{1/k} \le \overline{\lim_{k\to\infty}} \left(\binom{n+k-1}{n} \rho^n \right)^{1/k} = 1.$$

Hence the theorem is proved.

Corollary 2.3.22 *Identity (2.192) is analytic for $|t| < 1$ and so is (2.193) for $|t| < +\infty$.*

Example 2.3.23 *Evidently, the following power series*

$$G(t) = \frac{1}{\sqrt{1-4t}} = \sum_{k=0}^{\infty} \binom{2k}{k} t^k$$

converges absolutely for $|t| < 1/4$. As a consequence of (2.190) (see Exercise 2.8), we get the following exact formula

$$\sum_{k=0}^{\infty} \binom{2k}{k} S_n^k(f) t^k = \sum_{k=1}^{n} \frac{(2k)!}{k!} \frac{t^k}{(1-4t)^{k+1/2}} T_n^k(f) \tag{2.195}$$

for $|t| < 1/4$.

Example 2.3.24 *It is well known that the Fibonacci number sequence $(F_n)_{n\ge 0}$ is generated by the generating function*

$$G(t) = \frac{1}{1-t-t^2} = \sum_{k=0}^{\infty} F_k t^k,$$

which is convergent absolutely for $|t| < (\sqrt{5}-1)/2$. To ease notation, we let $a = (1+\sqrt{5})/2$ and $b = (1-\sqrt{5})/2$ and therefore have (see Exercise 2.9)

$$G(t) = \frac{1}{(1-at)(1-bt)} = \frac{1}{\sqrt{5}} \sum_{k=0}^{\infty} (a^{k+1} - b^{k+1}) t^k$$

and $g(k) = F_k = (a^{k+1} - b^{k+1})/\sqrt{5}$. Then from (2.190) we have

$$\sum_{k=0}^{\infty} F_k S_n^k(f) t^k = \sum_{k=1}^{n} \frac{k!}{\sqrt{5}} \left[\left(\frac{a}{1-at} \right)^{k+1} - \left(\frac{b}{1-bt} \right)^{k+1} \right] T_n^k(f) t^k, \tag{2.196}$$

which is an exact formula for $|t| < (\sqrt{5}-1)/2$.

Example 2.3.25 *Setting $G(t) = (1+t)^\alpha$ for $\alpha \in \mathbb{R}$ and $|t| < 1$ and substituting it into (2.190) yields (see Exercise 2.10)*

$$\sum_{k=0}^{\infty} \binom{\alpha}{k} S_n^k(f) t^k = \sum_{k=1}^{n} (\alpha)_k \frac{t^k}{(1+t)^{k-\alpha}} T_n^k(f). \tag{2.197}$$

The further substitution $\alpha \to -\alpha - 1$ leads us to

$$\sum_{k=0}^{\infty} \binom{\alpha+k}{k} S_n^k(f)(-t)^k = \sum_{k=1}^{n} (\alpha+k)_k \frac{(-t)^k}{(1+t)^{k+\alpha+1}} T_n^k(f). \tag{2.198}$$

Obviously, both (2.197) and (2.198) are valid for $|t| < 1$. It is also clear that (2.192) can be deduced from (2.197) via the substitutions $\alpha \to -1$ and $t \to -t$.

2.3.5 Some operator summation formulas from multifold convolutions

It is known that both D and Δ are delta operators so that by use of Mullin-Rota's substitution rule (cf. Section 3.1), we may deduce some special operator summation formulas from (2.193), (2.195), (2.197), and (2.198), respectively.

For any function $\phi(t)$ over $\mathbb{R}$, it is easy to check that

$$(1+\Delta)^\alpha \phi(t) = E^\alpha \phi(t) = \phi(t+\alpha).$$

Thus, under the substitution $t \to \Delta$, we see that (2.197) and (2.198) together yields a pair of Δ-type operator summation formulas. The results are as follows:

$$\sum_{k=0}^{\infty} \binom{\alpha}{k} S_n^k(f) \Delta^k \phi(0) = \sum_{k=1}^{n} (\alpha)_k T_n^k(f) \Delta^k \phi(\alpha - k), \tag{2.199}$$

$$\sum_{k=0}^{\infty} \binom{\alpha+k}{k} S_n^k(f)(-\Delta)^k \phi(0) = \sum_{k=1}^{n} (\alpha+k)_k T_n^k(f)(-\Delta)^k \phi(-\alpha - k - 1). \tag{2.200}$$

Alternatively, the substitution $t \to -\frac{1}{4}\Delta$ reduces (2.195) to a Δ-type summation formula of the form

$$\sum_{k=0}^{\infty} \frac{(-1)^k}{2^{2k}} \binom{2k}{k} S_n^k(f) \Delta^k \phi(0) = \sum_{k=1}^{n} \frac{(-1)^k (2k)!}{2^{2k} k!} T_n^k(f) \Delta^k \phi(-k - 1/2) \tag{2.201}$$

As above, substituting t by D in (2.193) and simplifying the result by the relations that $e^D = E = 1 + \Delta$, we come up with a D-type summation formula for $\phi(t) \in C^\infty$ (the set of infinitely differentiable real functions over $\mathbb{R}$) evaluated at $t = 0$:

$$\sum_{k=0}^{\infty} \frac{1}{k!} S_n^k(f) D^k \phi(0) = \sum_{k=1}^{n} T_n^k(f) D^k \phi(1). \tag{2.202}$$

In accordance with Example 2.3.25 in Subsection 2.3.4, it is clear that (2.199) and (2.200) are exact formulas under the condition

$$\overline{\lim_{k\to\infty}}\left|\Delta^k\phi(0)\right|^{1/k} < 1. \tag{2.203}$$

Actually condition (2.203) also ensures the validity of formula (2.201) inasmuch as it is just equivalent to the condition (see Example 2.3.23)

$$\overline{\lim_{k\to\infty}}\left|\frac{1}{4^k}\Delta^k\phi(0)\right|^{1/k} < \frac{1}{4}.$$

Moreover, Corollary 2.3.22 states that (2.202) is an exact formula under the condition

$$\overline{\lim_{k\to\infty}}\left|D^k\phi(0)\right|^{1/k} < +\infty. \tag{2.204}$$

It may happen that $\lim_{k\to\infty}\left|\Delta^k\phi(0)\right|^{1/k} = 1$. As such, the convergence conditions for (2.199), (2.200), and (2.201) should be investigated separately.

As one might expect, all formulas from (2.199) to (2.202) can be employed to produce various special formulas and identities, because that $f(x)$ and $\phi(t)$ are free to choose. In what follows we will detail how to find concrete identities by considering some interesting examples.

Example 2.3.26 *Substitute $\phi(t)$ by $\phi_1(t) = \binom{t+\beta}{m}$ and $\phi_2(t) = 1/(t+\beta)$ in turn, $m \in \mathbb{N}_+$ and $\beta > 0$. Then we have*

$$\Delta^k\phi_1(t) = \binom{t+\beta}{m-k}, \quad \Delta^k\phi_2(t) = \frac{(-1)^k}{t+\beta}\binom{t+\beta+k}{k}^{-1}.$$

From now on, we write briefly $\binom{t+\beta+k}{k}^{-1}$ for $1/\binom{t+\beta+k}{k}$. Under these two choices, it is easy to deduce the following six formulas, respectively from (2.199), (2.200), and (2.201):

$$\sum_{k=0}^{m}\binom{\alpha}{k}\binom{\beta}{m-k}S_n^k(f) = \sum_{k=1}^{n}(\alpha)_k\binom{\alpha+\beta-k}{m-k}T_n^k(f), \tag{2.205}$$

$$\sum_{k=0}^{m}(-1)^k\binom{\alpha+k}{k}\binom{\beta}{m-k}S_n^k(f) = \sum_{k=1}^{n}(-1)^k(\alpha+k)_k\binom{\beta-\alpha-k-1}{m-k}T_n^k(f), \tag{2.206}$$

$$\sum_{k=0}^{m}\frac{(-1)^k}{4^k}\binom{2k}{k}\binom{\beta}{m-k}S_n^k(f) = \sum_{k=1}^{n}\frac{(-1)^k(2k)!}{4^k k!}\binom{\beta-k-1/2}{m-k}T_n^k(f), \tag{2.207}$$

$$\sum_{k=0}^{\infty}(-1)^k\binom{\alpha}{k}\binom{\beta+k}{k}^{-1}S_n^k(f)=\beta\sum_{k=1}^{n}\frac{(-1)^k(\alpha)_k}{\alpha+\beta-k}\binom{\alpha+\beta}{k}^{-1}T_n^k(f),\tag{2.208}$$

$$\sum_{k=0}^{\infty}\binom{\alpha+k}{k}\binom{\beta+k}{k}^{-1}S_n^k(f)=\beta\sum_{k=1}^{n}\frac{(\alpha+k)_k}{\beta-\alpha-k-1}\binom{\beta-\alpha-1}{k}^{-1}T_n^k(f),\tag{2.209}$$

$$\sum_{k=0}^{\infty}\frac{1}{4^k}\binom{2k}{k}\binom{\beta+k}{k}^{-1}S_n^k(f)=\beta\sum_{k=1}^{n}\frac{(2k)!}{4^k k!(\beta-k-1/2)}\binom{\beta-1/2}{k}T_n^k(f).\tag{2.210}$$

We remark that all infinite series involved in (2.208), (2.209), and (2.210) are assumed to be convergent under suitable conditions for α,β *and* $S_n^k(f)$.

Example 2.3.27 *The most simple case of* $S_n^k(f)$ *is when* $n=1$ *with* $f(1)=1$. *In such a case, it is easy to check that*

$$S_1^k(f)=k, T_1^1(f)=1, T_1^k(f)=0 \text{ for } k>1.$$

Consequently, each identity from (2.205) to (2.210) yields correspondingly a special identity for $n=1$. *The results are stated as follows:*

$$\sum_{k=0}^{m}k\binom{\alpha}{k}\binom{\beta}{m-k}=\alpha\binom{\alpha+\beta-1}{m-1}\quad (\textit{Vandermonde}),\tag{2.211}$$

$$\sum_{k=0}^{m}(-1)^k k\binom{\alpha}{k}\binom{\beta}{m-k}=-(\alpha+1)\binom{\beta-\alpha-2}{m-1},\tag{2.212}$$

$$\sum_{k=0}^{m}\frac{(-1)^k k}{4^k}\binom{2k}{k}\binom{\beta}{m-k}=-\frac{1}{2}\binom{\beta-3/2}{m-1},\tag{2.213}$$

$$\sum_{k=0}^{\infty}(-1)^k k\binom{\alpha}{k}\binom{\beta+k}{k}^{-1}=-\frac{\alpha\beta}{(\beta+\alpha)(\alpha+\beta-1)},\tag{2.214}$$

$$\sum_{k=0}^{\infty}k\binom{\alpha+k}{k}\binom{\beta+k}{k}^{-1}=\frac{(1+\alpha)\beta}{(\beta-\alpha-1)(\beta-\alpha-2)},\tag{2.215}$$

$$\sum_{k=0}^{\infty}\frac{k}{4^k}\binom{2k}{k}\binom{\beta+k}{k}^{-1}=\frac{\beta}{2(\beta-1/2)(\beta-3/2)}.\tag{2.216}$$

Note that for $\alpha>0$ *and* $\beta>0$ *we have the estimates*

$$\left|\binom{\alpha}{k}\right|\leq\frac{\lfloor\alpha\rfloor+1}{k}=O(1/k),\ \binom{2k}{k}/4^k\sim\frac{1}{\sqrt{k\pi}}=O(1/\sqrt{k})$$

and

$$\binom{\beta+k}{k}\geq\binom{k+\lfloor\beta\rfloor}{\lfloor\beta\rfloor}=O(k^{\lfloor\beta\rfloor})\quad(k\to\infty),$$

where $|\binom{\alpha}{k}|$ *is the absolute value of* $\binom{\alpha}{k}$, $\lfloor x \rfloor$ *denotes the largest integer not greater than* x, *the big O notation takes the usual meaning of asymptotic. Thus all infinite series appearing in (2.214), (2.215), and (2.216) are absolutely convergent under their respective conditions, i.e.,*

$$(2.214) : \alpha > 0, \beta \geq 2; \quad (2.215) : \alpha > 0, \beta \geq \alpha + 3; \quad (2.216) : \beta \geq 2.$$

Both (2.214) and (2.215) are comparable with the series

$$\sum_{k=0}^{\infty} (-1)^k \binom{\alpha}{k} \binom{\beta + k}{k}^{-1} = \frac{\beta}{\alpha + \beta} \tag{2.217}$$

for $\alpha > 0, \beta \geq 1$. *It is of interest to note that the special case of (2.217) when* $\alpha > 0$ *is integer is recorded as a "theorem" in Wilf [216, p.134, Theorem], being employed as an example of the well-known W-Z algorithm (or WZ method). Thus we believe that (2.214), (2.215), and (2.216) may also be verified by means of the W-Z algorithm (cf. Subsubsection 5.3.2.2).*

Example 2.3.28 *By the multivariate Vandermonde convolution formula it is easily found that for* $\alpha \in \mathbb{R}$ *and* $r \in \mathbb{N}$, *there hold*

$$\begin{aligned} \sum_{[n,k,0]} \binom{\alpha}{x_1} \binom{\alpha}{x_2} \cdots \binom{\alpha}{x_k} &= \binom{k\alpha}{n}, \\ \sum_{[n,k,0]} \binom{x_1}{r} \binom{x_2}{r} \cdots \binom{x_k}{r} &= \binom{n+k-1}{kr+k-1}. \end{aligned}$$

Now, replace $f(x)$ *with* $\binom{\alpha}{x}$ *and* $\binom{r+x}{r}$ *in turn. For both cases, we therefore obtain*

$$S_n^k\left(\binom{\alpha}{x}\right) = \binom{k\alpha}{n}, \; T_n^k\left(\binom{\alpha}{x}\right) = \sum_{\sigma(n,k)} \frac{\binom{\alpha}{1}^{k_1} \binom{\alpha}{2}^{k_2} \cdots \binom{\alpha}{n}^{k_n}}{k_1! k_2! \cdots k_n!}, \tag{2.218}$$

$$S_n^k\left(\binom{r+x}{r}\right) = \binom{k(r+1)+n-1}{n}, \tag{2.219}$$

$$T_n^k\left(\binom{r+x}{r}\right) = \sum_{\sigma(n,k)} \frac{\binom{r+1}{r}^{k_1} \binom{r+2}{r}^{k_2} \cdots \binom{r+n}{r}^{k_n}}{k_1! k_2! \cdots k_n!}. \tag{2.220}$$

Accordingly, (2.190) leads us to a pair of combinatorial series as follows:

$$\sum_{k=0}^{\infty} g(k) \binom{k\alpha}{n} t^k = \sum_{k=1}^{n} G^{(k)}(t) t^k T_n^k\left(\binom{\alpha}{x}\right), \tag{2.221}$$

$$\sum_{k=0}^{\infty} g(k) \binom{k(r+1)+n-1}{n} t^k = \sum_{k=1}^{n} G^{(k)}(t) t^k T_n^k\left(\binom{r+x}{r}\right). \tag{2.222}$$

As is expected to be, a variety of special identities can be deduced from (2.205) and (2.210) with $S_n^k(f)$ and $T_n^k(f)$ being replaced by those of (2.218) and (2.220). For instance, by virtue of (2.209) and (2.218), we may find a combinatorial series

$$\sum_{k=0}^{\infty}\binom{\alpha+k}{k}\binom{\beta+k}{k}^{-1}\binom{k\delta}{n}=\sum_{k=1}^{n}\frac{(\alpha+k)_k\beta}{\beta-\alpha-k-1}\binom{\beta-\alpha-1}{k}^{-1}T_n^k\left(\binom{\delta}{x}\right). \tag{2.223}$$

A bit analysis shows that this series is absolutely convergent under the sufficient condition $\alpha, \delta > 0, \beta \geq \alpha + n + 2$. Due to space limitations, other identities are left to the interested reader to work out.

Exercises

2.1 Prove Lemma 2.1.5.

2.2 Derive the second Gauss symbolic expression and the second Gauss interpolation formula.

2.3 Prove series transformation formulas (2.98)–(2.110).

2.4 Prove summation formulas (2.116)–(2.124).

2.5 Prove the expansion of Bernoulli polynomial $B_k(x)$ shown in (2.159)

2.6 Prove the convolution sum (2.170).

2.7 Show $\overline{\lim}_{k\to\infty}\left(\binom{n+k-1}{n}\rho^n\right)^{1/k}=1$.

2.8 Accomplish Example 2.3.23 by using the hint:

$$G^{(k)}(t)=\frac{(2k)!}{k!}(1-4t)^{-k-1/2},\quad g(k)=\binom{2k}{k}.$$

2.9 Accomplish Example 2.3.24 by using simple computation

$$G^{(k)}(t)=\frac{1}{\sqrt{5}}D^k\left(\frac{a}{1-at}-\frac{b}{1-bt}\right)=\frac{k!}{\sqrt{5}}\left[\left(\frac{a}{1-at}\right)^{k+1}-\left(\frac{b}{1-bt}\right)^{k+1}\right].$$

2.10 Accomplish Example 2.3.25 by noticing for $G(t)=(1+t)^\alpha$ $g(k)=\binom{\alpha}{k}$ and $G^{(k)}(t)=(\alpha)_k(1+t)^{\alpha-k}$.

2.11 Prove formulas (2.205)–(2.210).

2.12 Prove formulas (2.211)–(2.216).

2.13 Prove formulas (2.218)–(2.223).

3

Source Formulas for Symbolic Methods

CONTENTS

It is known that the symbolic calculus with operators Δ (differencing), E (operation of displacement), and D (derivative) plays an important role in the Calculus of Finite Differences, which is often employed by statisticians and numerical analysts. Various well-known results can be found in some classical treatises, e.g., those by Jordan [141], Milne-Thomson [164], and some materials represented in Chapter 2. Since all the symbolic expressions used and operated in the calculus could be formally expressed as power series in Δ (or D or E) over the real or complex number field, it is clear that the theoretical basis of the calculus may be found within the general theory of the formal power series. Worth reading is a sketch of the theory of formal series that has been given briefly in Comtet [44] (see §1.12 and § 3.2–§ 3.5) (cf. Bourbaki [23] Chap. 4-5).

In the first section of this chapter, we present a general substitution rule called Mullin-Rota's substitution rule. Given a generating function or a formal power series expansion, a certain operational formula may be obtained. Twelve generating functions for the well-known sequences will be displayed.

DOI: 10.1201/9781003051305-3

Then 12 operational formulas will be found by using the substitution rule. Finally, some operator summation formulas from multifold convolutions are constructed similarly. With the aid of Mullin-Rota's substitution rule, we shall show in the second section that the Sheffer-type differential operators together with the delta operators Δ and D can be used to construct a pair of expansion formulas that imply a wide variety of summation formulas in the discrete analysis and combinatorics. A convergence theorem is established for a fruitful source formula that implies more than 20 noted classical formulas and identities. Numerous new formulas as also represented as illustrative examples. Finally, it is shown that a kind of lifting process can be used to produce certain chains of (∞^m) degree formulas for $m \geq 3$ with $m \equiv 1 \pmod 2$ and $m \equiv 1 \pmod 3$, respectively. The second section of this chapter represents a further investigation on a general source formula (GSF) that has been proved capable of deducing more than 30 classical and new formulas for series expansions and summations besides those given in the previous section. It will be shown that the pair of series transformation formulas found and utilized in Section 2.2 is also deducible from the GSF as consequences. Thus it is found that the GSF actually implies more than 50 special series expansions and summation formulas. Finally, several expository remarks relating to the $(\Sigma\Delta D)$ formula class are given at the end of Section 3.3.

3.1 An Application of Mullin-Rota's Theory of Binomial Enumeration

The results shown in Section 2.1 can be considered as special cases of the results in this section (see Remark 3.1.3). We shall show that a variety of formulas and identities containing famous number sequences, namely Bell, Bernoulli, Euler, Fibonacci, Genocchi, and Stirling, could be quickly derived by using a symbolic method with operators Δ, E, and D. The key idea is a suitable application of a certain symbolic substitution rule to the generating functions for those number sequences so that a number of symbolic expressions could be obtained, which then can be used as stepping-stones to yielding particular formulas or identities of interest.

Frequently we shall get formulas and identities involving infinite series expansions. Certainly, any convergence problems, if involved in the results, should be treated separately.

3.1.1 A substitution rule and its scope of applications

As usual, we denote by C^∞ the class of real functions, infinitely differentiable in $\mathbb{R} = (-\infty, \infty)$. We will make frequent use of the operators Δ, E, and D

which are known to be defined for all $f \in C^\infty$ via the relations

$$\Delta f(t) = f(t+1) - f(t), \quad Ef(t) = f(t+1), \quad Df(t) = \frac{d}{dt} f(t).$$

Consequently, they satisfy some simple symbolic relations such as

$$E = 1 + \Delta, \quad E = e^D, \quad \Delta = e^D - 1, \quad D = log(1+\Delta), \tag{3.1}$$

where the unity 1 serves as an identity operator I such that $If(t) = f(t) = 1f(t)$, and e^D and $log(1+D)$ are meaningful in the sense of formal power series expansions, namely

$$e^D = \sum_{k\geq 0} \frac{1}{k!} D^k, \qquad log(1+\Delta) = \sum_{k\geq 1} \frac{(-1)^{k-1}}{k} \Delta^k$$

so that $e^D f(t) = \sum_{k\geq 0} D^k f(t)/k! = f(t+1) = Ef(t)$, (cf. Jordan [141] and Section 2.1).

An operator T which commutes with the shift operator E is called a *shift-invariant operator* (cf. for example, [168]), i.e.,

$$TE^\alpha = E^\alpha T,$$

where $E^\alpha f(t) = f(t+\alpha)$ and $E^1 \equiv E$. Clearly, the identity operator 1, the differentiation operator D, and the difference operator Δ are all shift-invariant operators. A shift-invariant operator Q is called a *delta operator* if Qt is a non-zero constant. Obviously, the identity operator, the differentiation operator, the difference operator, and the backward difference, the central difference, Laguerre and Abel operators (cf. [168] and Section 2.1) are all delta operators.

Note that there are following two well-known operational formulas involving (signed) Stirling numbers of the first and second kinds, respectively, $(-1)^{n-m}\left[{n \atop m}\right]$ (or $S_1(n,m)$) and $\left\{{n \atop m}\right\}$ (or $S_2(n,m)$), where $\left[{n \atop m}\right]$ is known as the signless Stirling number of the first kind, which counts the number of ways to arrange n objects into m cycles instead of subsets, and $\left\{{n \atop m}\right\}$ counts the number of ways to partition a set of n objects into m non-empty subsets (cf. Remark 2.1.13).

$$D^m f(t) = \sum_{n\geq m} \frac{m!}{n!} (-1)^{n-m} \left[{n \atop m}\right] \Delta^n f(t) \tag{3.2}$$

$$\Delta^m f(t) = \sum_{n\geq m} \frac{m!}{n!} \left\{{n \atop m}\right\} D^n f(t). \tag{3.3}$$

These could be derived using the Newton interpolation series and Taylor series, respectively. (cf. [141] § 56 and § 67).

Certainly, according to (3.1), it is obvious that (3.2) and (3.3) may be viewed as direct consequences of the substitutions $t \to \Delta$ and $t \to D$ into the following generating functions, respectively

$$
\begin{aligned}
(log(1+t))^m &= \sum_{n \geq m} \frac{m!}{n!}(-1)^{n-m} \begin{bmatrix} n \\ m \end{bmatrix} t^n \\
\left(e^t - 1\right)^m &= \sum_{n \geq m} \frac{m!}{n!} \begin{Bmatrix} n \\ m \end{Bmatrix} t^n .
\end{aligned}
$$

Note that certain particular identities could be deduced from (3.2) and (3.3) with particular choices of $f(t)$ (cf. for example, [141], loc.cit).

The above description is an example of the following general substitution rule shown in Mullin and Rota [168] (cf. also in Loeb and Rota [150]).

Theorem 3.1.1 *[168] Let Q be a delta operator and let F be the ring of formal power series in the variable t, over the same field, then there exists an isomorphism from F onto the ring $\sum$ of shift-invariant operators, which carries*

$$f(t) = \sum_{k \geq 0} \frac{a_k}{k!} t^k \text{ into } f(Q) = \sum_{k \geq 0} \frac{a_k}{k!} Q^k .$$

From Theorem 3.1.1 we have the following general substitution rule for the formal power series expansion of the functions regarding e^t or $log(1+t)$.

Substitution Rule (w. r. t. operators D and Δ): Given a generating function or a formal power series expansion $F(t) = \sum_{k \geq 0} f_k t^k$, where $F(t)$ is expressed either in the form $G(t, e^t)$ or in the form $G(t, log(1+t))$, a certain operational formula may be obtained as follows.

(i) For the case $F(t) = G(t, e^t)$, the substitution $t \to D$ leads to the symbolic formula

$$F(D) = G(D, 1+\Delta) = \sum_{k \geq 0} f_n D^k . \tag{3.4}$$

(ii) For the case $F(t) = G(t, log(1+t))$, $t \to \Delta$ leads to the formula

$$F(\Delta) = G(\Delta, D) = \sum_{k \geq 0} f_k \Delta^k . \tag{3.5}$$

Of course, (3.4) and (3.5) can be deduced from (3.1). In what follows we display 12 generating functions for the sequences (W_k) (Bell numbers), $(B_k^{(n)})$ and $(B_k^{(-n)})$ (generalized Bernoulli numbers of the orders n and $-n$), $(E_k(t))$ (Euler polynomials), $\{e_k = E_k(0))$, $(\alpha_k(t))$ (Eulerian fractions) and (G_k) (Genocchi numbers), $(\phi_k(t))$ (Bernoulli polynomials of the first kind), $(\psi(t))$ (Bernoulli polynomials of the second kind), (b_k) (Bernoulli numbers of the second kind), respectively.

(G_1) $exp(e^t - 1) = \sum_{k\geq 0} \frac{1}{k!} W_k t^k$ (Comtet [44], p. 210),

(G_2) $\left(\frac{t}{e^t-1}\right)^n = \sum_{k\geq 0} \frac{1}{k!} B_k^{(n)} t^k$ (David-Barton [50], p. 287),

(G_3) $\left(\frac{e^t-1}{t}\right)^n = \sum_{k\geq 0} \frac{1}{k!} B_k^{(-n)} t^k$ (cf. [50], p. 287),

(G_4) $\frac{te^{xt}}{e^t-1} = \sum_{k\geq 0} \phi_k(x) t^k$, where $\phi_k(0) = B_k^{(1)}/k!$, (Jordan [141], p. 250),

(G_5) $\frac{2e^{xt}}{e^t+1} = \sum_{k\geq 0} E_k(x) t^k$ (Jordan [141], p. 309),

(G_6) $\frac{2}{e^t+1} = \sum_{k\geq 0} e_k t^k$, where $e_k = E_k(0)$, (cf. [141], p. 309),

(G_7) $\frac{1}{1-xe^t} = \sum_{k\geq 0} \frac{1}{k!} \alpha_k(x) t^k$ (Wang and Hsu [210], p.24),

(G_8) $\frac{2t}{e^t+1} = \sum_{k\geq 0} \frac{1}{k!} G_k t^k$ (Comtet [44], p. 49),

(G_9) $\frac{t(1+t)^x}{\log(1+t)} = \sum_{k\geq 0} \psi_k(x) t^k$ (Jordan [141], p. 279),

(G_{10}) $\frac{t}{\log(1+t)} = \sum_{k\geq 0} b_k t^k$, where $b_k = \psi_k(0)$, (cf. [141], p. 279),

(G_{11}) $\frac{1}{1-\Delta-\Delta^2} = \sum_{k\geq 0} F_k \Delta^k$, where $F_0 = F_1 = 1$ and $F_k = F_{k-1} + F_{k-2}$ for $k = 2, 3, \ldots$,

(G_{12}) $\frac{1}{(1-t)e^t} = \sum_{k\geq 0} \frac{\phi(0)}{k!} t^k$.

In comparison of (G_8) with (G_6), we see that Genocchi numbers G_{k+1} are equivalent to $(k+1)!e_k$ $(k = 0, 1, 2, \ldots)$, with $e_k = E_k(0)$. It is also known that $G_{2m+1} = 0$ and $G_{2m} = 2\left(1 - 2^{2m}\right) B_{2m}$, where $B_{2m} \equiv B_{2m}^{(1)}$ are Bernoulli numbers given by generating function (G_2) with $n = 1$. Surely, all the generating functions shown above could be found in comprehensive books on the Calculus of Finite Differences, in particular, e.g., in Jordan [141] (§78, §85, §95 and 96, §109). For (G_7), see [210].

Clearly, the substitution rule is applicable to each of the generating functions (G_1)–(G_{12}) so that 12 operational formulas could be obtained. This will be shown in the next section (§3.1.2).

3.1.2 Various symbolic operational formulas

Let us apply the substitution rule to each left-hand side (LHS) of (G_1)–(G_{12}). We easily find

$LHS(G_1)$: $exp(e^D - 1) = exp\Delta = \sum_{k\geq 0} \frac{1}{k!} \Delta^k$,

$LHS(G_2)$: $\left(\frac{D}{e^D-1}\right)^n = \frac{D^n}{\Delta^n}$,

$LHS(G_3)$: $\left(\frac{e^D-1}{D}\right)^n = \frac{\Delta^n}{D^n}$,

$LHS(G_4)$: $\frac{D(e^D)^x}{e^D-1} = \frac{DE^x}{\Delta}$,

$LHS(G_5)$: $\frac{2(e^D)^x}{e^D+1} = \frac{2E^x}{2+\Delta} = E^x \sum_{k\geq 0} (-1)^k \left(\frac{\Delta}{2}\right)^k$,

$LHS(G_6)$: $\frac{2}{e^D+1} = \frac{2}{2+\Delta} = \sum_{k\geq 0} (-1)^k \left(\frac{\Delta}{2}\right)^k$,

$LHS(G_7)$: $\frac{1}{1-xe^D} = \frac{1}{1-xE} = \sum_{k\geq 0} x^k E^k$,
$LHS(G_8)$: $\frac{2D}{e^D+1} = \frac{2D}{2+\Delta} = D\sum_{k\geq 0}(-1)^k \left(\frac{\Delta}{2}\right)^k$,
$LHS(G_9)$: $\frac{\Delta(1+\Delta)^x}{\log(1+\Delta)} = \frac{\Delta E^x}{D}$,
$LHS(G_{10})$: $\frac{\Delta}{\log(1+\Delta)} = \frac{\Delta}{D}$,
$LHS(G_{11})$: $\frac{1}{1-\Delta-\Delta^2} = \frac{1}{1-\Delta(\Delta+1)} = \frac{1}{1-\Delta E} = \sum_{k\geq 0} \Delta^k E^k$,
$LHS(G_{12})$: $\frac{1}{(1-D)E} = E^{-1}\sum_{k\geq 0} D^k$.

Thus, pairing each $LHS(G_i)$ with $RHS(G_i)$ ($i = 1, 2, \ldots, 12$), we can obtain formally 12 operational formulas for $f(t) \in C^\infty$ evaluated at $t = a$ or at $t = y$, namely

(O_1) $\sum_{k\geq 0} \frac{1}{k!}\Delta^k f(a) = \sum_{k\geq 0} \frac{W_k}{k!} D^k f(a)$,

(O_2) $D^n f(a) = \sum_{k\geq 0} \frac{B_k^{(n)}}{k!} \Delta^n D^k f(a)$,

(O_3) $\Delta^n f(a) = \sum_{k\geq 0} \frac{B_k^{(-n)}}{k!} D^{n+k} f(a)$,

(O_4) $Df(x+y) = \sum_{k\geq 0} \phi_k(x) D^k[f(y+1) - f(y)]$,

(O_5) $\sum_{k\geq 0} \left(-\frac{1}{2}\right)^k \Delta^k f(x) = \sum_{k\geq 0} E_k(x) D^k f(0)$,

(O_6) $\sum_{k\geq 0} \left(-\frac{1}{2}\right)^k \Delta^k f(a) = \sum_{k\geq 0} e_k D^k f(a)$,

(O_7) $\sum_{k\geq 0} f(a+k)x^k = \sum_{k\geq 0} \frac{\alpha_k(x)}{k!} D^k f(a)$,

(O_8) $\sum_{k\geq 0} \left(-\frac{1}{2}\right)^k \Delta^k f(a) = \sum_{k\geq 0} \frac{G_k}{k!} D^{k-1} f(a)$,

(O_9) $\Delta f(x+y) = \sum_{k\geq 0} \psi_k(x) \Delta^k Df(y)$,

(O_{10}) $\Delta f(a) = \sum_{k\geq 0} b_k \Delta^k Df(a)$,

(O_{11}) $\sum_{k\geq 0} \Delta^k f(k) = \sum_{k\geq 0} F_k \Delta^k f(0)$,

(O_{12}) $\sum_{k\geq 0} D^k f(a-1) = \sum_{k\geq 0} \frac{\phi(0)}{k!} D^k f(a)$.

We surmise that at least half of the above formulas may be new, or at least not easily found in the literature.

Certainly, all the series expansions shown above involve convergence problems for given functions, some of which will be considered in the next subsection. Note that (O_4) and (O_9) are well known, and their equivalent forms with applications have been fully expounded in Jordan [141]. Identity (O_7) appears to be an unfamiliar formula whose finite form with certain estimable remainders has been used as a summation formula for power series (cf. [210]).

As may be predicted, a considerable variety of particular identities containing some famous number sequences (or polynomial sequences) could be obtained from the formulas (O_1)–(O_{12}) with special choices of the functions $f(t) \in C^\infty$. This will be partially justified with selective examples in the last subsection.

Remark 3.1.2 *For other delta operators such as the backward difference, the central difference, Laguerre, Bernoulli, and Abel operators, we can construct many other symbolic sum formulas similarly, which are left for the interested readers.*

Remark 3.1.3 *We may construct symbolic sum formulas from some identities such as*

$$\frac{1}{1-x(1+t)} = \frac{1}{1-x}\frac{1}{1-\frac{x}{1-x}t}.$$

The series expansion of the above identity can be written formally as

$$\sum_{k\geq 0} x^k(1+t)^k = \sum_{k\geq 0}\frac{x^k}{(1-x)^{k+1}}t^k.$$

Hence, using the substitution rule for $t \to \Delta$ and noting $1+\Delta = E$ yields formally the following sum formula

$$\sum_{k\geq 0} x^k f(k) = \sum_{k\geq 0}\frac{x^k}{(1-x)^{k+1}}\Delta^k f(0),$$

which is the generalized Euler series transformation formula. Particularly, for $x=-1$, the above formula is reduced to the ordinary Euler series transformation formula shown in Corollary 1.3.12. The series was developed in [99], and its convergence conditions were established in He, Hsu, and Shiue [102].

3.1.3 Some theorems on Convergence

First, let us introduce a definition as follows:

Definition 3.1.4 *Let $\{f_k(t)\}$ and $\{g_k(t)\}$ be two sequences of functions. The commutator of $\{f_k(t)\}$ and $\{g_k(t)\}$ is defined as*

$$[f,g](x,y) \equiv [\{f_k\},\{g_k\}](x,y) := \sum_{k\geq 0}[f_k(x)g_k(y) - f_k(y)g_k(x)]. \tag{3.6}$$

If $[f,g]\equiv 0$, i.e., two sequences of functions $\{f_k(t)\}$ and $\{g_k(t)\}$ satisfy the formal equation/equality

$$\sum_{k\geq 0} f_x(x)g_k(y) = \sum_{k\geq 0} f_k(y)g_k(x), \tag{3.7}$$

we say that $\{f_k(t)\}$ and $\{g_k(t)\}$ have a symmetrical product summation property, or briefly a SPS-property.

From the definition, we immediately have $[f,g](x,y) = -[f,g](y,x)$ or $[f,g](x,y)+[f,g](y,x)=0$. Denote the Fourier transform of a function $h(t)$ as $\hat{h}(\xi)$, if it exists. If each function in sequences $\{f_k(t)\}$ and $\{g_k(t)\}$ has the Fourier transform (e.g., $f_k, g_k \in L_1$, $k \geq 0$), then $\widehat{[f,g]}(\xi,\eta) = [\hat{f},\hat{g}](\xi,\eta)$. Thus, $[f,g]=0$ iff $[\hat{f},\hat{g}]=0$; i.e., $\{f_k(t)\}$ and $\{g_k(t)\}$ have a SPS-property iff $\{\hat{f}(\xi)\}$ and $\{\hat{g}(\xi)\}$ have a SPS-property.

Rota's binomial-type functions (polynomials) are those characterized by the equation

$$f_n(x+y) = \sum_{k\geq 0} \binom{n}{k} f_k(x) f_{n-k}(y),$$

which may be rewritten as

$$\frac{1}{n!} f_n(x+y) = \sum_{k\geq 0} \frac{f_k(x)}{k!} \frac{f_{n-k}(y)}{(n-k)!}.$$

Thus, for fixed $n \geq 1$, the pair of sequences $\langle f_k(t)/k!, f_{n-k}(t)/(n-k)!\rangle$ has the SPS-property. Moreover, we have the following result.

Theorem 3.1.5 *The three pairs* $\langle \phi_k(t), \Delta D^k f(t)\rangle$, $\langle E_k(t), D^k f(t)\rangle$ *and* $\langle \psi_k(t), \Delta^k D f(t)\rangle$ *all have the SPS-property for* $f \in C^\infty$.

Proof. According to (O_4), (O_9), and (O_5), we have, respectively

$$Df(x+y) = \sum_{k\geq 0} \phi_k(x) \Delta D^k f(y), \tag{3.8}$$

$$\Delta f(x+y) = \sum_{k\geq 0} \psi_k(x) D\Delta^k f(y), \tag{3.9}$$

$$\sum_{k\geq 0} \left(-\frac{1}{2}\right)^k \Delta^k f(x+y) = \sum_{k\geq 0} E_k(x) D^k f(y). \tag{3.10}$$

As the LHS's remain the same when x and y are interchanged, we see that the theorem is true.

In what follows we will establish convergence conditions for the series expansions of (O_4), (O_5), (O_6), (O_8), (O_9), and (O_{11}). Convergence problems for other series expansions will be left to the interested reader for consideration (cf. Exercise 3.2). A general technique yielded from the convergence theorems on sum formulas (O_4)–(O_6), (O_8), (O_9), and (O_{11}) will be described in Remark 3.1.14 at the end of the section.

Theorem 3.1.6 *For given* $f \in C^\infty$ *and* $x, y \in \underline{R}$, *the absolute convergence of the series expansion (3.8) is ensured by the condition*

$$\overline{\lim_{k\to\infty}} \left|\Delta D^k f(y)\right|^{1/k} < 1. \tag{3.11}$$

Proof. The root test confirms that the *LHS* of (3.8) will be absolutely convergent provided that

$$\overline{\lim_{k\to\infty}} \left|\phi_k(x)\Delta D^k f(y)\right|^{1/k} < 1. \tag{3.12}$$

We shall show that

$$\overline{\lim_{k\to\infty}} \left|\phi(x)\right|^{1/k} \leq 1. \tag{3.13}$$

so that (3.11) plus (3.13) will imply (3.12).

Recall that the Bernoulli polynomial may be written in the form (cf. Jordan [141], §78-§82)

$$\phi_k(x) = \sum_{j=0}^{k} \frac{B_j}{(k-j)!j!} x^{k-j} = \sum_{j=0}^{k} \frac{x^{k-j}}{(k-j)!} \alpha_j,$$

where $\alpha_j = B_j/j! = B_j^{(1)}/j!$, and B_j are ordinary Bernoulli numbers. Note that $\alpha_0 = 1$, $\alpha_1 = -1/2$, $\alpha_{2m+1} = 0$ $(m \in \mathbb{N})$ and (cf. [141], p. 245)

$$|\alpha_{2m}| \leq \frac{1}{12(2\pi)^{2m-2}}, \quad (m = 0, 1, 2, \ldots).$$

It follows that

$$\begin{aligned} |\phi_k(x)| &\leq \frac{|x|^k}{k!} + \frac{|x|^{k-1}}{2(k-1)!} + \sum_{j=2}^{k} \left(\frac{1}{12(2\pi)^{j-2}}\right) \frac{|x^{k-j}|}{(k-j)!} \\ &< \sum_{j=0}^{k} \frac{|x|^j}{j!} \leq \sum_{r=0}^{\infty} \frac{|x|^r}{r!} = e^{|x|}, \quad (k \geq 2). \end{aligned}$$

Consequently, we get $|\phi_k(x)|^{1/k} < exp(|x|/k) \to 1$ as $k \to \infty$, and the assertion (3.13) is proved.

Theorem 3.1.7 *The absolute convergence of the series expansion (3.9) is ensured by the condition*

$$\overline{\lim_{k\to\infty}} \left|\Delta^k Df(y)\right|^{1/k} < 1. \tag{3.14}$$

Proof. Given condition (3.14). Using the root test again, we have to show that

$$\overline{\lim_{k\to\infty}} \left|\psi_k(x)\Delta^k Df(y)\right|^{1/k} < 1. \tag{3.15}$$

For this it suffices to prove that

$$\overline{\lim_{k\to\infty}} \left|\psi_k(x)\right|^{1/k} \leq 1. \tag{3.16}$$

Recall that there is an integral representation of $\psi_k(x)$, namely (cf. [141], p. 268)

$$\psi_k(x) = \int_0^1 \binom{x+t}{k} dt. \tag{3.17}$$

For $t \in [0,1]$ and for large k, we have the order estimation

$$\left|\binom{x+t}{k}\right| = \frac{|(x+t)_k|}{k!} = \frac{|(k-x-t-1)_k|}{k!} = o\left(\frac{(k+\lfloor |x| \rfloor)_k}{k!}\right) = o\left(k^{\lfloor |x| \rfloor}\right).$$

This means that there is a constant $M > 0$ such that

$$\max_{0 \le t \le 1} \left|\binom{x+t}{k}\right| < Mk^{\lfloor |x| \rfloor}.$$

Thus it follows that

$$\overline{\lim_{k\to\infty}} |\psi_k(x)|^{1/k} \le \overline{\lim_{k\to\infty}} \left(\int_0^1 \left|\binom{x+t}{k}\right| dt\right)^{1/k} \le \overline{\lim_{k\to\infty}} \left(Mk^{\lfloor |x| \rfloor}\right)^{1/k} = 1.$$

This is a verification of (3.16), and Theorem 3.1.7 is proved.

We need the following lemmas for discussing the convergence of the series in (3.10), (O_6) and (O_8).

Lemma 3.1.8 *Let $f \in C^\infty$. Then $\overline{\lim}_{k\to\infty} \left|D^k f(y)\right|^{1/k} < a$ implies*

$$\overline{\lim_{k\to\infty}} \left|\Delta^k f(y)\right|^{1/k} < e^a - 1.$$

Proof. Assume $\overline{\lim}_{n\to\infty} |D^n f(y)|^{1/n} < a$. Denote $\overline{\lim}_{n\to\infty} |D^n f(y)|^{1/n} = \theta$. Then there exists a number γ such that $\theta < \gamma < a$. Thus for large enough n we have $|D^n f(y)|^{1/n} < \gamma$ or $|D^n f(y)| < \gamma^n$.

From (3.3), noting $\left\{ {n \atop m} \right\} \ge 0$ and $|D^n f(y)| < \gamma^n$ yields

$$\begin{aligned} \left|\Delta^k f(y)\right| &= \left|\sum_{n\ge k} \frac{k!}{n!} \left\{ {n \atop k} \right\} D^n f(y)\right| \le \sum_{n \ge k} \frac{k!}{n!} \left\{ {n \atop k} \right\} |D^n f(y)| \\ &\le \sum_{n\ge k} \frac{k!}{n!} \left\{ {n \atop k} \right\} \gamma^n = (e^\gamma - 1)^k < (e^a - 1)^k. \end{aligned}$$

Here the rightmost equality is from Jordan [141] (cf. P. 176). This proves the lemma.

Lemma 3.1.9 *Let $f \in C^\infty$. If $\overline{\lim}_{k\to\infty} \left|D^k f(y)\right|^{1/k} < a$ for some y, then for any fixed t we have*

$$\overline{\lim_{k\to\infty}} \left|D^k f(t)\right|^{1/k} < a.$$

Proof. Denote $\overline{\lim}_{k\to\infty} \left|D^k f(y)\right|^{1/k} = \theta$. Then there exists a number γ such that $\theta < \gamma < a$. Thus, for large enough k $\left|D^k f(y)\right| < \gamma^k$. Denote $x = t - y$. Hence, for large k

$$\begin{aligned} \left|D^k f(t)\right| &= \left|D^k f(y+x)\right| = \left|D^k E^x f(y)\right| = \left|D^k e^{xD} f(y)\right| \\ &= \left|\sum_{j=0}^{\infty} \frac{x^j}{j!} D^{k+j} f(y)\right| \le \sum_{j=0}^{\infty} \frac{|x|^j}{j!} \left|D^{k+j} f(y)\right| \\ &\le \sum_{j=0}^{\infty} \frac{|x|^j}{j!} \gamma^{k+j} = \gamma^k e^{|x|\gamma}, \end{aligned}$$

which implies

$$\left|D^k f(t)\right|^{1/k} \le \gamma e^{|x|\gamma/k} \ (x = t - y).$$

For given t we choose large k such that

$$k > \frac{|x|\gamma}{\log a - \log \gamma} = \frac{|t-y|\gamma}{\log a - \log \gamma}.$$

Thus,

$$\left|D^k f(t)\right|^{1/k} \le \gamma e^{|x|\gamma/k} < \gamma \frac{a}{\gamma} = a.$$

This completes the proof of the lemma.

Theorem 3.1.10 *The absolute convergence of the series expansions involved in (3.10) is ensured by the condition*

$$\overline{\lim_{k\to\infty}} \left|D^k f(y)\right|^{1/k} < 1. \tag{3.18}$$

Proof. From condition (3.18), by using Lemmas 3.1.9 and 3.1.8, we have $\overline{\lim}_{k\to\infty} \left|D^k f(x+y)\right|^{1/k} < 1$ and

$$\left|\Delta^k f(y+x)\right|^{1/k} < e - 1 < 2,$$

respectively. Then, the above inequality implies the absolute convergence of the series on the LHS of (3.10) is obvious in view of the root test for convergence.

Given (3.18), the absolute convergence of the series on the RHS of (3.10) is implied by

$$\overline{\lim_{k\to\infty}} \left|E_k(x)\right|^{1/k} \le 1. \tag{3.19}$$

To prove (3.19), we note that Euler polynomial $E_k(x)$ can be written in the form

$$E_k(x) = \sum_{j=0}^{k} e_j \frac{x^{k-j}}{(k-j)!}, \ (e_0 = 1), \tag{3.20}$$

where $e_j = E_j(0)$, $e_{2m} = 0$ $(m = 1, 2, \ldots)$, and e_{2m-1} satisfies the inequality (cf. [141], p. 302)

$$|e_{2m-1}| < \frac{2}{3\pi^{2m-2}} < 1 \ (m = 1, 2, \ldots). \tag{3.21}$$

Thus we have the estimation

$$|E_k(x)| \leq \frac{|x|^k}{k!} + \sum_{j=1}^{k} |e_j| \frac{|x|^{k-j}}{(k-j)!} \leq \frac{|x|^k}{k!} + \sum_{j=1}^{k} \frac{|x|^{k-j}}{(k-j)!} < e^{|x|}.$$

Consequently, we get

$$\overline{\lim_{k\to\infty}} |E_k(x)|^{1/k} \leq \lim_{k\to\infty} \left(e^{|x|}\right)^{1/k} = 1.$$

Hence (3.19) is verified.

Remark 3.1.11 *From the LHS of (3.10), we recognize the condition*

$$\overline{\lim_{k\to\infty}} \left|\Delta^k f(x+y)\right|^{1/k} < 2 \tag{3.22}$$

implies the absolute convergence of the series. In the proof of Theorem 3.1.10, this condition is derived from condition (3.18). Hence, the reader may propose the question: Are conditions (3.18) and (3.22) equivalently? Namely, does (3.22) also imply (3.18)? The following example shows that the answer is negative.

Consider $f(t) = 2.8^t$. Let both x and y to be zero. Then,

$$\left|\Delta^k f(x+y)\right|^{1/k} = \left|\Delta^k f(0)\right|^{1/k} = \left[(2.8-1)^k\right]^{1/k} < 2.$$

However,

$$\left|D^k f(y)\right|^{1/k} = \left|D^k f(0)\right|^{1/k} = \left[(\log(2.8))^k\right]^{1/k} > 1.$$

Similar to Theorem 3.1.10, we obtain the following convergence results for (O_6) and (O_8).

Theorem 3.1.12 *The absolute convergence of the series expansions involved in (O_6) and (O_8) is ensured by the condition*

$$\overline{\lim_{k\to\infty}} \left|D^k f(a)\right|^{1/k} < 1.$$

Proof. From the given condition, by using Lemma 3.1.8, we have

$$\overline{\lim_{k\to\infty}} \left|\Delta^k f(a)\right|^{1/k} < e - 1 < 2.$$

Hence, the series on the *LHS* of (O_6) and (O_8) are absolutely convergent.

Note that $e_k = E_k(0)$ satisfies $|e_k| < e^0 = 1$ and $G_{k+1}/(k+1)! = e_k$, we immediately obtain the convergence of the series on the *RHS* of (O_6) and (O_8) from the given condition.

Let us consider the operational formula (O_{11}):

$$\sum_{k\geq 0} F_k \Delta^k f(0) = \sum_{k\geq 0} \Delta^k f(k), \tag{3.23}$$

where F_k has Binet expression $F_k = (\alpha^{k+1} - \beta^{k+1})/\sqrt{5}$ with $\alpha = (1+\sqrt{5})/2$ and $\beta = (1-\sqrt{5})/2$ so that $\alpha + \beta = 1$ and $\alpha|\beta| = 1$.

Theorem 3.1.13 *The following condition*

$$\overline{\lim_{k\to\infty}} \left|\Delta^k f(0)\right|^{1/k} < |\beta| = \frac{\sqrt{5}-1}{2} \tag{3.24}$$

ensures the absolute convergence of the series on both sides of (3.23).

Proof. Clearly, we have (see Wilf [216] Eq. (1.3.4))

$$\overline{\lim_{k\to\infty}} (F_k)^{1/k} = \alpha = \frac{\sqrt{5}+1}{2}.$$

Thus condition (3.24) implies that

$$\overline{\lim_{k\to\infty}} \left|F_k \Delta^k f(0)\right|^{1/k} < \alpha|\beta| = 1.$$

so that the series on the LHS of (3.23) is absolutely convergent.

Rewrite (3.24) in the form

$$\overline{\lim_{k\to\infty}} \left|\Delta^k f(0)\right|^{1/k} = \theta < |\beta|.$$

Choose a number γ such that $\theta < \gamma < |\beta|$. Then for large enough k we have $\left|\Delta^k f(0)\right|^{1/k} < \gamma$, i.e., $\left|\Delta^k f(0)\right| < \gamma^k$. Consequently,

$$\begin{aligned}
&\left|\Delta^k f(k)\right|^{1/k} = \left|\Delta^k E^k f(0)\right|^{1/k} = \left|\Delta^k (1+\Delta)^k f(0)\right|^{1/k} \\
= &\left|\sum_{j=0}^{k} \binom{k}{j} \Delta^{k+j} f(0)\right|^{1/k} \leq \left(\sum_{j=0}^{k} \binom{k}{j} \left|\Delta^{k+j} f(0)\right|\right)^{1/k} \\
< &\left(\sum_{j=0}^{k} \binom{k}{j} \gamma^{k+j}\right)^{1/k} = \left(\gamma^k (1+\gamma)^k\right)^{1/k} = \gamma(1+\gamma) < |\beta|\alpha = 1.
\end{aligned}$$

It follows that

$$\overline{\lim_{k\to\infty}} \left|\Delta^k f(k)\right|^{1/k} \leq \gamma(1+\gamma) < 1.$$

This shows that the series on the RHS of (3.23) is also absolutely convergent.

Remark 3.1.14 *We may sort sum formulas (O_1)–(O_{12}) into two classes. The first class includes only either the sum $\sum \gamma_k D^k f$ or the sum $\sum \eta_k E^k f$ in the formulas such as (O_2)–(O_4), (O_9), and (O_{10}). The second class includes the sums $\sum \gamma_k D^k f$ and/or $\sum \eta_k E^k f$ on both sides, such as (O_1), (O_5)–(O_8), (O_{11}), and (O_{12}). Similar to Theorems 3.1.6, 3.1.7 and 3.1.13, we may establish the convergence condition, $\overline{\lim}_{k\to\infty} \left|D^k f\right|^{1/k} < 1$ (or $\overline{\lim}_{k\to\infty} \left|E^k f\right|^{1/k} < 1$), for the first-class series expansions if we can determine $|\gamma_k| \le 1$ (or $|\eta_k| \le 1$). We may also establish the convergence condition, $\overline{\lim}_{k\to\infty} \left|D^k f\right|^{1/k} < 1$, for the second class series expansions, which is similar to Theorems 3.1.10 and 3.1.12. More precisely, if there exist $|\gamma_k| \le 1$ and $|\eta_k| \le (1/2)^k$, by using Lemmas 3.1.8 and 3.1.9, we get the convergence of the series expansions.*

3.1.4 Examples

Surely, the list of operational formulas (O_1)–(O_{11}) may provide a fruitful source of particular identities relating to some famous number sequences and polynomials involved in those formulas. In this subsection, we will present a number of particular identities or formulas as examples, in which $f(x)$'s are taken to be simple elementary functions.

First, let us mention several elementary functions with simpler differences and derivatives.

(i) For $f(x) = x^m$ $(m \ge 1)$ we have (with $k \le m$)

$$\Delta^k f(0) = \left[\Delta^k x^m\right]_{x=0} = k!\left\{ {m \atop k} \right\},$$
$$\left[D^k x^m\right]_{x=0} = \left[(m)_k x^{m-k}\right]_{x=0} = \delta_{m,k} m!,$$

where $\left\{ {m \atop k} \right\}$ is the Stirling number of the second kind (cf. [141], P. 168), and $\delta_{m,k}$ is the Kronecker symbol with $\delta_{m,k} = 1$ for $m = k$ and zero for $m \ne k$.

(ii) For $f(x) = \binom{x}{m}$ and $m \ge k \ge 0$ we have

$$\Delta^k f(0) = \binom{x}{m-k}_{x=0} = \binom{0}{m-k} = \delta_{m,k},$$
$$D^k f(0) = \left[D^k \frac{(x)_m}{m!}\right]_{x=0} = \frac{k!}{m!}(-1)^{m-k}\begin{bmatrix} m \\ k \end{bmatrix}.$$

(iii) For $f(x) = a^x$ $(a > 1)$, we have

$$\Delta^k a^x = (a-1)^k a^x, \quad D^k a^x = (\log a)^k a^x,$$
$$\left[\Delta^k a^x\right]_{x=0} = (a-1)^k, \quad \left[D^k a^x\right]_{x=0} = (\log a)^k.$$

(iv) For $f(x) = \frac{1}{1+x}$, we have

$$\Delta^k f(x) = \frac{(-1)^k k!}{(x+k+1)_{k+1}}, \quad D^k f(x) = \frac{(-1)^k k!}{(1+x)^{k+1}},$$

$$\Delta^k f(0) = \frac{(-1)^k}{k+1}, \quad D^k f(0) = (-1)^k k!.$$

(v) For $f(x) = e^{ix}$ and $g(x) = e^{-ix}$ $(i^2 = -1)$, we have

$$\Delta^k e^{\pm ix} = e^{\pm ix}(e^{\pm i} - 1)^k, \quad D^k e^{\pm ix} = (\pm i)^k e^{\pm ix},$$
$$\left[\Delta^k e^{\pm ix}\right]_{x=0} = (e^{\pm i} - 1)^k, \quad \left[D^k e^{\pm ix}\right]_{x=0} = (\pm i)^k.$$

(vi) For $f(x) = \cos x = \frac{1}{2}\left(e^{ix} + e^{-ix}\right)$ we have

$$\left[\Delta^k \cos x\right]_{x=0} = \frac{(e^i - 1)^k + (e^{-i} - 1)^k}{2} = \frac{(1 + (-1)^k e^{-ik})(e^i - 1)^k}{2},$$
$$\left[D^k \cos x\right]_{x=0} = \frac{i^k + (-i)^k}{2} = i^k \left(\frac{1 + (-1)^k}{2}\right) = i^k \delta_k,$$

where δ_k is the parity function, viz., $\delta_k = 0$ if k is an odd integer, and $\delta_k = 1$ if k is an even integer.

As an immediate generalization of (O_1), there exists

$(O_1)^*$ $\sum_{k\geq 0} \frac{x^k}{k!}\Delta^k f(a) = \sum_{k\geq 0} \tau_k(x) D^k f(a)$,

where $\tau_k(x)$ are known as Touchard polynomials generated by the expansion (cf. Hsu and Shiue [134], p. 186)

$$e^{x(e^t - 1)} = \sum_{k=0}^{\infty} \tau_k(x) t^k.$$

Clearly, $(O_1)^*$ is obtained via the substitution $t \to D$, namely

$$e^{x\Delta} f(a) = \sum_{k\geq 0} \frac{x^k \Delta^k}{k!} f(a) = \sum_{k\geq 0} \tau_k(x) D^k f(a).$$

Example 3.1.15 *Taking $f(t) = t^m$ $(m \geq 1)$, we see that the LHS of $(O_1)^*$ with $a = 0$ yields*

$$\sum_{k\geq 0} \frac{x^k}{k!}\left[\Delta^k t^m\right]_{t=0} = \sum_{k=0}^{m} x^k \left\{ m \atop k \right\}.$$

The RHS of $(O_1)^*$ gives

$$\sum_{k\geq 0} \tau_k(x)\left[D^k t^m\right]_{t=0} = m! \tau_m(x).$$

Hence we get an identity of the form

$$\sum_{k=0}^{m} \left\{ m \atop k \right\} x^k = m! \tau_m(x). \tag{3.25}$$

Usually the LHS of (3.25) is called "exponential polynomial" for Stirling numbers. Thus (3.25) shows that the exponential polynomial is precisely given by Touchard polynomial. It can be checked that the Touchard polynomial sequence is a binomial type polynomial sequence and has the SPS-property.

Example 3.1.16 *In view of the generating function of* $\{\tau_k(x)\}$ *we see that Bell numbers* W_k *are given by* $W_k = k!\tau_k(1)$. *Thus (3.25) with implies the well-known relation*

$$\sum_{k=0}^{m} \left\{ {m \atop k} \right\} = W_m. \tag{3.26}$$

Example 3.1.17 *Taking* $f(t) = (t)_m$ *with* $m \geq 1$, *we find*

$$\Delta^k (t)_m \big|_{t=0} = m!\Delta^k \binom{t}{m}\bigg|_{t=0} = m!\delta_{m,k},$$

$$D^k (t)_m \big|_{t=0} = k!(-1)^{m-k} \left[{m \atop k} \right],$$

where $(-1)^{m-k}\left[{m \atop k}\right]$ *is the Stirling number of the first kind. Using* $(O_1)^*$ *with* $x = 1$ *and* $a = 0$ *we see that its LHS and RHS are respectively* $\sum_{k\geq 0} \frac{1}{k!} m!\delta_{mk} = 1$ *and* $\sum_{k\geq 0} \frac{W_k}{k!} k!(-1)^{m-k}\left[{m \atop k}\right]$. *Thus we obtain an elegant identity of the form*

$$\sum_{k=0}^{m} (-1)^{m-k} W_k \left[{m \atop k} \right] = 1.$$

Substituting (3.25) with $x = 1$ *into the above identity yields identity (cf. [141], P. 183)*

$$\sum_{k=0}^{m} \sum_{j=0}^{k} (-1)^{m-k} \left[{m \atop k} \right] \left\{ {k \atop j} \right\} = 1.$$

Example 3.1.18 *Taking* $f(t) = x_m(t)$, *the mth degree Bernoulli polynomial of the 2nd kind, and recalling that (cf. [141], §89-§97)*

$$x_m(t) = \int_0^1 \binom{t+x}{m} dx = \sum_{j=0}^{m} b_j \binom{t}{m-j},$$

where $b_j = x_j(0) = \int_0^1 \binom{x}{j} dx$, *we have*

$$\Delta^k x_m(t) = x_{m-k}(t), \ (0 \leq k \leq m),$$

$$\Delta^k x_m(0) = x_{m-k}(0) = b_{m-k}, \ b_0 = 1.$$

Thus using (O_1) *we get a relation involving three kinds of special numbers as follows*

$$\sum_{k=1}^{m} \frac{1}{k} W_k (-1)^{m-k} \left[{m-1 \atop k-1} \right] = (m-1)! \sum_{k=1}^{m} \frac{1}{k!} b_{m-k}, \tag{3.27}$$

where b_j *are known as Bernoulli numbers of the 2nd kind, defined by* (G_{10}). *Note that there is a known formula for namely (cf. [141])*

$$\Delta^n = \sum_{k=0}^{\infty} n! \left\{ {n+k \atop n} \right\} \frac{D^{n+k}}{(n+k)!}.$$

Comparing this equation with (O_3), we get the well-known relation between Bernoulli numbers and Stirling numbers, viz.

$$B_k^{(-n)} = \binom{n+k}{k}^{-1} \left\{ {n+k \atop n} \right\}.$$

This implies that (O_3) is another form for the expression of Δ^n. Formula (O_2) appears to be not so familiar. However, the case $n = 1$ is well-known, and it leads to the classical Euler-Maclaurin summation formula.

Remark 3.1.19 *$x_m(t)$ can be written as $x_m(t) = J\binom{x}{m}$ symbolically, where J is the Bernoulli operator: $J : p(t) \mapsto \int_t^{t+1} p(x)dx$.*

Example 3.1.20 *Let $f(t) = t^m$ ($m > n \geq 1$) and $a = 0$. Then*

$$\Delta^n D^k f(0) = (m)_k \Delta^n t^{m-k}\big|_{t=0} = (m)_k n! \left\{ {m-k \atop n} \right\}.$$

We find the RHS of (O_2) is $\sum_{k=0}^{m-n} \binom{m}{k} B_k^{(n)} n! \left\{ {m-k \atop n} \right\}$. That is

$$\sum_{k=0}^{m-n} \binom{m}{k} B_k^{(n)} n! \left\{ {m-k \atop n} \right\} = D^n t^m\big|_{t=0} = 0. \tag{3.28}$$

In particular, the case $n = 1$ gives the well-known recurrence relations for Bernoulli numbers, viz.

$$\sum_{k=0}^{m-1} \binom{m}{k} B_k = (1+B)^m - B_m = 0, \tag{3.29}$$

where $(1 + B)^m$ is written in the sense of umbrel calculus, in which B^i must be substituted for B_i.

Example 3.1.21 *Take $f(t) = 2^t$, we find $D^k f(t) = 2^t(\log 2)^k$, $\Delta^k f(t) = 2^t$. Thus, formula (O_2) gives*

$$D^n f(t) = 2^t(\log 2)^n = \sum_{k=0}^{\infty} \frac{B_k^{(n)}}{k!} 2^t (\log 2)^k.$$

That is

$$\sum_{k=0}^{\infty} \frac{B_k^{(n)}}{k!} \frac{(\log 2)^k}{k!} = (\log 2)^n. \tag{3.30}$$

The reader may find various examples in Jordan [141] for the equivalent of (O_4) and (O_9). Here we supplement some other examples.

Example 3.1.22 *A case of* (O_4) *is the following expression:*

$$\sum_{k\geq 0}\phi_k(x)D^k[f(1)-f(0)] = Df(x).$$

Taking $f(t)=t^m$ $(m\geq 1)$ *we get*

$$\sum_{k=0}^{m-1}(m)_k\phi_k(x) = mx^{m-1}. \tag{3.31}$$

Example 3.1.23 *For* $f(t)=t^m$ $(m\geq 1)$ *formula* (O_9) *with* $y=0$ *yields*

$$\sum_{k\geq 0}\psi_k(x)m\left[\Delta^k y^{m-1}\right]_{y=0} = \sum_{k=0}^{m-1}\psi_k(x)m\cdot k!\left\{ {m-1 \atop k} \right\} = (x+1)^m - x^m.$$

This leads to the formula

$$\sum_{k=0}^{m-1}k!\psi_k(x)\left\{ {m-1 \atop k} \right\} = \frac{(x+1)^m-x^m}{m}. \tag{3.32}$$

Example 3.1.24 *Let* $f(t)=\binom{t}{m}$ $(m\geq 1)$. *Then (cf. [141], P.64)*

$$D\Delta^k f(y) = D_y\binom{y}{m-k}, \quad \Delta f(x+y) = \binom{x+y}{m-1}.$$

Thus it follows from (O_9) *the closed formula*

$$\sum_{k=0}^{m}\psi_k(x)\frac{d}{dy}\binom{y}{m-k} = \binom{x+y}{m-1}. \tag{3.33}$$

Example 3.1.25 *Replacing* $f(x)$ *by* $f(x+y)$ *in* (O_5), *we have*

$$\sum_{k\geq 0}E_k(x)D^k f(y) = \sum_{k\geq 0}\left(-\frac{1}{2}\right)^k \Delta^k f(x+y).$$

Taking $f(y)=\binom{y}{m}$ $(m\geq 1)$, *we find* $D^k f(y)\big|_{y=0} = (-1)^{m-k}\frac{k!}{m!}\left[{m \atop k} \right]$ *and*

$$\sum_{k=0}^{m}(-1)^{m-k}E_k(x)\frac{k!}{m!}\left[{m \atop k} \right] = \sum_{k=0}^{m}\left(-\frac{1}{2}\right)^k\binom{x}{m-k}. \tag{3.34}$$

Putting $x=m$ *and* $x=0$ *in (3.34), respectively, we easily obtain*

$$\sum_{k=0}^{m}(-1)^{m-k}\frac{k!}{m!}E_k(m)\left[{m \atop k} \right] = \left(\frac{1}{2}\right)^m \tag{3.35}$$

and

$$\sum_{k=0}^{m} k!e_k(-1)^k \begin{bmatrix} m \\ k \end{bmatrix} = \frac{m!}{2^m}, \tag{3.36}$$

where $e_k = E_k(0)$, *defined by* (G_6), *(3.36) may also be derived from* (O_6) *by setting* $a = 0$. *Most likely, identities (3.32)–(3.36) may be new or not easily found in classical literature.*

Example 3.1.26 *As mentioned before (cf. the last part of Subsection 3.1.2), a comparison of* (G_8) *with* (G_6) *leads to the relation* $G_{k+1} = (k+1)!e_k$. *Thus the equality* (3.36) *may also be written in terms of Genocchi numbers, viz.*

$$\sum_{k=1}^{m+1} (-1)^{k-1} \frac{G_k}{k} \begin{bmatrix} m \\ k-1 \end{bmatrix} = \frac{m!}{2^m}. \tag{3.37}$$

Example 3.1.27 *In* (O_7) *taking* $f(t) = t^m$ $(m \geq 1)$ *and* $a = 0$, *we get*

$$\sum_{k \geq 0} k^m x^k = \sum_{k \geq 0} \frac{\alpha_k(x)}{k!} m!\delta_{mk} = \alpha_m(x). \tag{3.38}$$

This is the classical formula of Euler for the arithmetic-geometric series (cf. Lemma 2.1.7 and Example 2.2.7).

Example 3.1.28 *Obviously* (O_7) *can be written in the form*

$$\sum_{k=a}^{\infty} f(k)x^k = \sum_{k=0}^{\infty} \frac{\alpha_k(x)}{k!} x^a f^{(k)}(a).$$

Thus for any given non-negative integers $a < b-1$ *we have*

$$\sum_{k=a}^{b-1} f(k)x^k = \sum_{k=0}^{\infty} \frac{\alpha_k(x)}{k!} \left[x^a f^{(k)}(a) - x^b f^{(k)}(b) \right]. \tag{3.39}$$

The partial sum of the RHS of (3.39) with a remainder can be used as a summation formula for the LHS. This problem has been treated much in detail by Wang and Hsu [210].

Example 3.1.29 *Recall that (3.38) may be rewritten in the form* (cf. *Comtet [44],* p. *243-5)*

$$\sum_{k=0}^{\infty} k^m x^k = \alpha_m(x) = \frac{A_m(x)}{(1-x)^{m+1}}, \ (|x| < 1), \tag{3.40}$$

where $A_m(x)$ *is the* m*th degree Eulerian polynomial given by the expression* $A_0(x) = 1$ *and*

$$A_m(x) = \sum_{k=1}^{m} A(m,k)x^k, \ (m \geq 1),$$

with $A(m,0)=0$ *and*

$$A(m,k)=\sum_{j=0}^{k}(-1)^j\binom{m+1}{j}(k-j)^m,\ (1\le k\le m),$$

$A(m,k)$ *being known as Eulerian numbers (cf. Subsection 2.1.1).*

Now, (3.40) can be symbolized in this way: Letting x *be substituted by* $E=1+\Delta$, *we have* $(x-1)^{m+1}\to\Delta^{m+1}$. *Thus (3.40) leads to the symbolic formula*

$$(O_{13}):\ \sum_{k=0}^{\infty}k^m\Delta^{m+1}f(k)=(-1)^{m+1}A_m(E)f(0).$$

This summation formula can be used to compute the series of the form as shown on the LHS of (O_{13}). *For instance, taking* $f(t)=1/(1+t)$, *we find*

$$\Delta^{m+1}f(k)=(-1)^{m+1}\frac{(m+1)!}{(m+k+2)_{m+2}}=\frac{(-1)^{m+1}}{m+2}\binom{m+k+2}{m+2}^{-1}.$$

Consequently, using formula (O_{13}) *we obtain*

$$\frac{1}{m+2}\sum_{k=0}^{\infty}k^m\Big/\binom{m+k+2}{m+2}=\sum_{k=1}^{m}A(m,k)/(k+1). \tag{3.41}$$

This appears to be a novel identity.

3.2 On a Pair of Operator Series Expansions Implying a Variety of Summation Formulas

In this section, we shall construct a pair of expansion formulas by using Mullin-Rota's substitution rule. A convergence theorem is established for a fruitful source formula

Generally, an operator T is called a shift-invariant operator (cf. [107]) if $TE^\alpha=E^\alpha T$ for every $\alpha\in\mathbf{R}$. Recall that if in addition, $Tt\ne 0$ (a non-zero constant), then T is called a delta operator as we defined before. Obviously, both Δ and D are delta operators.

We shall frequently utilize a general proposition due to Mullin and Rota as stated in Theorem 3.1.1 and Mullin-Rota's substitution rule.

For the fps $f(t)$, note that $D^kf(t)=f^{(k)}(t)$ means a kth order formal derivative of $f(t)$ so that $D^kf(0)=f^{(k)}(0)$, and that $f(t)$ can also be written as a formal Taylor series.

Also, we shall need three basic concepts as given by the following definitions.

Definition 3.2.1 *For any given fps $A(t)$ and $g(t)$ such that $A(0) = 1, g(0) = 0$ and $g'(0) = Dg(0) \neq 0$, the polynomials $p_k(z)(k \in \mathbf{N})$, defined by the generating function (GF)*

$$A(t)e^{zg(t)} = \sum_{k\geq 0} p_k(z)t^k \tag{3.42}$$

are called the Sheffer-type polynomials, where $p_0(z) = 1$. More explicitly, we may denote

$$p_k(z) \equiv p_k(z, A(t), g(t)) = [t^k]A(t)e^{zg(t)}, \tag{3.43}$$

where $[t^k]$ is the so-called extracting coefficient operator.

Accordingly, $p_k(D)$ with $D \equiv d/dt$ is called the Sheffer-type differential operator of degree k. In particular, $p_0(D) = 1$ is the identity operator.

Definition 3.2.2 *Any expansion formula or a summation formula in the theory of formal power series as well as in the computational analysis is called an (∞^m) degree formula, if it consists of m arbitrary functions that could be chosen in infinitely many ways, where m is called the freedom degree of the formula. For example, the expansion (3.42) is an (∞^2) degree formula.*

Definition 3.2.3 *For $A(t)$ and $g(t)$ as given in Definition 1.1, the numbers d_{kj} defined by (*cf. *[103, 97],[196] etc)*

$$d_{kj} = [t^k]A(t)(g(t))^j, \quad 0 \leq j \leq k \in \mathbb{N}_0 \tag{3.44}$$

are said to form a Riordan array (d_{kj}) which is denoted by $(A(t),g(t))$.

In substance, this section consists of two main parts plus a few remarks. The first part is concerned with a pair of (∞^4) degree expansion formulas that could lead to various specializations and examples. The object of the second part is to investigate a source formula in some detail. It will be shown that the source formula is really a particular consequence of an (∞^4) degree formula, and it becomes an exact formula under certain general convergence conditions. Finally, as one of the concluding remarks, we will explain why there could exist infinitely many (∞^m) formulas via a kind of lifting process.

3.2.1 A pair of (∞^4) degree formulas

The main result to be represented in this subsection is a basic theorem that involves a pair of (∞^4) degree expansion formulas.

Theorem 3.2.4 *Let $A(t), g(t)$, and $f(t)$ be an fps such that $A(0) = 1, g(0) = 0$ and $g'(0) = Dg(0) \neq 0$. Suppose that $p_k(D)(k \in N)$ are the Sheffer-type differential operators associated with $A(t)$ and $g(t)$. Then for any given $\phi(t) \in C^\infty$ there hold formally a pair of (∞^4) degree expansion formulas as follows:*

$$A(D)f(g(D))\phi(t) = \sum_{k\geq 0} (p_k(D)f(0))D^k\phi(t), \tag{3.45}$$

$$A(\Delta)f(g(\Delta))\phi(t) = \sum_{k\geq 0}(p_k(\Delta)f(0))\Delta^k\phi(t), \tag{3.46}$$

where $p_k(D)$ *has the explicit expression*

$$p_k(D) = \sum_{j=0}^{k}\left(\frac{1}{j!}d_{kj}\right)D^j, \tag{3.47}$$

with $(d_{kj}) = (A(t), g(t))$ *being the Riordan array. Moreover, if it is assumed that*

$$\theta = \overline{\lim_{k\to\infty}}\,|p_k(D)f(0)|^{1/k} > 0, \tag{3.48}$$

then the expansions (3.45) *and* (3.46) *are absolutely convergent at* $t = 0$*, under the following conditions, respectively*

$$\overline{\lim_{k\to\infty}}\,|D^k\phi(0)\,|^{1/k} < 1/\theta. \tag{3.49}$$

$$\overline{\lim_{k\to\infty}}\,|\Delta^k\phi(0)\,|^{1/k} < 1/\theta. \tag{3.50}$$

Proof. First, notice that the generating function for the Sheffer-type polynomial sequence $\{p_k(z)\}$ given by (3.42) with replacement $t \longmapsto x$ is an fps in x as well as in z:

$$A(x)\sum_{k\geq 0}\frac{(g(x))^k z^k}{k!} = \sum_{k\geq 0}p_k(z)x^k. \tag{3.51}$$

Actually, this is a formal identity involving arguments z and x. Thus one may apply Mullin-Rota's substitution rule to the formal series (3.51) with replacement $z \longmapsto D$. Accordingly, by letting the resultant operator series act on the $f(t)$ at $t = 0$, it gives formally

$$A(x)\sum_{k\geq 0}\frac{(g(x))^k}{k!}(D^k f(t))_{t=0} = \sum_{k\geq 0}x^k(p_k(D)f(t))_{t=0}. \tag{3.52}$$

Observe that the left-hand side (LHS) of (3.52) apart from $A(x)$ is just a formal Taylor series expansion of $f(g(x))$ in powers of $g(x)$. Consequently, (3.52) can be rewritten in the form

$$A(x)f(g(x)) = \sum_{k\geq 0}(p_k(D)f(0))x^k. \tag{3.53}$$

This is actually a known formula (*cf.* e.g., Theorems 2.3.2-2.3.3 of Roman's [186] and He, Hsu, and Shiue's Expression (3.48) of [103]. Using the conditions for $A(t), g(t)$, and $f(t)$ as stated in the theorem, one may see that Mullin-Rota's substitution rule can also be applied to the formal identity (3.53). Thus the expansion formulas (3.45) and (3.46) can be obtained with the substitutions $x \longmapsto D$ and $x \longmapsto \Delta$, respectively.

Moreover, it is clear that the condition $\theta|x| < 1$ for the absolute convergence of the expansion (3.53) just follows from Cauchy's root-test. Certainly, the similar argument also applies to the conditions (3.49) and (3.50).

It remains to verify the equality (3.47). Clearly, the condition $Dg(0) = g'(0) \neq 0$ implies that the formal power series expansion in t of the following formal series

$$\sum_{j>k} z^j (g(t))^j / j!$$

involves powers of t greater than k. Thus one may deduce from (3.42) that

$$p_k(z) = [t^k]A(t)\sum_{j=0}^{k} z^j (g(t))^j / j! = \sum_{j=0}^{k} [t^k](A(t)(g(t))^j / j!) z^j = \sum_{j=0}^{k} (d_{kj}/j!) z^j,$$

where $d_{kj} = [t^k]A(t)(g(t))^j$ with $0 \le j \le k \in \mathbb{N}$ just form the Riordan array $(d_{kj}) = (A(t), g(t))$. This completes the proof of the theorem.

It may be worth noticing that the formula (3.53) involved in the proof is deducible from either of (3.45) and (3.46) as a particular consequence. More precisely we have the following:

Corollary 3.2.5 *Either of the choices* $\phi(t) = e^{xt}$ *in* (3.45) *and* $\phi(t) = (1+x)^t$ *in* (3.46) *yields the* (∞^2) *degree formula* (3.53) *that is convergent absolutely under the condition* $|x| < 1/\theta$ *with* θ *being defined by* (3.48).

Obviously, one may find by easy computations

$$[D^k e^{xt}]_{t=o} = x^k, \quad [\Delta^k (1+x)^t]_{t=0} = x^k, \quad x \in \mathbf{R}, \quad k \in \mathbf{N}.$$

Note that $A(d)f(g(D))$ and $A(\Delta)f(g(\Delta))$ have formal power series expansions in D and Δ, respectively. Thus, it follows that

$$[A(D)f(g(D))e^{xt}]_{t=0} = A(x)f(g(x)),$$
$$[A(\Delta)f(g(\Delta))(1+x)^t]_{t=0} = A(x)f(g(x)).$$

Hence (3.53) is implied by either of (3.45) and (3.46), with Cauchy-type convergence condition $\theta|x| < 1$.

As suggested by two basic theorems of He, Hsu, and Yin's paper [107], we could give another corollary of Theorem 2.1. Let

$$G(x, y, z) = F_1(x, y, z)/F_2(x, y, z)$$

denote a rational function, namely, both F_1 and F_2 are polynomials involving the variables x, y and z. Then, as a consequence of Theorem 3.2.4, we have the following corollary.

Corollary 3.2.6 *Let $A(t), g(t)$, and $f(t)$ be given as in Theorem 2.1. Then we have two statements.*

(i) Suppose that $A(t)f(g(t))$ is expressible as a rational function in t, e^t and $e^{\alpha t}$ of the form

$$A(t)f(g(t)) = G(t, e^t, e^{\alpha t}), \quad \alpha \in \mathbf{R} \text{ or } \mathbf{C}.$$

Then there holds a formal expansion formula for any $\phi(t) \in C^\infty$:

$$G(D, E, E^\alpha)\phi(t) = \sum_{k\geq 0}(p_k(D)f(0))D^k\phi(t). \tag{3.54}$$

(ii) For the case $A(t)f(g(t)) = G(t, log(1+\alpha t), (1+\alpha t)^\beta)$ with $\alpha \neq 0$ and $\beta \in \mathbf{R}$, there holds a formal expansion for any $\phi(t) \in C^\infty$:

$$G\left(\frac{1}{\alpha}\Delta, D, E^\beta\right)\phi(t) = \sum_{k\geq 0}(p_k(D)f(0))\Delta^k\phi(t)/\alpha^k. \tag{3.55}$$

Note that (3.54)–(3.55) are parallel to the related formulas obtained in [107], but entirely different in formula structures. Certainly, $A(t), g(t)$, and $f(t)$ mentioned in the statements of the Corollary are not really arbitrary, as they have to satisfy the conditions imposed by (i) or (ii).

In the follow-up subsections, the reader will see that a great variety of particular consequences (involving new and old formulas) could be deduced from the expansion formulas displayed in this subsection.

3.2.2 Specializations and examples

Several specializations of (3.45)–(3.46) may be displayed in what follows. There are three pairs of (∞^3) degree formulas of the following forms (applicable to $\phi(t) \in C^\infty$) :

$$\begin{cases} f(g(D))\phi(t) = \sum_{k\geq 0}(p_k(D)f(0))D^k\phi(t), \\ f(g(\Delta))\phi(t) = \sum_{k\geq 0}(p_k(D)f(0))\Delta^k\phi(t). \end{cases} \tag{3.56}$$

$$\begin{cases} A(D)f(D)\phi(t) = \sum_{k\geq 0}(p_k(D)f(0))D^k\phi(t), \\ A(\Delta)f(\Delta)\phi(t) = \sum_{k\geq 0}(p_k(D)f(0))\Delta^k\phi(t). \end{cases} \tag{3.57}$$

and

$$\begin{cases} A(D)\exp(zg(D))\phi(t) = \sum_{k\geq 0}p_k(z)D^k\phi(t), \\ A(\Delta)\exp(zg(\Delta))\phi(t) = \sum_{k\geq 0}(p_k(z)\Delta^k\phi(t). \end{cases} \tag{3.58}$$

where $p_k(D)$ contained in (3.56) has the expression

$$p_k(D) \equiv p_k(D, 1, g(t)) = \sum_{j=0}^{k} \left(\frac{1}{j!} d_{kj}\right) D^j, \quad k \in \mathbf{N}, \tag{3.59}$$

with $d_{kj} = [t^k](g(t))^j$; the operator $p_k(D)$ in (3.57) takes the form

$$p_k(D) \equiv p_k(D, A(t), t) = \sum_{j=0}^{k} \left(\frac{1}{j!} d_{kj}\right) D^j, \quad k \in \mathbf{N}, \tag{3.60}$$

with $d_{kj} = [t^k](A(t), t^j) = [t^{k-j}]A(t), (0 \le j \le k)$; and $p_k(z)$ involved in (3.58) is a Sheffer-type polynomial, viz

$$p_k(z) = p_k(z, A(t), g(t)) = \sum_{j=0}^{k} \left(\frac{1}{j!} d_{kj}\right) z^k, \quad k \in \mathbf{N}, \tag{3.61}$$

with d_{kj} being given by the Riordan array $(d_{kj}) = ([t^k]A(t)(g(t))^j)_{0 \le j \le k}$.

Certainly one may get further specializations from the above expansions.

For instance, as special cases of (3.58) with $g(t) = t$, there are two (∞^2) degree formulas of the forms

$$\begin{cases} A(D)\exp(zD)\phi(t) = \sum_{k \ge 0} p_k(z) D^k \phi(t), \\ A(\Delta)\exp(z\Delta)\phi(t) = \sum_{k \ge 0} (p_k(z) \Delta^k \phi(t). \end{cases} \tag{3.62}$$

where

$$p_k(z) \equiv p_k(z, A(t), t) = [t^k](A(t)e^{zt})$$

may be written as

$$p_k(z) = \sum_{j=0}^{k} \left(\frac{1}{j!} d_{kj}\right) z^j, \quad k \in \mathbf{N}, \tag{3.63}$$

where the numbers d_{kj} being given by $d_{kj} = [t^{k-j}]A(t)$ $(0 \le j \le k)$.

By means of practice, one may find that various special formulas and identities could be deduced as consequences from (3.45)–(3.46) and some of the specializations mentioned above. In what follows we will give several selected examples.

The first three examples are related to the Touchard polynomials $\tau_k(z)$, the Toscano polynomials $(Tos)_k^{(\lambda)}(z)$ with $\lambda \ne 0$, and the generalized Laguerre polynomials $L_k^{(p-1)}(z)$ with $p > 0\,(k \in \mathbf{N}.)$ These polynomials are known as special Sheffer-type polynomials (*cf.* Boas and Buck [21]). In accordance with the denotion (3.43), we may denote them as follows:

$$\tau_k(z) = \tau_k(z, 1, e^t - 1),$$

$$(\mathrm{Tos})_k^{(\lambda)}(z) = (\mathrm{Tos})_k^{(\lambda)}(z, e^{\lambda t}, 1-e^t),$$
$$L_k^{(p-1)}(z) = L_k^{(p-1)}(z, (1-t)^{-p}, t/(t-1)).$$

Accordingly, we could get 3 expansion formulas involving special Sheffer-type operators $\tau_k(D), (\mathrm{Tos})_k^{(\lambda)}(D)$, and $L_k^{(p-1)}(D)$, respectively.

Example 3.2.7 *For the case $A(t) = 1$ and $g(t) = e^t - 1$ we have $A(D) = 1$ (identity operator) and $g(D) = e^D - 1 = E - 1 = \Delta$ so that in accordance with (3.45) or the first equation of (3.56), we may get an (∞^2) degree formula of the form (for $\phi(t) \in C^\infty$) :*

$$f(\Delta)\phi(t) = \sum_{k\geq 0}(\tau_k(D)f(0))\phi^{(k)}(t). \tag{3.64}$$

Herein the Touchard operator $\tau_k(D)$ is given by $\tau_0(D) = 1$ and

$$\tau_k(D) = \sum_{j=0}^{k}\left(\frac{1}{j!}[t^k](e^t-1)^j\right)D^j = \frac{1}{k!}\sum_{j=1}^{k}\left\{ {k \atop j} \right\}D^j, \quad (k \geq 1), \tag{3.65}$$

with $\left\{ {k \atop j} \right\}$ denoting the Stirling numbers of the second kind in Knuth's notation, viz.

$$\left\{ {k \atop j} \right\} = \frac{1}{j!}(\Delta^j t^k)_{t=0}.$$

Example 3.2.8 *For the case $A(t) = e^{\lambda t}$ and $g(t) = 1 - e^t$ we have*

$$A(D) = e^{\lambda D} = E^\lambda, \quad g(D) = 1 - e^D = 1 - E = -\Delta,$$

and

$$A(D)f(g(D))\phi(t) = E^\lambda f(-\Delta)\phi(t) = f(-\Delta)\phi(\lambda + t).$$

Accordingly, as a consequence of (3.45), we obtain an (∞^2) degree formula for $\phi(t) \in C^\infty$:

$$f(-\Delta)\phi(\lambda+t) = \sum_{k\geq 0}((Tos)_k^{(\lambda)}(D)f(0))\phi^{(k)}(t), \tag{3.66}$$

with the Toscano operator being given by the expression

$$(Tos)_k^{(\lambda)}(D) = \sum_{j=0}^{k}\frac{(-1)^j}{k!}S_2^{(\lambda)}(k,j)D^j. \tag{3.67}$$

Herein $S_2^{(\lambda)}(k,j)$ *are known as the second kind of weighted Stirling numbers due to Carlitz [32], which are defined by*

$$\frac{1}{j!}e^{\lambda t}(e^t-1)^j = \sum_{m=j}^{\infty} \frac{t^m}{m!} S_2^{(\lambda)}(m,j) \tag{3.68}$$

so that (3.67) *follows from* (3.47) *and* (3.68), *namely*

$$(\mathit{Tos})_k^{(\lambda)}(D) = \sum_{j=0}^{k} \left(\frac{1}{j!}[t^k]e^{\lambda t}(1-e^t)^j\right) D^j = \mathit{RHS} \text{ of } (3.67). \tag{3.69}$$

Note that the simplest case of (3.66) *with* $f(t) \equiv 1$ *just gives the Taylor expansion of* $\phi(\lambda+t)$ *in powers of* λ, *since* $S_2^{(\lambda)}(m,0) = \lambda^m$.

Example 3.2.9 *Given* $A(t) = (1-t)^{-p}, (p>0)$ *and* $g(t) = t/(t-1)$. *We have* $A(-\Delta) = (1+\Delta)^{-p} = E^{-p}, g(-\Delta) = (-\Delta)/(-\Delta-1) = \Delta/E$ *and* $A(-\Delta)f(g(-\Delta)) = E^{-p}f(\Delta/E)$. *Consequently, an application of* (3.46) *gives an* (∞^2) *degree formula of the form*

$$f\left(\frac{\Delta}{E}\right)\phi(t-p) = \sum_{k\geq 0}(L_k^{(p-1)}(D)f(0))(-1)^k\Delta^k\phi(t), \tag{3.70}$$

where the Laguerre operator has the expression

$$L_k^{(p-1)}(D) = \sum_{j=0}^{k} \frac{(-1)^j}{j!}\binom{k+p-1}{k-j} D^j. \tag{3.71}$$

Actually the numbers $(-1)^j\binom{n+p-1}{k-j} = d_{kj}$ *involved in* (3.71) *from the special Riordan array* $(d_{kj}) = ((1-t)^p, t/(t-1))$.

As is easily seen, the simplest case of (3.70) with $f(t) \equiv 1$ and $t=0$ gives the Newton interpolation series for $\phi(-p)$.

Surely it is possible to get some special formulas or identities from (3.64), (3.66), and (3.70) via suitable choices of $f(t)$ and $\phi(t)$. Here we shall mention an application of (3.64) to the derivation of some formulas in Enumerative Combinatorics.

Note that (3.65) leads us to consider the Bell number $\omega(k)$ as defined by

$$\omega(0) = 1, \ \ \omega(k) = \sum_{j=1}^{k} \left\{ {k \atop j} \right\} \quad (k \geq 1), \tag{3.72}$$

where $\omega(k)$ is known as the number of ways to partition a set of k things into nonempty subsets. Naturally, we have to take $f(t) = e^t$ in order to make (3.64) to yield a useful formula of the form

$$\sum_{k\geq 0} \frac{\Delta^k \phi(t)}{k!} = \sum_{k\geq 0} \frac{\omega(k)\phi^{(k)}(t)}{k!}. \tag{3.73}$$

This formula has some interesting special consequences.

(1) Choosing $\phi(t) = e^{xt}$ with x being a parameter, we have $\Delta^k \phi(0) = (\Delta^k e^{xt})_{t=0} = (e^x - 1)^k$ and $D^k \phi(0) = (D^k e^{xt})_{t=0} = x^k$. Thus we see that (3.73) yields the following equality at $t = 0$:

$$e^{(e^x-1)} = \sum_{k\geq 0} \omega(k) \frac{x^k}{k!}. \tag{3.74}$$

This gives the well-known exponential generating function for the sequence of Bell numbers.

(2) Taking $\phi(t) = m!\binom{t}{m} = (t)_m$ with $(t)_0 = 1$, we have

$$\Delta^k \phi(0) = m! \binom{t}{m-k}_{t=0} = m! \binom{0}{m-k} = m!\delta_{mk},$$

and

$$D^k \phi(0) = \left(D^k \sum_{j=0}^{m} (-1)^{m-j} \begin{bmatrix} m \\ j \end{bmatrix} t^j \right)_{t=0} = (-1)^{m-k} \begin{bmatrix} m \\ k \end{bmatrix} \cdot k!.$$

Consequently, (3.73) with $t = 0$ yields the following well-known identity

$$\sum_{k=0}^{m} (-1)^{m-k} \begin{bmatrix} m \\ k \end{bmatrix} \omega(k) = 1. \tag{3.75}$$

The interested readers may give a combinatorial interpretation for (3.75) (Exercise 3.10).

(3) Letting $\phi(t) = B_n(t)$ (Bernoulli polynomial of degree n) with $B_0(t) = 1$ and denoting $B_n = B_n(0)$ (n-th Bernoulli number) with $B_0 = 1$, we have (*cf.* [157])

$$D^k B_n(t) = (n)_k B_{n-k}(t), \quad D^k B_n(0) = k! \binom{n}{k} B_{n-k},$$

$$\Delta^k B_n(0) = (\Delta^k B_n(t))_{t=0} = \Delta^{k-1}(nt^{n-1})_{t=0} = (k-1)!n \begin{Bmatrix} n-1 \\ k-1 \end{Bmatrix}, \quad (k \geq 1).$$

Consequently, by using (3.73) we obtain

$$\sum_{k=1}^{n} \frac{n}{k} \begin{Bmatrix} n-1 \\ k-1 \end{Bmatrix} = \sum_{k=1}^{n} \binom{n}{k} B_{n-k}\omega(k). \tag{3.76}$$

This is a strange identity showing that a combinatorial convolution of Bell numbers and Bernoulli numbers can be expressed in terms of the Stirling numbers of the second kind.

(4) As may be observed, by taking $f(t) = e^{-t}$ one may find that (3.66) and (3.67) lead to an extension of (3.73), viz.

$$\sum_{k\geq 0} \frac{\Delta^k \phi(t+\lambda)}{k!} = \sum_{k\geq 0} \frac{\omega_\lambda(k) D^k \phi(t)}{k!}, \tag{3.77}$$

where $\omega_\lambda(k)(\lambda \in \mathbf{R}, k \in \mathbf{N})$ denotes a kind of generalized Bell number, viz.

$$\omega_\lambda(k) = \sum_{j=0}^{k} S_2^{(\lambda)}(k,j). \tag{3.78}$$

Also, it may be found that (3.77) yields an extension of (3.74) with $\phi(t) = e^{xt}$:

$$\exp(e^x + \lambda x - 1) = \sum_{k\geq 0} \omega_\lambda(k) \frac{x^k}{k!}. \tag{3.79}$$

This gives a generating function for the generalized Bell numbers.

Finding further examples of (3.66) and (3.67) and possible applications of (3.70) and (3.71) may be left to the interested reader.

Example 3.2.10 *Let $\psi(t)$ be a polynomial of degree r, and let $\phi(t) \in C^\infty$. Then the following summations formula* (cf. *[107])*

$$\sum_{k=0}^{\infty} \frac{\psi(k)\phi^{(k)}(0)}{k!} = \sum_{k=0}^{r} \frac{\Delta^k \psi(0)\phi^{(k)}(1)}{k!} \tag{3.80}$$

is a particular consequence of (3.45) *with $A(t) = e^t, g(t) = t$ and*

$$f(t) = \sum_{k=0}^{r} t^k \Delta^k \psi(0)/k!. \tag{3.81}$$

For proof, it suffices to show that the RHS of (3.80) and the LHS of (3.80) are given respectively by the LHS and the RHS of (3.45) for the given $A(t), g(t)$, and $f(t)$. Indeed, we have

$$A(D)f(g(D))\phi(t)|_{t=0} = \left(e^D \sum_{k=0}^{r} D^k \Delta^k \psi(0)/k! \right) \phi(t) \bigg|_{t=0} = \sum_{k=0}^{r} \frac{\Delta^k \psi(0)}{k!} \phi^{(k)}(1).$$

Moreover, the RHS of (3.45) gives

$$\sum_{k\geq 0} (p_k(D, e^t, t) f(0)) \phi^{(k)}(0).$$

Notice that

$$p_k(z) = p_k(z, e^t, t) = [t^k]e^t \cdot e^{zt} = (1+z)^k/k!$$

and that (3.81) may be rewritten as

$$f(t) = \sum_{k=0}^{\infty} t^k \Delta^k \psi(0)/k!$$

because $\Delta^k\psi(0) = 0$ for all $k > r$ so that $\Delta^k\psi(0)$ has meaning for all $k \geq 0$. Consequently, we see that the RHS of (3.45) can be evaluated as follows

$$\begin{aligned}
&\sum_{k\geq 0}(p_k(D)f(0))\phi^{(k)}(0) \\
&= \sum_{k\geq 0}\left(\frac{1}{k!}(1+D)^k f(0)\right)\phi^{(k)}(0) = \sum_{k\geq 0}\left(\frac{1}{k!}\sum_{j=0}^{k}\binom{k}{j}D^j f(0)\right)\phi^{(k)}(0) \\
&= \sum_{k\geq 0}\left(\frac{1}{k!}\sum_{j=0}^{k}\binom{k}{j}\Delta^j\psi(0)\right)\phi^{(k)}(0) = \sum_{k\geq 0}\frac{\psi(k)\cdot\phi^{(k)}(0)}{k!}.
\end{aligned}$$

Note that (3.80) has a different proof that appeared in [107], and that its particular case with $\psi(t) = t^r$ implies several interesting special identities. For instance, by means of the following special choices (1) $\phi(t) = e^t$, (2) $\phi(t) = 1 + t + \cdots + t^m$ $(m \geq 1)$ and (3) $\phi(t) = (1-tx)^{-1}$ with $|tx| < 1$, one may get, respectively, (1) the well-known formula of Dobinski for the Bell number $\omega(r)$, (2) the useful formula of Stirling for the arithmetic progression of higher order, and (3) the formula of Euler for the arithmetic-geometric series (cf. Lemma 2.1.7 and Example 2.2.7).

$$\sum_{k=0}^{\infty} k^r x^k.$$

Example 3.2.11 *Taking* $\psi(t) = \binom{t+r}{r}$, *it follows from* (3.80) *that*

$$\sum_{k=0}^{\infty}\binom{k+r}{r}\frac{\phi^{(k)}(0)}{k!} = \sum_{k=0}^{r}\binom{r}{k}\frac{\phi^{(k)}(1)}{k!}. \tag{3.82}$$

Recall that

$$D^k \cos t = \cos\left(t + \frac{k\pi}{2}\right) \text{ and } D^k \sin\, t = \sin\left(t + \frac{k\pi}{2}\right),$$

so that the special choices $\phi(t) = \cos t$ *and* $\phi(t) = \sin t$ *in* (3.82) *lead to a pair of combinatorial series sums:*

$$\sum_{k=0}^{\infty}\frac{(-1)^k}{(2k)!}\binom{2k+r}{r} = \sum_{k=0}^{r}\frac{1}{k!}\binom{r}{k}\cos\left(1+\frac{1}{2}k\pi\right), \tag{3.83}$$

$$\sum_{k=0}^{\infty} \frac{(-1)^k}{(2k+1)!} \binom{2k+r+1}{r} = \sum_{k=0}^{r} \frac{1}{k!} \binom{r}{k} \sin\left(1 + \frac{1}{2}k\pi\right) \quad (3.84)$$

Example 3.2.12 *Recall that the generalized m-th order Bernoulli polynomials and Euler polynomials are defined by* (cf. *[157]*)

$$e^{zt}\left(\frac{t}{e^t-1}\right)^m = \sum_{k=0}^{\infty} B_k^{(m)}(z)\frac{t^k}{k!} \qquad (|t| < 2\pi), \quad (3.85)$$

$$e^{zt}\left(\frac{2}{e^t+1}\right)^m = \sum_{k=0}^{\infty} E_k^{(m)}(z)\frac{t^k}{k!} \qquad (|t| < \pi), \quad (3.86)$$

where $z \in \mathbf{C}, m \in \mathbf{N}, m \geq 1$. *Evidently both* $B_k^{(m)}(z)/k!$ *and* $E_k^{(m)}(z)/k!$ *are Sheffer-type polynomials associated with the (GF) pairs* $\{A(t) = (t/(e^t - 1))^m, g(t) = t\}$ *and* $\{A(t) = (2/(e^t+1))^m, g(t) = t\}$, *respectively (cf. [59, 157]). Also, both the LHS of* (3.85) *and* (3.86) *may be written as*

$$A(t)f(g(t)) = A(t)f(t) = A(t)exp(zt),$$

where $f(t) = e^{zt}$, *which are rational functions in* t, e^t *and* e^{zt}. *Thus using the first equation of* (3.58) *and referring to Corollary 2.2, we find that* (3.85) *and* (3.86) *with* $t \longmapsto D$ *lead to the following expansions (for given* $\phi(t) \in C^\infty$*)*

$$\begin{aligned} & (D/(E-1))^m E^z \phi(t) = \Delta^{-m}\phi^{(m)}(z+t) \\ = & \sum_{k=0}^{\infty} \frac{1}{k!} B_k^{(m)}(z)\phi^{(k)}(t), \end{aligned} \quad (3.87)$$

$$\begin{aligned} & (2/(E+1))^m E^z \phi(t) = \nabla^{-m}\phi^{(m)}(z+t) \\ = & \sum_{k=0}^{\infty} \frac{1}{k!} E_k^{(m)}(z)\phi^{(k)}(t), \end{aligned} \quad (3.88)$$

where ∇ *is the mean-value operator (in Nörlund's notation) defined by*

$$\nabla = \frac{1}{2}(1+E) = 1 + \frac{1}{2}\Delta.$$

Now, letting the operators Δ^m *and* ∇^m *be applied to both sides of* (3.87) *and* (3.88) *(with evaluation at* $t = 0$*), respectively, we obtain*

$$\phi^{(m)}(z) = \sum_{k=0}^{\infty} \frac{\Delta^m \phi^{(k)}(0)}{k!} B_k^{(m)}(z), \quad (3.89)$$

$$\phi^{(m)}(z) = \sum_{k=0}^{\infty} \frac{\nabla^m \phi^{(k)}(0)}{k!} E_k^{(m)}(z). \quad (3.90)$$

These two expansions are absolutely convergent under the following conditions, respectively

$$\sup_k |\ \Delta^m \phi^{(k)}(0)|^{1/k} < 2\pi, \quad \sup_k |\nabla^m \phi(k)(0)|^{1/k} < \pi. \quad (3.91)$$

Obviously (3.91) *follows from Cauchy's root-test and the conditions contained in* (3.85) *and* (3.86), *respectively.*

Note that both (3.85) *and* (3.86) *reduce to the Taylor expansion for* $m = 0$. *In the case for* $m = 1$, *they provide the classical expansions of an analytic function* $\phi'(z)$ *in terms of the ordinary Bernoulli and Euler polynomials, respectively.*

3.2.3 A further investigation of a source formula

The so-called source formula is a basic result that has been successfully applied to yield a good many special formulas, including a variety of remarkable formulas and identities in Discrete Mathematics and Combinatorics. The object of this subsection is to prove two related results that are given by the following theorems.

Theorem 3.2.13 *Let* $\psi(t)$ *and* $\phi(t)$*be real-valued functions defined on* $\mathbb{Z}$ *and* $\mathbb{R}$, *respectively, and let* $\alpha \in \mathbf{R}$. *Then the source formula of* (∞^2) *degree of the form*

$$\sum_{k\geq 0} \binom{\alpha}{k} \psi(k)\Delta^k\phi(0) = \sum_{k\geq 0} \binom{\alpha}{k} \Delta^k\psi(0) \cdot \Delta^k\phi(\alpha - k) \tag{3.92}$$

is a consequence form of (3.46) *of Theorem 3.2.4 with* $A(t) = (1+t)^\alpha$, $g(t) = t/(t+1)$ *and*

$$f(t) = \sum_{k\geq 0} \binom{\alpha}{k} t^k\Delta^k\psi(0). \tag{3.93}$$

Proof. It requires to show that (3.46) of Theorem 3.2.4 yields (3.92) with the given $A(t), g(t)$, and $f(t)$. Clearly, the LHS of (3.46) (evaluated at $t = 0$) gives

$$A(\Delta)f(g(\Delta))\phi(0) = (1+\Delta)^\alpha \sum_{k\geq 0} \binom{\alpha}{k} \Delta^k\psi(0) \cdot (g(\Delta))^k\phi(t)|_{t=0}$$

$$= E^\alpha \sum_{k\geq 0} \binom{\alpha}{k} \Delta^k\psi(0) \left(\frac{\Delta}{E}\right)^k \phi(t)|_{t=0}$$

$$= \sum_{k\geq 0} \binom{\alpha}{k} \Delta^k\psi(0) \cdot \Delta^k\phi(\alpha - k) = RHS \text{of } (3.92).$$

On the other hand, the RHS of (3.46) (evaluated at $t = 0$) may be written as

$$\sum_{k\geq 0}(p_k(D)f(0))\Delta^k\phi(0) = \sum_{k\geq 0}\left(\sum_{j=0}^{k} \frac{1}{j!} d_{kj} D^j f(0)\right) \Delta^k\phi(0).$$

Herein d_{kj} is given by

$$d_{kj} = [t^k](1+t)^\alpha (t/(t+1))^j = [t^{k-j}](1+t)^{\alpha-j} = \binom{\alpha-j}{k-j}.$$

Then, according to (3.93), we have

$$\begin{aligned} p_k(D)f(0) &= \sum_{j=0}^{k} \frac{1}{j!}\binom{\alpha-j}{k-j} D^j f(0) = \sum_{j=0}^{k}\binom{\alpha}{j}\binom{\alpha-j}{k-j}\Delta^j\psi(0) \\ &= \binom{\alpha}{k}\sum_{j=0}^{k}\binom{k}{j}\Delta^j\psi(0) = \binom{\alpha}{k}\psi(k). \end{aligned}$$

Finally, the RHS of (3.46) gives

$$\sum_{k\geq 0}(p_k(D)f(0))\Delta^k\phi(0) = \sum_{k\geq 0}\binom{\alpha}{k}\psi(k)\Delta^k\phi(0) = LHS \text{of } (3.92).$$

Thus, we complete the proof.

Theorem 3.2.14 *Let $\psi(t)$ and $\phi(t)$ be defined as in Theorem 3.2.13. Denote*

$$\theta_1 = \overline{\lim_{k\to\infty}}|\Delta^k\phi(0)|^{1/k}, \quad \theta_2 = \overline{\lim_{k\to\infty}}|\Delta^k\psi(0)|^{1/k}. \tag{3.94}$$

Then (3.92) *becomes an exact equality for $\alpha \in \mathbf{R}$, provided that*

$$\theta_1(1+\theta_2) < 1. \tag{3.95}$$

Proof. Starting with the familiar equalities

$$\psi(k) = \sum_{j=0}^{k}\binom{R}{j}\Delta^j\psi(0) \quad \text{and} \quad \binom{\alpha}{k}\binom{k}{j} = \binom{\alpha}{j}\binom{\alpha-j}{k-j}, \tag{3.96}$$

which have been employed in the preceding proof, one may see that the LHS of (3.92) can be computed formally as follows:

$$\begin{aligned} \text{LHS of } (3.92) &= \sum_{k=0}^{\infty}\binom{\alpha}{k}\Delta^k\phi(0)\sum_{j=0}^{k}\binom{k}{j}\Delta^j\psi(0) \\ &= \sum_{j=0}^{\infty}\Delta^j\psi(0)\sum_{k=j}^{\infty}\binom{\alpha}{k}\binom{k}{j}\Delta^k\phi(0) \\ &= \sum_{j=0}^{\infty}\binom{\alpha}{j}\Delta^j\psi(0)\sum_{k=j}^{\infty}\binom{\alpha-j}{k-j}\Delta^k\phi(0) \end{aligned}$$

$$= \sum_{j=0}^{\infty} \binom{\alpha}{j} \Delta^j \psi(0)(1+\Delta)^{\alpha-j} \Delta^j \phi(0)$$

$$= \sum_{j=0}^{\infty} \binom{\alpha}{j} \Delta^j \psi(0) \Delta^j \phi(\alpha - j) = \text{RHS of (3.92)}.$$

Apparently, the key step (without justification) in the formal derivation is the exchange of orders of repeated summation. Thus in order to prove the theorem, it suffices to show that the condition (3.95) ensures the absolute convergence of the repeated summations involved. This can be done in what follows.

First, according to (3.95), one may choose $\delta_1 > 0$, $\delta_2 > 0$ so small that lead to

$$\overline{\theta}_1(1+\overline{\theta_2}) < 1 \text{ with } \overline{\theta}_1 = \theta_1 + \delta_1, \;\; \overline{\theta}_2 = \theta_2 + \delta_2.$$

The condition (3.94) implies that there exist integers $m > 0$ and $n > 0$, such that

$$|\Delta^k \phi(0)|^{1/k} < \overline{\theta}_1, \;\; |\Delta^j \psi(0)|^{1/j} < \overline{\theta}_2,$$

for $k \geq m$ and $j \geq n$, where it is no real restriction to assume that $m > n$. Consequently, it can be shown that the following series consisting of repeated summations

$$S_{m,n} := \sum_{k \geq m} \binom{\alpha}{k} \Delta^k \phi(0) \sum_{j \geq n} \binom{k}{j} \Delta^j \psi(0)$$

is absolutely convergent. Indeed, we have

$$|S_{m,n}| \leq \sum_{k \geq m} \left| \binom{\alpha}{k} \Delta^k \phi(0) \right| \sum_{j \geq n} \left| \binom{k}{j} \Delta^j \psi(0) \right|$$

$$\leq \sum_{k \geq m} \left| \binom{\alpha}{k} \overline{\theta}_1^k \right| \sum_{j \geq n} \binom{k}{j} \overline{\theta}_2^j$$

$$< \sum_{k \geq m} \left| \binom{\alpha}{k} \overline{\theta}_1^k (1+\overline{\theta}_2)^k \right| < \infty.$$

It remains to show that the absolute convergence of $S_{m,n}$ implies that of $S_{m,0}$ and so of $S_{0,0}$. Notice that

$$S_{m,0} = S_{m,n} + \sum_{k \geq m} \binom{\alpha}{k} \Delta^k \phi(0) \sum_{j=0}^{n-1} \binom{k}{j} \Delta^j \psi(0). \tag{3.97}$$

Clearly, the right-most summation on the RHS of (3.97) has an order estimate (in Landau's denotation) $\mathcal{O}(nk^{n-1})$ $(k \to \infty)$. Consequently, it is seen that the absolute convergence of the series (on the RHS of (3.97))

$$\sum_{k \geq m} \binom{\alpha}{k} \Delta^k \phi(0) \cdot \mathcal{O}(nk^{n-1})$$

follows from Cauchy's root test with

$$\overline{\lim_{k\to\infty}} \left| \binom{\alpha}{k} \mathcal{O}(nk^{n-1}) \right|^{1/k} = 1, \quad \overline{\lim_{k\to\infty}} |\Delta^k \phi(0)|^{1/k} = \theta_1 < 1.$$

Moreover, the absolute convergence of $S_{0,0}$ follows from that of $S_{m,0}$. Hence the theorem is proved.

As will be observed from many examples displayed in the next subsection that (3.92) is really a useful tool for finding various summation formulas and identities (including classical ones and new ones), and (3.94) and (3.95) provide a kind of general convergence condition which is more available than those intractable ones given in [128].

Corollary 3.2.15 *The formula* (3.92) *is an exact equality whenever* $\theta_1 + \theta_2 < 1$. *In particular, the condition may be replaced by* $\theta_1 < 1$ *if* $\psi(t)$ *is a polynomial in* t.

In fact, we have $\theta_2 = 0$ *if* $\psi(t)$ *is a polynomial.*

3.2.4 Various consequences of the source formula

In this subsection we will present 24 selected instances to illustrate the real capacity of (3.92) for finding or deriving special formulas and novel identities. It has been found that some particular choices of the triplet $\{\alpha, \psi(t), \phi(t)\}$ of (3.92) could lead to numerous interesting formulas and combinatorial identities. Some noticeable examples given in [128] may be briefly recalled for references.

First, let us mention that the following 24 special triplets (T_{21} includes two cases) yield, respectively, 24 notable formulas (Formula 21 includes two cases) shown below.

$$T_1 = \{\alpha = -1, \psi(t), \phi(t) = 2^t\},$$

$$T_2 = \left\{\alpha = x, \psi(t) = 1, \phi(t) = \binom{y+t}{n}\right\},$$

$$T_3 = \{\alpha = -1, \psi(t) = t^r, \phi(t) = (1-x)^t\},$$

$$T_4 = \left\{\alpha = -1, \psi(t) = t^r, \phi(t) = \binom{m-t}{m}\right\},$$

$$T_5 = \left\{\alpha = -s-1, \psi(t) = \binom{t}{j}, \phi(t) = \binom{n-t}{n}\right\},$$
$$T_6 = \{\alpha = n, \psi(t) = 1, \phi(t) = 1/(t+a)\},$$
$$T_7 = \left\{\alpha = n, \psi(t) = \binom{x+n-t}{n+m}, \phi(t) = \binom{m+t}{n}\right\},$$
$$T_8 = \left\{\alpha = n, \psi(t) = \binom{m+t}{m}, \phi(t) = \binom{p+m+t}{p}\right\},$$
$$T_9 = \left\{\alpha \longmapsto -(\alpha+1), \psi(t) = t^r, \phi(t) = \binom{n-t}{n}\right\},$$
$$T_{10} = \{\alpha = m, \psi(t) = a/(a+t), \phi(t) = (-1)^t\} \quad (a = -(2m+1)).$$
$$T_{11} = \{\alpha = -1, \psi(t) = H_t, \phi(t) = 1-x)^t\}$$
$$T_{12(i)} = \{\alpha = n, \psi(t) = t^r, \phi(t) = F_{s+t}\},$$
$$T_{12(ii)} = \{\alpha = n, \psi(t) = t^r, \phi(t) = F_{s+2t}\},$$
$$T_{13} = \left\{\alpha = x, \psi(t) = \binom{t+y}{y+b-a}, \phi(t) = \binom{y+b+t}{b}\right\}$$
(with $b = \min\{a, b\}$, $x, y, \in \mathbf{N}_j$ making$b \to (b-k)$, betting finally $x, y \in \mathbf{R}$),
$$T_{14} = \{\alpha = x, \psi(t) = t^m, \phi(t) = \cos(\alpha t + \beta)\},$$
$$T_{15} = \{\alpha = -1, \psi(t) \to \psi(t+1), \phi(t) = (1-x)^t\},$$
$$T_{16} = \{\alpha = -m-1, \psi(t) = H_{m+t} - H_m, \phi(t) = (1-x)^t\},$$
$$T_{17} = \Big\{\alpha = n, \psi(k) = a(kx-a)^{k-1}(kx+b)^{n-k} + (-a)^k b^{n-k},$$
$$\phi(t) = \binom{n-t}{n}\Big\},$$
$$T_{18} = \left\{\alpha, \psi(t) = t^n, \phi(t) = (1+\frac{1}{\alpha})(\cdots)t\right\},$$
$$T_{19} = \left\{\alpha = n-r, \psi(t) = \binom{2m+t}{n}, \phi(t) = 2^{-t}\right\},$$
$$T_{20} = \left\{\alpha, \psi(t) = t^r, \phi(t) = \binom{m-t}{m}\right\},$$
$$T_{21(i)} = \left\{\alpha = n, \psi(t) = H_{a+t}, \phi(t) = \binom{n-t}{n}\right\},$$
$$T_{21(ii)} = \left\{\alpha = n, \psi(t) = H_{a+t}, \phi(t) = \binom{t}{n}\right\},$$
$$T_{22} = \left\{\alpha = n, \psi(t) = H_{a+t}, \phi(t) = \binom{r+n-t}{n}\right\},$$
$$T_{23} = \left\{\alpha, \psi(t) = t^r, \phi(t) = \frac{1}{t+\lambda}\right\},$$
$$T_{24} = \left\{\alpha, \psi(t) = F_{r+3t}, \phi(t) = \binom{\alpha-t}{n}\right\},$$

1 Euler's formula for series transform (with $\psi(k) \downarrow 0$ as $k \to \infty$)

$$\sum_{k=0}^{\infty}(-1)^k\psi(k) = \sum_{j=0}^{\infty}(-1)^j\Delta^j\psi(0)/2^{j+1}.$$

2 Vandermonde's convolution formula (with $x, y \in \mathbf{R}$)

$$\sum_{k=0}^{n}\binom{x}{k}\binom{y}{n-k} = \binom{x+y}{n}.$$

3 Euler's formula for the arithmetic-geometric series (cf. Lemma 2.1.7 and Example 2.2.7)

$$\sum_{k=0}^{\infty}k^r x^k = \sum_{j=0}^{r}\frac{j!x^j}{(1-x)^{j+1}}\left\{ {r \atop j} \right\} (|x| < 1).$$

4 Stirling's formula for the arithmetic series of higher order

$$\sum_{k=0}^{m}k^r = \sum_{j=0}^{r}j!\left\{ {r \atop j} \right\}\binom{m+1}{j+1}.$$

5 Knuth's combinatorial identity ($s \in \mathbf{R}$)

$$\sum_{k=j}^{n}\binom{s+k}{k}\binom{k}{j} = \binom{s+j}{j}\binom{n+s+1}{n-j}.$$

6 A formula stated as a theorem in Wilf's [216] (with $\alpha \in \mathbf{R}$).

$$\sum_{k=0}^{n}(-1)^k\binom{n}{k}\Big/\binom{k+a}{k} = \frac{a}{n+a} \quad (n+a \neq 0)\ .$$

7 Riordan's identity (with $x \in \mathbf{R}$)

$$\sum_{k=0}^{n}\binom{n}{k}\binom{m}{n-k}\binom{x+n-k}{n+m} = \binom{x}{m}\binom{x}{n}.$$

8 An identity due to Li Shanlai

$$\sum_{k=0}^{n}\binom{n}{k}\binom{m}{k}\binom{p+n+m-k}{n+m} = \binom{p+n}{n}\binom{p+m}{m}.$$

9 An extended Stirling summation formula (with $\alpha \in \mathbf{R}$)

$$\sum_{k=0}^{n} \binom{\alpha+k}{k} k^r = \sum_{j=0}^{r} \binom{\alpha+j}{j} \binom{\alpha+n+1}{n-j} j! \left\{ {r \atop j} \right\}.$$

10 A so-called miraculous formula treated in Graham, Knuth, and Patashnik [79]

$$\sum_{k=0}^{m} \binom{m}{k} \frac{2m+1}{2m+1-k} (-2)^k = 1 \Big/ \binom{-1/2}{m}.$$

The above instances suggest that (3.92) may be continually used to deduce more formulas of some interest via various suitable choices of the triplet $\{\alpha, \psi, \phi\}$. Indeed, as further consequences of (3.92), we may mention in succession the following 14 verifiable special formulas.

11 The (GF) for harmonic numbers $H_0 = 0$, $H_k = 1 + \frac{1}{2} + \cdots + \frac{1}{k}$, $(k \geq 1)$

$$\sum_{k=1}^{\infty} H_k x^k = \frac{1}{1-x} \log \frac{1}{1-x} \quad (|x| < 1).$$

12 A pair of identities involving Fibonacci numbers defined by $F_0 = 0$, $F_1 = 1$, $F_n = F_{n-1} + F_{n-2}$, $n \in \{2, 3, \cdots\}$

(i) $$\sum_{k=0}^{n} \binom{n}{k} k^r F_{s-k} = \sum_{j=0}^{r} j! \left\{ {r \atop j} \right\} \binom{n}{j} F_{s+n-j} \quad (s \geq k).$$

(ii) $$\sum_{k=0}^{n} \binom{n}{k} k^r F_{s+k} = \sum_{j=0}^{r} j! \left\{ {r \atop j} \right\} \binom{n}{j} F_{s+2n-j} \quad (s \geq 0).$$

13 Stanley's identity (with $x, y \in \mathbf{R}$)

$$\sum_{k=0}^{\min\{a,b\}} \binom{x+y+k}{k} \binom{y}{a-k} \binom{x}{b-k} = \binom{x+a}{b} \binom{x+b}{a}.$$

14 A summation formula for a kind of trigonometric series involving 4 parameters $\alpha, \beta, x \in \mathbf{R}$ and $m \in \mathbf{N}$ (with $0 < |\alpha| < \pi/3$)

$$\sum_{k=0}^{\infty} \binom{x}{k} k^m \left(2\sin\frac{\alpha}{2}\right)^k \cos\left(\beta + \frac{k}{2}(\alpha+\pi)\right)$$
$$= \sum_{j=0}^{m} \binom{x}{j} j! \left\{ {m \atop j} \right\} \left(2\sin\frac{\alpha}{2}\right)^j \cos\left(\alpha x + \beta + \frac{j}{2}(\pi - \alpha)\right).$$

15 Montmort's series transform formula with $|x| < 1, |x/(1-x)| < 1$ (cf. Theorem 1.3.10)

$$\sum_{k=1}^{\infty} \psi(k)x^k = \sum_{j=0}^{\infty} \Delta^j \psi(1) \left(\frac{x}{1-x}\right)^{j+1}.$$

16 A formula due to D. A. Zave [80] (with $m \in \mathbf{N}, |x| < 1$)

$$\sum_{k=1}^{\infty} \binom{k+m}{m} (H_{k+m} - H_m)x^k = (1-x)^{-m-1} \log\left(\frac{1}{1-x}\right).$$

17 An equivalent form of Abel's identity

$$\sum_{k \geq 0} (-1)^k \binom{n}{k} \left[a(kx-a)^{k-1}(kx+b)^{n-k} + (-a)^k b^{n-k}\right] = 0.$$

18 A pre-limit form of Dobinski's formula for Bell numbers (with $\alpha > 1$)

$$\sum_{k=0}^{\infty} \binom{\alpha}{k} \left(\frac{1}{\alpha}\right)^k k^n = \sum_{j=0}^{n} \binom{\alpha}{j} \left(\frac{1}{\alpha}\right)^j j! \left\{ {n \atop j} \right\} \left(1 + \frac{1}{\alpha}\right)^{\alpha - j}.$$

19 Grosswald's formula (with $n > r > 0$ and $n + r = 2m$)

$$\sum_{k=0}^{n-r} (-1)^k \binom{n-r}{k} \binom{2m+k}{n} 2^{-k}$$
$$= (-1)^{(n-r)/2} 2^{r-n} \binom{n}{m} \binom{2m}{n} \binom{n}{r}^{-1}.$$

20 A generalized Stirling formula (with $\alpha \in \mathbf{R}$)

$$\sum_{k=0}^{m} \binom{\alpha}{k} (-1)^k k^r = \sum_{j=0}^{r} (-1)^j \binom{\alpha}{j} \binom{m-\alpha}{m-j} j! \left\{ {r \atop j} \right\}.$$

21 Two identities involving harmonic numbers

(i) $\displaystyle\sum_{k=0}^{n} (-1)^{k-1} \binom{n}{k} H_{a+k} = \frac{1}{a+1} \Big/ \binom{a+n}{n-1} \quad (n \geq 1, a \in \mathbf{N}).$

(ii) $\displaystyle\sum_{k=1}^{n} (-1)^{k-1} \binom{n}{k} \Big/ \binom{a+k}{k-1} = (a+1)(H_{a+n} - H_a.)$

22 A general combinatorial identity involving harmonic numbers

$$\sum_{k=0}^{n}\binom{n}{k}\binom{r+n-k}{r}(-1)^{k-1}H_{a+k}$$
$$=\frac{1}{a+1}\sum_{j=0}^{r}\binom{r}{j}\binom{n}{j}\Big/\binom{a+n-j}{a+1}.$$

23 A summation formula for a kind of combinatorial series

$$\sum_{k=0}^{\infty}(-1)^k\binom{\alpha}{k}\binom{\lambda+k}{k}^{-1}k^r$$
$$=\frac{\lambda}{\alpha+\lambda}\sum_{j=0}^{r}(-1)^j\binom{\alpha}{j}j!\left\{\begin{matrix}r\\ j\end{matrix}\right\}\binom{\alpha+\lambda-1}{j}^{-1},$$

where $\alpha,\lambda\in\mathbf{R}$ and $r\in\mathbf{N}$ such that $\alpha\geq-1$ and $\lambda\geq r+2$.

24 A miraculous formula involving Fibonacci numbers (with $\alpha\in\mathbf{R}, r\in\mathbf{Z}$)

$$\sum_{k=0}^{n}\binom{\alpha}{k}\binom{\alpha-k}{n-k}(-1)^kF_{r+3k}=\binom{\alpha}{n}(-2)^nF_{r+n}.$$

Note that suitable specializations of 22–24 could lead to several known formulas and identities, and that the convergence conditions $|x|<1$ and $|\alpha|<\pi/3$ involved in formulas 3 and 14, respectively, are both deducible from the condition $\theta_1<1$ given by Corollary 3.2.15.

Surely, for the sake of immediate verification (cf. Exercise 3.12), the reader may find that the following short table of 7 difference formulas and the different formulas shown in Subsection 3.1.4 are both needful and helpful. One may notes that some difference formulas given in Subsection 3.1.4 are the special cases of some formulas shown below.

(i) $\Delta^k\binom{a+t}{n}=\binom{a+t}{n-k}$ $(k\leq n)$; $\Delta^k\binom{a+t}{n}_0=\binom{a}{n-k}$;

(ii) $\Delta^k\binom{a-t}{n}=(-1)^k\binom{a-t-k}{n-k}$ $(k\leq n)$; $\Delta^k\binom{a-t}{n}_0=(-1)^k\binom{a-k}{n-k}$;

(iii) $\Delta^k\left(\frac{1}{t+a}\right)=\frac{(-1)^kk!}{(t+a)(t+a+1)\cdots(t+a+k)}$;

$\Delta^k\left(\frac{1}{t+a}\right)_0=\frac{(-1)^k}{a}\Big/\binom{k+a}{k}$ $(a\neq0)$;

(iv) $\Delta^k\cos(at+b)=(2\cdot\sin\frac{a}{2})^k\cos(at+b+\frac{k}{2}(a+\pi))$;

(v) $\Delta^k\sin(at+b)=(2\cdot\sin\frac{a}{2})^k\sin(at+b+\frac{k}{2}(a+\pi))$;

(vi) $\Delta^kH_t=\Delta^{k-1}\frac{1}{t+1}=\frac{(-1)^{k-1}(k-1)!}{(t+k)_k}$, $(\Delta^kH_t)_0=\frac{(-1)^{k-1}}{k}$;

(vii) $\Delta^kF_t=F_{t-k}$ $(k\in\mathbf{N}, t\geq k)$.

3.2.5 Lifting process and formula chains

For brevity, the formulas (3.45) and (3.46) may be rewritten as a single formula

$$A(\delta)f(g(\delta))\phi(t) = \sum_{k\geq 0}(p_k(D)f(0))\delta^k\phi(t), \tag{3.98}$$

where δ denotes either D or Δ. Also, we will make use of (3.53) with $x \longmapsto t$

$$A(t)f(g(t)) = \sum_{k\geq 0}(p_k(D)f(0))t^k. \tag{3.99}$$

In what follows we will show that both (3.98) and (3.99) could be used as basic formulas to produce chains of formulas having freedom degrees in increasing orders. The basic idea is to construct a kind of iteration process starting from a given initial formula with a form of either (3.98) or (3.99).

Let $\{A_m(t)\}_1^\infty, \{g_m(t)\}_1^\infty$ and $\{f_m(t)\}_1^\infty$ be an arbitrary fps or functions of the class C^∞ such that $A_m(0) = 1, g_m(0) = 0, g'_m(0) = Dg_m(0) \neq 0, (m = 1, 2, \cdots)$. Accordingly, for each $m \geq 1$ we may construct a Riordan array $(d_{kj}) = (A_m(t), g_m(t))$ with d_{kj} being defined by

$$d_{kj} = [t^k]A_m(t)(g_m(t))^j, \quad (j \geq k < \infty).$$

Consequently, we have Sheffer-type operators $p_k^{(m)}(D)$ defined by

$$p_k^{(m)}(D) = \sum_{j=0}^{k}(d_{kj}/j!)D^j.$$

This leads to the expansion formula

$$A_m(t)f_m(g_m(t)) = \sum_{k\geq 0}(p_k^{(m)}(D)f_m(0))t^k \tag{3.100}$$

and its allied operator series expansion formula for $\phi_m(t) \in C^\infty$

$$A_m(\delta)f_m(g_m(\delta))\phi_m(t) = \sum_{k\geq 0}(p_k^{(m)}(D)f_m(0))\delta^k\phi_m(t). \tag{3.101}$$

Now, suppose that for each $m \geq 1$, the RHS of (3.100) defines a function $f_{m+1}(t)$ iteratively, viz

$$\sum_{k\geq 0}(p_k^{(m)}(D)f_m(0))t^k = f_{m+1}(t). \tag{3.102}$$

Accordingly, the sequence $\{f_m(t)\}_2^\infty$ is created by the iteration process of (3.102) starting from A_1, g_1 and f_1. In this way we see that (3.100) gives a (∞^{2m+1}) degree formula, since it consists of arbitrary functions $A_j(t), g_j(t), (j = 1, 2, \cdots, m)$ and $f_1(t)$.

Moreover, start from (3.101) with $m = 1$ and the initial choices A_1, g_1, f_1 and ϕ_1 and let f_{m+1} be defined iteratively by the RHS of (3.101), viz

$$\sum_{k\geq 0}(p_k^{(m)}(D)f_m(0))\delta^k\phi_m(t) = f_{m+1}(t) \ \ (m \geq 1). \tag{3.103}$$

Then we see that (3.101) is a formula of (∞^{3m+1}) degree, as it consists of arbitrary functions $A_j, g_j, \phi_j, (j = 1, 2, \cdots, m)$ and f_1.

From that given above we may conclude that the iteration processes defined by (3.102) and (3.103) could produce two chains of formulas with increasing freedom degrees $(2m + 1)$ and $(3m + 1)$ $(m = 1, 2, \cdots)$, respectively. The iterations process may also be called the lifting process, for it could lift the freedom degrees of formulas successively.

Note that every member of the formula chains could be employed as a source formula to yield a formula family. Thus we may state a set-theoretic proposition belonging Cantor's category: "There exist countably infinitely many formula families (with various degrees)".

Certainly, the rather special and quite interesting formula family is the so-called $(\Sigma\Delta)$ class that consists of all exact formulas (series summations and expansions) deducible from (3.92) via proper triplets. Moreover, the set of all exact formulas deducible from (3.98) also gives an interesting family, say $[\Sigma\delta]$ family, which is much more comprehensive than $(\Sigma\Delta)$ class. Surely, more fruitful applications of both (3.92) and (3.98) to Computational Analysis are worthy of further discussion in Section 4.2.

3.3 $(\Sigma\Delta D)$ General Source Formula and Its Applications

Section 2.2 gives a symbolic operator method for the summation of power series based upon two series transformation formulas. One of these formulas involves an extension of Eulerian fractions. Numerous applications of these series transformation formulas are given as special cases. Meanwhile, many illustrative examples, including a Gegenbauer type series transformation formula, are represented. The last section of this chapter extends the results of Section 2.2. We now consider a more general source formula (GSF) called $(\Sigma\Delta D)$ source formula. As an illustration of the application of the source formula, more than 50 special series expansions and summation formulas have been given in the previous sections, and more formulas will be shown later.

3.3.1 $(\Sigma\Delta D)$ GSF

As may be seen, our previous sections have been concerned with finding some source formulas that could be used to draw various special formulas for series expansions and summations. The basic tools we employed are the symbolic operator calculus, the theory of formal power series (fps) and that of differential operators. What we have obtained and utilized are certain series transformation formulas involving the ordinary difference operators Δ^k (with increment 1 for Δ) and differential operators D^k (with $D = d/dt$), wherein $k \in \mathbb{N}$ and $\Delta^0 = D^0 = 1$, the identity operator. Some fruitful results may be recalled briefly as follows.

Let $\psi(t)$ and $\phi(t)$ be real-valued functions defined on $\mathbb{R}$, and let $x \in \mathbf{R}$. Then there holds a series expansion formula of the form

$$\sum_{k\geq 0} \binom{x}{k} \psi(k)\Delta^k \phi(0) = \sum_{k\geq 0} \binom{x}{k} \Delta^k \psi(0)\Delta^k \phi(x-k), \tag{3.1}$$

which is given in Theorem 3.2.13.

Also, if $\psi(t)$ and $\phi(t)$ are infinitely differentiable on $[0, \infty)$, there are two expansion formulas involving derivatives obtained in Theorem 2.2.2:

$$\sum_{k\geq 0} \psi(k)\phi^{(k)}(0)\frac{x^k}{k!} = \sum_{k\geq 0} \Delta^k \psi(0)\phi^{(k)}(x)\frac{x^k}{k!} \tag{3.2}$$

$$\sum_{k\geq 0} \psi(k)\phi^{(k)}(0)\frac{x^k}{k!} = \sum_{k\geq 0} \frac{1}{k!}\psi^{(k)}(0)A_k(x, \phi(x)), \tag{3.3}$$

where $A_m(x, g(x))$ is an extension of Euler fraction in terms of $g(x)$ defined as in (2.91), i.e.,

$$A_m(x, \phi(x)) = \sum_{j=0}^{m} \left\{ {m \atop j} \right\} \phi^{(j)}(x)x^j, \quad A_0(x, \phi(x)) = \phi(x), \tag{3.4}$$

where $\left\{ {m \atop j} \right\}$ are Stirling numbers of the second kind, i.e., $j!\left\{ {m \atop j} \right\} = \left[\Delta^j t^m\right]_{t=0}$, which is also denoted by $S(m, j)$.

As an analog of derivative operator D, Δ acting on $\{f(n)\}$ such that $\Delta f(n) = f(n+1) - f(n)$ and for $(n)_k = n(n-1)\cdots(n-k+1)$, $\Delta(n)_k = k(n)_{k-1}$. Hence

$$f(n) = \sum_{k\geq 0} \Delta^k f(0)\frac{(n)_k}{k!} = \sum_{k\geq 0} \Delta^k f(0)\binom{n}{k}. \tag{3.5}$$

Equation (3.5) can be proved by using the substitution of the known divided difference formula

$$\Delta^k f(0) = \sum_{j=0}^{k} (-1)^{k-j}\binom{k}{j} f(j)$$

into the leftmost side of (3.5) and switching the order of the summation to yield

$$\sum_{k\geq 0} \Delta^k f(0)\binom{n}{k} = \sum_{k\geq 0}\sum_{j=0}^{k} (-1)^{k-j}\binom{n}{k}\binom{k}{j} f(j)$$

$$= \sum_{j\geq 0}\left(\sum_{k=j}^{n}(-1)^{k-j}\binom{n}{k}\binom{k}{j}\right)f(j) = \sum_{j\geq 0}\left(\sum_{k=0}^{n-j}(-1)^{k}\binom{n}{k+j}\binom{k+j}{j}\right)f(j)$$

$$= \sum_{j\geq 0}\left(\sum_{k=0}^{n-j}(-1)^{k}\binom{n-j}{k}\right)\binom{n}{j}f(j) = f(n).$$

In the above sense, the series expansion formula (3.1) is an analog of series expansion formula (3.2). More details can be seen in Remark 3.3.5.

Apparently, when taking $\psi(t) = 1$, formulas (3.1)–(3.3) will be reduced to Newton's interpolation series and Maclaurin's expansion of $\phi(x)$, respectively. Series expansions (3.1)–(3.3) are the basic results given in Sections 2.2 and 3.2, and there have been already given plenty of examples showing that a variety of special formulas and identities accordingly. Hence, (3.1), (3.2), and (3.3) may be called, respectively, the 1st, 2nd and 3rd source formula, or denoted briefly as SF(1), SF(2), and SF(3). Certainly, each of these formulas is associated with a given triplet $\{x, \psi, \phi\}$, and all possible special formulas are deduced via suitable special choices of the triplets.

As was mentioned in Section 3.2, the pair of operator series expansions given by the following Theorem 3.3.1 could be rewritten as a single formula involving a delta operator δ. Also, it has been proved that the SF(1) is deducible from the single formula with $\delta = \Delta$ (cf. loc. cit) so that the so-called single formula (3.98) may be adopted as a general source formula (GSF). In this subsection we shall give a utilizable specialization of the GFS and will show that both SF(2) and SF(3) are included in the specialization as consequences. Thus as a conclusion, one may think that the GSF is really a common source for all the SF(i)'s ($i = 1, 2, 3$).

As in Section 3.2, $A(t), g(t), f(t), \phi(t)$, etc. always denote the fps or the functions in C^∞ defined in $\mathbb{R}$ or $\mathbb{C}$ (real or complex number field). All operators are assumed to be acting on the fps or functions of t, unless otherwise stated. As usual, the shift operator E is defined by $E^\alpha f(t) = f(t+\alpha)$. An operator Q is said to be shift-invariant if $QE^\alpha = E^\alpha Q$ for every α. Recall that a shift-invariant operator Q is called a delta operator whenever $Qt \neq 0$ (a non-zero constant). Obviously, $\Delta, D, \Delta, D, \Delta E^\alpha$, and DE^α are the most useful delta operators.

Let us now reformulate the basic result of Section 3.2 in a more general form as follows.

Theorem 3.3.1 *Let $A(t), g(t)$, and $f(t)$ be fps over $\mathbb{R}$ or $\mathbb{C}$ such that $A(0) = 1$, $g(0) = 0$ and $g'(0) = Dg(0) \neq 0$. Let $\phi(t) \in C^\infty$, and let δ be a delta operator. Then there holds formally an operator series expansion formula of the form*

$$A(\delta)f(g(\delta))\phi(t) = \sum_{k\geq 0}(p_k(D)f(0))\delta^k\phi(t). \tag{3.6}$$

Herein $p_k(D)$ ($k \in \mathbb{N}$) are Sheffer-type differential operators given by the expression

$$p_k(D) = \sum_{j=0}^{k} \left(\frac{1}{j!} d_{kj}\right) D^j \tag{3.7}$$

within which $(d_{kj}) = (A(t), g(t))$ is a Riordan array obtainable via the use of the extracting-coefficient operator $[t^k]$, namely

$$d_{kj} = [t^k] A(t)(g(t))^j, \quad 0 \le j \le k, \quad j,k \in \mathbb{N}. \tag{3.8}$$

Moreover, if it is assumed that

$$\theta = \overline{\lim_{k\to\infty}} |p_k(D) f(0)|^{1/k} > 0. \tag{3.9}$$

Then the expansion formula (3.6) *becomes an exact equality at $t = 0$, provided that*

$$\overline{\lim_{k\to\infty}} \left|\delta^k \phi(0)\right|^{1/k} < \frac{1}{\theta}. \tag{3.10}$$

Actually, (3.6) is obtained by twice the applications of Mullin-Rota's substitution rule. In what follows we will give a utilizable specialization of (3.6).

Corollary 3.3.2 *By taking $\delta = D$, $g(t) = t$ and $A(t) = e^{xt}$ in* (3.6) *with x being a given real of complex parameter, we may get a formal series expansion of the form*

$$f(D)\phi(x+t) = \sum_{k\ge 0} \frac{1}{k!}(x+D)^k f(0) D^k \phi(t), \tag{3.11}$$

where f and ϕ are fps or functions in C^∞ defined on $\mathbb{R}$ or $\mathbb{C}$.

Proof. From the given conditions we see that the LHS of (3.6) gives

$$e^{xD} f(D)\phi(t) = E^x f(D)\phi(t) = f(D)\phi(x+t) = \text{LHS of (3.11)}.$$

Also, in accordance with the RHS of (3.6), we have to compute $p_k(D)f(0)$. Using (3.7) and (3.8) we easily find

$$\begin{aligned} p_k(D)f(0) &= \sum_{j=0}^{k} \frac{1}{j!} d_{kj} D^j f(0) = \sum_{j=0}^{k} \frac{1}{j!} [t^k](e^{xt} \cdot t^j) D^j f(0) \\ &= \sum_{j=0}^{k} \frac{x^{k-j}}{j!(k-j)!} D^j f(0) \\ &= \frac{1}{k!} \sum_{j=0}^{k} \binom{k}{j} x^{k-j} D^j f(0) = \frac{1}{k!}(x+D)^k f(0). \end{aligned}$$

Hence the RHS of (3.6) yields the RHS of (3.11).

Evidently, Taylor's expansion formula is a particular case of (3.11) with $f(t) = 1$. In the next subsection we will give an important application of (3.11).

Example 3.3.3 *Recall that the Bernoulli polynomials $B_k(x)$ and Charlier polynomials $C_k^{(\alpha)}(x)$ $(k \in \mathbb{N})$, are Sheffer-type polynomials so that $B_k(D)$ and $C_k^{(\alpha)}(D)$ give Sheffer-type differential operators. Note that they are generated by $\{A(t) = t/(e^t - 1), g(t) = t\}$ and $\{A(t) = e^{-\alpha t}, g(t) = log(1+t)\}$, respectively. Accordingly, the GSF (3.6) yields the following two series expansions*

$$\delta/(e^\delta - 1)f(\delta)\phi(t) = \sum_{k\geq 0} \frac{1}{k!} B_k(D)f(0) \cdot \delta^k \phi(t), \tag{3.12}$$

$$e^{-\alpha\delta} f(\log(1+\delta))\phi(t) = \sum_{k\geq 0} C_k^{(\alpha)}(D)f(0) \cdot \delta^k \phi(t). \tag{3.13}$$

In particular, taking $\delta = D$ for (3.12) and $\delta = \Delta$ for (3.13), and noticing that $e^D - 1 = E - 1 = \Delta$ and $\log(1+\Delta) = D$, we find the following two formal expansions

$$f(D)\phi'(t) = \sum_{k\geq 0} \frac{1}{k!} B_k(D)f(0)(\phi^{(k)}(t+1) - \phi^{(k)}(t)), \tag{3.14}$$

$$f(D)\phi(t) = \sum_{k\geq 0} C_k^{(\alpha)}(D)f(0) \cdot e^{\alpha\Delta}\Delta^k \phi(t), \tag{3.15}$$

where the operator $1/\Delta$ involved in the LHS of (3.12) has been removed to the RHS, and a similar process has been applied to the equation (3.13) to get (3.15).

3.3.2 GSF implies SF(2) and SF(3)

To prove that GSF represented in the previous subsection implies SF(2) and SF(3), it suffices to show that SF(2) and SF(3) could be deduced from (3.11) with special choices of $f(t)$.

Proposition 3.3.4 *The formal expansion formulas (3.2) and (3.3), namely SF(2) and SF(3), are deducible from (3.11) with evaluation at t = 0 and with the following choices of $f(t)$*

$$f(t) = \sum_{k\geq 0} \frac{1}{k!} \Delta^k \psi(0) x^k t^k, \tag{3.16}$$

$$f(t) = \sum_{k\geq 0} \frac{1}{k!} \psi^k(0) \sum_{j=0}^{k} \left\{ {k \atop j} \right\} x^j t^j, \tag{3.17}$$

respectively, where x is a given parameter.

Proof. First, it may be seen that the LHS of (3.6) evaluated at $t = 0$ just provides the RHS of (3.2) and BHS of (3.3) with f(t) being defined by (3.16) and (3.17), respectively. Moreover, it is clear that for (3.16) we have formal derivatives

$$f^{(j)}(0) = D^j f(o) = x^j \Delta^j \psi(0), \quad j \in \mathbb{N}.$$

Also, for (3.17)we have

$$f^{(j)}(0) = \sum_{k\geq 0} \frac{1}{k!}\psi^{(k)}(0) j! \begin{Bmatrix} k \\ j \end{Bmatrix} x^j = x^j \sum_{k\geq 0} \frac{1}{k!}\psi^{(k)}(0)(\Delta^j t^k)_{t=0}$$

$$= x^j \left(\Delta^j \sum_{k\geq 0} \frac{1}{k!}\psi^{(k)}(0) t^k \right)_{t=0} = x^j (\Delta^j \psi(t))_{t=0} = x^j \Delta^j \psi(0).$$

This shows that for both (3.16) and (3.17) we have the same expression

$$(x+D)^k f(0) = \sum_{j=0}^{k} \binom{k}{j} x^{k-j} x^j \Delta^j \psi(0) = x^k \psi(k).$$

Hence both (3.16) and (3.17) are implied by (3.6) with the evaluation of $t = 0$.

Remark 3.3.5 *What is worth mentioning is the fact that SF(2) and SF(3) have been found before so that the choices of $f(t)$ for S(2) and S(3) appear to be a relatively easier matter. Also, it is known that SF(1) is a special case of* (3.1) *(viz. GSF) with $A(t) = (1+t)^x$, $g(t) = t/(t+1)$ and*

$$f(t) = \sum_{k\geq 0} \binom{x}{k} \Delta^k \psi(0) t^k. \tag{3.18}$$

Here (3.18) *is a discrete analog of* (3.1)*. This is quite natural since SF(1) is actually a discrete analog of SF(2).*

3.3.3 Embedding techniques and remarks

As shown in our previous Sections 2.2 and 3.2, the main technique used for the derivation of most special formulas (cf. examples in Sections 2.2 and 3.2) is to make suitable choices of the triplets $\{x, \psi, \phi\}$ involved in the source formulas. Certainly, the unified technique may be called "embedding technique". In this subsection we will present some additional examples and give some more explanations for a few selected instances exhibited previously.

Note that both SF(1) and SF(2) involve computations of higher order differences. We have reproduced a short table of difference formulas in Subsection 3.1.4 and at the end of Subsection 3.2.4.

Example 3.3.6 *What is worth mentioning is that there are* 3 *classical formulas due to Euler, all deducible from the SF(1) with special choices of the triplet $\{x, \psi, \phi\}$. The first two formulas are known as the transformation formula* *for the alternating series and the summation formula for the arithmetical-geometric series. The third formula is usually called Euler's finite difference theorem, which may be expressed by the following equality*

$$\sum_{k=0}^{n}(-1)^k\binom{n}{k}f(k) = (-1)^n\Delta^n f(0) = \begin{cases} 0, & 0 \le m \le n-1, \\ (-1)^n n! a_n, & m = n, \end{cases} \tag{3.19}$$

where $f(t)$ is a polynomial of degree m, namely

$$f(t) = \sum_{j=0}^{m} a_j t^j, \quad m \in \mathbb{N}. \tag{3.20}$$

As may be observed, (3.19) *follows from* (3.1) *(SF(1)) via embedding with the special triple $\{x = n, \psi(t) = f(t), \phi(t) = \binom{n-t}{n}\}$ since $\phi(t)$ gives*

$$\Delta^n\phi(t) = (-1)^k\binom{n-t-k}{n-k}, \quad \Delta^k\phi(0) = (-1)^k,$$

and

$$\Delta^k\phi(n-k) = (-1)^k\binom{0}{n-k} = (-1)^k\delta_{n,k} = \begin{cases} (-1)^n, & k = n, \\ 0, & k < n. \end{cases}$$

Moreover, $\Delta^n f(0) = n!a_n$ just follows from a simple computation when $m = n$.

Example 3.3.7 *It may be of interest to notice that Abel's famous identity*

$$(a+b)^n = \sum_{k=0}^{n}\binom{n}{k}a(a-kx)^{k-1}(b+kx)^{n-k} \tag{3.21}$$

is implied by Euler's formula (3.19) *as a consequence. Clearly,* (3.21) *is equivalent to the algebraic identity*

$$\sum_{k=0}^{n}(-1)^k\binom{n}{k}\left[a(kx-a)^{k-1}(kx+b)^{n-k} + (-a)^k b^{n-k}\right] = 0, \tag{3.22}$$

where the LHS is of the same form as that of (3.19) *with*

$$f(k) = a(kx-a)^{k-1}(kx+b)^{n-k} + (-a)^k b^{n-k}. \tag{3.23}$$

Thus in order to prove that (3.19) *implies* (3.22)*, it suffices to show that* (3.23) *can be expressed algebraically as a polynomial in k of degree $(n-1)$, for $0 \le k \le n$. First, it is easily seen that the RHS of* (3.23) *can be expanded into a polynomial in kx of degree $(k-1)+(n-k) = n-1$ with the last term*

$(-a)^k b^{n-k}$ being cancelled within the expression and $f(0) = 0$. Moreover, $f(k)$ can be expressed algebraically in the following form (with $1 \le k \le n$)

$$\begin{aligned} f(k) =& a(kx+b)^{n-k} \sum_{0 \le j \le (n-1)} \binom{k-1}{j} (kx+b)^{k-1-j}(-a-b)^j + (-a)^k b^{n-k} \\ =& a \sum_{0 \le j \le (n-1)} (kx+b)^{n-j-1} \binom{k-1}{j} (-a-b)^j + (-a)^k b^{n-k} \\ =& \text{polynomial in } k \text{ of degree} (n-j-1)+j = n-1, \end{aligned}$$

wherein the term $(-a)^k b^{n-k}$ is already cancelled. Hence (3.19) implies (3.22).

As known, two classical proofs of Abel's identity have been given by Lucas and Francon (cf. §3.1, p. 128–129, [44]).

Example 3.3.8 *In a recently published interesting book [180] by Quaintance and Gould, chapter 7 is entitled "Melzak's formula", in which several nice combinatorial identities have been derived as applications of the formula. Also represented in the chapter is an elaborate proof of the formula (cf. loc. cit., p. 79–82, [180]). What we have found here is that a simplified form of Melzak's formula could be embedded in the SF(2) (viz. (3.2)), thus leading to a short proof.*

Let $f(x)$ be a polynomial in x of degree n, and let $y \in \mathbb{C}$. Melzak's formula states

$$f(x+y) = y\binom{y+n}{n} \sum_{k=0}^{n} (-1)^k \binom{n}{k} \frac{f(x-k)}{y+k}, \tag{3.24}$$

where $y \ne 0, -1, -2, \ldots, -n$. Clearly, we may treat $F(y) = f(x+y)$ as a polynomial in y of degree n with coefficients involving the parameter x. Thus (3.24) may be rewritten as

$$\sum_{k=0}^{n} (-1)^k \binom{n}{k} \frac{F(-k)}{y+k} = \frac{F(y)}{y} \Big/ \binom{y+n}{n}. \tag{3.25}$$

As $F(y)$ is a linear combination of monomials $\alpha_m y^m$ $(0 \le m \le n)$, it suffices to verify (3.25) with taking $F(y) \to y^m$. We only need to show

$$\sum_{k=0}^{n} \frac{(-k)^m}{k+y} (-1)^k \binom{n}{k} = \frac{y^m}{y} \Big/ \binom{y+n}{n}. \tag{3.26}$$

This may be embedded in the particular formula SF(2) with $x = 1$, viz. (3.2) with $x = 1$

$$\sum_{k \ge 0} \frac{1}{k!} \psi(k) \phi^{(k)}(0) = \sum_{k \ge 0} \frac{1}{k!} \Delta^k \psi(0) \phi^{(k)}(1). \tag{3.27}$$

Indeed, taking $\psi(t) = (-t)^m/(t+y)$ and $\phi(t) = (1-t)^n$, we find that the LHS of (3.27) just gives the LHS of (3.26). Also, we have $\phi^{(k)}(1) =$

$D^k(1-t)^n\big|_{t=1} = 0$ $(0 \le k \le n)$, $\phi^{(n)}(1) = (-1)^n n!$, *and we find the RHS of* (3.27)$= (-1)^n \Delta^n \psi(0)$. *Now we have (noticing that* $(m-1) < n$*)*

$$\begin{aligned}\Delta_t^n \psi(0) =& \Delta^n \psi(t)_0 = (-1)^m \left\{ \Delta_t^n \left(\frac{t^m - (-y)^m}{t-(-y)} \right)_0 + \Delta_t^n \left(\frac{(-y)^m}{t-(-y)} \right)_0 \right\} \\ =& \Delta_t^n \left(\frac{y^m}{t+y} \right)_0 = (-1)^n \frac{y^m}{y} \Big/ \binom{n+y}{n}.\end{aligned}$$

Hence the RHS of (3.27) *gives the RHS of* (3.26).

Example 3.3.9 *The following formula due to Zave [224] (with* $m \in \mathbb{N}$ *and* $|x| < 1$*)*

$$\sum_{k=1}^{\infty} \binom{k+m}{m} (H_{k+m} - H_m) x^k = (1-x)^{-m-1} \log \left(\frac{1}{1-x} \right) \tag{3.28}$$

can be obtained from the SF(1) *(viz.* (3.1)*) with the chosen triplet*

$$\{x = -m-1, \psi(t) = H_{m+1} - H_m, \phi(t) = (1-x)^t\}.$$

Indeed, using (3.1) *one may find that*

$$\begin{aligned}LHS \text{ of } (3.28) =& \sum_{k \ge 1} \binom{-m-1}{k} (H_{m+k} - H_m) \left(\Delta^k (1-x)^t \right)_0 \\ =& \sum_{j \ge 1} \binom{-m-1}{j} \Delta^j (H_{m+t} - H_m)_0 \left(\Delta^j (1-x)^t \right)_{t=-m-1-j} \\ =& \sum_{j \ge 1} (-1)^j \binom{m+j}{j} \left(\Delta^{j-1} \frac{1}{m+1+t} \right)_0 \\ & \quad \left((-x)^j (1-x)^t \right)_{t=-m-1-j} \\ =& \sum_{j \ge 1} \frac{1}{j} \binom{m+j}{j-1} \frac{(-1)^{j-1}}{\binom{m+j}{j-1}} \left(\frac{x}{1-x} \right)^j (1-x)^{-m-1} \\ =& (1-x)^{-m-1} \sum_{j \ge 1} \frac{(-1)^{j-1}}{j} \left(\frac{x}{1-x} \right)^j = RHS \text{ of } (3.28).\end{aligned}$$

Example 3.3.10 *Recall that the well-known C-numbers first introduced and utilized by Charalambides and Koutras, may be defined by the following [33]*

$$C(n, k; a, b) = \frac{1}{k!} \Delta^k (at+b)_n \Big|_{t=0}, \tag{3.29}$$

wherein $(x)_n = x(x-1)\cdots(x-n+1)$ $(n \ge 1)$ *and* $(x)_0 = 1$. *As shown in [33], C-numbers are particularly useful for obtaining closed sum formulas for combinatorial identities involving* $\binom{ak+b}{m}$ *as a factor in the summands, wherein* $m, k \in \mathbb{N}$ *and* $a, b \in \mathbb{R}$. *Indeed, there are two related general formulas that have been represented and utilized in [33], namely the following*

$$\sum_{k\geq 0}\binom{ak+b}{m} f^{(k)}(0)\frac{t^k}{k!} = \frac{1}{m!}\sum_{j=0}^{m} C(m,j;a,b) f^{(j)}(t)t^j, \tag{3.30}$$

$$\sum_{k\geq 0}\binom{ak+b}{m}\Delta^k g(0)\binom{t}{k} = \frac{1}{m!}\sum_{j=0}^{m} C(m,j;a,b)\Delta^j g(t-j)(t)_j, \tag{3.31}$$

where $f \in C^\infty$ *and* $g(t)$ *is defined on* $\mathbb{Z}$. *Obviously,* (3.30) *and* (3.31) *could be deduced from the SF*(2) *and SF*(1) *with the following special triplets, respectively.*

$$\left\{x = t, \psi(t) = \binom{at+b}{m}, \phi(t) = f(t)\right\}, \left\{x = t, \psi(t) = \binom{at+b}{m}, \phi(t) = g(t)\right\}.$$

Remark 3.3.11 *It is easily seen that* (3.30) *is an exact formula for* $|t| < \rho$, *provided that* $f(t)$ *has a Maclaurin series expansion for* $|t| < \rho$. *Moreover, the absolute convergence of the series in* (3.31) *is ensured by the conditions* $|t| < \infty$ *and*

$$\overline{\lim_{k\to\infty}} |\Delta^k g(0)|^{1/k} < 1. \tag{3.32}$$

Also, note that a variety of special formulas and identities deducible from either (3.30) *and* (3.31) *can be found in [33] or elsewhere.*

At the end of this subsection, we now give a survey of the substitution techniques used above, where all the mathematical terminologies will be used in the ordinary sense in mathematical sciences. Let us give here the following two definitions.

Definition 3.3.12 *A mathematical formula is said to be deducible from the GSF, if it could be deduced formally from any of the SF*(i) ($i = 1, 2, 3$) *or from the GSF itself with a special choice of the quintuplet* $\{\delta, A(t), f(t), g(t), \phi(t)\}$.

Definition 3.3.13 *All the mathematical formulas which are deducible from the GSF are said to form a formula class, so-called Sigma-Delta-D* ($\Sigma\Delta D$) *class.*

As known from our former sections quoted in the preceding sections, more than 50 special formulas and identities are the members belonging to the ($\Sigma\Delta D$) class. What is worth noticing is the fact that the ($\Sigma\Delta D$) class includes as special members those well-known classical formulas due to, respectively, Newton, Taylor, Euler, Stirling, Vandermonde, Montmort, Riordan, Carlitz, Li Shanlai, Knuth, Grosswald, Rosenbaum, Stanley, Gould, Melzak, Zave, et al.

Remark 3.3.14 *Sometimes, certain members of the* $(\Sigma\Delta D)$ *class may have limits when some parameters tend to* ∞. *For instance, taking the special triplet of the SF*(1)

$$\left\{x = \alpha, \psi(t) = t^n, \phi(t) = \left(1 + \frac{1}{\alpha}\right)^t\right\}, \quad \alpha > 1,$$

one may get a special formula of the form

$$\sum_{k=0}^{\infty} \binom{\alpha}{k} \left(\frac{1}{\alpha}\right)^k k^n = \sum_{j=0}^{n} \binom{\alpha}{j} \left(\frac{1}{\alpha}\right)^j j! \left\{ n \atop j \right\} \left(1 + \frac{1}{\alpha}\right)^{\alpha - j}, \tag{3.33}$$

which belongs to the $(\Sigma\Delta D)$ *class. Obviously,* (3.33) *yields the following limit when* $\alpha \to \infty$:

$$\sum_{k=0}^{\infty} \frac{k^n}{k!} = e \sum_{j=0}^{n} \left\{ n \atop j \right\} = e\omega(n). \tag{3.34}$$

This is the well-known Dobinski formula for Bell numbers $\omega(n)$. *Thus, if the* $(\Sigma\Delta D)$ *class is extended to include all those limits of members as members, then the Dobinski formula is a member of the class. Similarly, observe that*

$$\lim_{\alpha \to \infty} a^{-m} C(m, k; a, b) = \frac{1}{k!} \Delta^k (t^m)_{t=0} = \left\{ m \atop k \right\}$$

and it is seen that the limit form of (3.30) *(with* $a \to \infty$*) yields Grunnert's formula*

$$\left(t\frac{d}{dt}\right)^m f(t) = \sum_{k=0}^{\infty} k^m f^{(k)}(0) \frac{t^k}{k!} = \sum_{j=0}^{m} \left\{ m \atop j \right\} f^{(j)}(t) t^j. \tag{3.35}$$

Consequently, Grunnert's formula is also a member of the extended $(\Sigma\Delta D)$ *class.*

Remark 3.3.15 *The wording 'deduced formally' used in the Definition 3.3.12 may be given a little more explanation. Clearly, in the previous sections, the so-called formal derivation used for getting special formulas from source formulas, generally consists of using (i) operations with fps, (ii) symbolic operations with* $\Delta, D,$ *and* E, *(iii) ordinary algebraic computations, (iv) ordinary computational methods in mathematical analysis (including uses of Taylor's series expansion and Newton's interpolation series), (v) operations with infinite series, and exchange of the orders of repeated series summations without considering convergence problems, and (vi) mathematical tables including short tables of difference formulas and derivative formulas.*

Remark 3.3.16 *Evidently, by adopting the multi-index notational system, it is easy to formulate the GSF and SF*(i) $(i = 1, 2, 3)$ *in multivariate forms.*

Certainly, such a higher dimensional extension may be worth giving in detail, if it could be found really useful in applications. As regards the problem, whether it is possible to extend the main results of this chapter to the cases of q*-analysis should be worthy of investigation.*

Remark 3.3.17 *As seen in Section 3.2, we have defined a formula chain via iterations of the GSF. More precisely, a chain of formulas with freedom-degrees* (∞^{3m+1}) $(m = 1, 2, 3, \ldots)$ *could be generated successively by the iteration formulas*

$$\sum_{k\geq 0}(p_k^{(m)}(D)f_m(0))\delta^k\phi_m(t) = f_{m+1}(t), \quad m \geq 1,$$

with start from A_1, g_1, f_1 *and* ϕ_1*. In this way, each formula of freedom-degree* (∞^{3m+1}) *could be used as a general source formula to yield a formula class, denoted by* $(\Sigma\Delta D)_{(3m+1)}$*. Consequently, we may get a sequence of formula classes with increasing freedom-degrees, viz.*

$$(\Sigma\Delta D)_{(4)} \subset (\Sigma\Delta D)_{(7)} \subset \cdots \subset (\Sigma\Delta D)_{(3m+1)} \subset \cdots .$$

Here the first one is just the $\Sigma\Delta D$ *class treated in this section, and it may be the most available class of formulas in the Discrete Analysis and Combinatorics.*

Exercises

3.1 Prove operational formulas (3.2) and (3.3).

3.2 Find the convergence conditions for the series expansions in (O_1)–(O_3), (O_7), (O_{10}), and (O_{12}).

3.3 For $f(x) = x^m$ $(m \geq 1)$, show the following difference and derivative are true.

$$\Delta^k f(0) = \left[\Delta^k x^m\right]_{x=0} = k!\left\{ {m \atop k} \right\},$$
$$\left[D^k x^m\right]_{x=0} = \left[(m)_k x^{m-k}\right]_{x=0} = \delta_{m,k} m!,$$

where $\left\{ {m \atop k} \right\}$ is the Stirling number of the second kind (cf. [141], P. 168), and $\delta_{m,k}$ is the Kronecker symbol with $\delta_{m,k} = 1$ for $m = k$ and zero for $m \neq k$.

3.4 For $f(x) = \binom{x}{m}$ and $m \geq k \geq 0$, , show the following difference and derivative are true.

$$\Delta^k f(0) = \left.\binom{x}{m-k}\right|_{x=0} = \binom{0}{m-k} = \delta_{m,k},$$
$$D^k f(0) = \left[D^k \frac{(x)_m}{m!}\right]_{x=0} = \frac{k!}{m!}(-1)^{m-k}\left[{m \atop k} \right].$$

3.5 For $f(x) = a^x$ $(a > 1)$, show the following difference and derivative are true.

$$\Delta^k a^x = (a-1)^k a^x, \quad D^k a^x = (\log a)^k a^x,$$
$$\left[\Delta^k a^x\right]_{x=0} = (a-1)^k, \quad \left[D^k a^x\right]_{x=0} = (\log a)^k.$$

3.6 For $f(x) = \frac{1}{1+x}$, show the following difference and derivative are true.

$$\Delta^k f(x) = \frac{(-1)^k k!}{(x+k+1)_{k+1}}, \quad D^k f(x) = \frac{(-1)^k k!}{(1+x)^{k+1}},$$
$$\Delta^k f(0) = \frac{(-1)^k}{k+1}, \quad D^k f(0) = (-1)^k k!.$$

3.7 For $f(x) = e^{ix}$ and $g(x) = e^{-ix}$ $(i^2 = -1)$, show the following difference and derivative are true.

$$\Delta^k e^{\pm ix} = e^{\pm ix}(e^{\pm i} - 1)^k, \quad D^k e^{\pm ix} = (\pm i)^k e^{\pm ix},$$
$$\left[\Delta^k e^{\pm ix}\right]_{x=0} = (e^{\pm i} - 1)^k, \quad \left[D^k e^{\pm ix}\right]_{x=0} = (\pm i)^k.$$

3.8 For $f(x) = \cos x = \frac{1}{2}\left(e^{ix} + e^{-ix}\right)$, show the following difference and derivative are true.

$$\left[\Delta^k \cos x\right]_{x=0} = \frac{(e^i - 1)^k + (e^{-i} - 1)^k}{2} = \frac{(1 + (-1)^k e^{-ik})(e^i - 1)^k}{2},$$
$$\left[D^k \cos x\right]_{x=0} = \frac{i^k + (-i)^k}{2} = i^k\left(\frac{1 + (-1)^k}{2}\right) = i^k \delta_k,$$

where δ_k is the parity function, viz., $\delta_k = 0$ if k is an odd integer, and $\delta_k = 1$ if k is an even integer.

3.9 Prove $\left[\Delta^k (1+x)^t\right]_{t=0} = x^k$ for $x \in \mathbb{R}$ and $k \in \mathbb{N}_0$.

3.10 Give a combinatorial interpretation for (3.75).

3.11 Prove 24 formulas given in Subsection 3.2.4.

3.12 Prove the difference formulas shown at the end of Subsection 3.2.4.

4

Methods of Using Special Function Sequences, Number Sequences, and Riordan Arrays

CONTENTS

In Chapters 2 and 3, we apply operators Σ, Δ, D and E to formal power series f and g (or ψ and ϕ) to construct identities and summation formulas. In the first two sections of this chapter, we shall continue this process for some special function sequences and number sequences such as the sequences of Bernoulli

DOI: 10.1201/9781003051305-4

functions, Stirling numbers, Fibonacci numbers, etc. In the third section, we construct identities and summation formulas for the function sequences and number sequences related to Riordan arrays. A Riordan array is an infinite lower triangular matrix, which columns are multiplication of certain power series g and f. The theory of Riordan arrays provides a modern method for classical umbra calculus, bringing new insights into many areas of combinatorial importance. The third section of this chapter gives introduction to the fields of interest for students and researchers, who seek novel ways of working in fields such as combinatorial identities, triangles for enumerating combinatorial numbers, special polynomial sequences, orthogonal polynomials, and the row summations of Riordan arrays.

4.1 Use of Stirling Numbers, Generalized Stirling Numbers, and Eulerian Numbers

Section 2.2 contains a pair of series transformation formulas with a variety of illustrative examples. As a theoretical completion or a substantial supplement to Section 2.2, we provide some convergence theorems for the transformation formulas under certain general conditions. We also show that these two transformation formulas subject to the convergence conditions can be further utilized to produce more than 30 special power series sums and combinatorial identities involving some well-known number sequences.

4.1.1 Basic convergence theorem

If $g(t)$ are infinitely differentiable on $[0,\infty)$, there are two expansion formulas involving derivatives obtained in Theorem 2.2.2:

$$\sum_{k=0}^{\infty} f(k)g^{(k)}(0)\frac{t^k}{k!} = \sum_{k=0}^{\infty} \Delta^k f(0)g^{(k)}(t)\frac{t^k}{k!} \tag{4.1}$$

$$\sum_{k=0}^{\infty} f(k)g^{(k)}(0)\frac{t^k}{k!} = \sum_{k=0}^{\infty} \frac{1}{k!} f^{(k)}(0)A_k(t,g(t)), \tag{4.2}$$

where $(f(k))$ in (4.1) is an arbitrary sequence of numbers in $\mathbb{C}$, $f(t)$ in (4.2) is infinitely differentiable at $t=0$, and $A_k(t,g(t))$ is an extension of Euler fraction in terms of $g(t)$ defined as in (2.91), i.e.,

$$A_k(t,g(t)) := \sum_{j=0}^{k} \left\{ k \atop j \right\} g^{(j)}(t)t^j, \tag{4.3}$$

where $\left\{ {k \atop j} \right\}$ being Stirling numbers of the second kind, i.e., $j!\left\{ {k \atop j} \right\} = \left[\Delta^j t^k\right]_{t=0}$, which is also denoted by $S(k, j)$. Formulas in Theorem 2.2.2 form the source formulas (3.2) and (3.3) in Chapter 3.

As was already shown in Section 2.2 more than 20 special power series summations and analytic/combinatorial identities have been derived from (4.1) and (4.2) without considering convergence problems. As one of main purposes of the present section, we will prove general convergence theorems for (4.1) and (4.2) by assuming $g(z)$ to be an analytic function and $f(z)$ to be functions satisfying certain general conditions. Generally speaking, the freedom of choices of g and f in (4.1) and (4.2) should imply the versatility of uses of the transformation formulas for establishing various series summation formulas including some analytic combinatorial identities. Here, several well-known results are mentioned for immediate reference.

(1) *Euler's formula for the arithmetic geometric series* with $p \in \mathbb{N}$

$$\sum_{k=0}^{\infty} k^p t^k = \sum_{k=0}^{p} \left\{ {p \atop k} \right\} \frac{k! t^k}{(1-t)^{k+1}} \quad (|t| < 1), \tag{4.4}$$

is a simple example of (4.1) with $f(t) = t^p$ and $g(t) = (1-t)^{-1}$.

(2) The Montmort series transformation formula

$$\sum_{k=1}^{\infty} f(k) t^k = \sum_{k=1}^{\infty} \Delta^{k-1} f(1) \left(\frac{t}{1-t} \right)^k, \quad (|t| < 1), \tag{4.5}$$

is a particular instance of (4.1) given by taking $g(t) = (1-t)^{-1}$ and replacing $f(t)$ and t^k by $f(t+1)$ and t^{k+1}, respectively.

(3) The choice $f(t) = t^p$ and $g(t) = e^t$ in (4.1) (with $t = 1$) leads to Dobinski's formula for the Bell number $\omega(p)$, namely

$$\frac{1}{e} \sum_{k=0}^{\infty} \frac{k^p}{k!} = \sum_{k=0}^{p} \left\{ {p \atop k} \right\} = \omega(p). \tag{4.6}$$

(4) The summation problem of the finite series

$$S_n(t) = \sum_{k=0}^{n} k^p t^k \tag{4.7}$$

has once been the subject of some papers of the Fibonacci Quarterly fore and after 1990s of the last century. As a matter of fact, either (4.1) or (4.2) essentially gives an explicit solution to the problem by just taking $f(t) = t^p$ and $g(t) = 1 + t + t^2 + \cdots + t^n$ or $g(t) = (1 - t^{n+1})/(1-t)$, $t \neq 1$. The details may be referred to Hsu and Shiue [133] (cf. Exercise 4.1).

(5) A summation formula of practical value, established by Wang and Hsu [210], for sectional power series using Eulerian fractions $\alpha_k(t) = A_k(t, (1 -$

$t)^{-1}$) is also a particular consequence of (4.2) with $g(t) = (1-t)^{-1}$, $|t| < 1$. Indeed, the principal results of [210] are built on

$$\sum_{k=a}^{b-1} f(k)t^k = \sum_{k=0}^{\infty} \frac{\alpha_k(t)}{k!}\left[t^a f^{(k)}(a) - t^b f^{(k)}(b)\right], \tag{4.8}$$

which, in turn, falls under the purview of (4.2), where $a, b \in \mathbb{N}$ with $a < b-1$ (cf. Exercise 4.2).

The rest of this subsection is wholly devoted to the analysis of convergency of (4.1) and (4.2). For those who are mainly interested in the computational applications of (4.1) and (4.2), it may be better first to look through following subsections §4.1.2–4.1.3.

Throughout the rest of this subsection, we will consider functions $g : \mathbb{C} \to \mathbb{C}$ and $f : \mathbb{C} \to \mathbb{C}$ so that we may write $g(z)$ and $f(z)$ instead of $g(t)$ and $f(t)$, respectively.

Theorem 4.1.1 *Let $g(z)$ be an analytic function with convergence radius $r > 0$, and let $(f(k))$ be a number sequence in $\mathbb{C}$ such that $\overline{\lim}_{k\to\infty} \sqrt[k]{|f(k)|} = \theta > 0$. Then, the expansion formula (4.1) with t being replaced by z is absolutely convergent for $|z| < r/(2+\theta)$.*

Corollary 4.1.2 *Let $g(z)$ be an entire function and $(f(k))$ be a number sequence in $\mathbb{C}$ such that $\overline{\lim}_{k\to\infty} \sqrt[k]{|f(k)|} < \infty$. Then, the formula (4.1) with $t \to z$ is absolutely convergent for every $z \in \mathbb{C}$.*

Proof. We will need to make the change of orders of summations so that justification of the absolute convergence property of certain series becomes an essential portion of our proof.

First, it is easily shown that the condition $|z| < r/(2+\theta)$ implies

$$\overline{\lim_{k\to\infty}} \left| f(k) g^{(k)}(0) \frac{z^k}{k!} \right|^{1/k} < 1,$$

so that the left-hand side (LHS) of (4.1) converges absolutely. In fact, by Cauchy-Hadamard's convergence radius formula

$$r = 1/\overline{\lim_{k\to\infty}} \sqrt[k]{|a_k|},$$

for $g(z) = \sum_{k\geq 0} a_k z^k$, we have

$$\begin{aligned}
&\overline{\lim_{k\to\infty}} \left| f(k) g^{(k)}(0) \frac{z^k}{k!} \right|^{1/k} \\
\leq &\overline{\lim_{k\to\infty}} |f(k)|^{1/k} \cdot \overline{\lim_{k\to\infty}} \left| \frac{g^{(k)}(0)}{k!} \right| \cdot |z| \\
= &\theta \cdot r^{-1} \cdot |z| < \theta \cdot r^{-1} \cdot \frac{r}{2+\theta} = \frac{\theta}{2+\theta} < 1.
\end{aligned}$$

For $|z| < r/(2+\theta)$, we may choose $\rho > \theta$ such that

$$|z| < \frac{r}{2+\rho} < \frac{r}{2+\theta}. \tag{4.9}$$

For such ρ, there can be found a number $K > 0$ such that $|f(k)|^{1/k} < \rho$, i.e. $|f(k)| < \rho^k$ whenever $k > K$. Consequently, there exists a sufficiently large M such that $|f(k)| < M\rho^k$ for all $k \in \mathbb{N}_0$.

Let us use the familiar formulas involving finite differences

$$f(m) = \sum_{k=0}^{m} \binom{m}{k} \Delta^k f(0), \quad \Delta^k f(0) = \sum_{j=0}^{k} (-1)^{k-j} \binom{k}{j} f(j).$$

With the help of these two formulas, we now expand the LHS of (4.1) into the form with $t \to z$:

$$\text{LHS of (4.1)} = \sum_{m=0}^{\infty} \frac{g^{(m)}(0)}{m!} z^m \sum_{k=0}^{m} \binom{m}{k} \Delta^k f(0). \tag{4.10}$$

Here, the rightmost sum in (4.10) can be estimated in absolute values as follows:

$$\begin{aligned}\sum_{k=0}^{m} \binom{m}{k} |\Delta^k f(0)| &\leq \sum_{k=0}^{m} \binom{m}{k} \sum_{j=0}^{k} \binom{k}{j} |f(j)| \\ &< \sum_{k=0}^{m} \binom{m}{k} \sum_{j=0}^{k} \binom{k}{j} M\rho^j \\ &= M \sum_{k=0}^{m} \binom{m}{k} (1+\rho)^k = M(2+\rho)^m.\end{aligned}$$

This states that the right-hand side (RHS) of (4.10) is bounded absolutely by

$$M \sum_{m=0}^{\infty} \frac{g^{(m)}}{m!} (2+\rho)^m |z|^m < \infty. \tag{4.11}$$

Since $g(z)$ is analytic within $\{z : |z| < r\}$ and by (4.9), $(2+\rho)|z| < r$, the absolute convergence of the LHS of (4.1) is ensured by (4.11).

Moreover, (4.11) also implies that the RHS of (4.10) can be simplified via the change of summation orders, namely

$$\text{RHS of (4.10)} = \sum_{k=0}^{\infty} \frac{\Delta^k f(0)}{k!} \sum_{m=k}^{\infty} \frac{g^{(m)}(0)}{(m-k)!} z^m = \sum_{k=0}^{\infty} \Delta^k f(0) g^{(k)}(z) \frac{z^k}{k!}.$$

This is precisely the RHS of (4.1) (with $t \to z$) whose absolute convergence is also ensured by (4.11). Hence, both the validity and the absolute convergence of (4.1) are proved under the condition $|z| < r/(2+\theta)$.

In particular, since $r = \infty$ for the entire function $g(z)$, we get the corollary of the theorem.

■

As usual, for any analytic function $f(z) = \sum_{k=0}^{\infty} b_k z^k$, we may take $\hat{f}(z) = \sum_{k=0}^{\infty} |b_k| z^k$ as a simple majorant series (function) of $f(z)$, which has the same convergence radius as that of $f(z)$.

For functions that can be represented by a power series, the term "majorant" is often given a more special meaning. In such cases the term denotes the sum of a power series with positive coefficients that are not less than the absolute values of the corresponding coefficients of the given series. If $h(x)$ is a majorant (in this special sense) of the function $g(x)$, then we write $h(x) >> g(x)$. For example, $x/(1-x) >> \ln(1+x)$ since

$$\ln(1+x) = x - \frac{x^2}{2} + \frac{x^3}{3} - \cdots + (-1)^{n-1}\frac{x^n}{n} + \cdots$$
$$\frac{x}{1-x} = x + x^2 + x^3 + \cdots + x^n + \cdots.$$

It is obvious that the function $f(x) = x$ is the majorant for $x > -1$ of the function $g(x) = \ln(1+x)$ since $x \geq \ln(1+x)$ for all values $x > -1$. However, in this (special) sense $f(x) = x$ is no longer the majorant of $\ln(1+x)$. The majorants of power series are widely used in the theory of differential equations.

Theorem 4.1.3 *Let $g(z)$ be an analytic function with convergence radius $r > 0$, and let $f(z)$ be an entire function with its simple majorant function $\hat{f}(z)$ satisfying the condition*

$$\overline{\lim_{k\to\infty}} \sqrt[k]{|\hat{f}(k)|} = \theta > 0.$$

Then, the series transformation formula (4.2) *with $t \to z$ and $f(k)$ being replaced by $\hat{f}(k)$ is absolutely convergent for $|z| < \min\{r, r/\theta\}$.*

Corollary 4.1.4 *Let both $g(z)$ and $f(z)$ be entire functions and let θ be given as in Theorem 4.1.3. Then, formula* (4.2) *is absolutely convergent for every $z \in \mathbb{C}$ whenever $0 < \theta < \infty$.*

Proof. It is known that m^k $(m, k \in \mathbb{N}_0)$ may be expressed in terms of finite differences of zero, or of Stirling numbers of the second kind, namely

$$m^k = \sum_{j=0}^{m} \binom{m}{j} \Delta^j 0^k = \sum_{j=0}^{k} (m)_j \left\{ {k \atop j} \right\},$$

where the jth falling factorial $(m)_j = m!/(m-j)!$ for $m \geq j \geq 0$ and 0 otherwise.

Since $\hat{f}(z)$ is also an entire function, it is clear that the following expansions are absolutely convergent:

$$\begin{aligned}\frac{g^{(m)}(0)}{m!}\hat{f}(m) =& \frac{g^{(m)}(0)}{m!}\sum_{k=0}^{\infty}|f^{(k)}(0)|\frac{m^k}{k!}\\ =& \sum_{k=0}^{\infty}\frac{|f^{(k)}(0)|}{k!}\sum_{j=0}^{k}\frac{(m)_j}{m!}\left\{ {k \atop j} \right\} g^{(m)}(0),\end{aligned}$$

for $m \in \mathbb{N}_0$.

On the other hand, for $|z| < \min\{r, r/\theta\}$, we have by Cauchy-Hadamard's formula for r:

$$\overline{\lim_{k\to\infty}}\left|\hat{f}(k)g^{(k)}(0)\frac{z^k}{k!}\right|^{1/k} \leq \theta \cdot r^{-1} \cdot |z| < \theta \cdot r^{-1} \cdot \frac{r}{\theta} = 1.$$

Thus by Cauchy's root test, we see that the simple majorant series of the LHS of (4.2) (with $t \to z$) with $f(k)$ being replaced by $\hat{f}(k)$ must be absolutely convergent. This guarantees that all the double series occurring below are absolutely convergent, and that the exchange of orders of double summations is permissible. Namely, we have

$$\begin{aligned}\sum_{m=0}^{\infty}\frac{g^{(m)}(0)}{m!}\hat{f}(m)z^m =& \sum_{m=0}^{\infty}\left(\sum_{k=0}^{\infty}\frac{|f^{(k)}(0)|}{k!}\sum_{j=0}^{k}\frac{(m)_j}{m!}\left\{ {k \atop j} \right\} g^{(m)}(0)\right) z^m\\ =& \sum_{k=0}^{\infty}\frac{|f^{(k)}(0)|}{k!}\sum_{m=0}^{\infty}\left(\sum_{j=0}^{k}\frac{(m)_j}{m!}\left\{ {k \atop j} \right\} g^{(m)}(0)\right) z^m\\ =& \sum_{k=0}^{\infty}\frac{|f^{(k)}(0)|}{k!}\sum_{j=0}^{k}\left\{ {k \atop j} \right\} z^j\left(\sum_{m=j}^{\infty}\frac{g^{(m)}(0)}{(m-j)!}z^{m-j}\right)\\ =& \sum_{k=0}^{\infty}\frac{|f^{(k)}(0)|}{k!}\sum_{j=0}^{k}\left\{ {k \atop j} \right\} g^{(j)}(z)z^j\\ =& \sum_{k=0}^{\infty}\frac{|f^{(k)}(0)|}{k!}A_k(z, g(z)).\end{aligned}$$

This shows that (4.2) holds for $\hat{f}(z)$ instead of $f(z)$. As may be observed, the above procedure can be performed similarly when $g(z)$ is replaced by its simple majorant series $\hat{g}(z)$ (accordingly $g^{(m)}(z)$ by $\hat{g}^{(m)}(z)$). Thus, we can conclude that the absolute convergence of (4.2) is always ensured by the condition $|z| < \min\{r, r/\theta)$, and the theorem is proved.

From our argument, Corollary 4.1.4 follows as a result for the special case $r = \infty$.

■

Let us now make use of Theorems 4.1.1 and 4.1.3 to furnish some convergence conditions for the following expansions. More applications to the convergence of the other source expansions will be seen in the next subsection.

Example 4.1.5 *Let $f(z)$ be a polynomial in z of degree $\partial f(t) = p$. Then, (4.1) and (4.2) yield the following closed sum formulas:*

$$\sum_{k=0}^{\infty} f(k)g^{(k)}(0)\frac{z^k}{k!} = \sum_{k=0}^{p} \Delta^k f(0)g^{(k)}(z)\frac{z^k}{k!} \tag{4.12}$$

$$\sum_{k=0}^{\infty} f(k)g^{(k)}(0)\frac{z^k}{k!} = \sum_{k=0}^{p} \frac{1}{k!}f^{(k)}(0)A_k(z, g(z)), \tag{4.13}$$

Under the known condition, $\overline{\lim}_{k\to\infty} \sqrt[k]{|f(k)|} = 1$. So, it follows that both series on the LHS of (4.12) and (4.13) are absolutely convergent for $|z| < r$ wherever $g(z)$ is analytic for $|z| < r$. Of course, the validity of (4.13) for $|z| < r$ also follows from Theorem 4.1.3. Moreover, it may be worth noticing that the conditions $\Delta^k f(0) = 0$ $(k > p)$ could make the proof of Theorem 4.1.1 much simplified so that the conclusion involving convergence condition $|z| < r/(2+\theta)$ may be extended to the form $|z| < r/\theta$, partly consistent with the conclusion of Theorem 4.1.3.

As will be seen, convergence condition mentioned in Example 4.1.5 could apply to all the infinite series appearing in the next two subsections.

4.1.2 Summation formulas involving Stirling numbers, Bernoulli numbers, and Fibonacci numbers

What we will introduce in this subsection are two specifications of (4.1) and an operational formula, which will prove to be very useful for obtaining various series summation formulas and combinatorial identities.

We begin by recalling that various Stirling-type numbers proposed by Riordan, Carlitz, Howard, Broder, Charalambides, Koutras, respectively (cf. He [88]), are included in the three-parameter *generalized Stirling numbers (GSN)* $S(n,k) \equiv S(n,k;\alpha,\beta,\gamma)$ which are defined as connection coefficients of a linear transformation between generalized factorials (cf. Hsu and Shiue [132] and [88]), namely

$$(t|\alpha)_n = \sum_{k=0}^{n} S(n,k;\alpha,\beta,\gamma)(t-\gamma|\beta)_k, \tag{4.14}$$

where $(t|\alpha)_0 = 1$ and $(t|\alpha)_n = t(t-\alpha)\cdots(t-n\alpha+\alpha)$ for $n \in \mathbb{N}$, and the parameters α, β, γ are any real or complex numbers with $(\alpha,\beta,\gamma) \neq (0,0,0)$. Clearly, $S(0,0,;\alpha,\beta,\gamma) = 1$ and $S(n,k;\alpha,\beta,\gamma) = 0$ for $k > n$. In particular, we have

$$S(n,k;1,0,0) = (-1)^{n-k}\begin{bmatrix} n \\ k \end{bmatrix}, \ S(n,k;0,1,0) = \begin{Bmatrix} n \\ k \end{Bmatrix}, \ S(n,k;0,0,1) = \binom{n}{k}, \tag{4.15}$$

where $(-1)^{n-k}\left[{n \atop k}\right]$, as relative to $\left\{{n \atop k}\right\}$, stands for the classical *Stirling numbers of the first kind.* Of all results concerning these two kinds of number sequences, the following orthogonality relation, called the *Stirling inversion*, is worthy of mention:

$$\sum_{k=j}^{n}(-1)^{n-k}\begin{bmatrix} n \\ k \end{bmatrix}\begin{Bmatrix} k \\ j \end{Bmatrix} = \sum_{k=j}^{n}(-1)^{k-j}\begin{Bmatrix} n \\ k \end{Bmatrix}\begin{bmatrix} k \\ j \end{bmatrix} = \delta_{n,j}, \tag{4.16}$$

where $\delta_{n,j}$ denotes the usual Kronecker delta.

For our purpose, $(-1)^{n-k}\left[{n \atop k}\right]$ may be defined by

$$(-1)^{n-k}\begin{bmatrix} n \\ k \end{bmatrix} = \frac{1}{k!}\left[D^k(t)_n\right]_{t=0}, \tag{4.17}$$

where $(t)_n = (t|1)_n$. In this form, it is easily found that

$$\frac{1}{k!}\left[D^k(t)_n\right]_{t=1} = \begin{bmatrix} n-1 \\ k \end{bmatrix} + \begin{bmatrix} n-1 \\ k-1 \end{bmatrix}, \quad (k \geq 1). \tag{4.18}$$

From now on, we always write $S(n,k)$ instead of $S(n,k;\alpha,\beta,\gamma)$ unless otherwise is stated. For the case $\beta \neq 0$, we see that

$$(\beta t + \gamma|\alpha)_n = \sum_{k=0}^{n} S(n,k)(\beta t|\beta)_k = \sum_{k=0}^{n}(t)_k S(n,k)\beta^k.$$

Thus, it follows that

$$S(n,k) = \frac{1}{k!\beta^k}\left[\Delta^k(\beta t+\gamma|\alpha)_n\right]_{t=0}. \tag{4.19}$$

Note that (4.1) becomes a closed sum formula for the series on the LHS of (4.1) if $f(t)$ is a polynomial with degree $\partial f(t) = p \in \mathbb{N}$. This suggests that two useful specializations may be achieved by taking $f(t)$ to be binomial coefficients and generalized factorials of the forms

$$f(t) = \binom{\alpha+t}{p} \text{ and } f(t) = (\beta t + \gamma|\alpha)_p$$

with $\partial f(t) = p$, where $\alpha, \beta, \gamma \in \mathbb{R}$ with $\beta \neq 0$. As rewarding, we get a pair of power series summation formulas as follows.

Theorem 4.1.6 *Let $g(z)$ be an analytic function with convergence radius $r > 0$. Then, for $z : |z| < r$, we have*

$$\sum_{k=0}^{\infty} \binom{\alpha+k}{p} g^{(k)}(0)\frac{z^k}{k!} = \sum_{k=0}^{p} \binom{\alpha}{p-k} g^{(k)}(z)\frac{z^k}{k!}, \tag{4.20}$$

$$\sum_{k=0}^{\infty} (\beta k+\gamma|\alpha)_p g^{(k)}(0)\frac{z^k}{k!} = \sum_{k=0}^{p} S(p,k) g^{(k)}(z)(\beta z)^k. \tag{4.21}$$

Evidently, under the same assumption as above, from formula (4.1) with $t \to z$ and $g(z) = (1+z)^\alpha$, we have

$$\sum_{k=0}^{\infty} \binom{\alpha}{k} f(k) z^k = \sum_{k=0}^{\infty} \binom{\alpha}{k} \Delta^k f(0) \frac{z^k}{(1+z)^{k-\alpha}}. \tag{4.22}$$

This not only gives a generating function for the sequence $(\binom{\alpha}{k} f(k))$ but also admits a *symbolization process* (cf. Section 3.1). To be more precise for the latter, let z be replaced by the difference operator Δ so that

$$\frac{z^k}{(1+z)^{k-\alpha}} \to \frac{\Delta^k}{(1+\Delta)^{k-\alpha}} = \Delta^k E^{\alpha-k},$$

and we see that the last equality turns out to be an operational formula applicable to any formal power series ϕ.

Theorem 4.1.7 *Let $f(t)$ be a polynomial with degree $\partial f(t) = p \in \mathbb{N}$. Then, we have formally*

$$\sum_{k=0}^{\infty} \binom{\alpha}{k} f(k) \Delta^k \phi(0) = \sum_{k=0}^{p} \binom{\alpha}{k} \Delta^k f(0) \Delta^k \phi(\alpha-k). \tag{4.23}$$

We now show that (4.20)–(4.23) can be used to produce a variety of power series sums and combinatorial identities. Some of these results involve *Bernoulli polynomials* $B_n(t)$, *Bernoulli numbers* $B_n = B_n(0)$, *Fibonacci numbers* F_n and GSN $S(n,k)$ as well as classical Stirling numbers $(-1)^{n-k}\left[{n \atop k}\right]$ and $\left\{{n \atop k}\right\}$.

For the latter application, one may recall that $B_n(t)$ ($n \in \mathbb{N}_0$) are given by

$$\frac{xd^{tx}}{e^x - 1} = \sum_{n=0}^{\infty} B_n(t) \frac{x^n}{n!}, \tag{4.24}$$

where $B_n(t)$ have the properties

$$B_n(t+1) - B_n(t) = nt^{n-1}, \tag{4.25}$$

$$B_n(1-t) = (-1)^n B_n(t), \tag{4.26}$$

$$B_n(t+s)=\sum_{k=0}^{n}\binom{n}{k}B_k(t)s^{n-k}, \tag{4.27}$$

$$D^kB_n(t)=(n)_kB_{n-k}(t), \tag{4.28}$$

$$D^kB_n(0)=(n)_kB_{n-k}. \tag{4.29}$$

Also, Fibonacci numbers F_k are given by

$$\frac{1}{1-t-t^2}=\sum_{k=0}^{\infty}F_kt^k,\quad (F_0=F_1=1). \tag{4.30}$$

It is well-known that

$$F_k=\frac{1}{\sqrt{5}}\left(a^{k+1}-b^{k+1}\right)$$

with $a=(1+\sqrt{5})/2$, $b=(1-\sqrt{5})/2$, and that

$$D^k\left(\frac{1}{1-t-t^2}\right)=\frac{k!}{\sqrt{5}}\left\{\left(\frac{a}{1-at}\right)^{k+1}-\left(\frac{b}{1-bt}\right)^{k+1}\right\}.$$

Next, making use of (4.18) and (4.19) and choosing $g(t)$ to be the easily computed functions such as

(*i*) $(1+t)^s$, (*ii*) $(1-t)^{s-1}$ $(s\in\mathbb{R})$,(*iii*) $1/\sqrt{1-4t}$, (*iv*) $(t)_n$ $(n\in\mathbb{N})$,
(*v*) $\log(1-t)$, (*vi*) a^t $(a>0, a\neq 1)$,(*vii*) $B_n(t)$, (*viii*) $\frac{1}{1-t-t^2}$,

successively, we can obtain a list of formulas and identities, which we state without going into details as the following and leave their proofs as Exercise 4.3:

$$\sum_{k=0}^{\infty}\binom{\alpha+k}{p}\binom{s}{k}t^k=\sum_{k=0}^{p}\binom{\alpha}{p-k}\binom{s}{k}\frac{t^k}{(1+t)^{k-s}}\quad (|t|<1), \tag{4.31}$$

$$\sum_{k=0}^{\infty}\binom{\alpha+k}{p}\binom{s+k}{k}t^k=\sum_{k=0}^{p}\binom{\alpha}{p-k}\binom{s+k}{k}\frac{t^k}{(1-t)^{s+k+1}}\quad (|t|<1), \tag{4.32}$$

$$\sum_{k=0}^{\infty}\binom{\alpha+k}{p}\binom{2k}{k}t^k=\sum_{k=0}^{p}\binom{\alpha}{p-k}\binom{2k}{k}\frac{t^k}{(1-4t)^{k+1/2}}\quad (|t|<1/4), \tag{4.33}$$

$$\sum_{k=0}^{\infty}\binom{\alpha+k}{p}(-1)^k\begin{bmatrix}n\\k\end{bmatrix}=\sum_{k=0}^{p}(-1)^{k+1}\binom{\alpha}{p-k}\left(\begin{bmatrix}n-1\\k\end{bmatrix}-\begin{bmatrix}n-1\\k-1\end{bmatrix}\right)\ (p\in\mathbb{N}), \tag{4.34}$$

$$\sum_{k=0}^{\infty}\binom{\alpha+k}{p}\frac{(t\log a)^k}{k!}=a^t\sum_{k=0}^{p}\binom{\alpha}{p-k}\frac{(t\log a)^k}{k!}\quad (t\in\mathbb{R}), \tag{4.35}$$

$$\sum_{k=0}^{\infty}\binom{\alpha+k}{p}\frac{t^k}{k}=\binom{\alpha}{p}\log\frac{1}{1-t}+\sum_{k=1}^{p}\binom{\alpha}{p-k}\frac{t^k}{k(1-t)^k}\quad(|t|<1),\qquad(4.36)$$

$$\sum_{k=0}^{n}\binom{\alpha+k}{p}\binom{n}{k}B_{n-k}t^k=\sum_{k=0}^{p}\binom{\alpha}{p-k}\binom{n}{k}B_{n-k}t^k,\qquad(4.37)$$

$$\sum_{k=0}^{n}\binom{k}{p}\binom{n}{k}B_{n-k}t^k=B_{n-p}(t)\binom{n}{p}t^p,\qquad(4.38)$$

$$\sum_{k=0}^{\infty}\binom{\alpha+k}{p}F_kt^k=\frac{1}{\sqrt{5}}\sum_{k=0}^{p}\binom{\alpha}{p-k}\left\{\left(\frac{a}{1-at}\right)^{k+1}-\left(\frac{b}{1-bt}\right)^{k+1}\right\}t^k,\qquad(4.39)$$

$$\sum_{k=0}^{\infty}\binom{k}{p}F_kt^k=\frac{1}{\sqrt{5}}\left\{\left(\frac{a}{1-at}\right)^{p+1}-\left(\frac{b}{1-bt}\right)^{p+1}\right\}t^p,\qquad(4.40)$$

Both (4.39) and (4.40) are valid for $|t|<(\sqrt{5}-1)/2$.

$$\sum_{k=0}^{\infty}(\beta k+\gamma|\alpha)_p\binom{s}{k}t^k=\sum_{k=0}^{p}S(p,k)\frac{(s)_k(\beta t)^k}{(1+t)^{k-s}}\quad(|t|<1),\qquad(4.41)$$

$$\sum_{k=0}^{\infty}(\beta k+\gamma|\alpha)_p\binom{s+k}{k}t^k=\sum_{k=0}^{p}S(p,k)\frac{(s+k)_k(\beta t)^k}{(1+t)^{k+s+1}}\quad(|t|<1),\qquad(4.42)$$

$$\sum_{k=0}^{\infty}(\beta k+\gamma|\alpha)_p\binom{2k}{k}t^k=\sum_{k=0}^{p}S(p,k)\frac{(2k)_k(\beta t)^k}{(1-4t)^{k+1/2}}\quad(|t|<1/4),\qquad(4.43)$$

$$\sum_{k=0}^{\infty}(-1)^k(\beta k+\gamma|\alpha)_p\begin{bmatrix}n\\k\end{bmatrix}=\sum_{k=1}^{p}(-1)^{k+1}S(p,k)\left(\begin{bmatrix}n-1\\k\end{bmatrix}-\begin{bmatrix}n-1\\k-1\end{bmatrix}\right)k!\beta^k,\qquad(4.44)$$

$$\sum_{k=0}^{\infty}(\beta k+\gamma|\alpha)_p\frac{(t\log a)^k}{k!}=a^t\sum_{k=0}^{p}S(p,k)(\beta t\log a)^k\quad(t\in\mathbb{R}),\qquad(4.45)$$

$$\sum_{k=0}^{\infty}(\beta k+\gamma|\alpha)_p\frac{t^k}{k!}=e^t\sum_{k=0}^{p}S(p,k)(\beta t)^k\quad(t\in\mathbb{R})\qquad(4.46)$$

$$\sum_{k=0}^{\infty}(\beta k+\gamma|\alpha)_p\frac{t^k}{k!}=(\gamma|\alpha)_p\log\frac{1}{1-t}+\sum_{k=1}^{p}S(p,k)\frac{(k-1)!(\beta t)^k}{(1-t)^k}\quad(|t|<1)\qquad(4.47)$$

$$\sum_{k=0}^{\infty}(\beta k+\gamma|\alpha)_p\binom{n}{k}B_{n-k}t^k=\sum_{k=0}^{p}S(p,k)B_{n-k}(t)(\beta t)^k,\qquad(4.48)$$

$$\sum_{k=0}^{n}k^p\binom{n}{k}B_{n-k}t^k=\sum_{k=0}^{p}B_{n-k}(t)\left\{{p\atop k}\right\}(n)_kt^k,\qquad(4.49)$$

$$\sum_{k=0}^{\infty}(\beta k+\gamma|\alpha)_p F_k t^k = \sum_{k=0}^{p}\frac{k!S(p,k)}{\sqrt{5}\beta}\left\{\left(\frac{a\beta}{1-at}\right)^{k+1}-\left(\frac{b\beta}{1-bt}\right)^{k+1}\right\}t^k, \tag{4.50}$$

$$\sum_{k=0}^{\infty}k^p F_k t^k = \sum_{k=0}^{p}\left\{ {p \atop k} \right\}\frac{k!}{\sqrt{5}}\left\{\left(\frac{a}{1-at}\right)^{k+1}-\left(\frac{b}{1-bt}\right)^{k+1}\right\}t^k, \tag{4.51}$$

Both (4.50) and (4.51) are valid for $|t| < (\sqrt{5}-1)/2$.

$$\sum_{k=p}^{n}(-1)^k \begin{bmatrix} n \\ k \end{bmatrix}\binom{k}{p} = (-1)^{p+1}\left(\begin{bmatrix} n-1 \\ p \end{bmatrix}-\begin{bmatrix} n-1 \\ p-1 \end{bmatrix}\right) \quad (1 \le p \le n), \tag{4.52}$$

$$\sum_{k=1}^{n}(-1)^k k^p \begin{bmatrix} n \\ k \end{bmatrix} = \sum_{k=1}^{p}(-1)^{k+1}k!\left(\begin{bmatrix} n-1 \\ k \end{bmatrix}-\begin{bmatrix} n-1 \\ k-1 \end{bmatrix}\right)\left\{ {p \atop k} \right\} \quad (p \ge 1), \tag{4.53}$$

$$\sum_{k=0}^{n}k^p\binom{n}{k} = \sum_{k=0}^{p}(n)_k\left\{ {p \atop k} \right\}2^{n-k} \quad (p \ge 0). \tag{4.54}$$

Surely, most of the formulas displayed in the above list are not given before. However, the last three identities (4.52)–(4.54) appear to be not so strange and are included as particular cases of (4.34), (4.44), and (4.41), respectively. It may be also of interest to notice that Euler's formula (4.4) and Dobinski's equality (4.6) are special consequences implied by (4.42) (with $s = 0$, $(\alpha, \beta, \gamma) = (0, 1, 0)$) and (4.46) (with $t = 1$, $(\alpha, \beta, \gamma) = (0, 1, 0)$), respectively. Apart from these, it is especially noteworthy that (4.52) can be inverted via the Stirling inversion (4.16) to get the following elegant and peculiar identity involving three kinds of elementary counting numbers (cf. (4.15)) simultaneously:

$$\sum_{k=p}^{n}\left\{ {n \atop k} \right\}\left(\begin{bmatrix} k-1 \\ p \end{bmatrix}+\begin{bmatrix} k-1 \\ p-1 \end{bmatrix}\right) = \binom{n}{p}, \quad (1 \le p \le n). \tag{4.55}$$

4.1.3 Summation formulas involving generalized Eulerian functions

Let $f(z)$ be an entire function with its simple majorant function $\hat{f}(z)$ satisfying the condition that $\overline{\lim}_{k\to\infty}\sqrt[k]{|\hat{f}(k)|} = \theta > 0$. Then, the choices $g(z) = (1+z)^\alpha$ and $g(z) = (1-z)^{-\alpha-1}$ ($\alpha \in \mathbb{R}$) in (4.2) yield the following closed expansions (cf. Section 2.2)

$$\sum_{k=0}^{\infty}\binom{\alpha}{k}f(k)z^k = \sum_{k=0}^{\infty}\frac{1}{k!}f^{(k)}(0)A_k(z,(1+z)^\alpha), \tag{4.56}$$

$$\sum_{k=0}^{\infty}\binom{\alpha+k}{k}f(k)z^k=\sum_{k=0}^{\infty}\frac{1}{k!}f^{(k)}(0)A_k(z,(1-z)^{-\alpha-1}), \tag{4.57}$$

$$\sum_{k=m}^{\infty}\binom{k}{m}f(k)z^k=\sum_{k=0}^{\infty}\frac{z^m}{k!}f^{(k)}(m)A_k(z,(1-z)^{-m-1}). \tag{4.58}$$

In view of Theorem 4.1.7, we see that the series expansions involved in (4.56)-(4.58) are all convergent absolutely for $|z|<\min\{1,1/\theta\}$.

We now go to the real field. Let g(t) be a given formal power series. As previously, a direct application of (4.2) to $f(t)=(t)_p$ ($p\in\mathbb{N}$) yields an identity of the form

$$\sum_{k=1}^{p}\begin{bmatrix}p\\k\end{bmatrix}A_k(t,g(t))=t^pg^{(p)}(t). \tag{4.59}$$

This can be verified directly as follows:

$$\begin{aligned}t^pg^{(p)}(t)=&t^pD^p\left(\sum_{k=0}^{\infty}g^{(k)}(0)\frac{t^k}{k!}\right)=\sum_{k=p}^{\infty}(k)_pg^{(k)}(0)\frac{t^k}{k!}\\=&\sum_{k=0}^{\infty}\frac{1}{k!}[D^k(t)_p]_{t=0}A_k(t,g(t))=\sum_{k=1}^{p}\begin{bmatrix}p\\k\end{bmatrix}A_k(t,g(t)),\end{aligned}$$

where $A_k(t,g(t))$ is given by (4.3). In fact, on taking the Stirling inversion (4.16) into account, we readily see that (4.59) is equivalent to (4.3).

In particular, for the classical Eulerian fraction $A_k(t,(1-t)^{-1})$, it is easily found that

$$\sum_{k=1}^{p}\begin{bmatrix}p\\k\end{bmatrix}A_k(t,(1-t)^{-1})=\frac{t^pp!}{(1-t)^{p+1}}. \tag{4.60}$$

Alternatively, by taking $f(t)=B_p(t)$ ($p\in\mathbb{N}$) in (4.2), we obtain immediately

$$\sum_{k=0}^{\infty}B_p(k)g^{(k)}(0)\frac{t^k}{k!}=\sum_{k=0}^{p}\binom{p}{k}B_{p-k}A_k(t,g(t)). \tag{4.61}$$

It still covers some particular identities as represented previously (cf. Section 2.2). Owing to that $\overline{\lim}_{k\to\infty}\sqrt[k]{|B_p(k)|}=1$, the series on the LHS (4.61) converges absolutely for $|t|<r$, provided that $g(z)$ is analytic for $|z|<r$ (cf. the beginning of this subsection).

In what follows we will proceed to show how the operational formula (4.23) can be utilized to construct several infinite and finite identities of combinatorial type. A typical example is when

$$\phi(t)=\frac{1}{t+\beta}\quad(\beta\in\mathbb{R},\beta>0).$$

We have via a brief computation

$$\Delta^k\phi(t)=\frac{(-1)^k}{t+\beta}\Big/\binom{t+\beta+k}{k}. \tag{4.62}$$

In particular, for $\alpha+\beta \neq k$, set $t=\alpha-k$ and 0 in (4.62), respectively, to get

$$\Delta^k\phi(\alpha-k)=\frac{(-1)^k}{\alpha+\beta-k}\Big/\binom{\alpha+\beta}{k},$$
$$\Delta^k\phi(0)=\frac{(-1)^k}{\beta}\Big/\binom{\beta+k}{k}.$$

Then, for any given polynomial $f(t)$ with degree $\partial f(t)=p$, we find that (4.23) yields

$$\sum_{k=0}^{\infty}(-1)^k f(k)\binom{\alpha}{k}\Big/\binom{\beta+k}{k}=\sum_{k=0}^{p}\Delta^k f(0)\frac{(-1)^k\beta}{\alpha+\beta-k}\binom{\alpha}{k}\Big/\binom{\alpha+\beta}{k}, \tag{4.63}$$

where $\alpha+\beta\neq k$ for integers k with $0\le k\le p$. As for its convergence, we have the following:

Proposition 4.1.8 *For $\alpha,\beta>0$, the infinite series involved in (4.63) is absolutely convergent for $[\beta]\ge p+1$, where $[\beta]$ denotes the integer part of b.*

Actually, for $\alpha,\beta>0$, we have the estimates

$$\left|\binom{\alpha}{k}\right|\le\frac{[\alpha]+1}{k}=O(1/k),\quad \binom{\beta+k}{k}\ge\binom{k+[\beta]}{[\beta]}=O(k^{[\beta]}),\quad (k\to\infty).$$

Thus, it follows that

$$\left|f(k)\binom{\alpha}{k}\right|\Big/\binom{\beta+k}{k}=O(k^{p-1-[\beta]}),\quad (k\to\infty).$$

This estimate justifies that the infinite series of (4.63) converges absolutely for $[\beta]\ge p+1$.

As direct applications, (4.63) contains some infinite-type (analytic) combinatorial identities with $f(t)=t^p$, $\binom{t+\lambda}{p}$, etc. as follows:

$$\sum_{k=0}^{\infty}(-1)^k k^p\binom{\alpha}{k}\Big/\binom{\beta+k}{k}=\sum_{k=0}^{p}\frac{(-1)^k\beta k!}{\alpha+\beta-k}\left\{ {p \atop k} \right\}\binom{\alpha}{k}\Big/\binom{\alpha+\beta}{k}, \tag{4.64}$$

$$\sum_{k=0}^{\infty}(-1)^k\binom{\lambda+k}{p}\binom{\alpha}{k}\Big/\binom{\beta+k}{k}=\sum_{k=0}^{p}\frac{(-1)^k\beta}{\alpha+\beta-k}\binom{\lambda}{p-k}\binom{\alpha}{k}\Big/\binom{\alpha+\beta}{k}, \tag{4.65}$$

$$\sum_{k=0}^{\infty}k^p\Big/\binom{\beta+k}{k}=\sum_{k=0}^{p}\frac{\beta k!}{\beta-k-1}\left\{ {p \atop k} \right\}\Big/\binom{\beta-1}{k},\ (\beta\ge p+2), \tag{4.66}$$

$$\sum_{k=0}^{\infty}(-1)^k\binom{\alpha}{k}\Big/\binom{\beta+k}{k}=\frac{\beta}{\alpha+\beta},\ (\alpha>0,\ \beta\ge1), \tag{4.67}$$

$$\sum_{k=0}^{\infty} \binom{\beta+k}{k}^{-1} = \frac{\beta}{\beta-1}, \ (\beta \geq 2). \tag{4.68}$$

Here, it may be of interest to note that identity (4.67) with $\alpha = n \in \mathbb{N}$ just gives a theorem in Section 4.4 of Wilf [216], which has served as a distinctive example proved by the well-known *WZ-method* [177].

With a bit of surprise, we find that numerous combinatorial identities of the [3/0] type as displayed in Egorychev's treatise [58] (pp. 78-85) and Gould's formulary [73] can be rediscovered or verified by use of (4.23) with suitable choices of $f(t)$ and $\phi(t)$. Typical examples include the following finite-type identities:

$$\sum_{k=0}^{\min\{m,n\}} \binom{x}{k}\binom{x-k}{m-k}\binom{y-m}{n-k} = \binom{x}{m}\binom{y}{n}, \ (x, y \in \mathbb{R}), \tag{4.69}$$

$$\sum_{k=0}^{n} \binom{n}{k}\binom{m}{n-k}\binom{x+n-k}{m+n} = \binom{x}{m}\binom{x}{n}, \ (x \in \mathbb{R}), \tag{4.70}$$

$$\sum_{k=0}^{\min\{a,b\}} \binom{x+y+k}{k}\binom{x}{b-k}\binom{y}{a-k} = \binom{x+a}{b}\binom{y+b}{a}, \tag{4.71}$$

In particular, it holds

$$\sum_{k=0}^{n} \binom{x}{k}\binom{y}{k}\binom{x+y+n-k}{n-k} = \binom{x+n}{n}\binom{y+n}{n}, \ (x, y \in \mathbb{R}). \tag{4.72}$$

It is worth mentioning that (4.71) is recorded by Koepf [146] (p. 41, Exer. 3.4) as *Stanley's identity*, which obviously contains (4.72) as the special case $a = b = n$. Identity (4.72) has been attributed as a well-known identity to the old Chinese mathematician *Li Shanlai* who lived in Qing dynasty of China. It was re-proved by a number of mathematicians including *P. Turan*, *L. K. Hua*, et al. in various tricky ways.

Observe that both sides of each of the identities (4.69)–(4.72) consist of algebraic polynomials with the same degrees in the real arguments x and y. Thus, it suffices to verify them for the cases where the arguments are restricted to be positive integers. In other words, we need only to consider the cases $x, y \in \mathbb{N}$.

Along this line, in order to deduce (4.69) from (4.23), we may take

$$f(t) = \binom{t+\beta}{n}, \quad \phi_m(t) = \binom{t}{m},$$

and that set $\alpha = x$, $\beta = y - m$. For these two functions, it is easy to calculate

$$\Delta^k f(0) = \binom{\beta}{\beta-k} = \binom{\beta}{k}, \quad \Delta^k f_m(\alpha - k) = \binom{\alpha-k}{m-k},$$

in particular, $\Delta^k \phi_m(0) = \delta_{m,k}$. In this case, the RHS of (4.23) reduces to

$$\sum_{k=0}^{\min\{m,n\}} \binom{x}{k}\binom{x-k}{m-k}\binom{y-m}{n-k}$$

and the LHS of (4.23) yields $\binom{x}{m}\binom{y}{n}$, respectively. Thus (4.69) is established.

For the derivation of (4.70) from (4.23), it suffices to choose

$$f(t) = \binom{t+m}{n}, \quad \phi(t) = \binom{x+t}{x-m}, \quad (x \in \mathbb{N}).$$

As for the verification of (4.71), it is no real restriction to assume $b = \min\{a, b\}$ and replace $b - k$ by k in the LHS. Thus, (4.71) may be rewritten as

$$\sum_{k=0}^{b} \binom{x}{k}\binom{y}{y+b-a-k}\binom{x+y+b-k}{b-k} = \binom{x+a}{b}\binom{y+b}{a}. \tag{4.73}$$

Then, the choice $\alpha = x$, and

$$f(t) = \binom{y+t}{y+b-a}, \quad \phi(t) = \binom{y+b+t}{b}$$

in (4.23), after the similar computation, shows that the both sides of (4.23) just generate identity (4.73), which is equivalent to (4.71).

Clearly, all summation formulas represented in this section just show that (4.1) and (4.2) with their specializations really form the sources of many particular formulas and identities, and as sources they should still be possible to be utilized for getting more types of special formulas and identities than those already displayed. Moreover, it may be worth predicting that certain proper q-analog of (4.1) and (4.2) could be constructed so that various novel q-identities might be derived therefrom. Meanwhile the author hope that WZ-style proofs for the aforementioned combinatorial identities will be found soon. Anyway, readers may suppose that much remains to be done after this present section.

4.2 Summation of Series Involving Other Famous Number Sequences

It has been seen that suitable manipulations with the operators Δ, E, and D (differentiation) could yield a variety of summation formulas and identities in the computational mathematics and combinatorics. In Section 3.3 we present three source formulas (3.1)–(3.3). From the previous section, §4.1, we have seen that formulas (3.2) and (3.3) form formulas (4.1) and (4.2), for which the general convergence theorems are given by assuming $g(z)$ to be an analytic function and $f(z)$ to be functions satisfying certain conditions. Generally speaking, the freedom of choices of g and f in (3.2) and (3.3) (i.e., (4.1) and (4.2)) should imply the versatility of uses of the transformation formulas for establishing various series summation formulas including some analytic combinatorial identities. Similarly, what we are considering in this section are the following two Δ-operational transformation formulas for Newton-type series from formula (3.1).

$$\sum_{k\geq 0}\binom{\alpha}{k}f(k)\Delta^k g(0)=\sum_{k\geq 0}\binom{\alpha}{k}\Delta^k f(0)\Delta^k g(\alpha-k), \tag{4.74}$$

$$\sum_{k\geq 0}\binom{\alpha+k}{k}f(k)(-\Delta)^k g(0)=\sum_{k\geq 0}\binom{\alpha+k}{k}\Delta^k f(0)(-\Delta)^k g(-\alpha-k-1). \tag{4.75}$$

In this section, we will use the formulas (4.74) and (4.75) to study two formulaic classes consisting of various combinatorial algebraic identities and series summation formulas. The basic ideas include utilizing properly the Δ-operator and Fibonacci numbers and Stirling numbers for some series transformations. A variety of classic formulas and remarkable identities are shown to be the members of the classes.

As may be observed, (4.74) and (4.75) could be deduced from each other by letting $\alpha \to -(\alpha+1)$. Thus (4.74) is the simpler basic formula to be concerned. In fact, the proof of the formal identity (4.74) is given in Theorem 3.2.13 and certain of its nice applications have been briefly represented in Section 3.3. Moreover, a further scrutiny of (4.74) reveals itself that it could even be served as a source for drawing a variety of classic formulas and novel results. Accordingly, (4.74) may be used as a source formula for defining a certain class of formulas and identities. All of these together with other related things will be expounded in this sections.

4.2.1 Convergence theorem and examples

Similar to Section 4.1, in order to be able to get various exact results from (4.74) or (4.75) we need a convergence theorem as follows.

Theorem 4.2.1 *Let $f(t)$ and $g(t)$ be real entire functions satisfying the following two conditions*

$$\overline{\lim_{k\to\infty}}\left|f(k)\Delta^k g(0)\right|^{1/k}<1, \tag{4.76}$$

$$\overline{\lim_{k\to\infty}}\left|\Delta^k f(0)\Delta^k g(\alpha-k)\right|^{1/k}<1, \quad \alpha\in\mathbb{R}. \tag{4.77}$$

Then all the series involved in (4.74) and (4.75) are absolutely convergent so that they give exact finite results.

Obviously, for the special case $f(t)$ being a polynomial, the right-hand sides (RHS) of (4.74) and (4.75) just give finite series so that the condition (4.77) certainly holds and may be omitted. Similarly both (4.76) and (4.77) can be omitted in case $g(t)$ is a polynomial. In what follows we will give a proof for the general case.

Proof. It suffices to establish (4.74) with conditions (4.76) and (4.77). For the sake of completeness and as a first step in the proof, a formal process may be reproduced here to justify the formal equality (4.74) quite straightforwardly as follows:

$$\begin{aligned}\sum_{k\geq 0}\binom{\alpha}{k}f(k)x^k =&\sum_{k\geq 0}\binom{\alpha}{k}x^k E^k f(0) = \sum_{k\geq 0}\binom{\alpha}{k}(xE)^k f(0) = (1+xE)^\alpha f(0)\\ =&(1+x+x\Delta)^\alpha f(0) = (1+x)^\alpha \left(1+\frac{x}{1+x}\Delta\right)^\alpha f(0)\\ =&\sum_{k\geq 0}\binom{\alpha}{k}x^k(1+x)^{\alpha-k}\Delta^k f(0).\end{aligned}$$

Having considered the above series as a formal power series in x, one may utilize the well-known substitution rule $x \to \Delta$ to get the formal operational equality of the form (cf. Section 2.2)

$$\begin{aligned}\sum_{k\geq 0}\binom{\alpha}{k}f(k)\Delta^k g(0) =&\sum_{k\geq 0}\binom{\alpha}{k}\Delta^k f(0)\Delta^k(1+\Delta)^{\alpha-k}g(0)\\ =&\sum_{k\geq 0}\binom{\alpha}{k}\Delta^k f(0)\Delta^k E^{\alpha-k}g(0)\\ =&\sum_{k\geq 0}\binom{\alpha}{k}\Delta^k f(0)\Delta^k g(\alpha-k).\end{aligned}$$

The second step is to show that both sides of the above resultant equation are absolutely convergent series. Note that

$$\overline{\lim_{k\to\infty}}\left|\binom{\alpha}{k}\right|^{1/k} = 1, \quad \alpha\in\mathbb{R}.$$

Combining this with the conditions (4.76) and (4.77), we see that Cauchy's root-test is applicable to the series on the both sides of (4.74). Hence the theorem is proved. ∎

Remark 4.2.2 *It is worth mentioning that Theorem 4.2.1 could be restated in a more general form in which the entire function $f(t)$ is replaced by the sequence $(f(k))$ $(k\in\mathbb{N}_0)$ so that the existence of the required sequence $(\Delta^k f(0))$ is also assumed.*

Here we will give a number of examples showing how to get various remarkable formulas or identities from (4.74) via special choices of α, $f(t)$, and $g(t)$. Some examples are given in Subsection 3.2.4 and/or Subsection 4.1.1 by using different symbolic operational source formulas. The table for the calculation of Δ^k given at the beginning of Subsection 3.3.3 is helpful in the following examples.

(1) The classical *Newton interpolation formula* is given by (4.74) with $f(t) = 1$. Also, the binomial theorem for the expansion of $(1 + x)^\alpha$ is given by (4.74) with $f(t) = 1$ and $g(t) = (1 + x)^t$ so that $\Delta^k g(0) = x^k$. Convergence conditions are given by (4.76), viz. $\sup |\Delta^k g(0)|^{1/k} < 1$ and $|x| < 1$, respectively.

(2) Euler's transform for alternating series is given by (4.74) with $\alpha = -1$ and $g(t) = 2^t$ (so that $\Delta^k g(0) = 1$ and $\Delta^k g(-k-1) = 1/2^{k+1}$), namely,

$$\sum_{k\geq 0}(-1)^k f(k) = \sum_{k\geq 0} \frac{(-1)^k}{2^{k+1}} \Delta^k f(0),$$

called the Euler's series transformation formula (cf. Example 2.2.6).

(3) *Vandermonde's convolution* is given by (4.74) with $\alpha \to x$, $f(t) = 1$, and $g(t) = \binom{t+y}{n}$ (so that $\Delta^k g(t) = \binom{t+y}{n-k}$), namely,

$$\sum_{k=0}^{n} \binom{x}{k}\binom{y}{n-k} = \binom{t+y}{n}_{t=x} = \binom{x+y}{n}.$$

(4) Euler's formula for arithmetic-geometric series (cf. Lemma 2.1.7 and Example 2.2.7) is given by (4.74) with $\alpha = -1$, $f(t) = t^r$ $(r \in \mathbb{N}_0)$, and $g(t) = (1-x)^t$ (so that $\Delta^k g(t) = (-x)^k(1-x)^t$), namely

$$\sum_{k\geq 0} k^r x^k = \sum_{j=0}^{r} \frac{j!x^j}{(1-x)^{j+1}} \left\{ {r \atop j} \right\}, \quad |x| < 1,$$

where $\left\{ {r \atop j} \right\}$ are Stirling numbers of the second kind.

(5) *Stirling's summation formula* for the higher order arithmetic progression is given by (4.74) with $\alpha = -1$, $f(t) = t^r$, and $g(t) = \binom{m-t}{m}$ (so that $\Delta^k g(t) = (-1)^k \binom{m-t-k}{m-k}$), namely

$$\sum_{k=0}^{m} k^r = \sum_{j=0}^{r} j! \left\{ {r \atop j} \right\} \binom{m+1}{j+1}.$$

(6) An identity due to Knuth [79] (cf. pp. 3-155) can be obtained from (4.74) by setting $\alpha = -s-1$, $f(t) = \binom{t}{j}$, and $g(t) = \binom{n-t}{n}$ so that

$$\binom{\alpha}{k} = \binom{-s-1}{k} = (-1)^k \binom{s+k}{k},$$

$$\Delta^k g(t) = (-1)^k \binom{n-t-k}{n-k}, \quad \Delta^k g(0) = (-1)^k,$$

for $0 \le k \le n$. More precisely, it can be written as

$$\sum_{k=j}^{n} \binom{s+k}{k}\binom{k}{j} = \binom{s+j}{j}\binom{n+s+1}{n-j} = \binom{n+1}{j}\binom{n+s+1}{s}\frac{n+1-j}{s+1+j}.$$

(7) A theorem stated in §4.4 of Wilf [216] is the identity

$$\sum_{k\ge 0} (-1)^k \binom{n}{k} \Big/ \binom{k+a}{k} = \frac{a}{n+a}, \quad n \ge 0.$$

This can be obtained from (4.74) by letting $\alpha \to n$, $f(t) = 1$, and $g(t) = 1/(t+a)$. In fact one may find

$$\Delta^k g(t) = (-1)^k k!/(t+a+k)_{k+1}, \quad \Delta^k g(0) = (-1)^k \Big/ \left(a\binom{k+a}{k}\right),$$

and $\Delta^k g(n-k) = (-1)^k/((n+a-k)\binom{n+a}{k})$.

(8) An identity of Riordan is the following (cf. [58])

$$\sum_{k=0}^{n} \binom{n}{k}\binom{m}{n-k}\binom{x+n-k}{n+m} = \binom{x}{m}\binom{x}{n}.$$

It is easily seen that the left-hand side (LHS) of the identity may be embedded in the LHS of (4.74) by letting $\alpha \to n$ and taking $f(t) = \binom{x+n-t}{n+m}$ and $g(t) = \binom{m+t}{n}$ so that

$$\Delta^k f(t) = (-1)^k \binom{x+n-t-k}{m+n-k}, \quad \Delta^k g(t) = \binom{m+t}{n-k}.$$

Correspondingly, the RHS of (4.74) gives

$$\begin{aligned}
&\sum_{k\ge 0}(-1)^k \binom{n}{k}\binom{x+n-k}{m+n-k}\binom{m+n-k}{n-k} \\
&= \binom{x}{m}\sum_{k\ge 0}(-1)^k \binom{n}{k}\binom{x+n-k}{n-k} = \binom{x}{m}\binom{x}{n}.
\end{aligned}$$

(9) The well-known identity of Li Shanlai (cf. Section 4.1) is the following

$$\sum_{k\ge 0} \binom{n}{k}\binom{m}{k}\binom{p+n+m-k}{n+m} = \binom{p+n}{n}\binom{p+m}{m}.$$

where the LHS may be embedded in the RHS of (4.74) by letting

$$\alpha \to n, \quad f(t) = \binom{m+t}{m}, \quad g(t) = \binom{p+m+t}{p}.$$

Obviously the LHS of (4.74) gives

$$\sum_{k\geq 0}\binom{n}{k}\binom{m+k}{m}\binom{p+m}{m+k}=\binom{p+m}{m}\sum_{k\geq 0}\binom{n}{k}\binom{p}{k}=\binom{p+m}{m}\binom{p+n}{n}.$$

Recall that (4.74) may also be written as (4.75) with the simple replacement $\alpha\to-(\alpha+1)$, and conversely.

(10) For the pair of *Rosenbaum's identities* (cf. [58], p. 83)

$$\sum_{k=0}^{m}(-1)^k\binom{k+n-1}{k}\binom{n}{m-k}=0,\quad n\geq m\geq 1,$$

$$\sum_{k=0}^{m}(-1)^k\binom{m}{k}\binom{k+n-1}{m-1}=0,\quad n>m\geq 1,$$

it is easily observed that the LHS's of the above two identities could be embedded in the LHS of (4.75) by taking

$$\left\{\alpha\to(n-1),\quad f(t)=1,\quad g(t)=\binom{t+n}{m}\right\},$$

and

$$\left\{\alpha\to-(m+1),\quad f(t)=\binom{t+n-1}{m-1},\quad g(t)=\binom{m-t}{m}\right\},$$

respectively. Correspondingly, the RHS of (4.75) gives, respectively

$$g(-n-k)|_{k=0}=\binom{n+t}{m}_{t=-n}=\binom{0}{m}=0,\quad m\geq 1,$$

$$\sum_{k=0}^{m-1}\binom{-m-1+k}{k}\binom{n-1}{m-1-k}\binom{0}{m-k}=0,\quad n>m\geq 1.$$

(11) As an extension of Stirling's formula in the above (5) there holds the identity

$$\sum_{k=0}^{n}\binom{\alpha+k}{k}k^r=\sum_{j=0}^{r}j!\binom{\alpha+j}{j}\binom{\alpha+n+1}{n-j}\left\{{r \atop j}\right\}.$$

It is seen that the LHS of the identity could be embedded in the LHS of (4.75) by letting $f(t)=t^r$ and $g(t)=\binom{n-t}{n}$ so that $(-\Delta)^k g(0)=\binom{n-k}{n-k}=1$ $(0\leq k\leq n)$. Obviously, the RHS of (4.75) just yields the RHS of the identity.

(12) The following identity is called a *miraculous formula* in Graham-Knuth-Patashnik's book [79] (cf. (5.104))

$$\sum_{k\geq 0}\binom{m}{k}\frac{2m+1}{2m+1-k}(-2)^k=(-1)^m 2^{2m}\Big/\binom{2m}{m}=1\Big/\binom{-1/2}{m}.$$

The LHS may be written as

$$\sum_{k\geq 0}\binom{m}{k}\frac{a}{k+a}\Delta^k(-1)^t\Big|_{t=0},\quad a=-(2m+1).$$

Thus the LHS may be embedded in the LHS of (4.74) by taking $\alpha\to m$, $f(t)=a/(t+a)$, and $g(t)=(-1)^t$. Note that

$$\Delta^k\left(\frac{a}{t+a}\right)_{t=0}=(-1)^k\Big/\binom{k+a}{k},\ \ \Delta^k g(m-k)=(-2)^k(-1)^{m-k}=(-1)^m2^k.$$

Accordingly, the RHS of (4.74) gives

$$\begin{aligned}&\sum_{k\geq 0}\binom{m}{k}(-1)^k\binom{k-2m-1}{k}^{-1}2^k(-1)^m\\&=(-1)^m\left(\sum_{k\geq 0}\binom{m}{k}\binom{2m}{k}^{-1}\binom{2m}{m}2^k\right)\Big/\binom{2m}{m}\\&=(-1)^m\left(\sum_{k\geq 0}\binom{2m-k}{m}2^k\right)\Big/\binom{2m}{m}=(-1)^m2^{2m}\Big/\binom{2m}{m},\end{aligned}$$

where the last sum follows easily from the fact that

$$\begin{aligned}&\sum_{k=0}^{m}\binom{2m-k}{m}2^k=\sum_{k=0}^{m}\binom{2m-k}{m}\sum_{j=0}^{k}\binom{k}{j}\\&=\sum_{k=0}^{m}\binom{2m-k}{m}\sum_{j=0}^{m}\binom{k}{j}=\sum_{j=0}^{m}\sum_{k=0}^{m}\binom{2m-k}{m}\binom{k}{j}\\&=\sum_{j=0}^{m}\binom{2m+1}{m+j+1}=\sum_{j=1}^{m+1}\binom{2m+1}{m+j}=2^{2m},\end{aligned}\tag{4.78}$$

where in the inner summation the *Chu-Vandermonde identity*

$$\sum_{k=0}^{s}\binom{s-k}{m}\binom{k}{j}=\binom{s+1}{m+j+1}\tag{4.79}$$

is used for $s=2m$.

(13) As is known, there are higher difference formulas for trigonometric functions, viz.

$$\begin{aligned}\Delta^k\cos(\alpha t+\beta)&=\left(2\sin\frac{\alpha}{2}\right)^k\cos\left(\alpha t+\beta+\frac{k}{2}(\alpha+\pi)\right),\\\Delta^k\sin(\alpha t+\beta)&=\left(2\sin\frac{\alpha}{2}\right)^k\sin\left(\alpha t+\beta+\frac{k}{2}(\alpha+\pi)\right),\quad\alpha,\beta\in\mathbb{R}.\end{aligned}$$

Thus the special choice $\{\alpha = x, f(t) = t^m, g(t) = \cos(\alpha t + \beta)\}$ in (4.74) leads to a particular summation formula as follows:

$$\sum_{k=0}^{\infty} \binom{x}{k} k^m \left(2\sin\frac{\alpha}{2}\right)^k \cos\left(\beta + \frac{k}{2}(\alpha+\pi)\right)$$
$$= \sum_{k=0}^{m} \binom{x}{k} k! \left\{ {m \atop k} \right\} \left(2\sin\frac{\alpha}{2}\right)^k \cos\left(\alpha x + \beta + \frac{k}{2}(\pi - \alpha)\right)$$

$$\sum_{k=0}^{\infty} \binom{x}{k} k^m \left(2\sin\frac{\alpha}{2}\right)^k \sin\left(\beta + \frac{k}{2}(\alpha+\pi)\right)$$
$$= \sum_{k=0}^{m} \binom{x}{k} k! \left\{ {m \atop k} \right\} \left(2\sin\frac{\alpha}{2}\right)^k \sin\left(\alpha x + \beta + \frac{k}{2}(\pi - \alpha)\right)$$

where the infinite series is absolutely convergent for $0 < \alpha < \pi$ so that $|2\sin\alpha| < 1$.

As we have seen, all what given above has shown that formula (4.74) could be really useful as a source for obtaining a considerable variety of formulas and identities, including classic ones and new ones. This fact may justify that one could useful (4.74) as a "source formula" to define a significant class of formulas and identities. Actually this will be one of the two formulaic classes to be considered in the next subsection.

4.2.2 More summation formulas involving Fibonacci numbers and generalized Stirling numbers

It is known that the ordinary Fibonacci numbers F_n $(n \in \mathbb{N}_0)$ can be extended to the case with negative indices. Actually, the basic recurrence relation originally defined for F_n $(n \in \mathbb{N}_0)$,

$$F_0 = 0, \quad F_1 = 1, \quad F_n = F_{n-1} + F_{n-2}, \quad n > 1$$

can be generally assumed to hold for all $n \in \mathbb{Z}$, so that one may find by induction the simple relation (cf. [79])

$$F_{-n} = (-1)^{n-1} F_n, \quad n \in \mathbb{Z}.$$

Consequently, we have

$$\Delta F_t = F_{t+1} - F_t = F_{t-1}, \quad \Delta^k F_t = F_{t-k}, \quad k \in \mathbb{N}_0,\ t \in \mathbb{Z}.$$

This suggests that formula (4.74) can be well utilized for obtaining combinatorial identities involving Fibonacci numbers. Using (4.74) with the particular choice

$$\left\{ f(t) = F_{r+t},\ g(t) = \binom{\alpha + s - t}{m} \right\}, \quad r, t \in \mathbb{Z},\ s \in \mathbb{N}_0,$$

we have

$$\Delta^k g(t) = (-1)^k \binom{\alpha + s - t - k}{m - k}, \quad \Delta^k g(\alpha - k) = (-1)^k \binom{s}{m - k}.$$

Thus the RHS of (4.74) gives

$$\sum_{k=0}^{m} \binom{\alpha}{k} \binom{\alpha + s - k}{m - k} (-1)^k F_{r+k} = \sum_{j=0}^{s} \binom{\alpha}{j} \binom{s}{m - j} (-1)^j F_{r-j},$$

which may be rewritten via $j \to m - j$ and $r \to -r$

$$\sum_{k=0}^{m} \binom{\alpha}{k} \binom{\alpha + s - k}{m - k} F_{r-k} = \sum_{j=0}^{s} \binom{s}{j} \binom{\alpha}{m - j} F_{m+r-j}.$$

This formula holds for $\alpha \in \mathbb{R}$, $r \in \mathbb{Z}$ and $s \in \mathbb{N}_0$, and may be specialized in various ways, e.g., the following special identities (including a few familiar ones):

$$\sum_{k=0}^{m} \binom{\alpha}{k} \binom{\alpha + 1 - k}{m - k} F_{r-k} = \binom{\alpha}{m} F_{m+r} + \binom{\alpha}{m - 1} F_{m+r-1},$$

$$\sum_{k=0}^{m} \binom{\alpha}{k} \binom{\alpha - k}{m - k} F_{r-k} = \binom{\alpha}{m} F_{m+r},$$

$$\sum_{k=0}^{m} \binom{\alpha}{k} \binom{\alpha - k}{m - k} (-1)^k F_k = -\binom{\alpha}{m} F_m,$$

$$\sum_{k=0}^{m} \binom{m}{k} (-1)^k F_{m+k} = 0, \quad \sum_{k=0}^{m} \binom{m}{k} F_{r-k} = F_{r+m}.$$

Using (4.74) with the following three choices

$$\{\alpha = n,\ f(t) = t^r,\ g(t) = F_{s+t}\}, \quad n, r \in \mathbb{N}_0,\ s,\ t \in \mathbb{Z},$$

$$\left\{\alpha = n,\ f(t) = \binom{\alpha + t}{r},\ g(t) = F_{s+t}\right\}, \quad r \in \mathbb{N}_0,\ s,\ t \in \mathbb{Z},$$

$$\{\alpha = n,\ f(t) = F_{r+t},\ g(t) = F_{s+t}\}, \quad n,\ s,\ t \in \mathbb{Z},$$

we may obtain the following three formulas, respectively,

$$\sum_{k=0}^{n} \binom{n}{k} k^r F_{s-k} = \sum_{j=0}^{r} j! \left\{ {r \atop j} \right\} \binom{n}{j} F_{s+n-2j},$$

$$\sum_{k=0}^{n} \binom{n}{k} \binom{\alpha + k}{r} F_{s-k} = \sum_{j=0}^{r} \binom{n}{j} \binom{\alpha}{r - j} F_{s+n-2j},$$

$$\sum_{k=0}^{n}\binom{n}{k}F_{r+k}F_{s-k}=\sum_{k=0}^{n}\binom{n}{k}F_{r-k}F_{s+n-2k}.$$

Certainly these identities also imply various special identities, e.g., the third identity has the following consequences

$$\sum_{k=0}^{n}\binom{n}{k}F_{n-k}(F_{n+k}+F_{2k-2n})=0,$$
$$\sum_{k=0}^{n}\binom{n}{k}F_{k}(F_{2k}-F_{2n-k})=0,$$
$$\sum_{k=0}^{n}\binom{n}{k}(-1)^{k}F_{k}(F_{k}+(-1)^{n}F_{2k-n})=0.$$

Taking $g(t)=F_{s+2t}$ $(s,\,t\in\mathbb{Z})$, we find

$$\Delta g(t)=F_{s+2t+2}-F_{s+2t}=F_{s+1+2t},$$
$$\Delta^{k}g(t)=\Delta^{k}F_{s+2t}=F_{s+k+2t}.$$

Thus, as an application of (4.74), the following choices

$$\{\alpha=n,\ f(t)=t^{r},\ g(t)=F_{s+2t}\},$$
$$\{\alpha=n,\ f(t)=F_{r+t},\ g(t)=F_{s+2t}\},$$
$$\left\{\alpha=n,\ f(t)=F_{r+2t},\ g(t)=\binom{n-t}{n}\right\},$$

can lead to the following three formulas, respectively,

$$\sum_{k=0}^{n}\binom{n}{k}k^{r}F_{s+k}=\sum_{j=0}^{r}j!\left\{{r \atop j}\right\}\binom{n}{j}F_{s+2n-j},$$
$$\sum_{k=0}^{n}\binom{n}{k}F_{r+k}F_{s+k}=\sum_{k=0}^{n}\binom{n}{k}F_{r-k}F_{s+2n-k},$$
$$\sum_{k=0}^{n}(-1)^{k}F_{r+2k}=(-1)^{n}F_{r+n},\quad r\in\mathbb{Z}.$$

These formulas also have various special consequences, e.g. the second formula implies the following identity with $s=r$:

$$\sum_{k=0}^{n}\binom{n}{k}F_{r+k}^{2}=\sum_{k=0}^{n}\binom{n}{k}F_{r-k}F_{r+2n-k}.$$

The simplest case of this identity is given by $n=1$, namely

$$F_{r}^{2}+F_{r+1}^{2}=F_{r}F_{r+2}+F_{r-1}F_{r+1}.$$

This may be written as a recurrence relation for r as follows:

$$F_rF_{r+2} - F_{r+1}^2 = -(F_{r-1}F_{r+1} - F_r^2), \quad r \geq 1.$$

Thus it follows that

$$F_rF_{r+2} - F_{r+1}^2 = (-1)^r(F_0F_2 - F_1^2) = (-1)^{r+1}.$$

This is known as one of the oldest theorems about Fibonacci numbers, and may be more precisely called *Cassini-Kepler identity* (cf. §6.6 of [79]).

The above examples suggest two combinatorial formulaic classes which may be briefly called *Sigma-Delta class* ($\Sigma\Delta$ *class)* and *Sigma-summation class* (ΣS *class)*. We introduce ($\Sigma\Delta$) class in this subsection and (ΣS) class in the next subsection. Some relations between these two classes will also be discussed later. Let us give the following definitions.

Definition 4.2.3 *The formula* (4.74) *is called a valid source formula whenever the involved sequence* $(f(k))$ $(k \in \mathbb{N}_0)$ *and the entire function* $g(t)$ $(t \in \mathbb{R})$ *satisfy the conditions* (4.76) *and* (4.77) *of Theorem 4.2.1.*

Definition 4.2.4 *Any special choice of the component set* $\{\alpha,\ f(k)$ *or* $f(t),\ g(t)\}$ *is called a "valid choice" if it could be embedded in the valid source formula* (4.74) *to get an exact formula or a valid identity.*

Definition 4.2.5 *Any exact formula or a valid identity is said to be a member of the* ($\Sigma\Delta$) *class, if it can be obtained from* (4.74) *via a valid choice of the component set* $\{\alpha,\ f,\ g\}$. *In particular, the* 3 *specialized conditions* (*I*) $\alpha \to \mathbb{N}_0$, (*II*) $f(t)$ *is a polynomial, and* (*III*) $g(t)$ *is a polynomial just lead to* 3 *subclasses of the* ($\Sigma\Delta$) *class, respectively.*

Evidently, all the formulas and identities given in this subsection are mostly noticeable special examples of the ($\Sigma\Delta$) class. We think that still much remains to be found or constructed successively.

The second formulaic class we wish to introduce is a set consisting of pairs of equivalent formulas involving *generalized binomial coefficients* and generalized Stirling numbers (GSN) defined. We will adopt some familiar notations as follows:

$$(t|d)_n = t(t-d)\cdots(t-nd+d), \quad (t|d)_0 = 1,$$
$$\binom{t}{k}_d = (t|d)_k/k!, \quad \binom{t}{k}_1 = \binom{t}{k}, \quad \binom{t}{k}_0 = t^k/k!.$$

It is known that the 3-parameter GSN's denoted by $S(n,k;\alpha,\beta,\gamma)$ with $(\alpha,\beta,\gamma) \neq (0,0,0)$ are defined by the following (cf. (4.14))

$$(t|\alpha)_n = \sum_{k=0}^{n} S(n,k;\alpha,\beta,\gamma)(t-\gamma|\beta)_k, \tag{4.80}$$

where α, β, and γ may be real or complex numbers. But in our present discussion we only need the GSN's with two parameters, namely, $S(n,k;\alpha,\beta) = S(n,k;\alpha,\beta,0)$, where $(\alpha,\beta) \in \mathbb{R}^2$. Thus, for $\alpha \neq \beta$, the numbers $\binom{t}{p}_\alpha$ and $\binom{t}{p}_\beta$ $(p \in \mathbb{N}_0)$ are related by the equations

$$\binom{t}{p}_\alpha = \sum_{j=0}^{p} \frac{j!}{p!} S(p,j;\alpha,\beta) \binom{t}{j}_\beta , \tag{4.81}$$

$$\binom{t}{p}_\beta = \sum_{j=0}^{p} \frac{j!}{p!} S(p,j;\beta,\alpha) \binom{t}{j}_\alpha . \tag{4.82}$$

These relations are symmetrical with respect to α and β. In particular, for the case $\alpha = \beta$ it is seen that

$$S(p,j;\alpha,\alpha) = \delta_{pj} = \begin{cases} 0, & for\ j < p, \\ 1, & for\ j = p. \end{cases}$$

Now, replacing the variable t of (4.80) by $\beta t + \gamma$, and applying Newton's interpolation formula to the polynomial $f(t) = (\beta t + \gamma|\alpha)_n$, we may find that the GSN as coefficients contained in (4.80) can be expressed using differences (cf. (4.19)), viz.,

$$S(n,k;\alpha,\beta,\gamma) = \frac{1}{\beta^k k!} \left[\Delta^k (\beta t + \gamma|\alpha)_n\right]_{t=0} , \tag{4.83}$$

Certainly an explicit expression of $S(n,k;\alpha,\beta,\gamma)$ can be deduced from (4.83). Of particular usefulness for our evaluation of sums are the numbers $S(n,k;\theta,1,j)$, $S(n,k;0,1,j)$, and $S(n,k;\theta,1,0)$, which are known as *Howard's degenerate weighted Stirling numbers*, *Carlitz's weighted Stirling numbers*, and *degenerate Stirling numbers of the second kind*, respectively (cf. [31, 119]). As an example, we consider the case of $(\alpha,\beta,\gamma) = (0,\beta,\gamma)$. Then

$$S(n,k;0,\beta,\gamma) = \frac{1}{\beta^j j!} \left.\Delta^j (\beta t + \gamma)^k\right|_{t=0}$$

$$= \frac{1}{\beta^j j!} \Delta^j \left(\sum_{i=0}^{k} \binom{k}{i} \gamma^{k-i} \beta^i t^i \right)_{t=0} = \frac{1}{\beta^j j!} \sum_{i=0}^{k} \binom{k}{i} \gamma^{k-i} \beta^i \left(\Delta^j t^i\right)_{t=0}$$

$$= \frac{1}{\beta^j j!} \sum_{i=0}^{k} \binom{k}{i} \gamma^{k-i} d^i j! \left\{ {i \atop j} \right\} = \sum_{i=0}^{k} \binom{k}{i} \left\{ {i \atop j} \right\} \gamma^{k-i} \beta^{i-j}.$$

Hence, we obtain an identity from (4.80)

$$t^n = \sum_{k=0}^{n} \left(\sum_{i=0}^{k} \binom{k}{i} \left\{ {i \atop j} \right\} \gamma^{k-i} \beta^{i-j} \right) (t - \gamma|\beta)_k . \tag{4.84}$$

Moreover, substituting (4.81) into (4.82) or conversely, one may get the orthogonality relations

$$\sum_{i\le j\le p} S(p,j;\alpha,\beta)S(j,i;\beta,\alpha) = \sum_{i\le j\le p} S(p,j;\beta,\alpha)S(j,i;\alpha,\beta) = \delta_{pi}.$$

Making use of these relations, one may find that the "Extended Summation Rule" as formulated in Hsu [127] can be restated in a refined form as follows.

Theorem 4.2.6 *Let α and β be distinct real numbers, and let $F(n,k)$ be a bivariate function defined for integers $n,k \ge 0$. If there can be found a summation formula or a combinatorial algebraic identity such as*

$$\sum_{k=0}^{n} F(n,k)\binom{k}{p}_{\alpha} = \phi(n,p,\alpha), \quad p \ge 0, \tag{4.85}$$

then this equality implies the following formula

$$\sum_{k=0}^{n} F(n,k)\binom{k}{p}_{\beta} = \sum_{j=0}^{p} \frac{j!}{p!} S(p,j;\beta,\alpha)\phi(n,j,\alpha). \tag{4.86}$$

Conversely, (4.86) *also implies* (4.85)*. In brief,* (4.85) *and* (4.86) *are equivalent.*

Proof. The implication (4.85) $\Rightarrow$ (4.86) is simply from the substitution of (4.82) into the LHS of (4.86) and then a application of (4.85) to derive the equivalence of both sides of (4.86), which is left as an exercise (cf. Exercise 4.5). Here, it suffice to verify (4.86) $\Rightarrow$ (4.85). Indeed, using (4.81) and one of the orthogonality relations of $S(p,j;\alpha,\beta)$ and $S(p,j;\beta,\alpha)$, we have via (4.86)

$$\begin{aligned}
\text{LHS of (4.85)} &= \sum_{j=0}^{p} \frac{j!}{p!} S(p,j;\alpha,\beta) \sum_{k=0}^{n} F(n,k)\binom{k}{j}_{\beta} \\
&= \sum_{j=0}^{p} \frac{j!}{p!} S(p,j;\alpha,\beta) \sum_{i=0}^{j} \frac{i!}{j!} S(j,i;\beta,\alpha)\phi(n,i,\alpha) \\
&= \sum_{i=0}^{p} \phi(n,i,\alpha) \sum_{j=i}^{p} \frac{j!}{p!} S(p,j;\alpha,\beta)\frac{i!}{j!} S(j,i;\beta,\alpha) \\
&= \sum_{i=0}^{p} \frac{i!}{p!} \phi(n,i,\alpha)\delta_{pi} = \phi(n,p,\alpha) = \text{RHS of (4.85)}.
\end{aligned}$$

Hence (4.85) and (4.86) are really equivalent to each other. ∎

Completely similarly, for the case of infinite series there holds the following theorem.

Theorem 4.2.7 *Let $(F(k))$ be a sequence in $\mathbb{R}$ such that*

$$\overline{\lim_{k\to\infty}}|F(k)|^{1/k} < 1. \tag{4.87}$$

Then the following pair of summation formulas are equivalent:

$$\sum_{k=0}^{\infty} F(k)\binom{k}{p}_{\alpha} = \phi(p,\alpha), \quad p \geq 0, \tag{4.88}$$

$$\sum_{k=0}^{\infty} F(k)\binom{k}{p}_{\beta} = \sum_{j=0}^{p} \frac{j!}{p!} S(p,j;\beta,\alpha)\phi(j,\alpha). \tag{4.89}$$

The proof follows from the similar lines as that of Theorem 4.2.6, and is left for Exercise 4.6.

Note that the ordinary Stirling numbers and *Lah's numbers* (in absolute value) are special cases of GSN's, namely,

$$S(n,k;1,0) = S_1(n,k) = \begin{bmatrix} n \\ k \end{bmatrix},$$

$$S(n,k;0,1) = S_2(n,k) = \begin{Bmatrix} n \\ k \end{Bmatrix},$$

$$S(n,k;-1,1) = \frac{n!}{k!}\binom{n-1}{k-1}.$$

Thus it is easily seen that Theorems 4.2.6 and 4.2.7 imply two corollaries as follows (by taking $(\alpha,\beta) = (1,0)$ and $(\alpha,\beta) = (1,-1)$, respectively).

Corollary 4.2.8 *The following three summation formulas are equivalent to each other:*

$$\sum_{k=0}^{n} F(n,k)\binom{k}{j} = \phi(n,j), \quad j \geq 0, \tag{4.90}$$

$$\sum_{k=0}^{n} F(n,k)k^p = \sum_{j=0}^{p} j!\phi(n,j)\begin{Bmatrix} p \\ j \end{Bmatrix}, \tag{4.91}$$

$$\sum_{k=0}^{n} F(n,k)\binom{k+p-1}{p} = \sum_{j=0}^{p} \phi(n,j)\binom{p-1}{j-1}. \tag{4.92}$$

Corollary 4.2.9 *Suppose that (4.87) holds. Then the following three summation formulas are equivalent to each other:*

$$\sum_{k=0}^{n} F(n,k)\binom{k}{j} = \Phi(j), \quad j \geq 0, \tag{4.93}$$

$$\sum_{k=0}^{n} F(n,k)k^p = \sum_{j=0}^{p} j!\Phi(j)\begin{Bmatrix} p \\ j \end{Bmatrix}, \tag{4.94}$$

$$\sum_{k=0}^{n} F(n,k)\binom{k+p-1}{p} = \sum_{j=0}^{p} \Phi(j)\binom{p-1}{j-1}. \tag{4.95}$$

Surely, formulas contained in these two corollaries can be used to obtain a variety of special formulas or identities in combinatorics. Indeed, for instance, some identities of the types (4.90) and (4.91) will be displayed below.

For the simplest case $F(n,k) = 1$, from (4.90) we find the corresponding $\phi(n,j)$ as

$$\phi(n,j) = \sum_{k=j}^{n} \binom{k}{j} = \binom{n+1}{j+1}. \tag{4.96}$$

This leads to the familiar formula from (4.91)

$$\sum_{k=1}^{n} k^p = \sum_{j=1}^{p} j!\binom{n+1}{j+1}\left\{ {p \atop j} \right\}. \tag{4.97}$$

Actually there are many known identities of type (4.90) in which $F(n,k)$ may consist of a binomial coefficient or a product of binomial coefficients. See, e.g., Egorychev [58], Gould [73], and Riordan [183]. Consequently, we may find various special summation formulas via (4.91). We now list a dozen formulas, as follows:

$$\sum_{k=0}^{n} k^p \binom{n}{k} u^k v^{n-k} = \sum_{j=0}^{p} j! u^j \binom{n}{j}\left\{ {p \atop j} \right\}, \quad u+v=1,\ u>0, \tag{4.98}$$

$$\sum_{k=0}^{[n/2]} k^p \binom{n}{2k} = \sum_{j=0}^{p} j! 2^{n-2j-1} \frac{n}{n-j}\binom{n-j}{j}\left\{ {p \atop j} \right\}, \tag{4.99}$$

$$\sum_{k=0}^{[n/2]} k^p \binom{n+1}{2k+1} = \sum_{j=0}^{p} 2^{n-2j} j! \binom{n-j}{j}\left\{ {p \atop j} \right\}, \tag{4.100}$$

$$\sum_{k=0}^{n} k^p \binom{n-k}{s} = \sum_{j=0}^{p} j! \binom{n+1}{s+j+1}\left\{ {p \atop j} \right\}, \tag{4.101}$$

$$\sum_{k=0}^{n} k^p \binom{s+k}{s} = \sum_{j=0}^{p} j! \frac{n+1-j}{s+1+j}\binom{n+1}{j}\binom{n+s+1}{s}\left\{ {p \atop j} \right\}, \tag{4.102}$$

$$\sum_{k=0}^{n} (-4)^k k^p \binom{n+k}{2k} = \sum_{j=0}^{p} (-1)^n 2^{2j} j! \frac{2n+1}{2j+1}\binom{n+j}{2j}\left\{ {p \atop j} \right\}, \tag{4.103}$$

$$\sum_{k=0}^{n} (-4)^k k^p \frac{n}{n+k}\binom{n+k}{2k} = \sum_{j=0}^{p} (-1)^n 2^{2j} j! \frac{n}{n+j}\binom{n+j}{2j}\left\{ {p \atop j} \right\}, \tag{4.104}$$

$$\sum_{k=0}^{[n/2]} (-1)^k 2^{n-2k} k^p \binom{n-k}{k} = \sum_{j=0}^{p} (-1)^j j! \binom{n+1}{2j+1}\left\{ {p \atop j} \right\}, \tag{4.105}$$

$$\sum_{k=0}^{n} k^p \binom{\alpha}{k}\binom{\beta}{n-k} = \sum_{j=0}^{p} j!\binom{\alpha}{j}\binom{\alpha+\beta-j}{n-j}\left\{ {p \atop j} \right\}, \quad \alpha,\beta \in \mathbb{R}, \tag{4.106}$$

$$\sum_{k=0}^{n} (-1)^k k^p \binom{n}{k}\binom{2n-k}{n} = \sum_{j=0}^{p} (-1)^j j! \binom{n}{j}^2 \left\{ {p \atop j} \right\}, \tag{4.107}$$

$$\sum_{k=0}^{[n/2]} 2^{n-2k} k^p \binom{n}{2k}\binom{2k}{k} = \sum_{j=0}^{p} j!\binom{n}{j}\binom{2n-2j}{n}\left\{ {p \atop j} \right\}, \tag{4.108}$$

$$\sum_{k=1}^{n} k^p H_k = \sum_{j=1}^{p} j!\binom{n+1}{j+1}\left(H_{n+1} - \frac{1}{j+1}\right)\left\{ {p \atop j} \right\}, \tag{4.109}$$

where $H_k = 1 + (1/2) + \cdots + (1/k)$, $(k \geq 1)$, are harmonic numbers.

Though most of the above formulas (except (4.98)) appear unfamiliar, or are difficult to find in the literature, they are actually companion formulas of some known identities. In fact, (4.98) is known as the pth moment of the binomial distribution of a discrete random variable. Formulas (4.99) and (4.100) represent companion formulas of the pair of *Moriarty identities* (cf. (2.73) and (2.74) of [58] and (3.120) and (3.121) of [73]). Also, (4.102) and (4.105) are just companion formulas of the following identities:

$$\sum_{k=j}^{n} \binom{k}{j}\binom{k+s}{s} = \frac{n+1-j}{s+1+j}\binom{n+1}{j}\binom{n+s+1}{s} \tag{4.110}$$

and

$$\sum_{k=j}^{[n/2]} (-1)^k 2^{n-2k} \binom{k}{j}\binom{n-k}{k} = (-1)^j \binom{n+1}{2j+1} \tag{4.111}$$

due to Knuth and Ascher, respectively (cf. (3.155) and (3.179) of [73]). Moreover, (4.109) may be inferred from the known relation (cf, e.g., pp. 98-99 in Ch. 3 of [2]):

$$\sum_{k=j}^{n} H_k \binom{k}{j} = \binom{n+1}{j+1}\left(H_{n+1} - \frac{1}{j+1}\right). \tag{4.112}$$

Evidently, both (4.101) and (4.102) imply (4.97) with $s = 0$, and (4.106) yields the Vandermonde convolution identity when $p = 0$. Moreover, it is easily found that (4.109) leads to an asymptotic relation, for $n \to \infty$ of the following,

$$\sum_{k=1}^{n} k^p H_k \sim \frac{n^{p+1}}{p+1}\left(\log n + \gamma - \frac{1}{p+1}\right), \tag{4.113}$$

where $\gamma := \lim_{n\to\infty}(H_n - \log n) = 0.5772...$ is *Euler's constant*. More precisely, whenever n is much bigger than p, say $n >> p^2$, the asymptotic behavior of the combinatorial sums in (4.109) is mainly determined by those principal terms (i.e., the terms with $j = p$). Hence, $\left\{ {p \atop p} \right\} = 1$, $H_{n+1} \sim \log n + \gamma$, and

$$p!\binom{n+1}{p+1} = p!\frac{(n+1)_{p+1}}{(p+1)!} \sim \frac{n^{p+1}}{p+1},$$

which implies (4.113).

4.2.3 Summation formulas of (ΣS) class

The examples (4), (5), (6), and (11) in Subsection 4.2.1 and the identities and summation formulas in corollaries 4.2.8 and 4.2.9 and their examples (4.98)–(4.112) suggest that it is reasonable to define a formulaic class by using (4.85)–(4.86) and (4.88)–(4.89) as source formulas.

Definition 4.2.10 *Any summation formula or a valid identity is said to be a member of the* (ΣS) *class, if it could be expressed in the same form as one of the formulas* (4.85)–(4.86) *and* (4.88)–(4.89), *in which* $F(n, k)$ *or* $F(k)$ *is a real definite function and* $\phi(n, p, \alpha)$ *or* $\phi(p, \alpha)$ *is an algebraic expression without containing any summation having number of terms unbounded.*

Evidently, all the formulas given by the examples (4), (5), (6), and (11) of Subsection 4.2.1 belong to both $(\Sigma\Delta)$ and (ΣS) so that the intersection set $(\Sigma\Delta) \cup (\Sigma S)$ is non-empty. Moreover, it is easily observed that the pair of Moriarty identities

$$\sum_{k=0}^{[n/2]} \binom{k}{j}\binom{n}{2k} = 2^{n-2j-1}\frac{n}{n-j}\binom{n-j}{j}, \quad \sum_{k=0}^{[n/2]} \binom{k}{j}\binom{n}{2k+1} = 2^{n-2j}\binom{n-j}{j} \tag{4.114}$$

cannot be embedded in (4.74) so that they are not the members of the $(\Sigma\Delta)$ class. However, they belong to the (ΣS) class. This means that (ΣS) is not a subset of $(\Sigma\Delta)$.

As typical members of (ΣS) we will mention here several pairs of equivalent formulas that are implied by the equivalence relations (4.90) $\Longleftrightarrow$ (4.91) and (4.93) $\Longleftrightarrow$ (4.94). In certain pairs given below the notations $\omega(p)$, H_k, and F_k denote, respectively, the *Bell numbers*, harmonic numbers and Fibonacci numbers, namely

$$\omega(p) = \sum_{j=1}^{p} \left\{ {p \atop j} \right\}, \quad H_k = 1 + \frac{1}{2} + \cdots + \frac{1}{k}, \quad k \geq 1,$$

and $F_k = (a^{k+1} - b^{k+1})/\sqrt{5}$ with $a = (1+\sqrt{5})/2$ and $b = (1-\sqrt{5})/2$.

$$\begin{cases} \sum_{k=0}^{n} \binom{k}{j} = \binom{n+1}{j+1} & \text{(Zhu Shijie identity)}, \\ \sum_{k=0}^{n} k^p = \sum_{j=0}^{p} j!\binom{n+1}{j+1}\left\{ {p \atop j} \right\} & \text{(Stirling's formula)}, \end{cases} \tag{4.115}$$

$$\begin{cases} \sum_{k=0}^{n} \frac{1}{k!}\binom{k}{j} = \frac{1}{j!}\sum_{k=0}^{\infty}\frac{(k)_j}{k!} = \frac{e}{j!}, \\ \sum_{k=0}^{n} \frac{1}{k!}k^p = e\sum_{j=0}^{p}\left\{ {p \atop j} \right\} = e\omega(p) \quad \text{(Dobinski's formula)}, \end{cases} \tag{4.116}$$

$$\begin{cases} \sum_{k=0}^{n} \binom{k}{j}x^k = \frac{x^j}{(1-x)^{j+1}}, & |x| < 1, \\ \sum_{k=0}^{n} k^p x^k = \sum_{j=0}^{p} j!\frac{x^j}{(1-x)^{j+1}}\left\{ {p \atop j} \right\} & \text{(Euler's formula)}, \end{cases} \tag{4.117}$$

$$\begin{cases} \sum_{k=0}^{n} \binom{s+k}{s}\binom{k}{j} = \binom{s+j}{j}\binom{n+s+1}{n-j} \quad \text{(Knuth's identity)}, \\ \sum_{k=0}^{n} \binom{s+k}{s}k^p = \sum_{j=0}^{p} j!\binom{s+j}{j}\binom{n+s+1}{n-j}\left\{ {p \atop j} \right\}, \end{cases} \tag{4.118}$$

$$\begin{cases} \sum_{k=0}^{n} \binom{k}{j}H_k = \binom{n+1}{j+1}\left(H_{n+1} - \frac{1}{j+1}\right), \\ \sum_{k=0}^{n} k^p H_k = \sum_{j=0}^{p} j!\binom{n+1}{j+1}\left(H_{n+1} - \frac{1}{j+1}\right)\left\{ {p \atop j} \right\}, \end{cases} \tag{4.119}$$

$$\begin{cases} \sum_{k=0}^{n} \binom{k}{j}F_k x^{k+1} = \frac{1}{\sqrt{5}}\left[\left(\frac{ax}{1-ax}\right)^{j+1} - \left(\frac{bx}{1-bx}\right)^{j+1}\right], \\ \sum_{k=0}^{n} k^p F_k x^{k+1} = \frac{1}{\sqrt{5}}\sum_{j=0}^{p} j!\left[\left(\frac{ax}{1-ax}\right)^{j+1} - \left(\frac{bx}{1-bx}\right)^{j+1}\right]\left\{ {p \atop j} \right\}, \\ \text{for} \quad |x| < 0.618. \end{cases} \tag{4.120}$$

As may be observed, the pairs (4.115), (4.116), and (4.117) just reveal the fact that certain somewhat harder summation formulas are actually equivalent to some rather simple equalities. Generally, the real situation is that once Stirling numbers have been introduced into play, everything regarding summation problems becomes apparently simplified and more easily treated.

We now make use of Knuth's identity (4.110) and the GSN of the form $S(n, k; \theta, 1, \lambda + j)$ to get some summation formulas. The first result is the following.

Theorem 4.2.11 *For any given integers $j \geq 0$ and $p \geq 0$, we have a summation formula as follows:*

$$\sum_{k=j}^{n}\binom{k}{j}(k+\lambda|\theta)_p = \binom{n+1}{j}\sum_{r=0}^{p}\frac{(n+1-j)_{r+1}}{r+j+1}S(p, r; \theta, 1, \lambda + j), \tag{4.121}$$

where λ and θ are real or complex numbers.

Proof. Let us start with the relation (4.80) which is actually an algebraic identity involving variables α, β, γ and t. Making substitutions $\alpha \to \theta$, $\beta \to 1$, $\gamma \to \lambda + j$, and $t \to k + \lambda$, we find that (4.80) may be rewritten in the form

$$(k+\lambda|\theta)_p = \sum_{r=0}^{p} r!S(p, r; \theta, 1, \lambda + j)\binom{k-j}{r}. \tag{4.122}$$

Using Knuth's identity (4.110) with substitutions $s \to j$ and $k \to k - j$, we have

$$\sum_{k=j}^{n} \binom{k-j}{r}\binom{k}{j} = \sum_{k=0}^{n-j} \binom{k}{r}\binom{k+j}{j} = \sum_{k=r}^{n-j} \binom{k}{r}\binom{k+j}{j}$$
$$= \frac{n+1-j-r}{r+j+1}\binom{n+1-j}{r}\binom{n+1}{j}$$
$$= \frac{r+1}{r+j+1}\binom{n+1-j}{r+1}\binom{n+1}{j}. \tag{4.123}$$

Thus, making use of (4.122) and (4.123), we find

$$\sum_{k=j}^{n} \binom{k}{j}(k+\lambda|\theta)_p = \sum_{r=0}^{p} r! S(p,r;\theta,1,\lambda+j) \sum_{k=j}^{n} \binom{k-j}{r}\binom{k}{j}$$
$$= \binom{n+1}{j} \sum_{r=0}^{p} r! \frac{r+1}{r+j+1}\binom{n+1-j}{r+1} S(p,r;\theta,1,\lambda+j)$$
$$= \binom{n+1}{j} \sum_{r=0}^{p} \frac{(n+1-j)_{r+1}}{r+j+1} S(p,r;\theta,1,\lambda+j). \tag{4.124}$$

This is what we desired.

■

Note that the GSNs $S(p,r;\theta,1,\lambda+j)$ may be expressed in the forms

$$S(p,r;\theta,1,\lambda+j) = \frac{1}{r!}\left[\Delta^r (t+\lambda+j|\theta)_p\right]_{t=0}, \tag{4.125}$$
$$S(p,r;\theta,1,\lambda+j) = \frac{1}{r!}\sum_{i=0}^{r}(-1)^{r-j}\binom{r}{j}(i+\lambda+j|\theta)_p. \tag{4.126}$$

Note also that (4.125) is implied by (4.83) and that (4.126) just follows from the well-known expression for higher differences. It is clear that (4.126) is a formula of rank 1 since it consists of only a single summation involving elementary terms. Consequently, formula (4.121) is of rank 2.

Remark 4.2.12 *In fact, that formula* (4.121) *is like a dual to formula (35) of [132], which states*

$$\sum_{k=0}^{n} \binom{n}{k}(k+\lambda|\theta)_p = \sum_{r=0}^{p} r! 2^{n-r}\binom{n}{r} S(p,r;\theta,1,\lambda), \tag{4.127}$$

where $S(p,r;\theta,1,\lambda)s$ *are given by* (4.126) *with* $j = 0$. *So,* (4.127) *is also a formula of rank* 2. *We can derive formula* (4.127) *using our method developed here, which is less complicated than that used in [132]. Indeed, from formula (4.122) and the identity*

$$\binom{n}{k}\binom{k}{r} = \binom{n}{r}\binom{n-r}{k-r}, \tag{4.128}$$

we have

$$\begin{aligned}\sum_{k=0}^{n}\binom{n}{k}(k+\lambda|\theta)_p &= \sum_{r=0}^{p} r!S(p,r;\theta,1,\lambda)\sum_{k=0}^{n}\binom{n}{k}\binom{k}{r}\\ &= \sum_{r=0}^{p} r!S(p,r;\theta,1,\lambda)\binom{n}{r}\sum_{k=r}^{n}\binom{n-r}{k-r}\\ &= \sum_{r=0}^{p} r!\binom{n}{r}S(p,r;\theta,1,\lambda)\sum_{k=0}^{n-r}\binom{n-r}{k}\\ &= \sum_{r=0}^{p} r!2^{n-r}\binom{n}{r}S(p,r;\theta,1,\lambda).\end{aligned} \tag{4.129}$$

A number of corollaries giving some previously known and unknown formulas may be stated as consequences of (4.121) and (4.126) as follows.

Corollary 4.2.13 *There holds the formula*

$$\sum_{k=j}^{n}\binom{k}{j}(k|\theta)_p = \binom{n+1}{j}\sum_{r=0}^{p}\frac{(n+1-j)_{r+1}}{r+j+1}S(p,r;\theta,1,j), \tag{4.130}$$

where Howard's GSN $S(p,r;\theta,1,j)$ (cf. [119]) is written in the form

$$S(p,r;\theta,1,j) = \frac{1}{r!}\sum_{i=0}^{r}(-1)^{r-j}\binom{r}{i}(i+j|\theta)_p. \tag{4.131}$$

In particular, if $\theta = 0$, then (4.130) is reduced to

$$\sum_{k=j}^{n}\binom{k}{j}k^p = \binom{n+1}{j}\sum_{r=0}^{p}\frac{(n+1-j)_{r+1}}{r+j+1}S(p,r;0,1,j), \tag{4.132}$$

where Carlitz's weighted Stirling numbers $S(p,r;0,1,j)$ (cf. [31]) is expressed in the form

$$S(p,r;0,1,j) = \frac{1}{r!}\sum_{i=0}^{r}(-1)^{r-j}\binom{r}{i}(i+j)^p. \tag{4.133}$$

Note that (4.130) and (4.132) are different from the main results of Gould and Wetweerapong [78], nevertheless they are comparable with each other.

By using residue method, Huang [136] provided an identity for (4.132) with $p = 2$. Jones [140] also used a telescoping series to derive a different formula of (4.132), namely,

$$\sum_{k=j}^{n} \binom{k}{j} k^p = j^p \binom{n+1}{j+1} + \sum_{i=j+1}^{n} (i^p - (i-1)^p) \left(\binom{n-1}{j+1} - \binom{i}{j+1} \right). \tag{4.134}$$

If $\theta = 1$ in Corollary 4.2.13, then we have the following:

Corollary 4.2.14 *We have a pair of formulas of rank* 1 *as follows:*

$$\sum_{k=j}^{n} \binom{k}{j}\binom{k}{p} = \binom{n+1}{j+1} \sum_{r=0}^{p} \frac{j+1}{r+j+1} \binom{n-j}{r} \binom{j}{p-r}, \tag{4.135}$$

$$\sum_{k=j}^{n} \binom{k}{p}^2 = \binom{n+1}{p+1} \sum_{r=0}^{p} \frac{p+1}{r+p+1} \binom{n-p}{r} \binom{p}{r}. \tag{4.136}$$

Proof. Obviously, (4.136) is implied by (4.135) with $j = p$. Notice that formula (4.130) with $\theta = 1$ implies that

$$\sum_{k=j}^{n} \binom{k}{j}\binom{k}{p} = \frac{1}{p!} \binom{n+1}{j} \sum_{r=0}^{p} \frac{(n+1-j)_{r+1}}{r+j+1} S(p, r; 1, 1, j). \tag{4.137}$$

Here, using (4.131) and the higher difference formula, that we easily find

$$\begin{aligned} S(p, r; 1, 1, j) &= \frac{p!}{r!} \sum_{i=0}^{r} (-1)^{r-j} \binom{r}{i} \binom{i+j}{p} \\ &= \frac{p!}{r!} \left[\Delta^r \binom{t+j}{p} \right]_{t=0} = \frac{p!}{r!} \binom{j}{p-r} \end{aligned} \tag{4.138}$$

for $p \geq r$. Thus, by substitution of the above resulting expression into the right-hand side of (4.137), we will attain the desired expression (4.135) after simple computations.

■

Corollary 4.2.15 *There holds*

$$\sum_{k=0}^{n} (k|\theta)_p = \sum_{r=0}^{p} r! \binom{n+1}{r+1} S(p, r; \theta, 1, 0), \tag{4.139}$$

where Carlitz's degenerate Stirling numbers $S(p, r; \theta, 1, 0)$ are expressed in the form

$$S(p, r; \theta, 1, 0) = \frac{1}{r!} \sum_{i=0}^{r} (-1)^{r-i} \binom{r}{i} (i|\theta)_p. \tag{4.140}$$

In particular, if $\theta = 0$, there holds the classical formula

$$\sum_{k=0}^{n} k^p = \sum_{r=0}^{p} r! \binom{n+1}{r+1} \left\{ {p \atop r} \right\}, \tag{4.141}$$

where $\left\{ {p \atop r} \right\} = S(p, r; 0, 1, 0)$ are the (ordinary) Stirling numbers of the second kind.

Corollary 4.2.16 *For any given real or complex number λ,*

$$\sum_{k=j}^{n} \binom{k}{j} = \binom{n+1}{j+1}. \tag{4.142}$$

where $S(p, r; 0, 1, \lambda)$ can be computed by (4.133) with $j = \lambda$.

Note that (4.142) implies (4.141) with $\lambda = 0$.

Corollary 4.2.17 *There holds probably the most simple identity*

$$\sum_{k=j}^{n} \binom{k}{j} = \binom{n+1}{j+1}. \tag{4.143}$$

Identity (4.143) (i.e., (4.115)) is known as the most old formula found in the 14th century by ancient Chinese mathematician Zhu Shijie. Indeed, (4.143) appeared in the second mathematics book of Zhu which was published in 1303 AD. Certainly, (4.143) is a formula of rank 0.

There is a closed formula for other types of combinatorial sums involving generalized factorials. For example, we have the following known results (cf. [132]):

$$\sum_{k=0}^{n} \binom{n}{k}^2 (k + \lambda|\theta)_p = \sum_{r=0}^{p} (n)_r \binom{2n-r}{n} S(p, r; \theta, 1, \lambda), \tag{4.144}$$

$$\sum_{k=0}^{[n/2]} \binom{n}{2k} (k + \lambda|\theta)_p = \sum_{r=0}^{p} r! 2^{n-2r-1} \frac{n}{n-r} \binom{n-r}{r} S(p, r; \theta, 1, \lambda), \tag{4.145}$$

where $S(p, r; \theta, 1, \lambda)$ are given by (4.131) with $j = \lambda$ so that (4.144) and (4.145) are all formulas of rank 2. Formula (4.145) is an extension of (4.99) for the classical Stirling numbers of the second kind. (4.144) and (4.145) can also be derived by using the represented method easily.

For proving (4.144), we note that Chu-Vandermonde's identity

$$\sum_{k=0}^{n} \binom{x}{k} \binom{y}{n-k} = \binom{x+y}{n}. \tag{4.146}$$

Then, using formula (4.122), we have

$$\begin{aligned}\sum_{k=0}^{n}\binom{n}{k}^2(k+\lambda|\theta)_p &= \sum_{r=0}^{p} r!S(p,r;\theta,1,\lambda)\sum_{k=0}^{n}\binom{n}{k}^2\binom{k}{r}\\ &= \sum_{r=0}^{p} r!S(p,r;\theta,1,\lambda)\binom{n}{r}\sum_{k=0}^{n}\binom{n}{k}\binom{n-r}{k-r}\\ &= \sum_{r=0}^{p}(n)_r S(p,r;\theta,1,\lambda)\sum_{k=0}^{n}\binom{n}{k}\binom{n-r}{k-r}. \end{aligned}\tag{4.147}$$

Taking $x = n$ and $y = n - r$ in (4.146), we have formula (4.144).

For proving (4.145), we have, from (4.122),

$$\sum_{k=0}^{[n/2]}\binom{n}{2k}^2(k+\lambda|\theta)_p = \sum_{r=0}^{p} r!S(p,r;\theta,1,\lambda)\sum_{k=0}^{[n/2]}\binom{n}{2k}\binom{k}{r}. \tag{4.148}$$

Since the first Moriarty identity (4.114)

$$\sum_{k=r}^{[n/2]}\binom{k}{r}\binom{n}{2k} = 2^{n-2r-1}\frac{n}{n-r}\binom{n-r}{r},$$

formula (4.145) follows immediately.

The Li Shanlan (1811 – 1882) identity is the well-known classical identity

$$\sum_{k=0}^{n}\binom{n}{k}^2\binom{m+n+k}{2n} = \binom{m+n}{n}^2, \tag{4.149}$$

which appeared in Li's writings in 1860s and was given several proofs 50 years ago by a number of authors including Paul Turan, Loo-Keng Hua, et al. The related references are too numerous to be given. Here it may be worth mentioning that this identity is a particular consequence of (4.144). Recall that

$$\binom{s}{m}\binom{m}{n} = \binom{s}{n}\binom{s-n}{m-n}. \tag{4.150}$$

From (4.138),we have $j!S(2n,j;1,1,m+n) = (2n)!\binom{m+n}{2n-j}$. Thus, we see that from formula (4.144) with $\lambda = m+n$, $\theta = 1$, and $p = 2n$ there holds

$$\begin{aligned}\text{LHS of (4.149)} &= \frac{1}{2n!}\sum_{k=0}^{n}\binom{n}{k}^2(k+m+n|1)_{2n}\\ &= \frac{1}{2n!}\sum_{j=0}^{n} j!\binom{2n-j}{n}\binom{n}{j}S(2n,j;1,1,m+n)\\ &= \sum_{j=0}^{n}\binom{n}{j}\binom{2n-j}{n}\binom{m+n}{2n-j}\end{aligned}$$

$$= \sum_{j=0}^{n} \binom{n}{j} \binom{m+n}{n} \binom{m}{n-j} \quad (by\ (4.150))$$

$$= \binom{m+n}{n} \binom{m+n}{n} = \text{RHS of (4.149).} \tag{4.151}$$

It is clear that formula (4.121) is useful for practical computations whenever n is much bigger than p, denoted by $n >> p^2$. Observe that for large n, the asymptotic behavior of the combinatorial sums in (4.121) is mainly determined by those principal terms (i.e., the terms with $r = p$) within the closed formula. Also note that $S(p, p; \cdot, \cdot, \cdot) = 1$. Thus, we easily obtain a simple asymptotic relation for $n \to \infty$ as follows:

$$\begin{aligned}\sum_{k=j}^{n} \binom{k}{j} (k + \lambda|\theta)_p =& \frac{(n+1)_j}{j!} \sum_{r=0}^{p} \frac{(n+1-j)_{r+1}}{r+j+1} S(p, r; \theta, 1, \lambda + j) \\ \sim& \frac{n^j}{j!} \frac{n^{p+1}}{p+j+1} = \frac{n^{p+j+1}}{j!(p+j+1)}.\end{aligned} \tag{4.152}$$

Certainly, the asymptotic estimate given by (4.152) can be refined by taking into account those terms with $r = p-1$, $r = p-2$, and so forth, within the closed formula. And accordingly, values of $S(p, p-1; \cdot, \cdot, \cdot)$, $S(p, p-2; \cdot, \cdot, \cdot)$, and so forth are required to be evaluated.

In what follows we will show how to make use of the GSN of the form $S(n, k, 1/m, 1)$ $(m \in \mathbb{N})$ to get some formulas with closed forms for the following sums

$$\sum_{k=0}^{n} \binom{s+k}{s} \binom{mk}{p}, \ \sum_{k=0}^{[n/2]} \binom{n+1}{2k+1} \binom{mk}{p}, \ \sum_{k=0}^{\infty} \frac{1}{k!} \binom{mk}{p}, \ \sum_{k=0}^{n} \binom{mk}{p} H_k.$$

Note that $\binom{k}{p}_{1/m} = \binom{mk}{p}/m^p$. Thus, by using Theorems 4.2.6 and 4.2.7 with $(\alpha, \beta) = (1, 1/m)$, we may easily obtain four sum formulas as follows

$$\sum_{k=0}^{n} \binom{s+k}{s} \binom{mk}{p} = m^p \sum_{j=0}^{p} \frac{j!}{p!} \binom{s+j}{j} \binom{n+s+1}{n-j} S\left(p, j; \frac{1}{m}, 1\right), \tag{4.153}$$

$$\sum_{k=0}^{[n/2]} \binom{n+1}{2k+1} \binom{mk}{p} = m^p \sum_{j=0}^{p} 2^{n-2j} \frac{j!}{p!} \binom{n-j}{j} S\left(p, j; \frac{1}{m}, 1\right), \tag{4.154}$$

$$\frac{1}{m^p} \sum_{k=0}^{\infty} \frac{1}{k!} \binom{mk}{p} = \frac{e}{p!} \sum_{j=0}^{p} S\left(p, j; \frac{1}{m}, 1\right), \tag{4.155}$$

$$\sum_{k=0}^{n} \binom{mk}{p} H_k = m^p \sum_{j=0}^{p} \frac{j!}{p!} \binom{n+1}{j+1} \left(H_{n+1} - \frac{1}{j+1}\right) S\left(p, j; \frac{1}{m}, 1\right). \tag{4.156}$$

Certainly, the above-mentioned sums with $m \geq 2$ cannot be evaluated into closed forms without using GSN's. Formulas (4.153), (4.154), and (4.156) would be practically useful when n is much bigger than p. Also, it is clear that the limiting case of (4.155) with $m \to \infty$ just yields Dobinski's identity since $S(p, j; 0, 1) = j$.

We will consider a new summation problem and an enlargement of the class (ΣS). As may be imagined, the sum formulas (4.153)–(4.156) would suggest the question of whether there are explicit sum formulas for those sums with the factor $\binom{mk}{p}$ of the summands being replaced by $\binom{mk+q}{p}$ $(q \in \mathbb{N})$. More generally, the problem may be stated in this form: Given a real function $F(n, k)$ for $n, k \geq 0$, and a sequence $(F(k))$ with

$$\overline{\lim_{k\to\infty}} |F(k)|^{1/k} < 1.$$

Suppose that there have been given known summation formulas such as (4.90) and (4.93). One may ask whether there could be constructed certain explicit sum formulas for the sums of the forms

$$\sum_{k=0}^{n} F(n,k)\binom{mk+q}{p} \quad \text{and} \quad \sum_{k=0}^{\infty} F(k)\binom{mk+q}{p}.$$

Certainly, the existence of solution is out of question since $\binom{mk+q}{p}$ is a polynomial in k of degree p and can be expressed as a linear combination of $\binom{k}{j}'$s. However, the real problem is to find constructive expressions. Recall that for the 3-parameter GSN's we have the basic relations

$$\binom{t}{p}_{\alpha} = \sum_{j=0}^{p} \frac{j!}{p!} S(p, j; \alpha, \beta, -r)\binom{t+r}{j}_{\beta}, \tag{4.157}$$

$$\binom{t+r}{p}_{\beta} = \sum_{j=0}^{p} \frac{j!}{p!} S(p, j; \beta, \alpha, r)\binom{t}{j}_{\alpha}. \tag{4.158}$$

Letting $t \to k$ and taking $\beta = 1/m$ $(m \in \mathbb{N})$ we have

$$\binom{k+r}{p}_{1/m} = \frac{1}{m^p}\binom{mk+rm}{p}.$$

Taking $\alpha = 1$ and setting $q = rm$ so that $r = q/m$, we have $S(p, j; \beta, \alpha, r) = S(p, j; 1/m, 1, q/m)$. Thus (4.158) gives

$$\frac{1}{m^p}\binom{mk+q}{p} = \sum_{j=0}^{p} \frac{j!}{p!} S\left(p, j; \frac{1}{m}, 1, \frac{q}{m}\right)\binom{k}{j}$$

and it follows from (4.90) and (4.93) that

$$\frac{1}{m^p}\sum_{k=0}^{n} F(n,k)\binom{mk+q}{p} = \sum_{j=0}^{p} \frac{j!}{p!} S\left(p, j; \frac{1}{m}, 1, \frac{q}{m}\right)\phi(n, j), \tag{4.159}$$

$$\frac{1}{m^p}\sum_{k=0}^{n}F(k)\binom{mk+q}{p}=\sum_{j=0}^{p}\frac{j!}{p!}S\left(p,j;\frac{1}{m},1,\frac{q}{m}\right)\Phi(j). \tag{4.160}$$

These two summation formulas are really equivalent to the simpler formulas (4.90) and (4.93), respectively, and this can be verified by using the orthogonality relations between $S(p,j;\alpha,\beta,-r)$ and $S(j,i;\beta,\alpha,r)$ with $(\alpha,\beta,-r)=(1,1/m,-q/m)$ (cf. Proof of Theorem 4.2.6).

Certainly, (4.159) and (4.160) may be used to generalize (4.153)–(4.156) to the forms in which p and $S(p,j,1/m,1)$ are replaced by $\binom{mk+p}{p}$ and $S(p,j;1/m,1,q/m)$, respectively. Of particular interest may be the extension for Dobinski's identity, namely

$$\frac{1}{em^p}\sum_{k=0}^{n}\frac{1}{k!}\binom{mk+q}{p}=\frac{1}{p!}\sum_{j=0}^{p}S\left(p,j;\frac{1}{m},1,\frac{q}{m}\right),$$

where

$$\lim_{m\to\infty}S\left(p,j;\frac{1}{m},1,\frac{q}{m}\right)=S(p,j;0,1,0)=\left\{ {p \atop j} \right\}.$$

As regards the practical computation of GSN $S(n,k)\equiv S(n,k;\alpha,\beta,r)$, the following recurrence relation may be utilized [132]:

$$S(n,k)=S(n-1,k-1)+(k\beta-n\alpha+r)S(n-1,k),$$

where $n\geq k\geq 1$, $S(0,0)=S(n,n)=1$, $S(1,0)=r$ and $S(n,0)=(r|\alpha)_n$.

Finally, as a significant thing to be noticed, both (4.159) and (4.160) can be also used as source formulas for getting various special formulas and identities other than those obtainable by (4.85)–(4.86), (4.88)–(4.89), (4.90)–(4.95). Accordingly, the formulaic class (ΣS) can be enlarged by admitting into it all those formulas and identities that are obtainable or deducible from (4.159) and (4.160), provided that $\phi(n,j)$ and $\Phi(j)$ are given by (4.90) and (4.93), respectively.

Note that formula (4.74) is also a consequence of the following series transformation formula via taking $g(t)=(1+t)^\alpha$ and taking $t\to\Delta$ (cf. Section 2.2)

$$\sum_{k\geq 0}f(k)g^{(k)}(0)\frac{t^k}{k!}=\sum_{k\geq 0}\Delta^k f(0)g^{(k)}(t)\frac{t^k}{k!}. \tag{4.161}$$

Accordingly, (4.74) may be called an associating operational formula of (4.161). Obviously, (4.161) implies Taylor's expansion with $f(k)=1$ ($k\in\mathbb{N}_0$). However, Taylor's expansion formula is not included in the $(\Sigma\Delta)$ class. Thus one may see that if (4.161) is taken to be the most basic source formula in which the substitutions $t\to\Delta$ and $t\to D$ (differential operator) are admitted, then all those formulas and identities that can be deduced from (4.161) or its associating operational formulas will form a formulaic class much bigger than $(\Sigma\Delta)$, say $(\Sigma\Delta D)$ class. Here, any exact formula or an identity is said to be a member of the class $(\Sigma\Delta D)$, if it can be obtained or deduced from (4.161) or its associating operational formulas. The rigorous definition of $(\Sigma\Delta D)$ class is given in Definition 3.3.13. However, in smaller classes $(\Sigma\Delta)$ and (ΣS), the summation formulas and identities seem to be constructed more efficiently.

From this definition it is clear that $(\Sigma\Delta)$ is a subset of $(\Sigma\Delta D)$. Also apparently, all the power-series expansions as well as numerous identities displayed in Section 2.2 are special members of the $(\Sigma\Delta D)$ class.

It is worth mentioning that the methods of *nonstandard analysis (NSA)* have been proved useful for combinatorics and graph theory since the years 1980's. Some related earlier works were done by Hurd [137], Hsu [124], and Hirshfeld [118], et al.. In connection with the present book, we just mention that both (4.74) and (4.161) can get unified into a single formula of the form in (NSA):

$$\sum_{k=0}^{\omega} \binom{t}{k}_{\delta} f(k) \, \nabla_{\delta}^{k}\, g(0) = \sum_{k=0}^{\omega} \binom{t}{k}_{\delta} \Delta^{k} f(0) \, \nabla_{\delta}^{k}\, g(t-k\delta), \tag{4.162}$$

where $t \in^* \mathbb{R}$, $0 < \delta \in^* \mathbb{R}$ (the nonstandard real number field), $\nabla_\delta = \delta^{-1}\Delta_\delta$ denotes the difference- quotient operator with increment δ, f, and g are maps (functions) of the class Map$(^*\mathbb{R}, ^*\mathbb{R})$ and $\omega \in^* \mathbb{N}_\infty$ (the set of star-finite positive integers). Actually, a formal derivation of (4.162) is quite similar to that of (4.74), but needs using some basic concepts in NSA. In particular, one may let $0 < \delta \in m(0)$ (the monad of zero) so that δ is a positive infinitesimal. Now by taking standard parts (so-called st operation) we have

$$st\binom{t}{k}_{\delta} = t^k/k!, \quad st(\nabla_{\delta}^{k} g(0)) = g^{(k)}(0), \quad st(\omega) = \infty, \quad etc.$$

Thus it follows that when taking standard parts for the cases of $\delta = 1$ and of $0 < \delta \in m(0)$ in (4.162) we will get (4.74) and (4.161), respectively. It is hopeful that (4.162) and its variants may be used to get various star-finite formulas and identities, some of which may be utilizable for dealing with some counting problems in the nonstandard combinatorics and infinite-graph theory. Clearly, much remains to be considered and investigated.

4.3 Summation Formulas Related to Riordan Arrays

The concept of a (proper) Riordan array (or Riordan matrix) is introduced as a generalization of the *Pascal triangle (or Pascal matrix)*

$$\begin{bmatrix} 1 & 0 & 0 & 0 & \cdots \\ 1 & 1 & 0 & 0 & \\ 1 & 2 & 1 & 0 & \\ 1 & 3 & 3 & 1 & \\ & & \cdots & & \ddots \end{bmatrix}.$$

From Definition 3.2.3, a Riordan array is a special type of infinite lower-triangular matrix. The set of all Riordan matrices forms a group called the *Riordan group*, which was first defined in 1991 by Shapiro, Getu, Woan, and Woodson [192]. Some of the results on the *Riordan group*, particularly, its applications to the combinatorial sums and identities can be found in [171], [195]-[196], [227], and [76]. In particular, in the work by Sprugnoli (*cf.* [201] and [202]).

More precisely, a Riordan array is an infinite, lower triangular matrices defined by a pair of formal power series $\{g(t), f(t)\}$ in $\mathbb{K}[\![t]\!]$, where $\mathbb{K}$ is either the field $\mathbb{R}$ or $\mathbb{C}$, with $g(0) \neq 0$, $f(0) = 0$, and $f'(0) \neq 0$. The usual way to represent the Riordan array, denoted by $(g(t), f(t))$ is by means of the infinite matrix $(d_{n,k})_{n,k\geq 0}$ with its generic element being

$$d_{n,k} = [t^n]g(t)f(t)^k, \tag{4.163}$$

where $[t^n]\sum_{k=0}^{\infty} h_k t^k = h_n$, the exact coefficient of t^n of a given formal power series. Hence, $[t^n] : \mathbb{K}[\![t]\!] \to \mathbb{R}$ is a functional satisfying $[t^n]t^k = \delta_{n,k}$. $[t^n]$ evidently posses the following properties (cf. Merlini, Sprugnoli, and Verri [160]).

$$[t^n](af(t) + bg(t)) = a[t^n]f(t) + b[t^n]g(t), \tag{4.164}$$

$$[t^n]tf(t) = [t^{n-1}]f(t), \tag{4.165}$$

$$[t^n]f'(t) = (n+1)[t^{n+1}]f(t), \tag{4.166}$$

$$[t^n]f(t)g(t) = \sum_{k=0}^{n}([t^n]f(t))([t^{n-k}]g(t)), \tag{4.167}$$

$$[t^n]f(g(t)) = \sum_{k\geq 0}([t^k]f(t))[t^n]g(t)^k. \tag{4.168}$$

Thus, the pascal matrix is the Riordan array $(1/(1-t), t/(1-t))$. The operation of the Riordan group is the regular matrix multiplication, and the identity matrix is

$$(1,t) = \begin{bmatrix} 1 & 0 & 0 & 0 & & \\ 0 & 1 & 0 & 0 & & \\ 0 & 0 & 1 & 0 & \cdots & \\ 0 & 0 & 0 & 1 & & \\ & & \cdots & & \ddots & \end{bmatrix}.$$

In this section, we will see that Riordan arrays provide a wealth of tools to construct summation formulas and identities.

4.3.1 Riordan arrays, the Riordan group, and their sequence characterizations

To present summation formulas related to Riordan arrays, we start from the rigorous definition of Riordan arrays and the Riordan group. More formally in mathematics, let us consider the set of formal power series ring $\mathcal{F} = \mathbb{K}[\![t]\!]$, where $\mathbb{K}$ is the field of $\mathbb{R}$ or $\mathbb{C}$. The *order* of $f(t) \in \mathcal{F}$, $f(t) = \sum_{k=0}^{\infty} f_k t^k$ ($f_k \in \mathbb{K}$), is the minimum number $r \in \mathbb{N}$ such that $f_r \neq 0$; $\mathcal{F}_r$ is the set of formal power series of order r. It is known that $\mathcal{F}_0$ is the set of invertible fps and $\mathcal{F}_1$ is the set of compositionally invertible fps or delta series, that is, the fps $f(t)$ for which the *compositional inverse* $\bar{f}(t)$ exists such that $f(\bar{f}(t)) = \bar{f}(f(t)) = t$. Let $g(t) \in \mathcal{F}_0$ and $f(t) \in \mathcal{F}_1$; the pair (g, f) defines the *Riordan array or Riordan matrix* $D = (d_{n,k})_{n,k \in \mathbb{N}_0} = (g, f)$ having $d_{n,k}$ shown in (4.163), or equivalently, having gf^k as the generating function of the kth column of the matrix (g, f). Some literatures call the Riordan array (g, f) with $f'(0) \neq 0$ a proper Riordan array. Hereinafter, unless otherwise specified, all Riordan arrays refer to proper Riordan arrays. Essentially the columns of the Riordan matrix can be thought of as a geometric sequence with g as the lead term and f as the multiplier term. Two examples of Riordan arrays are the Pascal matrix and the identity matrix shown above.

The Riordan group acts on the set of column vectors by matrix multiplication. In terms of generating functions we let $d(t) = d_0 + d_1 t + d_2 t^2 + \cdots$ and $h(t) = h_0 + h_1 t + h_2 t^2 + \cdots$. If $[d_0, d_1, d_2, \cdots]^T$ and $[h_0, h_1, h_2, \cdots]^T$ are the corresponding column vectors we observe that

$$(g, f)[h_0, h_1, h_2, \cdots]^T = [d_0, d_1, d_2, \cdots]^T$$

translates to

$$h_0 g(t) + h_1 g(t) f(t) + h_2 g(t) f(t)^2 + \cdots = g(t) \cdot h(f(t)) = d(t).$$

This simple observation is called the *Fundamental theorem of Riordan Arrays* and is abbreviated as FTRA. The general form of the FTRA can be written as

$$(g, f)h(t) = gh(f) \quad \text{or equivalently,} \quad \sum_{k\geq 0} d_{n,k} h_k = [t^n] g(t) h(f(t)). \quad (4.169)$$

Particularly, if $(g, f) = (1/(1-t), t/(1-t))$, then the FTRA (4.169) is the following well-known Euler transformation:

$$\sum_{k\geq 0} \binom{n}{k} h_k = [t^n] \frac{1}{1-t} h\left(\frac{t}{1-t}\right). \quad (4.170)$$

The first application of the fundamental theorem is to set $h(t) = G(t) F(t)^k$ so that

$$d(t) = g(t) \cdot G(f(t)) F(f(t))^k.$$

As k ranges over $0, 1, 2, \cdots$ the multiplication rule for Riordan arrays emerges:

$$(g, f)(G, F) = (g(G \circ f), (F \circ f)). \quad (4.171)$$

Thus, the *Riordan group* is the set of all pairs (g, f) together with the multiplication operation shown in (4.171). If we denote the compositional inverse of f as $\bar{f}$, then

$$(g, f)^{-1} = \left(\frac{1}{g \circ \bar{f}}\ \bar{f}\right).$$

As an example we return to the Pascal matrix where $f = \frac{t}{1-t}$. The inverse is $\overline{f} = \frac{t}{1+t}$, $g\left(\overline{f}\right) = \frac{1}{1-\left(\frac{t}{1+t}\right)} = 1 + t$ and the inverse matrix starts

$$\left(\frac{1}{1+t}, \frac{t}{1+t}\right) = \begin{bmatrix} 1 & 0 & 0 & 0 & 0 \\ -1 & 1 & 0 & 0 & 0 \\ 1 & -2 & 1 & 0 & 0 \\ -1 & 3 & -3 & 1 & 0 \\ 1 & -4 & 6 & -4 & 1 \end{bmatrix}.$$

Here is a list of six important subgroups of the Riordan group (cf. [196]).

- the *Appell subgroup* $\{(g(t), t)\}$.
- the *Lagrange (associated) subgroup* $\{(1, f(t))\}$.
- the *k-Bell subgroup* $\{(g(t), t(g(t))^k)\}$, where k is a fixed positive integer.
- the *hitting-time subgroup* $\{(tf'(t)/f(t), f(t))\}$.
- the *derivative subgroup* $\{(f'(t), f(t))\}$.
- the *checkerboard subgroup* $\{(g(t), f(t))\}$, where g is an even function and f is an odd function.

The 1-Bell subgroup is referred to as the Bell subgroup for short, and the Appell subgroup can be considered as the 0-Bell subgroup if we allow $k = 0$ to be included in the definition of the k-Bell subgroup. The proof of that the above six subsets are subgroups of the Riordan group is left as Exercise 4.6.

Some of the main results on the Riordan group and its application to combinatorial sums and identities can be found in [6], [14]-[17], [36]-[37], [38]–[42], [54]–[55], [65], [76], [84]–[95], [97], [105], [113], [108]-[110], [138], [151]-[153], [158]-[161], [170]-[171], [178], [174]-[175], [183]-[184], [193]-[195], [198]-[199], [201]–[202], [220]-[223], [209], and [227].

Theorem 4.3.1 *An infinite lower triangular array $D = (d_{n,k})_{n,k\in\mathbb{N}_0}$ is a Riordan array if and only if a sequence $A = (a_0, a_1, a_2, \ldots)$ exists such that for every $n, k \in \mathbb{N}_0$ there holds*

$$d_{n+1,k+1} = a_0 d_{n,k} + a_1 d_{n,k+1} + a_2 d_{n,k+2} + \cdots = \sum_{j=0}^{\infty} a_j d_{n,k+j}, \tag{4.172}$$

where the sum is actually finite since $d_{n,k} = 0$ *for all* $k > n$. *Hence, for* $D = (g(t), f(t))$

$$f(t) = tA(f(t)) \quad or \quad A(t) = \frac{t}{\bar{f}(t)}, \tag{4.173}$$

where $\bar{f}(t)$ *is the compositional inverse of* $f(t)$. *Moreover, we have the Lagrange inversion formula*

$$[t^n](f(t))^k = \frac{k}{n}[t^{n-k}](A(t))^n, \tag{4.174}$$

or equivalently,

$$[t^n](f(t))^k = \frac{k}{n}[t^{n-k}]\left(\frac{\bar{f}(t)}{t}\right)^{-n}. \tag{4.175}$$

Particularly, for $k = 1$,

$$[t^n]f(t) = \frac{1}{n}[t^{-1}]\frac{1}{\bar{f}(t)^n} = \frac{1}{n}[t^{-1}]A(t)^n. \tag{4.176}$$

Proof. Suppose that D is the Riordan array $(g(t), f(t))$. We consider the Riordan array $(g(t)f(t)/t, f(t))$ and define the Riordan array $(A(t), \tilde{A}(t))$ by the relation:

$$(A(t), \tilde{A}(t)) = (g(t), f(t))^{-1}\left(g(t)\frac{f(t)}{t}, f(t)\right)$$

or

$$(g(t), f(t))(A(t), \tilde{A}(t)) = \left(g(t)\frac{f(t)}{t}, f(t)\right).$$

By performing the Riordan array product, we find

$$g(t)A(f(t)) = g(t)\frac{f(t)}{t} \quad \text{and} \quad \tilde{A}(f(t)) = f(t). \tag{4.177}$$

The second identity of (4.177) implies $\tilde{A}(t) = t$. Therefore we have $(g(t), f(t))(A(t), t) = (g(t)f(t)/t, f(t))$. The (n, k) element of the left-hand resulting matrix after multiplication is

$$\sum_{j=0}^{\infty} d_{n,j}a_{j-k} = \sum_{j=0}^{\infty} a_j d_{n,k+j},$$

if as usual we interpret a_{j-k} as 0 when $j < k$. The same element in the right-hand member is

$$[t^n]\frac{g(t)f(t)}{t}f(t)^k = [t^{n+1}]g(t)f(t)^{k+1} = d_{n+1,k+1}.$$

By equating these two quantities, we have the identity (4.172). We remark that the first relation in (4.177) is equivalent to $tA(h(t)) = f(t)$.

For the converse, let us observe that (4.172) uniquely defines the array D when the elements of column 0, $(d_{0,0}, d_{1,0}, d_{2,0}, \ldots)$, are given. Let $g(t)$ be the generating function of the column 0, $A(t)$ the generating function of the sequence A. We now define $f(t)$ as the solution of the functional equation $f(t) = tA(f(t))$. It can be seen that f is uniquely determined because of the hypothesis $a_0 \neq 0$. We can therefore consider the Riordan array $\hat{D} = (g(t), f(t))$, which satisfies relation (4.172) for every $n, k \in \mathbb{N}_0$ by the proved first part of the theorem. Hence, $\hat{D}$ must be the same as D.

The equivalence between formulas (4.174) and (4.175) is obvious because of $A(t) = t/\bar{f}(t)$. The formula (4.175) results from (4.173) and the following process. Denote by $\hat{d}_{n,k}$ the (n, k) entry of the product of the matrix with the (n, k) entry shown on the right-hand side of (4.175) and the Riordan array with the (n, k) entry as $[t^n]\bar{f}(t)^k$. Then,

$$\begin{aligned}
\hat{d}_{n,k} &= \sum_{k\le \ell\le n} \frac{\ell}{n}[t^{n-\ell}]\left(\frac{\bar{f}(t)}{t}\right)^{-n}[t^\ell]\bar{f}(t)^k \\
&= \sum_{k\le \ell\le n} \frac{1}{n}[t^{n-\ell}]\left(\frac{\bar{f}(t)}{t}\right)^{-n}[t^\ell](tD(\bar{f}(t)^k) \\
&= \sum_{k\le \ell\le n} \frac{k}{n}[t^{n-\ell}]\left(\frac{\bar{f}(t)}{t}\right)^{-n}[t^\ell]t\bar{f}(t)^{k-1}\bar{f}'(t) \\
&= \frac{k}{n}[t^n]\left(t^{n+1}\bar{f}(t)^{-n+k-1}\bar{f}'(t)\right) \\
&= \frac{k}{n}[t^0]\left(t\bar{f}(t)^{-n+k-1}\bar{f}'(t)\right).
\end{aligned}$$

Consequently, for $k = n$ and $\bar{f}(t) = \sum_{j\ge 1}\bar{f}_j t^j$, we have

$$\hat{d}_{n,n} = [t^0]t\frac{\bar{f}'(t)}{\bar{f}(t)} = [t^0]t\frac{\sum_{j\ge 1} j\bar{f}_j t^{j-1}}{t\sum_{j\ge 1}\bar{f}_j t^{j-1}} = 1.$$

For $k < n$ and $\bar{f}(t) = \sum_{j\ge 1}\bar{f}_j t^j$, we have

$$\hat{d}_{n,k} = \frac{k}{n}[t^0]\left(tD\left(\frac{\bar{f}(t)^{-n+k}}{-n+k}\right)\right) = 0$$

because in the derivative of the Laurent series following sign D term t^{-1} cannot occur. Hence, $(\hat{d}_{n,k})_{n,k\ge 0} = (1, t)$, the identity Riordan array, which implies that the matrix with the (n, k) entry shown on the right-hand side of (4.175) is the inverse of $(1, \bar{f})$. Since the unique inverse of $(1, \bar{f})$ is $(1, f)$, we obtain (4.175). Consequently, (4.176) follows for $k = 1$.

■

Stanley [203] gives three proofs of the Lagrange inversion formula (4.174) (or (4.175)): direct algebraic argument by using Laurent series, ordinary generating function argument, and exponential generating function argument, while Theorem 4.3.1 gives a proof by using Riordan array argument.

We now use Riordan array approach to unify the Lagrange inversion formulas shown in Theorem A of [44] (cf. (4.175) below), Theorem 5.1 of [216] (cf. (4.179) below), and Formula $K6'$ of [160], respectively.

Corollary 4.3.2 *Let $f(t)\mathcal{F}_1$, and let $F(t)\in\mathcal{F}_0$. Denote the generating functions of the A-sequences of Riordan arrays $(1, f(t))$ and $(1, f(t))^{-1} = (1, \bar{f}(t))$ by $A(t)$ and $A^*(t)$, respectively. Then the Lagrange inversion formula* (4.175) *has the following equivalent forms.*

$$[t^n](\bar{f}(t))^k = \frac{k}{n}[t^{n-k}](A^*(t))^n = \frac{k}{n}[t^{n-k}]\left(\frac{f(t)}{t}\right)^{-n}, \tag{4.178}$$

$$[t^n]F(f(t)) = \frac{1}{n}[t^{n-1}]F'(t)A(t)^n, \tag{4.179}$$

$$[t^n]F(f(t)) = [t^n]F(t)A(t)^{n-1}(A(t) - tA'(t)), \tag{4.180}$$

where $A(t) = t/\bar{f}(t)$. Furthermore, the Lagrange inversion formulas (4.175), (4.179), *and* (4.180) *hold for any $f(t)\in\mathcal{F}_1$ and $A(t)\in\mathcal{F}_0$ provided they satisfy $f(t) = tA(f(t))$.*

Proof. Replacing into (4.174) and (4.175) $f(t)$ and $A(t)$ by $\bar{f}(t)$ and $A^*(t)$, respectively, and noting $A^*(t) = t/f$, we can get (4.178). Conversely, from (4.178), we can get (4.175).

For $F(t)\in\mathcal{F}_0$ with $F(t) = \sum_{k\geq 0} F_k t^k$, from (4.174) we have

$$\begin{aligned}
[t^n]F(f(t)) =& \sum_{k\geq 0} F_k [t^n] f(t)^k = \sum_{k\geq 0} F_k \frac{k}{n}[t^{n-k}]A(t)^n\\
=&\frac{1}{n}[t^{n-k}]\sum_{k\geq 0} kF_k A(t)^n = \frac{1}{n}[t^{n-1}]\left(\sum_{k\geq 0} kt^{k-1}F_k\right)A(t)^n\\
=&\frac{1}{n}[t^{n-1}]F'(t)A(t)^n.
\end{aligned}$$

Thus, we obtain (4.179) from (4.174). Obviously, for $F(t) = t^k$ we find (4.174) again from (4.179).

Secondly, we show that (4.179) implies (4.180). From (4.179) and noting $[t^{n-1}]f'(t) = n[t^n]f(t)$, we have

$$\begin{aligned}
[t^n]F(t)A(t)^n =& \frac{1}{n}[t^{n-1}](F(t)A(t)^n)'\\
=&\frac{1}{n}[t^{n-1}]F'(t)A(t)^n + [t^{n-1}]F(t)A(t)^{n-1}A'(t)\\
=&[t^n]F(f(t)) + [t^{n-1}]F(t)A(t)^{n-1}A'(t)
\end{aligned}$$

which implies

$$[t^n]F(t)A(t)^{n-1}(A(t)-tA'(t))=[t^n]F(f(t)),$$

i.e., (4.180).

Finally, we show that (4.180) implies (4.174) and (4.175). Taking derivative on the both sides of $f(t)=tA(f(t))$, we get

$$f'(t)=\frac{A(f(t))}{1-tA'(f(t))}=\frac{A(f)^2}{A(f)-fA'(f)},$$

where we have substituted $t=f/A(f)$. By using (4.180) and the last expression for $f'(t)$, we have

$$\begin{aligned}[t^n]tf(t)^{k-1}f'(t)=&[t^n]\frac{A(f)}{A(f)-fA'(f)}f^k\\ =&[t^n]\frac{t^kA(t)}{A(t)-tA'(t)}A(t)^{n-1}(A(t)-tA'(t))\\ =&[t^n]t^kA(t)^n=[t^{2n-k}](tA(t))^n.\end{aligned}$$

We are now ready to establish (4.174) and (4.175) from (4.180) as follows.

$$\begin{aligned}&\frac{k}{n}[t^{n-k}]\left(\frac{t}{\bar{f}(t)}\right)^n=\frac{k}{n}[t^{n-k}]A(t)^n=\frac{k}{n}[t^{2n-k}](tA(t))^n\\ =&\frac{k}{n}[t^n]tf(t)^{k-1}f'(t)=\frac{1}{n}[t^{n-1}](f(t))^k=[t^n]f(t)^k,\end{aligned}$$

which completes the proof of the corollary.

■

The Lagrange inversion formulas (4.174), (4.175), and (4.178)–(4.180) provide a class of source formulas to construct summations and identities by using generating functions.

Example 4.3.3 *As an example, we consider the generating function of the Fuss-Catalan number sequence* $(\frac{r}{mn+r}\binom{mn+r}{n})_{n\geq 0}$ $(m,n\in\mathbb{N}_0, r\in\mathbb{N})$, *using which many summation formulas will be constructed in Subsection 4.3.1. A closed form of the generating function of the Fuss-Catalan number sequence can be found by using an equivalent form of the Lagrange inversion formula (4.179). Using (4.179) and noting* $f(t)=tA(f(t))$ $(f\in\mathcal{F}_1, A\in\mathcal{F}_0)$, *we have*

$$\begin{aligned}&[t^n]F(t)A(t)^n=[t^{n-1}]\frac{F(t)}{t}A(t)^n=n[t^n]\int\frac{F(f(t))}{f(t)}df(t)\\ =&[t^{n-1}]\frac{d}{dt}\int\frac{F(f(t))}{f(t)}f'(t)dt=[t^{n-1}]\frac{F(f(t))}{f(t)}f'(t).\end{aligned}$$

Since $f(t) = tA(f(t))$ implies $f'(t) = A(f(t))/(1 - tA'(f(t)))$, we may substitute the expression of $f'(t)$ into the rightmost side of the last equations to obtain (cf. Merlini, Sprugnoli, and Verri [160])

$$[t^n]F(t)A(t)^n = [t^{n-1}]\frac{F(f(t))}{f(t)}\frac{A(f(t))}{1 - tA'(f(t))}$$
$$=[t^{n-1}]\frac{F(f(t))}{t(1 - tA'(f(t))} = [t^n]\frac{F(f(t))}{1 - tA'(f(t))}. \tag{4.181}$$

Using formula (4.181) and letting $A(t) = (1+t)^m$, the generating function of the Fuss-Catalan number sequence $(\frac{r}{mn+r}\binom{mn+r}{n})_{n\geq 0}$ can be written as

$$\sum_{n\geq 0}\frac{r}{mn+r}\binom{mn+r}{n}t^n = \sum_{n\geq 0}\left(\binom{mn+r}{n} - m\binom{mn+r-1}{n-1}\right)t^n$$
$$=\sum_{n\geq 0}[t^n]\left((1+t)^{mn+r} - (m-1)(1+t)^{mn+r-1}\right)t^n$$
$$=\sum_{n\geq 0}[t^n]\left((1-(m-1)t)(1+t)^{r-1}\left((1+t)^m\right)^n\right)t^n$$
$$=\sum_{n\geq 0}[t^n]\left(\frac{(1-(m-1)f(t))(1+f(t))^{r-1}}{1 - tm(1+f(t))^{m-1}}\right)t^n$$
$$=\frac{(1-(m-1)f(t))(1+f(t))^{r-1}}{1 - tm(1+f(t))^{m-1}}$$
$$=(1+f(t))^r\frac{1-(m-1)f(t)}{1+f(t) - tm(1+f(t))^m} = (1+f(t))^r, \tag{4.182}$$

where $f(t) = tA(f(t)) = t(1+f(t))^m$ is applied in the last step. Denote by $F_m(t)$ the generating function of the Fuss-Catalan number sequence $(\frac{1}{mn+1}\binom{mn+1}{n})_{n\geq 0}$. Then $F_m^r(t)$ is the generating function of the Fuss-Catalan number sequence $(\frac{r}{mn+r}\binom{mn+r}{n})_{n\geq 0}$ because $F_m(t) = 1 + f(t)$ and $F_m^r(t) = (1+f(t))^r$. From $f(t) = tA(f(t) = t(1+f(t))^m$, we obtain $F_m(t) = 1+F_m^m(t)$. If $m = 0$, then $f(t) = t$ and $F_0(t) = 1+t$. If $m = 1$, then $f(t) = t/(1-t)$ and $F_1(t) = 1/(1-t)$. If $m = 2$, then $f(t) = (1-2t-\sqrt{1-4t})/(2t)$ and $F_2(t) = (1-\sqrt{1-4t})/(2t)$. For the last case of $m = 2$, $(c_n = \frac{1}{2n+1}\binom{2n+1}{n})_{n\geq 0}$ is the Catalan number sequence, and $F_2(t)$ is the generating function $C(t)$ of the Catalan number sequence $(c_n)_{n\geq 0}$. The history, combinatorial interpretation, more properties, and applications of Fuss-Catalan number sequences will be presented in Subsection 4.3.6.

Although the two functions $g(t)$ and $A(t)$ completely characterize a Riordan array, we are mainly interested in another type of characterization. Let us consider the following result:

Theorem 4.3.4 *Let $D = (d_{n,k})_{n,k\in\mathbb{N}_0}$ be any infinite, lower triangular array with $d_{n,n} \neq 0$ for all $n \in \mathbb{N}_0$ (in particular, let it be a proper Riordan array);*

then a unique sequence $Z = (z_k)_{k\in\mathbb{N}_0}$ *exists such that every element in column 0 can be expressed as a linear combination of all the elements in the preceding row, i.e.:*

$$d_{n+1,0} = z_0 d_{n,0} + z_1 d_{n,1} + z_2 d_{n,2} + \cdots = \sum_{j=0}^{\infty} z_j d_{n,j}. \tag{4.183}$$

A proof can be given similarly, which is left as an exercise (cf. Exercise 4.7).

The sequence Z is called the Z-sequence for the Riordan array, which characterizes its column 0, except for the element $d_{0,0}$. Therefore, we can say that the triple $(d_{0,0}, A(t), Z(t))$ completely characterizes a proper Riordan array $(g(t), f(t)$. Usually, we denote $d_{0,0}$ as $g_0 = g(0)$.

To see how the Z-sequence is obtained by starting with the usual definition of a Riordan array, let us prove the following result.

Theorem 4.3.5 *Let* $(g(t), f(t))$ *be a Riordan array, and let* $Z(t)$ *be the generating function of the corresponding* Z*-sequence. We have:*

$$g(t) = \frac{g_0}{1 - tZ(f(t))} \quad \text{or} \quad Z(t) = \frac{g(\bar{f}(t)) - g_0}{\bar{f}(t)d(\bar{f}(t))}, \tag{4.184}$$

where $\bar{f}$ *is the compositional inverse of* f.

Proof. From (4.183) we have

$$[t^{n+1}]g(t) = \sum_{j=0}^{\infty} z_j [t^n]g(t)f(t)^j = [t^n]g(t)Z(f(t)).$$

Hence,

$$g(t) = g_0 + tg(t)Z(f(t)),$$

which implies the first equation of (4.184). Substituting $t = \bar{f}(t)$ into the first equation, we obtain

$$g(\bar{f}(t)) = \frac{g_0}{1 - \bar{f}(t)Z(t)},$$

which gives the second equation of (4.184).

∎

Example 4.3.6 *Clearly, in the Pascal triangle*

$$\left(\frac{1}{1-t}, \frac{t}{1-t}\right) = \left(\binom{n}{k}\right)_{n,k\in\mathbb{N}_0},$$

we have $A = (1, 1, 0, \ldots)$ *and* $Z = (1, 0, 0, 0, \ldots)$ *and so* $A(t) = 1 + t$ *and* $Z(t) = 1$. *Formulas (4.177) and (4.184) give immediately* $f(t) = t/(1-t)$ *and* $g(t) = 1/(1-t)$, *respectively.*

Example 4.3.7 *Let us consider a Riordan array (g, f) defined by $g(t) = (1-t-t^2)^{-1}$ so that column 0 composed by Fibonacci numbers. Besides, the A-sequence is $A = (1, 1, 1, 1, \ldots)$, that is, any element $d_{n+1,k+1}$ is obtained by summing all the elements in the previous row, starting from column $k \geq 1$. The generating function of the A-sequence is $1/(1-t)$. The Riordan array $(g(t), f(t))$ can be easily constructed. Formula (4.177) allows us to compute the function $f(t)$ of this array:*

$$f(t) = tA(f(t)) = \frac{t}{1-f(t)}.$$

This equation has two solutions, but we know that $A(0) \neq 0$ so that we should consider the solution with the minus sign:

$$f(t) = \frac{1-\sqrt{1-4t}}{2} = t + t^2 + 2t^3 + 5t^4 + 14t^5 + 42t^6 + 132t^7 + \cdots = tC(t),$$

where $C(t)$ is the generating function of the well-known Catalan numbers. The Catalan numbers occur in many counting situations. The book Enumerative Combinatorics: Volume 2 by Stanley [203] contains a set of exercises which describe 66 different interpretations of the Catalan numbers. His recent book [204], Catalan Numbers, contains more interesting applications. Other important resources on Catalan numbers can be found in Aigner [2]-[5], etc. The parametric Catalan numbers and Catalan triangles are studied in [86]. Therefore we have:

$$(g(t), f(t)) = \left(\frac{1}{1-t-t^2}, \frac{1-\sqrt{1-4t}}{2}\right),$$

which is called Fibonacci triangle. The Z-sequence for the Fibonacci triangle is a bit more complicated. If we set

$$y = f(t) = \frac{1-\sqrt{1-4t}}{2},$$

we can invert the function $f(t)$ and find $t = y - y^2$. Hence, $\bar{f}(t) = t - t^2$. We now substitute this expression in the second equation of (4.184) and find

$$\begin{aligned} Z(t) =& \frac{g(\bar{f}(t)) - g_0}{\bar{f}(t)d(\bar{f}(t))} \\ =& \left(\frac{1}{1-t-t^2} - 1\right)\frac{1-t-t^2}{t}\bigg|_{t\to t-t^2} \\ =& (1+t)|_{t\to t-t^2} = 1 + t - t^2. \end{aligned}$$

Hence, the Z-sequence is $Z = (1, 1, -1, 0, \ldots)$

We can characterize the main subgroups of the Riordan group by means of their A- and/or Z-sequences. In fact we have the following results.

Theorem 4.3.8 *Let us consider a Riordan array* $D = (g(t), f(t))$. *(1) D belongs to the Appell subgroup* $\mathcal{A}$ *if and only if its A-sequence satisfies* $A(t) = 1$. *Besides, in that case, we also have*

$$Z(t) = \frac{g(t) - g_0}{tg(t)}.$$

(2) D belongs to the Lagrange subgroup $\mathcal{L}$ *if and only if its Z-sequence satisfies* $Z(t) = 0$ *and* $g_0 = 1$. *(3) D belongs to the checkerboard subgroup* $\mathcal{C}$ *if and only if the generating functions* $A(t)$ *of its A-sequence is even and* $Z(t)$ *of its Z-sequence is odd. (4) D belongs to the Bell subgroup* $\mathcal{B}$ *if and only if its A- and Z- sequences satisfy* $A(t) = 1 + tZ(t)$. *(5) D belongs to the hitting-time subgroup if and only if its A- and Z- sequences satisfy* $Z(t) = A'(t)$.

Proof. The definitions of the subgroups are shown Subsection 4.3.1 (or see Exercise 4.6). The proof of the theorem follows the definitions of the subgroups and is left for an exercise (cf. Exercise 4.8).

■

Let us consider two Riordan arrays $D_1 = (g_1(t), f_1(t))$ and $D_2 = (g_2(t), f_2(t))$ and their product,

$$D_3 = D_1 D_2 = (g_1(t) g_2(f_1(t)), f_2(f_1(t)))$$

satisfies

$$g_3(t) = g_1(t) g_2(f_1(t)) \quad \text{and} \quad f_3(t) = f_2(f_1(t)).$$

Theorem 4.3.9 *(cf. Theorems 3.3 and 3.4 [113]) Let* D_1, D_2, *and* $D_3 = D_1 D_2$ *be the Riordan arrays shown above, and let* $A_i(t)$ *and* $Z_i(t)$ *be the generating functions of the A-sequence and Z-sequence of* D_i. *Then*

$$A_3(t) = A_2(t) A_1\left(\frac{t}{A_2(t)}\right),$$

$$Z_3(t) - \left(1 - \frac{t}{A_2(t)} Z_2(t)\right) Z_1\left(\frac{t}{A_2(t)}\right) + A_1\left(\frac{t}{A_2(t)}\right) z_2(t).$$

The proof of Theorem 4.3.9 is left as Exercise 4.9.

Theorem 4.3.10 *(cf. Theorems 4.1 and 4.2 [113]) Let* $D = (g, f)$ *be a Riordan array. Then the A-sequence of the inverse Riordan array* D^{-1} *is*

$$A^*(t) = \frac{1}{A(f(t))} \tag{4.185}$$

and the Z-sequence of the inverse Riordan array D^{-1} *is*

$$Z^*(t) = \frac{-tZ(f(t))}{f(t)(1 - tZ(f(t)))}. \tag{4.186}$$

The proof is left as an exercise (cf. Exercise 4.10).

4.3.2 Identities generated by using extended Riordan arrays and Faà di Bruno's formula

We now consider an extension of Riordan arrays called one-pth Riordan arrays, which are a useful extension of Riordan arrays. We start from the extensions of Riordan arrays that are referred to as the half Riordan arrays. It is known that the entries of a Riordan array have a multitude of interesting combinatorial explanations. The central entries play a significant role. For instance, the central entries of the Pascal matrix $(1/(1-z), z/(1-z))$ are the central binomial coefficients $\binom{2n}{n}$ (cf. the sequence A000984 in OEIS [197]) that can be explained as the number of ordered trees with a distinguished point. In addition, its exponential generating function is a modified Bessel function of the first kind. Similarly, the central entries of the Delannoy matrix $(1/(1-z), z(1+z)/(1-z))$, called the Pascal-like Riordan array, are the central Delannoy numbers $\sum_{k=0}^{n}\binom{n}{k}^2 2^k$ (cf. the sequence A001850 in OEIS [197]). The central Delannoy numbers can be explained as the number of paths from $(0,0)$ to (n,n) in an $n\times n$ grid using only steps north, northeast and east (i.e., steps $(1,0)$, $(1,1)$, and $(0,1)$). In addition, the nth central Delannoy numbers is the nth Legendre polynomial's value at 3. It is interesting, therefore, to be able to give generating functions of such central terms in a systematic way. In recent papers [16, 14, 220, 221, 222, 223], it has been shown how to find generating functions of the central entries of some Riordan arrays.

Yang, Zheng, Yuan, and the author [223] give the definition of half Riordan arrays (HRAs), which are called vertical half Riordan arrays in Barry [14]. Another type of halves of Riordan arrays, called horizontal Riordan arrays, are defined in [14]. The sequences characterizations and some related transformation and preserving properties of both type of halves of Riordan arrays are studied in a recent paper by the author in [95]. More precisely, we present the definitions of the Half Riordan arrays as follows. We will see they are special cases of one-pth Riordan arrays. Hence, the properties of the half Riordan arrays will be given as special cases of the properties of one-pth Riordan arrays later.

Definition 4.3.11 *Let $(g,f)=(d_{n,k})_{n,k\geq 0}$ be a Riordan array. Its related half Riordan array $(v_{n,k})_{n,k\geq 0}$, called the vertical half Riordan array (VHRA), is defined by*

$$v_{n,k}=d_{2n-k,n}. \tag{4.187}$$

Another related half Riordan array $(h_{n,k})_{n,k\geq 0}$, called the horizontal half Riordan array (HHRA), is defined by

$$h_{n,k}=d_{2n,n+k}. \tag{4.188}$$

As extensions of half Riordan arrays, the one-pth Riordan arrays of a given Riordan array (g,f) will be defined and constructed in the following theorems by using the Lagrange inversion formula (4.180). For the sake of convenience, we represent (4.180) with applicable notations. Let $F(t)$ be any formal power series, and let $\phi(t)$ and $u(t) = f(t)/t$ satisfy $\phi = tu(\phi)$. Then, we have the following Lagrange inversion formula that comes from (4.180) by replacing its f and A by ϕ and u, respectively.

$$[t^n]F(\phi(t)) = [t^n]F(t)u(t)^{n-1}(u(t) - tu'(t)). \tag{4.189}$$

Theorem 4.3.12 *Given a Riordan array $(d_{n,k})_{n,k\geq 0} = (g, f)$, for any integers $p \geq 1$ and $r \geq 0$, $(\widehat{d}_{n,k} = d_{pn+r-k,(p-1)n+r})_{n,k\geq 0}$ defines a new Riordan array, called the one-pth vertical Riordan array of (g, f) with respect to r, which can be written as*

$$\left(\frac{t\phi'(t)g(\phi)f(\phi)^r}{\phi^{r+1}}, \phi\right), \quad \text{where} \quad \phi(t) = \overline{\frac{t^p}{f(t)^{p-1}}}, \tag{4.190}$$

the compositional inverse of $t^p/f(t)^{p-1}$. Particularly, if $p = 1$ and $r = 0$, then $(\widehat{d}_{n,k} = d_{n-k,0})_{n,k\geq 0}$ is the Toeplitz matrix (or diagonal-constant matrix) of the 0th column of $(d_{n,k})_{n,k\geq 0}$. If $p = 2$ and $r = 0$, then $(\widehat{d}_{n,k} = d_{2n-k,n})_{n,k\geq 0}$ is the VHRA of the Riordan array $(d_{n,k})_{n,k\geq 0} = (t\phi'(t)g(\phi)/\phi, \phi)$, where $\phi = \overline{t^2/f(t)}$.

Moreover, the generating function of the A-sequence of the one-pth vertical Riordan array is $(A(f))^{p-1}$, where $A(t)$ is the generating function of the A-sequence of the given Riordan array $(d_{n,k})_{n,k\geq 0}$.

Proof. From $\phi(t) = \overline{t^p/f(t)^{p-1}}$ we have $\bar{\phi}(t) = t^p/f(t)^{p-1}$ and consequently, $t = \phi(t)^p/f(\phi(t))^{p-1}$. Hence, we may write

$$\phi = tu(\phi) \quad \text{where} \quad u(t) = \left(\frac{f(t)}{t}\right)^{p-1}.$$

Taking derivative on the both sides of $\phi = tu(\phi)$ and noting the definition of $u(t)$, we obtain

$$\phi'(t) = \left(\frac{f(\phi)}{\phi}\right)^{p-1} + t(p-1)\left(\frac{f(\phi)}{\phi}\right)^{p-2}\frac{f'(\phi)\phi'(t)\phi - \phi'(t)f(\phi)}{\phi^2},$$

which yields

$$\phi'(t) = \left(\frac{f(\phi)}{\phi}\right)^{p-1} \Big/ \left(1 - t(p-1)\left(\frac{f(\phi)}{\phi}\right)^{p-2}\frac{f'(\phi)\phi - f(\phi)}{\phi^2}\right).$$

Noting $t = \phi/u(\phi) = \phi^p/f(\phi)^{p-1}$, the last expression devotes

$$\begin{aligned}\phi'(t) &= \left(\frac{f(\phi)}{\phi}\right)^{p-1} \Big/ \left(1 - \frac{p-1}{f(\phi)}(f'(\phi)\phi - f(\phi))\right) \\ &= \frac{(f(\phi))^p}{\phi^{p-1}(f(\phi) - (p-1)(\phi f'(\phi) - f(\phi)))}\end{aligned} \tag{4.191}$$

We now use (4.191), $t = \phi^p/f(\phi)^{p-1}$ and the LIF shown in (4.189) to calculate $\widehat{d}_{n,k}$ for $n, k \geq 0$

$$\begin{aligned}
\widehat{d}_{n,k} &= [t^n]\frac{t\phi'(t)g(\phi)f(\phi)^r}{\phi^{r+1}}(\phi)^k \\
&= [t^n]\frac{\phi^p}{(f(\phi))^{p-1}}\frac{\phi^k(f(\phi))^{p+r}g(\phi)}{\phi^{p+r}(f(\phi)-(p-1)(\phi f'(\phi)-f(\phi)))} \\
&= [t^n]\frac{(f(\phi))^{r+1}g(\phi)}{\phi^{r-k}(f(\phi)-(p-1)(\phi f'(\phi)-f(\phi)))} \\
&= [t^n]\frac{(f(t))^{r+1}g(t)}{t^{r-k}(f(t)-(p-1)(tf'(t)-f(t)))}u(t)^{n-1}(u(t)-tu'(t)),
\end{aligned}$$

where $u(t) = \left(\frac{f(t)}{t}\right)^{p-1}$ and

$$u'(t) = (p-1)\left(\frac{f(t)}{t}\right)^{p-2}\frac{tf'(t)-f(t)}{t^2}.$$

Substituting the expressions of $u(t)$ and $u'(t)$ into the rightmost expression of $\widehat{d}_{n,k}$, we have

$$\begin{aligned}
\widehat{d}_{n,k} &= [t^n]\frac{(f(t))^{r+1}g(t)}{t^{r-k}(f(t)-(p-1)(tf'(t)-f(t)))}\frac{(f(t))^{(p-1)(n-1)}}{t^{(p-1)(n-1)}} \\
&\quad\times\left(\frac{(f(t)^{p-1}}{t^{p-1}}-t(p-1)\frac{(f(t))^{p-2}}{t^{p-2}}\frac{tf'(t)-f(t)}{t^2}\right) \\
&= [t^n]\frac{(f(t)^{(p-1)(n-1)+r+1}g(t)}{t^{(p-1)(n-1)+r-k}(f(t)-(p-1)(tf'(t)-f(t)))} \\
&\quad\times\frac{(f(t))^{p-2}}{t^{p-1}}(f(t)-(p-1)(tf'(t)-f(t))) \\
&= [t^n]g(t)\frac{(f(t))^{(p-1)n+r}}{t^{(p-1)n+r-k}} = [t^{pn+r-k}]g(t)(f(t))^{(p-1)n+r} \\
&= d_{pn+r-k,(p-1)n+r}.
\end{aligned}$$

Particularly, if $p = 1$ and $r = 0$, then $(\widehat{d}_{n,k} = d_{n-k,0})_{n,k}$ is the Toeplitz matrix of the 0th column of (g, f). If $p = 2$ and $r = 0$, then $(\widehat{d}_{n,k} = d_{2n-k,n})_{n,k\geq 0}$ is the VHRA of (g, f), and consequently, $(d_{n,k})_{n,k\geq 0} = (t\phi'(t)g(\phi)/\phi, \phi)$ with $\phi = \overline{t^2/f(t)}$.

As for the $\widehat{A}_p$, the generating function of the A-sequence of $(\widehat{d}_{n,k})_{n,k\geq 0}$, we have $t\widehat{A}_p(\phi) = \phi$, which implies $\widehat{A}_p(t) = t/(t^p/f^{p-1})$, or equivalently,

$$\widehat{A}_p(\bar{f}) = \left(\frac{t}{\bar{f}}\right)^{p-1} = (A(t))^{p-1}.$$

Hence, $\widehat{A}_p(t) = (A(f))^{p-1}$, completing the proof of the theorem.

Theorem 4.3.13 *Given a Riordan array $(d_{n,k})_{n,k\geq 0} = (g,f)$, for any integers $p \geq 1$ and $r \geq 0$, $(\tilde{d}_{n,k} = d_{pn+r,(p-1)n+r+k})_{n,k\geq 0}$ defines a new Riordan array, called the one-pth horizontal Riordan array of (g,f) with respect to r, which can be written as*

$$\left(\frac{t\phi'(t)g(\phi)f(\phi)^r}{\phi^{r+1}}, f(\phi)\right), \quad \text{where} \quad \phi(t) = \overline{\frac{t^p}{f(t)^{p-1}}}, \tag{4.192}$$

the compositional inverse of $t^p/f(t)^{p-1}$. Particularly, if $p = 1$ and $r = 0$, the one-pth Riordan array reduces to the given Riordan array $(\tilde{d}_{n,k})_{n,k\geq 0} = (g,f)$. If $p = 2$ and $r = 0$, the one-pth Riordan array is the HHRA of the given Riordan array $(\tilde{d}_{n,k})_{n,k\geq 0} = (t\phi'(t)g(\phi)/\phi, f(\phi))$, where $\phi = \overline{t^2/f(t)}$.

Moreover, the generating function of the A-sequence of the one-pth Riordan array is $(A(t))^p$, where $A(t)$ is the generating function of the A-sequence of the given Riordan array $(d_{n,k})_{n,k\geq 0}$.

Proof. We now use (4.191) and $t = \phi^p/f(\phi)^{p-1}$ and the LIF shown in (4.189) to calculate $\tilde{d}_{n,k}$ for $n,k \geq 0$.

$$\begin{aligned}
\tilde{d}_{n,k} =& [t^n]\frac{t\phi'(t)g(\phi)f(\phi)^r}{\phi^{r+1}}(f(\phi))^k \\
=& [t^n]\frac{\phi^p}{(f(\phi))^{p-1}}\frac{(f(\phi))^{p+r+k}g(\phi)}{\phi^{p+r}(f(\phi)-(p-1)(\phi f'(\phi)-f(\phi)))} \\
=& [t^n]\frac{(f(\phi))^{r+k+1}g(\phi)}{\phi^r(f(\phi)-(p-1)(\phi f'(\phi)-f(\phi)))} \\
=& [t^n]\frac{(f(t))^{r+k+1}g(t)}{t^r(f(t)-(p-1)(tf'(t)-f(t)))}u(t)^{n-1}(u(t)-tu'(t)),
\end{aligned}$$

where $u(t) = \left(\frac{f(t)}{t}\right)^{p-1}$. From the proof of Theorem 4.3.13

$$u'(t) = (p-1)\left(\frac{f(t)}{t}\right)^{p-2}\frac{tf'(t)-f(t)}{t^2}.$$

Substituting the expressions of $u(t)$ and $u'(t)$ into the rightmost expression of $\tilde{d}_{n,k}$, we have

$$\begin{aligned}
\tilde{d}_{n,k} =& [t^n]\frac{(f(t))^{r+k+1}g(t)}{t^r(f(t)-(p-1)(tf'(t)-f(t)))}\frac{(f(t))^{(p-1)(n-1)}}{t^{(p-1)(n-1)}} \\
&\times\left(\frac{(f(t)^{p-1}}{t^{p-1}} - t(p-1)\frac{(f(t))^{p-2}}{t^{p-2}}\frac{tf'(t)-f(t)}{t^2}\right) \\
=& [t^n]\frac{(f(t))^{(p-1)(n-1)+r+k+1}g(t)}{t^{(p-1)(n-1)+r}(f(t)-(p-1)(tf'(t)-f(t)))} \\
&\times\frac{(f(t))^{p-2}}{t^{p-1}}(f(t)-(p-1)(tf'(t)-f(t)))
\end{aligned}$$

$$=[t^n]g(t)\frac{(f(t))^{(p-1)n+r+k}}{t^{(p-1)n+r}} = [t^{pn+r}]g(t)(f(t))^{(p-1)n+r+k}$$
$$=d_{pn+r,(p-1)n+r+k}.$$

Particularly, if $p = 1$ and $r = 0$, then $\tilde{d}_{n,k} = d_{n,k}$. If $p = 2$ and $r = 0$, then $\tilde{d}_{n,k} = d_{2n,n+k}$, the (n, k) entry of the HHRA of (g, f), and consequently, $(\tilde{d}_{n,k})_{n,k\geq 0} = (t\phi'(t)g(\phi)/\phi, f(\phi))$ with $\phi = \overline{t^2/f(t)}$.

Let $A(t)$ be the generating function of the A-sequence of the given Riordan array (g, f). Then $A(f(t)) = f(t)/t$. Let $A_p(t)$ be the generating function of the A-sequence of the one-pth horizontal Riordan array in (4.192). Then $A_p(f(\phi)) = \frac{f(\phi)}{t}$. Substituting $t = \overline{\phi}(t)$ into the last equation yields

$$A_p(f) = \frac{f(t)}{\overline{\phi}(t)} = \frac{f(t)}{t^p/(f(t))^{p-1}} = \left(\frac{f(t)}{t}\right)^p = (A(f))^p,$$

i.e., $A_p(t) = (A(t))^p$ completing the proof.

We now show a half Riordan array approach in the construction of the Lagrange inversion formula (4.178). This new argument is not only simple but also stimulates a new form of the Lagrange inversion formula.

Theorem 4.3.14 *Let $f \in \mathcal{F}_1$, and let $\bar{f}$ be the compositional inverse of f. Then we have the following two Lagrange inversion formulas derived from VHRA and HHRA of Riordan array $(1, f)$, respectively.*

$$[t^n](f(t))^k = \frac{k}{n}[t^{n-k}]\left(\frac{\bar{f}(t)}{t}\right)^{-n} = \frac{k}{n}[t^{n-k}](A(t))^n, \tag{4.193}$$

$$[t^n](tf(t)^2)^k = \frac{2k}{n+k}[t^{n-k}]\left(\frac{\bar{f}(t)}{t}\right)^{-n-k} = \frac{2k}{n+k}[t^{n-k}](A(t))^{n+k}, \tag{4.194}$$

where $A(t) = t/\bar{f}(t)$ is the generating function of the A-sequence of $(1, f)$, and (4.193) *is* (4.178) *shown in Corollary 4.3.2.*

Proof. From equations (4.187) and (4.190) with $g = 1$, $p = 2$, and $r = 0$, we have

$$[t^{2n-k}]f^n = [t^n]\frac{t\phi'(t)}{\phi}\phi^k = [t^n]t\phi'(t)\phi^{k-1}(t)$$
$$=\frac{1}{k}[t^{n-1}]\frac{d}{dt}\phi^k(t) = \frac{n}{k}[t^n]\phi^k,$$

where $\phi = \overline{t^2/f}$. We substitute $f \to t^2/\bar{f}$ into the leftmost and rightmost sides of the equations and use the fact

$$\phi(t) = \overline{\frac{t^2}{t^2/\bar{f}(t)}} = f(t)$$

to obtain

$$[t^{2n-k}]\left(\frac{t^2}{\bar{f}}\right)^n = \frac{n}{k}[t^n]\phi^k = \frac{n}{k}[t^n]f^k,$$

which implies (4.178) and (4.193).

Similarly, from equations (4.188) and (4.192) with $g=1$, $p=2$, and $r=0$, we have

$$[t^{2n}]f^{n+k} = [t^n]\frac{t\phi'(t)}{\phi}f(\phi(t))^k,$$

where $\phi = \overline{t^2/f}$. Substituting $f \to t^2/\bar{f}$ into the above equation and noting $\phi(t) = f(t)$, we have

$$[t^{2n}]\left(\frac{t^2}{\bar{f}(t)}\right)^{n+k} = [t^n]\frac{tf'(t)}{f(t)}\left(\frac{f^2}{t}\right)^k = [t^{n+k-1}]f'f^{2k-1}$$
$$= \frac{1}{2k}[t^{n+k-1}]\frac{d}{dt}f^{2k} = \frac{n+k}{2k}[t^{n+k}]f^{2k},$$

which implies (4.194).

■

Corollary 4.3.15 *Let $d^H_{n,k}$, $d^D_{n,k}$, and $d^L_{n,k}$ be the (n,k) entries of the hitting-time Riordan array $(tf'(t)/f(t), f(t))$, the derivative Riordan array $(f'(t), f(t))$, and the associated Riordan array $(1, f(t))$, respectively. Then,*

$$\frac{k}{n}d^H_{n,k} = d^L_{n,k} \quad \frac{k+1}{n+1}d^D_{n,k} = d^L_{n+1,k+1}, \tag{4.195}$$

which implies an obvious relation $d^D_{n,k} = d^H_{n+1,k+1}$, $n,k \geq 0$.

Proof. From the proof of Theorem 4.3.14 (or the proof of Corollary 4.3.2), we obtain the first formula of (4.195). The second formula can be proved similarly as follows.

$$d^D_{n,k} = [t^n]f'(t)f(t)^k = \frac{1}{k+1}[t^n](f(t)^{k+1})' = \frac{n+1}{k+1}[t^{n+1}]f(t)^{k+1},$$

which implies the desired result.

■

We may use Theorems 4.3.12 and 4.3.13 and Faà di Bruno's formula to establish a class of summation formulas.

Faà di Bruno's formula is an identity generating the chain rule to higher derivatives. Though it is named after Francesco Faà di Bruno (1855, 1857), he was not the first to state or prove the formula. In 1800, more than 50 years before Faà di Bruno, the French mathematician Louis François Antoine Arbogast had stated the formula in a calculus textbook [12], which is considered to be the first published reference on the subject.

Perhaps the most well-known form of Faà di Bruno's formula says that (cf. Section 3.4 of [44])

$$\frac{d^n}{dt^n} f(g(t)) = \sum_{\sigma(n)} \frac{n!}{m_1!\,1!^{k_1}\,k_2!\,2!^{k_2}\,\cdots\,k_n!\,n!^{k_n}} \cdot f^k(g(t)) \cdot \prod_{j=1}^{n} \left(g^{(j)}(t)\right)^{m_j},$$

where the sum is over the set $\sigma(n)$ of all partitions of n, that is, over the set of all non-negative integral solutions $(k_1, k_2, \ldots, k_n)$ of the equations $k_1 + 2k_2 + \cdots + nk_n = n$ and $k_1 + k_2 + \cdots + k_n = k$, $k = 1, 2, \ldots, n$. Each solution $(k_1, k_2, \ldots, k_n)$ of the equations is called a partition of n with k parts and is denoted by $\sigma(n, k)$. Hence, the set $\sigma(n)$ is the union of all subsets $\sigma(n, k)$, $k = 1, 2, \ldots, n$.

Sometimes, to give it a memorable pattern, it is written in a way in which the coefficients that have the combinatorial interpretation discussed below are less explicit:

$$\frac{d^n}{dt^n} f(g(t)) = \sum_{\sigma(n)} \frac{n!}{k_1!\,k_2!\,\cdots\,k_n!} \cdot f^k(g(t)) \cdot \prod_{j=1}^{n} \left(\frac{g^{(j)}(t)}{j!}\right)^{k_j}. \tag{4.196}$$

Combining the terms with the same value of $k_1 + k_2 + \cdots + k_n = k$ and noticing that k_j has to be zero for $j > n - k + 1$ leads to a somewhat simpler formula expressed in terms of *incomplete exponential Bell polynomials*:

$$\frac{d^n}{dt^n} f(g(t)) = \sum_{k=1}^{n} f^{(k)}(g(t)) \cdot B_{n,k}\left(g'(t), g''(t), \ldots, g^{(n-k+1)}(t)\right). \tag{4.197}$$

Here, the incomplete exponential Bell polynomials $B_{n,k}(t_1, t_2, \ldots, t_{n-k+1})$ are defined by (cf. Section 3.3 [44])

$$\begin{aligned} &B_{n,k}(t_1, t_2, \ldots, t_{n-k+1}) \\ =&\sum \frac{n!}{j_1! j_2! \cdots j_{n-k+1}!} \left(\frac{t_1}{1!}\right)^{j_1} \left(\frac{t_2}{2!}\right)^{j_2} \cdots \left(\frac{t_{n-k+1}}{(n-k+1)!}\right)^{j_{n-k+1}}, \end{aligned} \tag{4.198}$$

where the sum is taken over all sequences $j_1, j_2, \ldots, j_{n-k+1}$ of non-negative integers such that these two conditions are satisfied:

$$j_1 + j_2 + \cdots + j_{n-k+1} = k, \;\; j_1 + 2j_2 + 3j_3 + \cdots + (n-k+1)j_{n-k+1} = n.$$

The sum

$$B_n(t_1, \ldots, t_n) = \sum_{k=1}^{n} B_{n,k}(t_1, t_2, \ldots, t_{n-k+1})$$

is called the nth *complete exponential Bell polynomial.*

Clearly, $B_{n,k}(t_1, t_2, \ldots, t_{n-k+1})$ can also be written as

$$B_{n,k}(t_1, t_2, \ldots, t_{n-k+1}) = \sum_{\sigma(n,k)} \frac{n!}{k_1! k_2! \cdots} \left(\frac{t_1}{1!}\right)^{k_1} \left(\frac{t_2}{2!}\right)^{k_2} \cdots \tag{4.199}$$

and $\sigma(n,k)$ as shown before is the set of the solution of the partition equations for a given k $(1 \leq k \leq n)$.

For the exponential formal power series $f(t) = \sum_{j\geq 0} f_j t^j/j!$ and $g(t) = \sum_{j\geq 1} g_j t^j/j!$, the Faà di Bruno's formula (4.197) for the composition $f(g(t)) = h(t) = \sum_{j\geq 0} h_j t^j/j!$ at $t = 0$ can be written as

$$\begin{aligned}\frac{d^n}{dt^n} f(g(0)) &= \sum_{\sigma(n)} \frac{n!}{k_1!\, k_2! \, \cdots \, k_n!} \cdot f^{(k)}(g(0)) \cdot \prod_{j=1}^{n} \left(\frac{g^{(j)}(0)}{j!}\right)^{k_j} \\ &= \sum_{k=1}^{n} f^{(k)}(g(0)) \cdot B_{n,k}\left(g'(0), g''(0), \ldots, g^{(n-k+1)}(0)\right),\end{aligned}$$

or equivalently,

$$h_n = \left[\frac{t^n}{n!}\right] f(g(t)) = \sum_{k=1}^{n} f_k B_{n,k}(g_1, g_2, \ldots, g_{n-k+1}), \quad h_0 = f_0, \tag{4.200}$$

where we use the facts $\frac{d^n}{dt^n} f(g(0)) = h_n = n![t^n]f(g(t))$, $f^{(k)}(g(0)) = f^{(k)}(0) = f_k = k![t^k]f(t)$, and $g^{(j)}(0) = g_j = j![t^j]g(t)$ for $j = 1, 2, \ldots, n-k+1$. Particularly, for $f(t) = t^k/k!$ and $g(t) = \sum_{j\geq 1} g_j t^j/j!$ we have

$$[t^n]\frac{1}{k!}(g(t))^k = \sum_{j=1}^{n} f_j B_{n,j} = B_{n,k},$$

which implies $B_{n,k} = [t^n/n!](g(t))^k/k!$. Hence, $B_{n,k} = B_{n,k}(t_1, t_2, \ldots)$ can be defined by (cf. Section 3.3 of [44]):

$$\frac{1}{k!}(g(t))^k = \sum_{n=k}^{\infty} B_{n,k} \frac{t^n}{n!}, \tag{4.201}$$

which implies that the matrix $B_{n,k}$ is the exponential Riordan array $(1, f(t))$. In [44], a matrix with the form of $(1, f(t))$ is called a *iteration matrix*.

The following important property of iteration matrices (cf. Theorem A on p. 145 of Comtet [44], Roman [186], and Roman and Rota [187])

$$B(f(g(t))) = B(g(t))B(f(t))$$

is trivial in the context of the theory of Riordan arrays, i.e.,

$$(1, f(g(t))) = (1, g(t))(1, f(t));$$

and the Faà di Bruno formula derived from the above property of the iteration matrix is an application of the FTRA.

Likewise, the partial ordinary Bell polynomial, in contrast to the usual incomplete exponential Bell polynomial defined before, is given by

$$\hat{B}_{n,k}(t_1, t_2, \ldots, t_{n-k+1}) = \sum \frac{k!}{j_1! j_2! \cdots j_{n-k+1}!} t_1^{j_1} t_2^{j_2} \cdots t_{n-k+1}^{j_{n-k+1}}, \qquad (4.202)$$

where the sum runs over all sequences $j_1, j_2, \ldots, j_{n-k+1}$ of non-negative integers such that

$$j_1 + j_2 + \cdots + j_{n-k+1} = k, \; j_1 + 2j_2 + \cdots + (n-k+1) j_{n-k+1} = n.$$

Clearly, the ordinary Bell polynomials can be expressed in the terms of incomplete exponential Bell polynomials:

$$\hat{B}_{n,k}(t_1, t_2, \ldots, t_{n-k+1}) = \frac{k!}{n!} B_{n,k}(1! \cdot t_1, 2! \cdot t_2, \ldots, (n-k+1)! \cdot t_{n-k+1}).$$

In general, Bell polynomial refers to the incomplete exponential Bell polynomial, unless otherwise explicitly stated.

For the ordinary formal power series $\hat{f}(t) = \sum_{j\geq 0} \hat{f}_j t^j$ and $\hat{g}(t) = \sum_{j\geq 1} \hat{g}_j t^j$, the Faà di Bruno's formula (4.197) for the composition $\hat{f}(\hat{g}(t)) = \hat{h}(t) = \sum_{j\geq 0} \hat{h}_j t^j$ at $t = 0$ can be written as

$$\begin{aligned}\frac{d^n}{dt^n} \hat{f}(\hat{g}(0)) &= \sum_{\sigma(n)} \frac{n!}{k_1!\, k_2! \,\cdots\, k_n!} \cdot \hat{f}^{(k)}(\hat{g}(0)) \cdot \prod_{j=1}^{n} \left(\frac{\hat{g}^{(j)}(0)}{j!} \right)^{k_j} \\ &= \frac{n!}{k!} \sum_{k=1}^{n} \hat{f}^{(k)}(g(0)) \cdot \hat{B}_{n,k} \left(\frac{\hat{g}'(0)}{1!}, \frac{\hat{g}''(0)}{2!}, \ldots, \frac{\hat{g}^{(n-k+1)}(0)}{(n-k+1)!} \right) \\ &= \frac{n!}{k!} \sum_{k=1}^{n} k! \hat{f}_k \hat{B}_{n,k} \left(\hat{g}_1, \hat{g}_2, \ldots, \hat{g}_{n-k+1} \right),\end{aligned}$$

where we use $\hat{f}^{(k)}(g(0)) = \hat{f}^{(k)}(0) = k!\hat{f}_k$ and $\hat{g}^{(j)}(0) = j!\hat{g}_j$ for $j = 1, 2, \ldots, n-k+1$. Noting $(d^n/dt^n)\hat{f}(\hat{g}(0)) = (d^n/dt^n)\hat{h}(0) = n!\hat{h}_n$, we may rewrite the last expression as

$$\hat{h}_n = \sum_{k=1}^{n} \hat{f}_k \hat{B}_{n,k} \left(\hat{g}_1, \hat{g}_2, \ldots, \hat{g}_{n-k+1} \right). \qquad (4.203)$$

Since $\hat{h}_n = h_n/n!$, $\hat{f}_k = f_k/k!$, and $\hat{g}_j = g_j/j!$ for $j = 1, 2, \ldots, n-k+1$, one may see the equivalence between (4.200) and (4.203).

Let $h(t) = \sum_{n=0}^{\infty} \alpha_n t^n$ be a formal power series in $\mathbb{R}$ with the case $h(0) = \alpha_0 = a \neq 0$. Assume that $f(t)$ has a ordinary formal power series expansion in t. Then the composition of f and h still possesses a formal series expansion in t, namely,

$$(f \circ h)(t) = f\left(a + \sum_{n=1}^{\infty} \alpha_n t^n\right)$$
$$= f(a) + \sum_{n=1}^{\infty} \left([t^n](f \circ h)(t)\right) t^n. \qquad (4.204)$$

Let $f^{(k)}(a)$ denote the kth derivative of $f(t)$ at $t = a$, i.e.,

$$f^{(k)}(a) = (d^k/dt^k) f(t)|_{t=a}.$$

Applying Faà di Bruno's formula (cf. FDB-1) to $(f \circ h)(t)$ and noting that the kth derivative of $f(t)$ at $t = 0$, $f^{(k)}(0)$, is $k![t^k]f(t)$, we have

$$[t^n](f \circ h) = \sum_{\sigma(n)} f^{(k)}(h(0)) \Pi_{j=1}^{n} \frac{1}{k_j!} \left([t^j]h\right)^{k_j}, \qquad (4.205)$$

where the summation ranges over the set $\sigma(n)$ of all partitions of n as we present in (4.196).

Let $\beta_n = [t^n](f \circ h)(t)$ and $h(0) = \alpha_0 = a$. Then there exists a pair of reciprocal relations

$$\beta_n = \sum_{\sigma(n)} f^{(k)}(a) \frac{\alpha_1^{k_1} \cdots \alpha_n^{k_n}}{k_1! \cdots k_n!}, \qquad (4.206)$$

$$\alpha_n = \sum_{\sigma(n)} \bar{f}^{(k)}(f(a)) \frac{\beta_1^{k_1} \cdots \beta_n^{k_n}}{k_1! \cdots k_n!}, \qquad (4.207)$$

where $\bar{f}$ denotes the compositional inverse of f. In fact, from (4.204) the given conditions ensure that there holds the pair of formal series expansions

$$f\left(a + \sum_{n\geq 1} \alpha_n t^n\right) = f(a) + \sum_{n\geq 1} \beta_n t^n, \qquad (4.208)$$

$$\bar{f}\left(f(a) + \sum_{n\geq 1} \beta_n t^n\right) = a + \sum_{n\geq 1} \alpha_n t^n. \qquad (4.209)$$

Thus, an application of Faà di Bruno's formula (4.205) to $(f \circ h)(t)$, on the LHS of (4.208) yields the expression (4.206) with $[t^i]h = \alpha_i$, $[t^n](f \circ h) = \beta_n$, and $h(0) = a$. Note that the LHS of (4.209) may be expressed as $h(t) = ((\bar{f} \circ f) \circ h)(t) = (\bar{f} \circ (f \circ h))(t)$ so that in a like manner an application of Faà di Bruno's formula to the LHS of (4.209) gives precisely the equality (4.207).

Replacing α_n by $x_n/n!$ and β_n by $y_n/n!$, we see that (4.206) and (4.207) can be expressed in terms of the incomplete exponential Bell polynomials (cf. (4.198)), namely,

$$y_n = \sum_{k=1}^{n} f^{(k)}(a) B_{n,k}(x_1, x_2, \ldots, x_{n-k+1}), \tag{4.210}$$

$$x_n = \sum_{k=1}^{n} \bar{f}^{(k)}(a) B_{n,k}(y_1, y_2, \ldots, y_{n-k+1}), \tag{4.211}$$

where $B_{n,k}(\ldots)$ are defined by (4.199).

Let $f(x) = x^p$ $(p \neq 0)$. Then $\bar{f}(x) = x^{1/p}$ with $f^{(k)}(1) = (p)_k$ and $\bar{f}^{(k)}(1) = (1/p)_k$, where $(p)_k = p(p-1)\ldots(p-k+1)$ and $(p)_0 = 1$. Hence, we obtain the special cases of (4.206) and (4.207):

$$\beta_n = \sum_{\sigma(n)} (\alpha)_k \frac{\alpha_1^{k_1} \cdots \alpha_n^{k_n}}{k_1! \cdots k_n!}, \tag{4.212}$$

$$\alpha_n = \sum_{\sigma(n)} (1/\alpha)_k \frac{\beta_1^{k_1} \cdots \beta_n^{k_n}}{k_1! \cdots k_n!}. \tag{4.213}$$

The above Faà di Bruno's relations have the associated relations

$$\left(1 + \sum_{n=1}^{\infty} \alpha_n t^n\right)^p = 1 + \sum_{n=1}^{\infty} \beta_n t^n, \tag{4.214}$$

$$\left(1 + \sum_{n=1}^{\infty} \beta_n t^n\right)^{1/p} = 1 + \sum_{n=1}^{\infty} \alpha_n t^n. \tag{4.215}$$

As example, if $h = a_0 + a_1 t$ and $f(t) = t^p$, then $f(h(t)) = a_0^p (1 + \alpha_1 t)^p$, where $\alpha_1 = a_1/a_0$. From (4.214) we have

$$(a_0 + a_1 t)^p = a_0^p (1 + \alpha_1 t)^p = a_0^p \left(1 + \sum_{j=1}^{\infty} \beta_j t^j\right),$$

where

$$\beta_j = \sum_{\sigma(j)} (\alpha)_k \frac{\alpha_1^{k_1} \cdots \alpha_n^{k_n}}{k_1! \cdots k_n!} = (p)_j \frac{\alpha_1^j}{j!} = \binom{p}{j} \alpha_1^j,$$

which presents the obvious expression $(a_0 + a_1 t)^p = a_0^p + \sum_{j=1}^{p} \binom{p}{j} a_0^{p-j} a_1^j t^j$.

Similarly, if $h = a_0 + a_1 t + a_2 t^2$, $a_0 \neq 0$, then

$$(a_0 + a_1 t + a_2 t^2)^p = a_0^p \left(1 + \frac{a_1}{a_0} t + \frac{a_2}{a_0} t^2\right)^p = a_0^p \left(1 + \sum_{j=1}^{p} \beta_j t^j\right),$$

where

$$\beta_j = \sum_{\sigma(j)} (p)_j \frac{1}{j_i!j_2!}\left(\frac{a_1}{a_0}\right)^{j_1}\left(\frac{a_2}{a_0}\right)^{j_2} = \sum_{j_i=0}^{j} \binom{p}{j}\binom{j}{j_1}\left(\frac{a_1}{a_0}\right)^{j_1}\left(\frac{a_2}{a_0}\right)^{j-j_1}.$$

Theorem 4.3.16 *Let $A(t) = \sum_{n\geq 0} a_n t^n$ ($a_0 \neq 0$) be the generating function of the A-sequence of the given Riordan array $(d_{n,k})_{n,k\geq 0} = (g, f)$, and let $(\tilde{d}_{n,k} = d_{pn+r,(p-1)n+r+k})_{n,k\geq 0}$ be the one-pth Riordan array of (g, f) with respect to r. Then from (4.204) there exists the following summation formula:*

$$d_{p(n+1)+r,(p-1)(n+1)+r+k+1} = \sum_{j=0}^{n-k} \beta_j d_{pn+r,(p-1)n+r+k+j}, \tag{4.216}$$

where by denoting $(p)_j = p(p-1)\ldots(p-j+1)$, $\beta_0 = a_0^p$, and $\alpha_i = a_i/a_0$, for $n \geq 1$

$$\begin{aligned}\beta_j =& a_0^p[t^j](A(t))^p = \sum_{\sigma(j)} (p)_j \frac{\alpha_1^{k_1}\cdots\alpha_j^{k_j}}{k_1!\cdots k_j!}\\ =& \sum_{i=1}^{j}\sum_{\sigma(j,i)} \binom{p}{j}\frac{j!}{k_1!k_2!\ldots}(\alpha_1)^{k_1}(\alpha_2)^{k_2}\ldots .\end{aligned} \tag{4.217}$$

Particularly, for $A(t) = a_0 + a_1 t$ and $A(t) = a_0 + a_1 t + a_2 t^2$, we have

$$\beta_j = \binom{p}{j} a_0^{p-j} a_1^j \quad \textit{and}$$

$$\beta_j = \sum_{i=0}^{j} \binom{p}{j}\binom{j}{i} a_0^{p-j} a_1^{j-i} a_2^i,$$

respectively.

The proof is left as Exercise 4.11.

Using (4.216) in Theorem 4.3.16, one may obtain many identities.

Example 4.3.17 *For Pascal matrix $(1/(1-t), t/(1-t))$, from Example 4.3.6, its A-sequence generating function is $A(t) = 1 + t$. Applying (4.216), we have*

$$\binom{p(n+1)+r}{(p-1)(n+1)+r+k+1} = \sum_{j=0}^{\min\{p,n-k\}} \binom{p}{j}\binom{pn+r}{(p-1)n+r+k+j}. \tag{4.218}$$

If $p = 1$ and $r = 0$, the above identity reduces to the well-known identity $\binom{n+1}{k+1} = \binom{n}{k} + \binom{n}{k+1}$.

The Riordan array $(1/(1-t-t^2), tC(t))$ *was considered in Example 4.3.7, where* $C(t) = \sum_{n=0}^{\infty} \binom{2n}{n} t^n/(n+1) = (1-\sqrt{1-4t})/(2t)$ *is the Catalan function. It was also shown in Example 4.3.7 that the A-sequence of the Riordan array* $(1/(1-t-t^2), tC(t))$ *is* $(1,1,1,\ldots)$, *i.e., the generating function of the A-sequence is* $A(t) = 1/(1-t)$. *From [79, 109] we have*

$$C(t)^k = \sum_{n=0}^{\infty} \frac{k}{2n+k} \binom{2n+k}{n} t^n. \tag{4.219}$$

Thus, the (n,k) *entry of the Riordan array* $(1/(1-t-t^2), tC(t))$ *is*

$$\begin{aligned} d_{n,k} &= [t^n] \frac{1}{1-t-t^2} (tC(t))^k \\ &= [t^{n-k}] \left(\sum_{i\geq 0} F_i t^i \right) \left(\sum_{j\geq 0} \frac{k}{2j+k} \binom{2j+k}{j} t^j \right) \\ &= [t^{n-k}] \sum_{i\geq 0} \left(\sum_{j=0}^{i} F_{i-j} \frac{k}{2j+k} \binom{2j+k}{j} \right) t^i \\ &= \sum_{j=0}^{n-k} F_{n-k-j} \frac{k}{2j+k} \binom{2j+k}{j}. \end{aligned}$$

Since

$$(A(t))^p = (1-t)^{-p} = \sum_{i\geq 0} \binom{-p}{i} (-t)^i = \sum_{i\geq 0} \binom{p+i-1}{i} t^i,$$

From (4.216) *there holds the identity*

$$\begin{aligned} &\sum_{j=0}^{n-k} F_{n-k-j} \frac{(p-1)(n+1)+r+k+1}{2j+(p-1)(n+1)+r+k+1} \\ &\times \binom{2j+(p-1)(n+1)+r+k+1}{j} \\ = &\sum_{i\geq 0} \binom{p+i-1}{i} \sum_{j=0}^{n-k-i} F_{n-k-i-j} \frac{(p-1)n+r+k+i}{2j+(p-1)n+r+k+i} \\ &\times \binom{2j+(p-1)n+r+k+i}{j}. \end{aligned}$$

Similarly, for the Riordan array $(C(t), tC(t))$, its (n,k) entry is

$$\begin{aligned} d_{n,k} =& [t^n]t^k(C(t))^{k+1} \\ =& [t^{n-k}]\sum_{j\geq 0}\frac{k+1}{2j+k+1}\binom{2j+k+1}{j}t^j \\ =& \frac{k+1}{2n-k+1}\binom{2n-k+1}{n-k}. \end{aligned}$$

Hence, from (4.216) *we may derive the identity*

$$\begin{aligned} &\frac{(p-1)(n+1)+r+k+2}{(p+1)(n+1)+r-k}\binom{(p+1)(n+1)+r-k}{n-k} \\ &=\sum_{j=0}^{n-k}\frac{(p-1)n+r+k+j+1}{(p+1)n+r-k-j+1}\binom{p+j-1}{j}\binom{(p+1)n+r-k-j+1}{n-k-j}. \end{aligned}$$

Faà di Bruno's relations (4.206) and (4.207) and their special cases such as (4.212) and (4.213) are also applicable rules to construct numerous identities. For instance, for the Catalan numbers, $c_n = \binom{2n}{n}/(n+1)$, and their generating function, $C(t) = (1-\sqrt{1-4t})/(2t)$, we have the identity

$$C_n = \frac{1}{n+1}\binom{2n}{n} = \frac{1}{n+1}\sum_{[n/2]\leq k\leq n}\left(-\frac{1}{2}\right)_k\frac{(-4)^{n-k}}{(2k-n)!(n-k)!}. \tag{4.220}$$

In fact, let $f(t) = t^{-1/2}$ and $\phi(t) = 1-4t$. Then $(f\circ\phi)(t)$ can be expanded as

$$(f\circ\phi)(t) = \frac{1}{\sqrt{1-4t}} = \sum_{n\geq 0}\binom{2n}{n}t^n. \tag{4.221}$$

Thus, substituting $\beta_n = \binom{2n}{n}$, $\alpha_1 = 1$, and $\alpha_2 = -4$ into the Faá di Bruno's relation (4.212) and noting $k_1 = 2k-n$ and $k_2 = n-k$, we obtain (4.220).

Catalan numbers provide solutions to many counting problems. The same is true for Motzkin numbers. For each non-negative integer n, the Motzkin number m_n is defined to be the counting number of all possible ways of connecting any subset of n points on a circle by nonintersecting chords. Let $M(t) = \sum_{n=0}^{\infty} m_n t^n$ be the generating function of the Motzkin numbers. It is known (cf. [18]) that $M(t)$ satisfies $t^2(M(t))^2 + (t-1)M(t) + 1 = 0$. In fact, $M(t)$ is of the form

$$M(t) = \frac{1-t-\sqrt{1-2t-3t^2}}{2t^2}. \tag{4.222}$$

Write $\sqrt{1-2t-3t^2} = 1 + \sum_{n=1}^{\infty} e_n t^n$ and take $\alpha_1 = -2$ and $\alpha_2 = -3$. From (4.212) and noting

$$\frac{d^k}{dx^k}\,x^{1/2}\bigg|_{x=1} = (-1)^{k-1}\frac{(2k-2)!}{2^{2k-1}(k-1)!}$$

for all $k \geq 1$, we have, after simplification,

$$e_n = \sum_{\sigma(n)} (-1)^{k-1} \frac{(2k-2)!}{2^{2k-1}(k-1)!} \frac{\alpha_1^{k_1} \cdots \alpha_n^{k_n}}{k_1! \cdots k_n!}$$
$$= -\frac{1}{2^{n-1}} \sum_{k=0}^{[n/2]} \frac{(2n-2k-2)! 3^k}{(n-k-1)!(n-2k)!k!}. \quad (4.223)$$

From (4.222), we have

$$M(t) = \sum_{n=0}^{\infty} \frac{-e^n + 2}{2} t^n = \sum_{n=0}^{\infty} \frac{1}{2^{n+2}} \left(\sum_{k=0}^{[(n+2)/2]} \frac{(2n-2k+3)! 3^k}{(n-2k+2)!(n-k+1)!k!} \right) t^n.$$

So,

$$m_n = \frac{1}{2^{n+2}(n+1)} \sum_{k=0}^{[(n+2)/2]} \binom{2n-2k+2}{n} \binom{n+1}{k} 3^k \quad (4.224)$$

for $n \geq 0$. It is known (cf. [18]) that

$$m_n = \sum_{k=0}^{[n/2]} \frac{1}{k+1} \binom{n}{2k} \binom{2k}{k}.$$

Jointing the last equation and equation 4.224 gives a new identity

$$\frac{1}{2^{n+2}(n+1)} \sum_{k=0}^{[(n+2)/2]} \binom{2n-2k+2}{n} \binom{n+1}{k} 3^k = \sum_{k=0}^{[n/2]} \frac{1}{k+1} \binom{n}{2k} \binom{2k}{k}.$$

For any non-negative integer n, the Riordan number r_n can be viewed as the number of plane tree of order $n+2$ in which the root has degree one and no vertex has degree two (cf. [18]). Let $R(t) = \sum_{n=0}^{\infty} r_n t^n$ be the generating function of Riordan numbers. It is known (cf. [18]) that $R(t)$ satisfies $(t + t^2)(R(t))^2 - (1+t)R(t) + 1 = 0$. So,

$$R(t) = \frac{1 + t - \sqrt{1 - 2t - 3t^2}}{2t(1+t)}.$$

From the above example of Motzkin numbers, using $\sqrt{1-2t-3t^2} = 1 + \sum_{n=1}^{\infty} e_n t^n$, where e_n are in (4.223), we have

$$R(t) = 1 + \sum_{n=2}^{\infty} t^n \sum_{m=2}^{n} \frac{(-1)^{n-m}}{2^{m+1} m} \sum_{k=0}^{[(m+1)/2]} \binom{2m-2k}{m-1} \binom{m}{k} 3^k.$$

Therefore $r_0 = 1$, $r_1 = 0$, and

$$r_n = \sum_{m=2}^{n} \frac{(-1)^{n-m}}{2^{m+1} m} \sum_{k=0}^{[(m+1)/2]} \binom{2m-2k}{m-1} \binom{m}{k} 3^k$$

for all $n \geq 2$. It is known (cf. [18]) that

$$r_n = \frac{1}{n+1}\sum_{m=0}^{n}(-1)^m\binom{n+1}{m}\binom{2n-2m}{n-m},$$

which implies the following new identity

$$\begin{aligned}&\sum_{m=2}^{n}\frac{(-1)^{n-m}}{2^{m+1}m}\sum_{k=0}^{[(m+1)/2]}\binom{2m-2k}{m-1}\binom{m}{k}3^k\\ =&\frac{1}{n+1}\sum_{m=0}^{n}(-1)^m\binom{n+1}{m}\binom{2n-2m}{n-m}.\end{aligned}$$

For each non-negative integer n, the Fine number f_n is considered to be the number of rooted trees of order n with root of even degree (cf. [56]). Let $F(t) = \sum_{n=0}^{\infty} f_n t^n$ be the generating function of the Fine numbers. It is known (cf. [56]) that

$$F(t) = \frac{1+2t-\sqrt{1-4t}}{2t^2+4t}.$$

From (4.214),

$$\begin{aligned}1+2t-\sqrt{1-4t} =&1+2t-\left(1+\sum_{n=1}^{\infty}(-1)^{n-1}\frac{(2n-2)!}{2^{2n-1}(n-1)!}\frac{(-4)^n}{n!}t^n\right)\\ =&4t+2t^2+\sum_{n=3}^{\infty}\frac{2}{n}\binom{2(n-1)}{n-1}t^n.\end{aligned}$$

So the expression of $F(t)$ can be written as

$$\begin{aligned}F(t) =&\frac{4+2t+\sum_{n=3}^{\infty}\frac{2}{n}\binom{2(n-1)}{n-1}t^{n-1}}{4+2t}\\ =&1+\frac{1}{2}\left(\sum_{n=2}^{\infty}\frac{1}{n+1}\binom{2n}{n}t^n\right)\left(\sum_{n=0}^{\infty}\left(\frac{-1}{2}\right)^n t^n\right)\\ =&1+\sum_{n=2}^{\infty}\frac{t^n}{2}\sum_{k=2}^{n}\frac{(-1)^{n-k}}{(k+1)2^{n-k}}\binom{2k}{k}.\end{aligned}$$

So, Finn numbers $f_0 = 1$, $f_1 = 0$, and

$$f_n = \frac{1}{2}\sum_{k=2}^{n}\frac{(-1)^{n-k}}{(k+1)2^{n-k}}\binom{2k}{k}$$

for all $n \geq 2$. This formula was considered in [53] using a different approach.

4.3.3 Various row sums of Riordan arrays

From FTRA, for a Riordan array $(g, f) = (d_{n,k})_{n,k\geq 0}$, $g \in \mathcal{F}_0$, $f \in \mathcal{F}_1$, and any formal power series $h = \sum_{k\geq 0} h_k t^k$ in $\mathcal{F} = \mathbb{K}[\![t]\!]$, we have

$$\sum_{k\geq 0} d_{n,k} h_k = [t^n] g(t) h(f(t)), \tag{4.225}$$

which can be written in short as

$$(g, f)h = g(h \circ f). \tag{4.226}$$

By choosing different h in (4.225) such as $h(t) = 1/(1-t)$, $1/(1+t)$, $t/(1-t)^2$, etc., we obtain the summation formula rules listed as follows:

$$\sum_{k\geq 0} d_{n,k} = [t^n] \frac{g(t)}{1 - f(t)} \quad \text{(row sums } R^+\text{)}$$

$$\sum_{k\geq 0} (-1)^k d_{n,k} = [t^n] \frac{g(t)}{1 + f(t)} \quad \text{(alternating row sums } R^-\text{)}$$

$$\sum_{k\geq 0} k d_{n,k} = [t^n] \frac{g(t) f(t)}{(1 - f(t))^2} \quad \text{(weighted row sums)} \ldots.$$

More precisely, if $h = 1/(1-t)$, then (4.226) presents the generating function, called the *sum function*, of the *row sum sequence* of the Riordan array $(g(t), f(t))$. We denote the sum function by $R^+(t)$. Hence,

$$R^+(t) \quad := \quad (g(t), f(t)) \frac{1}{1-t} = \frac{g(t)}{1 - f(t)}. \tag{4.227}$$

More briefly this can be written as

$$R^+ \quad = \quad \frac{g}{1-f} \tag{4.228}$$

Similarly, the generating function, denoted by R^-, of the *alternating sum sequence* of the Riordan array $(g(t), f(t))$ is called the *alternating sum function*, namely, defined as

$$R^-(t) := (g(t), f(t)) \frac{1}{1+t} = \frac{g(t)}{1 + f(t)}. \text{ i.e. } R^- = \frac{g}{1+f}. \tag{4.229}$$

It is easy to see that we have, reminiscent of Pythagorean triples,

$$g = \frac{2R^+ R^-}{R^+ + R^-}, \quad f = \frac{R^+ - R^-}{R^+ + R^-}. \tag{4.230}$$

Thus we can use the sum function R^+ and the alternating sum function R^- to characterize Riordan arrays.

Here is a list of several subgroups of $\mathcal{R}$ together with their R^+ and R^- functions

- For the *Appell subgroup* (which is a normal subgroup of the Riordan group), $A = \{(g(t), t) : g \in \mathcal{F}_0\}$, we have

$$R^+ = \frac{1+t}{1-t}R^-, \; R^- = \frac{1-t}{1+t}R^+ \quad \text{and} \quad g = R^+ - tR^+ = R^- + tR^-. \tag{4.231}$$

- For the *Associate subgroup (or Lagrange subgroup)*, $L = \{(1, f(t)) : f \in \mathcal{F}_1\}$, we have

$$R^+ = \frac{R^-}{2R^- - 1}, \; R^- = \frac{R^+}{2R^+ - 1} \quad \text{and} \quad f(t) = \frac{R^+ - 1}{R^+} = \frac{1 - R^-}{R^-}. \tag{4.232}$$

- For the *Bell subgroup*, $B = \{(g(t), tg(t)) : g \in \mathcal{F}_0\}$, we have

$$R^+ = \frac{R^-}{1 - 2tR^-} \quad \textit{and} \quad R^- = \frac{R^+}{1 + 2tR^+} \quad \text{and} \quad g = \frac{R^+}{1 + tR^+} = \frac{R^-}{1 - tR^-}. \tag{4.233}$$

- For the *checkerboard subgroup*, $\mathcal{C} = \{(g(t), f(t)) : g \in \mathcal{F}_0 \textit{ is even}, f \in \mathcal{F}_1 \textit{ is odd}\}$, we have

$$R^+(t) = R^-(-t). \tag{4.234}$$

- For the *stochastic subgroup*, $\mathcal{S} = \{(g(t), \; f(t))\}$, which row sums are one, we have

$$R^+ = 1/(1-t), \quad R^- = \frac{g}{2 - (1-t)g} \quad \text{or} \quad \frac{1-f}{(1-t)(1+f)}. \tag{4.235}$$

- For the *hitting-time subgroup* , $\mathcal{H} = \{(tf'(t)/(f(t)), f(t)) : f \in \mathcal{F}_1\}$, we have

$$R^+ = \frac{tf'}{f(1-f)} \quad \textit{and} \quad R^- = \frac{tf'}{f(1+f)}, \tag{4.236}$$

which implies

$$t\frac{D_t R^+}{R^+} - R^+ = t\frac{D_t R^-}{R^-} - R^-.$$

- For the *derivative subgroup*, $\mathcal{D} = \{(f'(t), f(t)) : f \in \mathcal{F}_1\}$, we have

$$R^+ = \frac{f'}{1-f} = -D_t \ln|1 - f(t)| \quad \textit{and} \quad R^- = \frac{f'}{1+f} = D_t \ln|1 + f(t)|. \tag{4.237}$$

This implies that

$$2 = e^{\int R^+(t)dt} + e^{\int R^-(t)dt}.$$

We survey the above characterizations of subgroups of Riordan group as follows.

Proposition 4.3.18 *Let $R^+(t)$ and $R^-(t)$ be the sum and alternating sum functions of a Riordan array $(d(t), h(t))$. If $(d(t), h(t))$ is an element of the subgroups of Appell, associate, Bell, checkboard, stochastic, hitting-time, and derivative, then the characterizations of $R^+(t)$ and $R^-(t)$ of $(d(t), h(t))$ are shown in* (4.231), (4.232), (4.233), (4.234), (4.235), (4.236), *and* (4.237), *respectively.*

As a heuristic principle two pieces of information will determine an element in the Riordan group. As examples we have: $d(t)$ and $h(t)$, R^+ and R^-, $A(t)$ and $Z(t)$. Next are some examples where given a subgroup and R^+ gives a unique element. For all these examples

$$\begin{aligned} R^+ =& C(t) = C = 1 + tC^2 = \frac{1-\sqrt{1-4t}}{2t} \\ =& \sum_{n\geq 0} \frac{1}{n+1}\binom{2n}{n} t^n = 1 + t + 2t^2 + 5t^3 + 14t^4 + \cdots, \end{aligned}$$

where $[t^n]C(t) = \binom{2n}{n}/(n+1)$ are the Catalan numbers. Formula (4.219) shows

$$C(t)^k = \sum_{n=0}^{\infty} \frac{k}{2n+k}\binom{2n+k}{n} t^n.$$

In the Appell subgroup, the element is

$$(C(1-t), t) = \begin{bmatrix} 1 & 0 & 0 & 0 & 0 & 0 & \\ 0 & 1 & 0 & 0 & 0 & 0 & \\ 1 & 0 & 1 & 0 & 0 & 0 & \\ 3 & 1 & 0 & 1 & 0 & 0 & \cdots \\ 9 & 3 & 1 & 0 & 1 & 0 & \\ 28 & 9 & 3 & 1 & 0 & 1 & \\ & & & \cdots & & & \end{bmatrix}$$

In the Lagrange subgroup, the element is

$$(1, tC) = \begin{bmatrix} 1 & 0 & 0 & 0 & 0 & 0 & \\ 0 & 1 & 0 & 0 & 0 & 0 & \\ 0 & 1 & 1 & 0 & 0 & 0 & \\ 0 & 2 & 2 & 1 & 0 & 0 & \cdots \\ 0 & 5 & 5 & 3 & 1 & 0 & \\ 0 & 14 & 14 & 9 & 4 & 1 & \\ & & & \cdots & & & \end{bmatrix}$$

Similarly, in the Bell subgroup, the element is (F, tF), where $F = \frac{C}{1+tC}$ is the generating function for the Fine numbers. In the Checkerboard subgroup, the element is $\left(\frac{2C(t)C(-t)}{C(t)+C(-t)}, \frac{C(t)-C(-t)}{C(t)+C(-t)}\right)$. In the Derivative subgroup, the element is $\left(\frac{C}{1+(C-1)e^{1-C}}, \frac{(C-1)e^{1-C}}{1+(C-1)e^{1-C}}\right)$. In the Hitting time subgroup, the element is $(C^2e^{-2tC}, 1-Ce^{-2tC})$.

The Riordan arrays (B, tC) and $\left(\frac{B}{C}, tC\right)$ have row sum generating functions BC and B, respectively. These are of combinatorial significance since B is the generating function for ordered trees with a marked vertex while $\frac{B}{C}$ is the generating function for marked leaves. Similar results hold for other kinds of ordered tress which have the same possibilities for updegrees at every vertex.

From the above Riordan arrays and the FTRA, some summation formulas can be found. For instance, one may consider the generic terms or the entries of the Riordan array $(C(t)^\ell, tC(t)^r)$, which includes its special cases of the Appell Riordan array $(C(t)^\ell, t)$ and the Lagrange Riordan array $(1, tC(t)^r)$. The (n,k) entry of $(C(t)^\ell, tC(t)^r)$ is

$$\begin{aligned}d_{n,k} =&[t^n]C^\ell(tC^r)^k = [t^{n-k}]C(t)^{rk+\ell}\\ =&[t^{n-k}]\sum_{j\geq 0}\frac{rk+\ell}{2j+rk+\ell}\binom{2j+rk+\ell}{j}t^j\\ =&\frac{rk+\ell}{2n+(r-2)k+\ell}\binom{2n+(r-2)k+\ell}{n-k}.\end{aligned}$$

From (4.226),

$$\left(C^\ell, tC^r\right)\frac{1}{1-t}=\frac{C^\ell}{1-tC^r}.$$

Particularly, if $\ell=0$ and $r=1$, noting $1-tC=1/C$ we may use (4.225) to obtain

$$\sum_{k=0}^n d_{n,k}|_{(0,1)}=[t^n]\frac{1}{1-tC(t)}=[t^n]C(t),$$

where $d_{n,k}|_{(0,1)}$ stands for the (n,k) entry of the Riordan array $(1, tC)$. Thus,

$$\sum_{k=0}^n\frac{k}{2n-k}\binom{2n-k}{n-k}=\frac{1}{n+1}\binom{2n}{n}.$$

Similarly, if $r=2$ and $\ell\in\mathbb{Z}$, then

$$\begin{aligned}&[t^n]\left(C^\ell, tC^2\right)\frac{1}{1-t}=[t^n]\frac{C^\ell}{1-tC^2}\\ =&[t^n]\frac{C^\ell}{2-C}=\frac{1}{2}\sum_{j\geq 0}[t^n]C^\ell\left(\frac{C}{2}\right)^j\\ =&\sum_{j\geq 0}\frac{1}{2^{j+1}}\frac{j+\ell}{2n+j+\ell}\binom{2n+j+\ell}{n}.\end{aligned}$$

By using (4.225) we get

$$\sum_{k=0}^{n} d_{n,k}|_{(\ell,r)=(\ell,2)} = [t^n]\left(C^\ell, tC^2\right)\frac{1}{1-t},$$

which implies

$$\sum_{k=0}^{n} \frac{2k+\ell}{2n+\ell}\binom{2n+\ell}{n-k} = \sum_{j\geq 0} \frac{1}{2^{j+1}}\frac{j+\ell}{2n+j+\ell}\binom{2n+j+\ell}{n}.$$

The above summation formula yields many identities. For instance, if $\ell = 0$, then the above summation formula is specified as

$$\sum_{k=0}^{n} \frac{k}{n}\binom{2n}{n-k} = \sum_{j\geq 0} \frac{j}{(2n+j)2^{j+1}}\binom{2n+j}{n},$$

which implies some summations such as

$$\sum_{j\geq 0}\frac{j}{2^{j+1}} = 1, \quad \sum_{j\geq 0}\frac{j(j+3)}{2^{j+2}} = 3,$$

etc. for the cases of $n = 1, 2, \ldots$, respectively.

Finally, if $h(t) = t/(1-t)^2$, then (4.226) presents the generating function, R^E, called the *weighted row sum function*, of the *weighted row sum sequence* of the Riordan array $(g(t), f(t))$ with the weights $\{0, 1, 2, \ldots\}$. Hence, R^E is also called the *expected value sum function* of $(g(t), f(t))$. More precisely, R^E is defined by

$$R^E \equiv R^E(t) = (g(t), f(t))\frac{t}{(1-t)^2} = \frac{g(t)f(t)}{(1-f(t))^2}. \tag{4.238}$$

The above expression can be written as

$$R^E = \frac{gf}{(1-f)^2} \tag{4.239}$$

It is easy to see that

$$g = \frac{(R^+)^2}{R^+ + R^E} \quad \text{and} \quad f = \frac{R^E}{R^+ + R^E}. \tag{4.240}$$

Thus we can use its sum function R^+ and the expected value sum function R^E to characterize a Riordan array.

We also have the following inverse relationship between (R^+, R^-) and (R^+, R^E).

$$R^- = \frac{(R^+)^2}{R^+ + 2R^E} \quad R^E = \frac{(R^+)^2 - R^+R^-}{2R^-}. \tag{4.241}$$

Therefore, from the characterizations of subgroups of Riordan group characterized by (R^+, R^-), we may obtain the characterizations of the subgroups with respect to (R^+, R^E) by busing the relationship (4.241) (cf. examples in Exercise 4.12).

Using the summation formula (4.238), one may establish numerous summation formulas and identities. For instance, for the Riordan array $(1, tC)$, from (4.238) we have

$$(1, tC)\frac{t}{(1-tC)^2} = \frac{tC}{(1-tC)^2} = tC(t)^3.$$

Noting

$$[t^n]tC(t)^3 = [t^{n-1}]\sum_{k\geq 0}\frac{3}{2n+3}\binom{2n+3}{n}t^n = \frac{3}{2n+1}\binom{2n+1}{n-1}$$

and

$$(1, tC)\frac{t}{(1-tC)^2} = \sum_{k=0}^{n} k[t^n](tC(t))^k$$
$$= \sum_{k=0}^{n} k[t^{n-k}]\frac{k}{2n+k}\binom{2n+k}{n} = \sum_{k=0}^{n}\frac{k^2}{2n-k}\binom{2n-k}{n-k},$$

we obtain the identity

$$\sum_{k=0}^{n}\frac{k^2}{2n-k}\binom{2n-k}{n-k} = \frac{3}{2n+1}\binom{2n+1}{n-1}.$$

We will give two different views to the *Bernoulli polynomials* and *Euler polynomial sequences* with the help of the row sums and alternating sums of Riordan arrays. First, the concepts of sum function and alternating sum function of a Riordan array can be extended to any array $(g(t), f(t)) = \left([t^n]g(t)f(t)^k\right)_{n,k\geq 0}$, where it is not necessary to have $g(t) \in \mathcal{F}_0$ and $f(t) \in \mathcal{F}_1$. We still define sum function and alternating sum function of $(g(t), f(t))$ as

$$R^+ \equiv R^+(t) := \frac{g(t)}{1-f(t)} \quad \text{and} \quad R^- \equiv R^-(t) := \frac{g(t)}{1+f(t)}. \tag{4.242}$$

This extension allow us to reconsider some well-known polynomial sequences and number sequences in a different way. Here, we show some examples related to the Bernoulli polynomial and the Euler polynomial sequences.

It is well-known that Bernoulli polynomials $B_n(x)$ and Euler polynomials $E_n(x)$, $n \in \mathbb{N}_0$, are defined by the power series

$$\frac{te^{xt}}{e^t-1} = \sum_{n=0}^{\infty} B_n(x)\frac{t^n}{n!} \quad \text{and} \quad \frac{2e^{xt}}{e^t+1} = \sum_{n=0}^{\infty} E_n(x)\frac{t^n}{n!}. \tag{4.243}$$

Hence, we may consider the function

$$R^+(t;x) := \sum_{n=0}^{\infty} B_n(x)\frac{t^n}{n!} \quad \text{and}$$

$$R^-(t;x) := -\frac{t}{2}\sum_{n=0}^{\infty} E_n(x)\frac{t^n}{n!} = \sum_{n=1}^{\infty} -\frac{n}{2}E_{n-1}(x)\frac{t^n}{n!} \tag{4.244}$$

as the sum function and the alternating sum function of the array $(-te^{xt}, e^t)$, respectively. Therefore, $B_n(x)$ is nth row sum of the array, while $-nE_{n-1}(x)/2$ is the nth alternating row sum of the array.

The definition of the sum functions shown in (4.242) provides

$$e^t = \frac{R^+(t;x) - R^-(t;x)}{R^+(t;x) + R^-(t;x)}$$

or equivalently,

$$R^+(t;x) - R^-(t;x) = e^t\left(R^+(t;x) + R^-(t;x)\right).$$

The last equation yields the identity

$$\sum_{k=1}^{n}\binom{n}{k}B_{n-k}(x) = nE_{n-1}(x) + \sum_{k=1}^{n-1}\frac{n-k}{2}\binom{n}{k}E_{n-k-1}(x). \tag{4.245}$$

More investigation on this approach to Bernoulli and Euler polynomials is left for the interested readers.

Secondly, we may introduce proper Riordan arrays for Bernoulli and Euler polynomial sequences by using sum functions. In order to do it, we re-write (4.243) as

$$\frac{te^{xt}}{e^t-1} = \frac{e^{xt}}{1-\frac{1+t-e^t}{t}} = \sum_{n=0}^{\infty} B_n(x)\frac{t^n}{n!} \quad \text{and} \tag{4.246}$$

$$\frac{2e^{xt}}{e^t+1} = \frac{e^{xt}}{1-\frac{1-e^t}{2}} = \sum_{n=0}^{\infty} E_n(x)\frac{t^n}{n!}. \tag{4.247}$$

Hence, $\sum_{n=0}^{\infty} B_n(x)\frac{t^n}{n!}$ is the generating function of the row sums (i.e., the sum function) of the exponential Riordan array $(e^{xt}, (1+t-e^t)/t)$. The corresponding generating function of the alternating sums (i.e., the alternating sum function) is

$$\frac{e^{xt}}{1+\frac{1+t-e^t}{t}} = \sum_{n=0}^{\infty} \tilde{B}_n(x)\frac{t^n}{n!}, \tag{4.248}$$

where $(\tilde{B}_n)_{n\geq 0}$ is called conjugate Bernoulli polynomial sequence, and its first few terms can be represented as

$$\begin{aligned}
\tilde{B}_0(x) &= 1,\\
\tilde{B}_1(x) &= x+\frac{1}{2},\\
\tilde{B}_2(x) &= x^2+x+\frac{5}{6},\\
\tilde{B}_3(x) &= x^3+\frac{3}{2}x^2+\frac{5}{2}x+2,\\
\tilde{B}_4(x) &= x^4+2x^3+5x^2+8x+\frac{191}{30},\\
\tilde{B}_5(x) &= x^5+\frac{5}{2}x^4+\frac{25}{3}x^3+20x^2+\frac{191}{6}x+\frac{67}{3}, \textit{etc.}
\end{aligned}$$

Similarly, $\sum_{n=0}^{\infty} E_n(x)\frac{t^n}{n!}$ is the generating function of the sums of the exponential Riordan array $(e^{xt}, (1-e^t)/2)$. The corresponding generating function of the alternating sums is

$$\frac{e^{xt}}{1-\frac{1-e^t}{2}} = \sum_{n=0}^{\infty} \tilde{E}_n(x)\frac{t^n}{n!}, \tag{4.249}$$

where $(\tilde{E}_n)_{n\geq 0}$ is called the conjugate Euler polynomial sequence, and its first few terms are

$$\begin{aligned}
\tilde{E}_0(x) &= 1,\\
\tilde{E}_1(x) &= x+\frac{1}{2},\\
\tilde{E}_2(x) &= x^2+x+1,\\
\tilde{E}_3(x) &= x^3+\frac{3}{2}x^2+3x+\frac{11}{4},\\
\tilde{E}_4(x) &= x^4+2x^3+6x^2+11x+10,\\
\tilde{E}_5(x) &= x^5+\frac{5}{2}x^4+10x^3+\frac{55}{2}x^2+50x+\frac{91}{2}, \textit{etc.}
\end{aligned}$$

Some properties of the conjugate Bernoulli and conjugate Euler polynomial sequences similar as those of Bernoulli and Euler polynomial sequences can be found easily such as

$$\tilde{B}_n'(x) = n\tilde{B}_{n-1}(x) \quad \text{and} \quad \tilde{E}_n'(x) = n\tilde{E}_{n-1}(x).$$

In addition, from the definitions of sum functions R^+ and R^-, we may find the relationship between Bernoulli polynomial sequence and $(\tilde{B}_n(x))_{n\geq 0}$ and the relationship between Euler polynomial sequence and $(\tilde{E}_n)_{n\geq 0}$. From the first equation of (4.230), there holds

$$e^{xt}\left(\sum_{n\geq 0} B_n(x)\frac{t^n}{n!} + \sum_{n\geq 0}\tilde{B}_n(x)\frac{t^n}{n!}\right)$$

$$= \sum_{k\geq 0}\frac{(xt)^k}{k!}\left(\sum_{n\geq 0}B_n(x)\frac{t^n}{n!}+\sum_{n\geq 0}\tilde{B}_n(x)\frac{t^n}{n!}\right)$$
$$= 2\sum_{n\geq 0}\sum_{k\geq 0}B_n(x)\tilde{B}_k(x)\frac{t^{n+k}}{n!k!},$$

which implies

$$\sum_{n\geq 0}\left(\sum_{k=0}^{n}\binom{n}{k}x^kB_{n-k}(x)\right)\frac{t^n}{n!}+\sum_{n\geq 0}\left(\sum_{k=0}^{n}\binom{n}{k}x^k\tilde{B}_{n-k}(x)\right)\frac{t^n}{n!}$$
$$= 2\sum_{n\geq 0}\left(\sum_{k=0}^{n}\binom{n}{k}B_{n-k}(x)\tilde{B}_k(x)\right)\frac{t^n}{n!}.$$

Comparing the coefficients on the two sides of the above equation, we have the identities

$$\sum_{k=0}^{n}\binom{n}{k}x^k\left(B_{n-k}(x)+\tilde{B}_{n-k}(x)\right)=2\sum_{k=0}^{n}\binom{n}{k}B_{n-k}(x)\tilde{B}_k(x) \tag{4.250}$$

for all $n\geq 0$.

Similarly, applying the first equation of (4.230) to Euler polynomial sequence and $(\tilde{E}_n(x))_{n\geq 0}$, we have

$$\sum_{k=0}^{n}\binom{n}{k}x^k\left(E_{n-k}(x)+\tilde{E}_{n-k}(x)\right)=2\sum_{k=0}^{n}\binom{n}{k}E_{n-k}(x)\tilde{E}_k(x) \tag{4.251}$$

for all $n\geq 0$.

From the second equation of (4.230), there holds

$$(1+t-e^t)\left(\sum_{n\geq 0}B_n(x)\frac{t^n}{n!}+\sum_{n\geq 0}\tilde{B}_n(x)\frac{t^n}{n!}\right)=t\left(\sum_{n\geq 0}B_n(x)\frac{t^n}{n!}-\sum_{n\geq 0}\tilde{B}_n(x)\frac{t^n}{n!}\right),$$

or equivalently,

$$\sum_{n\geq 0}\left(B_n(x)+\tilde{B}_n(x)\right)\frac{t^n}{n!}+\sum_{n\geq 1}n\left(B_{n-1}(x)+\tilde{B}_{n-1}(x)\right)\frac{t^n}{n!}$$
$$-\sum_{n\geq 0}\left(\sum_{k=0}^{n}\binom{n}{k}\left(B_{n-k}(x)+\tilde{B}_{n-k}(x)\right)\right)\frac{t^n}{n!}$$
$$=\sum_{n\geq 0}n\left(B_{n-1}(x)-\tilde{B}_{n-1}(x)\right)\frac{t^n}{n!}.$$

Comparing the coefficients on both sides of the above equation yields

$$2n\tilde{B}_{n-1}(x) = \sum_{k=1}^{n} \binom{n}{k} \left(B_{n-k}(x) + \tilde{B}_{n-k}(x)\right) \tag{4.252}$$

for all $n \geq 1$.

Similarly, we may applying the second equation of (4.230) to the Euler polynomial sequence and $(\tilde{E}_n(x))_{n\geq 0}$ to obtain the identity

$$2\tilde{E}_n(x) = 2E_n(x) + \sum_{k=1}^{n} \binom{n}{k} \left(E_{n-k}(x) + \tilde{E}_{n-k}(x)\right) \tag{4.253}$$

for all $n \geq 1$. More results on the conjugate Bernoulli polynomials and conjugate Euler polynomials can be found in [92].

4.3.4 Identities generated by using improper or non-regular Riordan arrays

The idea on the various row sums of Riordan arrays shown in the previous subsection can be extended to an improper Riordan array (g, f) with $f(0) \neq 0$, or an non-regular Riordan array (g, f) with $f(0) = f'(0) = 0$. The FTRA, $(g, f)h = gh(f)$, remains valid for the improper Riordan array (g, f) with $f(0) \neq 0$, under more restrictive conditions such as when the sequence $(h_k)_{k\geq 0}$ is actually finite and hence its generating function $h(t)$ reduces to a polynomial. For instance, for a Riordan array (g, f), (g, tf) is an non-regular Riordan array because $[t]tf = 0$. The rows of (g, tf) are the diagonals of (g, f). Hence, from (4.225) we have a summation rule

$$\sum_{k=0}^{[n/2]} d_{n-k,k} = [t^n]\frac{g(t)}{1 - tf(t)} \quad \text{(diagonal sums).} \tag{4.254}$$

The diagonal sums can be extended to the slope k row sums of a Riordan array $(g(t), f(t)) = (d_{n,k})_{n,k\geq 0}$ defined by

$$S_n^k := \sum_{j=0}^{[n/(k+1)]} d_{n-k\,j,j} \tag{4.255}$$

for $n \geq 0$, i.e.,

$$S_n^k = d_{n,0} + d_{n-k,1} + d_{n-2k,2} + \cdots.$$

The slop k row sums of (g, f) are the row sums of the non-regular Riordan array $(g, t^k f)$. Particularly, if $k = 1$, the slope 1 row sums of (g, f) are its diagonal sums and the row sums of the non-regular Riordan array (g, tf). For an example, the slope 1 sums of the Pascal array

$$\begin{bmatrix} 1 & 0 & 0 & 0 & 0 & 0 & \\ 1 & 1 & 0 & 0 & 0 & 0 & \\ 1 & 2 & 1 & 0 & 0 & 0 & \\ 1 & 3 & 3 & 1 & 0 & 0 & \cdots \\ 1 & 4 & 6 & 4 & 1 & 0 & \\ 1 & 5 & 10 & 10 & 5 & 1 & \\ & & & \cdots & & & \end{bmatrix}$$

are

$$\begin{aligned} &1\\ &1\\ &1+1=2\\ &1+2=3\\ &1+3+1=5\\ &1+4+3=8\\ &1+5+6+1=13,\cdots, \end{aligned}$$

which give the Fibonacci number sequence. In general, for the diagonal sums of $(1/(1-t), t/(1-t))$ or the row sums of $(1/(1-t), t^2/(1-t))$ we have

$$\sum_{k=0}^{\infty}\binom{n-k}{k} = [t^n]\frac{1}{1-t}\frac{1}{1-t^2/(1-t)} = [t^n]\frac{1}{1-t-t^2} = F_n.$$

We now present an alternative view of R_k^+ in the following theorem.

Theorem 4.3.19 *Let S_n^k be the slope k row sums defined by (4.255). Then their ordinary generating function $\sum_{n\geq 0} S_n^k t^n$, denoted by $R_k^+(t)$, is the sum function of the non-regular Riordan array $(g(t), t^k f(t))$, i.e.,*

$$R_k^+ \equiv R_k^+(t) = \frac{g}{1-t^k f}. \tag{4.256}$$

If $(g(t), f(t))$ is a Bell-type Riordan array with $f = tg$, then for a given R_k^+ there exists a unique

$$g \equiv g(t) = \frac{R_k^+}{1+t^{k+1}R_k^+} \tag{4.257}$$

such that R_k^+ is the sum function of $(g, t^k f)$ or the slope k sum function of (g, tg).

Proof. Since

$$\sum_{j\geq 0} d_{n-kj,j} = \sum_{j\geq 0}[t^{n-kj}]gf^j = [t^n]\sum_{j\geq 0} g\left(t^k f\right)^j$$

and

$$d_{n-k\,j,j} = 0$$

when $j > n/(k+1)$, we immediately obtain

$$R_k^+(t) = \sum_{n\geq 0} S_n^k t^n = \sum_{n\geq 0} [t^n] \left(\sum_{j\geq 0} g\left(t^k f\right)^j \right) t^n$$
$$= \sum_{j\geq 0} g\left(t^k f\right)^j = \frac{g}{1-t^k f}.$$

If (g, f) is a Bell-type Riordan array and its slope k sum function R_k^+ is given, then $f = tg$ and from (4.256) there holds

$$\frac{g}{1-t^k(tg)} = R_k^+,$$

which has the unique solution g as

$$g = \frac{R_k^+}{1+t^{k+1}R_k^+}.$$

We complete the proof.

■

Example 4.3.20 *As an example, let $k = 1$ and $R_1^+ = 1/(1-t-t^2)$, the generating function of Fibonacci sequence, then from (4.257) we have $g = 1/(1-t)$. Hence, the Pascal array*

$$\left(\frac{1}{1-t}, \frac{t}{1-t}\right)$$

is the unique Riordan array that has slope 1 sum function $R_1^+ = \frac{1}{1-t-t^2}$.

From the above various row sums, one may see the summation rule (4.225) makes an important role in the construction of identities related to Riordan arrays. The key in using this rule to find the sum $\sum_{k\geq 0} d_{n,k}h_k$ is to recognize the generating function of the sequence $(h_k)_{k\geq 0}$ and the generating functions g and f satisfying $d_{n,k} = [t^n]gf^k$. Theorem 2.1 of [201] gives a basic theorem on this topic, which can be represented as follows.

Theorem 4.3.21 *Let $\binom{n+ak}{m+bk}$ be the (n,k) entry with a parameter m of an infinity array denoted by $(\binom{n+ak}{m+bk})_{n,k\geq 0}$. If $b > a$ and $b - a$ is an integer, then $D = (\binom{n+ak}{m+bk})_{n,k\geq 0}$ is a Riordan array. If $b < 0$ is an integer, then $\widehat{D} = (\binom{n+ak}{m+bk})_{n,k\geq 0}$ is a Riordan array. Furthermore,*

$$D = \left(\frac{t^m}{(1-t)^{m+1}}, \frac{t^{b-a}}{(1-t)^b}\right) \quad \text{and} \quad \widehat{D} = \left((1+t)^n, \frac{t^{-b}}{(1+t)^{-a}}\right). \tag{4.258}$$

Proof. From the well-known properties of binomial coefficients, for $b > a$ and $b - a$ is an integer we have

$$\begin{aligned}\binom{n+ak}{m+bk} &= \binom{n+ak}{n-m+ak-bk}\\ &= \binom{-n-ak+n-m+ak-bk-1}{n-m+ak-bk}(-1)^{n-m+ak-bk}\\ &= \binom{-m-bk-1}{(n-m)+(a-b)k}(-1)^{n-m+ak-bk}\\ &= [t^{n-m+ak-bk}]\frac{1}{(1-t)^{m+1+bk}} = [t^n]\frac{t^m}{(1-t)^{m+1}}\left(\frac{t^{b-a}}{(1-t)^b}\right)^k.\end{aligned}$$

For an integer $b < 0$ we find

$$\binom{n+ak}{m+bk} = [t^{m+bk}](1+t)^{n+ak} = [t^m](1+t)^n(t^{-b}(1+t)^a)^k.$$

Thus, the theorem is a consequence of Newton's rule.

■

For $m = a = 0$ and $b = 1$, the Riordan array is the Pascal triangle. For $b = a$, D is an improper Riordan array, while so $\widehat{D}$ is for $b = 0$.

Example 4.3.22 *We use Theorem 4.3.19 for $a = -1$ and $b = 0$ and FTRA to find the sum*

$$\sum_{k\geq 0}\binom{n-k}{m} = [t^n]\frac{t^m}{(1-t)^{m+1}}\frac{1}{1-t} = [t^{n-m}]\frac{1}{(1-t)^{m+2}} = \binom{n+1}{m+1}.$$

Since

$$\binom{n+k}{m+2k} = [t^n]\frac{t^m}{(1-t)^{m+1}}\left(\frac{t}{(1-t)^2}\right)^k$$

and

$$[t^k]C(-t) = [t^k]\frac{1-\sqrt{1+4t}}{-2t} = \frac{(-1)^k}{k+1}\binom{2k}{k},$$

we use FTRA to have

$$\begin{aligned}&\sum_{k\geq 0}\binom{n+k}{m+2k}\binom{2k}{k}\frac{(-1)^k}{k+1} = [t^n]\frac{t^m}{(1-t)^{m+1}}C\left(-\frac{t}{(1-t)^2}\right)\\ =\ & [t^{n-m}]\frac{1}{(1-t)^{m+1}}\frac{\sqrt{1+4\frac{t}{(1-t)^2}}-1}{2\frac{t}{(1-t)^2}}\\ =\ & [t^{n-m}]\frac{1}{(1-t)^m} = \binom{n-1}{m-1}.\end{aligned}$$

Similarly, from $\binom{n}{k} = [t^n](1/(1-t))(t/(1-t))^k$ *and the Stirling number of the second kind*

$$\left\{ {k \atop m} \right\} = [t^k] \frac{t^m}{(1-t)(1-2t)\cdots(1-mt)},$$

we have

$$\sum_{k\geq 0} \binom{n}{k} \left\{ {k \atop m} \right\} = [t^n] \frac{1}{1-t} \frac{(t/(1-t))^m}{(1-t/(1-t))(1-2t/(1-t))\cdots(1-mt/(1-t))}$$
$$= [t^{n+1}] \frac{t^{m+1}}{(1-t)(1-2t)\cdots(1-(m+1)t)} = \left\{ {n+1 \atop m+1} \right\}.$$

In the sum $\sum_{k\geq 0} \binom{n+k}{k}\binom{n}{k} k(-1)^{k-1}$ *we have that the generating function of the sequence* $(\binom{n}{k} k(-1)^{k-1})_{k\geq 0}$ *is* $h(t) = nt(1-t)^{n-1}$*, which is a polynomial. Hence, the FTRA remains valid for the improper Riordan array* $((1+t)^n, 1+t)$*, where*

$$[t^n](1+t)^n(1+t)^k = \binom{n+k}{n} = \binom{n+k}{k}.$$

Hence, we have

$$\sum_{k\geq 0} (-1)^{k-1} k \binom{n}{k} \binom{n+k}{k} = [t^n](1+t)^n h(1+t)$$
$$= n[t^n](1+t)^{n+1}(-t)^{n-1}$$
$$= (-1)^{n-1} n [t](1+t)^{n+1} = (-1)^{n-1} n(n+1).$$

Let $h(t) = \sum_{k\geq 0} h_k t^k$ and

$$H(t) = \sum_{k\geq 0} \frac{h_k}{k} t^k = \int \frac{h(t)-h(0)}{t} dt$$

with $H(0) = 0$. It is clear that

$$\binom{n-k}{k} = \frac{n-k}{k} \binom{n-k-1}{k-1} = \frac{n-k}{k} [t^n] \left(\frac{t^2}{1-t} \right),$$

except for $k = 0$, when the LHS of the last equation is 1 and the RHS is undefined. Hence, by FTRA

$$\sum_{k\geq 0} \frac{n}{n-k} \binom{n-k}{k} h_k = h_0 + n \sum_{k=1}^{\infty} \binom{n-k-1}{k-1} \frac{h_k}{k} = h_0 + n[t^n] H\left(\frac{t^2}{1-t} \right). \tag{4.259}$$

Formula (4.259) gives an immediately proof of Hardy's identity:

$$\begin{aligned}\sum_{k\geq 0}(-1)^k\frac{n}{n-k}\binom{n-k}{k} &=[t^n]\log\frac{1+t}{1+t^3}\\ &=[t^n]\left(\log\frac{1}{1+t^3}-\log\frac{1}{1+t}\right)\\ &=\begin{cases}(-1)^n 2/n, & if\ 3|n,\\ (-1)^{n-1}/n, & otherwise.\end{cases}\end{aligned}$$

Similarly, one may prove a generalization of Hardy's identity

$$\sum_{k\geq 0}\frac{n}{n-k}\binom{n-k}{k}(a+b)^{n-2k}(-ab)^k = a^n+b^n, \tag{4.260}$$

or equivalently,

$$\sum_{k\geq 0}\frac{n}{n-k}\binom{n-k}{k}x^{n-2k}(-1)^k = \frac{(x+\sqrt{x^2-4})^n+(x-\sqrt{x^2-4})^n}{2^n},$$

where $a=(x+\sqrt{x^2-4})/2$ and $b=(x-\sqrt{x^2-4})/2$.

Let $d(t)=\sum_{k=0}^{\infty}d_{2k}t^{2k}$, $h_1(t)=\sum_{k=0}^{\infty}h_{1,2k+1}t^{2k+1}$, and $h_2(t)=\sum_{k=0}^{\infty}h_{2,2k+1}t^{2k+1}$. Then the double Riordan matrix in terms of $d(t)$, $h_1(t)$, and $h_2(t)$, denoted by $(d;h_1,h_2)$, is defined by the generating function of its columns as (cf. [49, 93])

$$(d,\ dh_1,\ dh_1h_2,\ dh_1^2h_2,\ dh_1^2h_2^2,\ldots).$$

There two cases of the *first fundamental theorem of double Riordan arrays (FTDRA)* (cf. [49]):

$$(d;h_1,h_2)A(t)=B(t), \tag{4.261}$$

where for $A(t)=\sum_{k\geq 0}a_{2k}t^{2k}$ and $A(t)=\sum_{k\geq 0}a_{2k+1}t^{2k+1}$, we have $B(t)=dA(\sqrt{h_1h_2})$ and $B(t)=d\sqrt{h_1/h_2}A(\sqrt{h_1h_2})$, respectively. Based on the fundamental theorem of double Riordan arrays, we may define a multiplication of two double Riordan arrays as

$$(d;h_1,h_2)(g;f_1,f_2)=(dg(\sqrt{h_1h_2});\sqrt{h_1/h_2}f_1(\sqrt{h_1h_2}),\sqrt{h_2/h_1}f_2(\sqrt{h_1h_2})), \tag{4.262}$$

where $g(t)=\sum_{k=0}^{\infty}g_{2k}t^{2k}$, $f_1(t)=\sum_{k=0}^{\infty}f_{1,2k+1}t^{2k+1}$, and $f_2(t)=\sum_{k=0}^{\infty}f_{2,2k+1}\ t^{2k+1}$. The collection of all double Riordan arrays associated with the multiplication defined above forms a group, which is called double Riordan group and is denoted by $\mathcal{DR}$, where the identity of the group is $(1;t,t)$.

Let n (or m) be a variable, and let m (resp. n), a, and b be parameters. Then the following result may provides many pairs of sums.

Proposition 4.3.23 *Suppose $b > a$, $c > d$ and $b - a$ and $c - d$ are integers, then the double Riordan array (or improper double Riordan array when $m \neq 0$) $(d_{n,k})_{n,k\geq 0} = (t^{2m}/(1-t^2)^{m+1}, t^{2b-2a+1}/(1-t^2)^b, t^{2c-2d+1}/(1-t^2)^d)$ possesses its entries as follows:*

$$d_{2n,2k} = \binom{n + (a + c - 1)k}{m + (b + d)k} \quad \text{and} \quad d_{2n+1,2k+1} = \binom{n + a(k + 1) + (c - 1)k}{m + b(k + 1) + dk} \tag{4.263}$$

for $n, k = 0, 1, , 2, \ldots$, and 0 otherwise. Particularly, if $b = a = 0$, then

$$d_{2n,2k} = d_{2n+1,2k+1} = \binom{n + (c - 1)k}{m + dk}$$

Proof. We first give $d_{2n,2k}$ of (4.263). From the definition of double Riordan array, we have

$$\begin{aligned}
d_{2n,2k} &= [t^{2n}]\frac{t^{2m}}{(1 - t^2)^{m+1}}\left(\frac{t^{2b-2a+2d-2c+2}}{(1 - t^2)^{b+d}}\right)^k \\
&= [t^{2n-2m+2(a-b+c-d-1)k}]\frac{1}{(1 - t^2)^{m+1+(b+d)k}} \\
&= [t^{2n-2m+2(a-b+c-d-1)k}]\sum_{j\geq 0}\binom{-m - 1 - (b + d)k}{j}(-t^2)^j \\
&= \binom{-m - 1 - (b + d)k}{n - m + (a - b + c - d - 1)k}(-1)^{n-m+(a-b+c-d-1)k} \\
&= \binom{n + (a + c - 1)k}{n - m + (a - b + c - d - 1)k} = \binom{n + (a + c - 1)k}{m + (b + d)k}.
\end{aligned}$$

We now derive $d_{2n+1,2k+1}$ of (4.263) using a similar argument. From the definition of double Riordan array,

$$\begin{aligned}
d_{2n+1,2k+1} &= [t^{2n+1}]\frac{t^{2m}}{(1 - t^2)^{m+1}}\frac{t^{2b-2a+1}}{(1 - t^2)^b}\left(\frac{t^{2b-2a+2d-2c+2}}{(1 - t^2)^{b+d}}\right)^k \\
&= [t^{2n-2m+2a-2b+2(a-b+c-d-1)k}]\frac{1}{(1 - t^2)^{m+1-b+(b+d)k}} \\
&= [t^{2n-2m+2a-2b+2(a-b+c-d-1)k}]\sum_{j\geq 0}\binom{-m - 1 - b - (b + d)k}{j}(-t^2)^j \\
&= \binom{-m - 1 - b - (b + d)k}{n - m + a - b + (a - b + c - d - 1)k}(-1)^{n-m+a-b+(a-b+c-d-1)k} \\
&= \binom{n + a + (a + c - 1)k}{n - m + a - b + (a - b + c - d - 1)k} = \binom{n + a(k + 1) + (c - 1)k}{m + b(k + 1) + dk},
\end{aligned}$$

which completes the proof.

Example 4.3.24 *Suppose $a = b = c = d = 1$, Proposition 4.3.23 gives the following entries of $(t^{2m}/(1-t^2)^{m+1}, t/(1-t^2), t/(1-t^2))$:*

$$d_{2n,2k} = \binom{n+k}{m+2k} \quad \text{and} \quad d_{2n+1,2k+1} = \binom{n+k+1}{m+2k+1}. \tag{4.264}$$

From the generating function of Catalan numbers

$$\sum_{k\geq 0} \frac{1}{k+1}\binom{2k}{k} t^k = \frac{1-\sqrt{1-4t}}{2t}$$

we have

$$\sum_{k\geq 0} \frac{(-1)^k}{k+1}\binom{2k}{k} t^{2k} = \frac{\sqrt{1+4t^2}-1}{2t^2} \quad \text{and} \quad \sum_{k\geq 0} \frac{(-1)^k}{k+1}\binom{2k}{k} t^{2k+1} = \frac{\sqrt{1+4t^2}-1}{2t}. \tag{4.265}$$

Hence,

$$\begin{aligned} &\left(\frac{t^{2m}}{(1-t^2)^{m+1}}, \frac{t}{1-t^2}, \frac{t}{1-t^2}\right) \frac{\sqrt{1+4t^2}-1}{2t^2} \\ = &\ \frac{t^{2m}}{(1-t^2)^{m+1}} \left.\frac{\sqrt{1+4y^2}-1}{2y^2}\right|_{y=t/(1-t^2)} \\ = &\ \frac{t^{2m}}{(1-t^2)^{m+1}} \frac{\frac{1+t^2}{1-t^2}-1}{\frac{2t^2}{(1-t^2)^2}} = \frac{t^{2m}}{(1-t^2)^m}. \end{aligned}$$

The above equations imply

$$\sum_{k\geq 0} d_{2n,2k} \frac{(-1)^k}{k+1}\binom{2k}{k} = [t^{2n}] \frac{t^{2m}}{(1-t^2)^m},$$

i.e.,

$$\sum_{k\geq 0} \frac{(-1)^k}{k+1}\binom{2k}{k}\binom{n+k}{m+2k} = \binom{n-1}{m-1},$$

4.3.5 Identities related to recursive sequences of order 2 and Girard-Waring identities

We now use the summation formula rules shown in the last part of the previous subsection to establish some identities. We start from the following consideration: What can be done by generalizing the ring of integers $\mathbb{Z}$ to the ring of polynomials $\mathbb{Z}[x]$ or $\mathbb{Z}[x, y]$ or $\mathbb{Z}[x_1, x_2, \cdots, x_n]$? There is a survey paper of Henry Gould giving the history on Girard-Waring identities in the rings of polynomials $\mathbb{Z}[x,y]$ and $\mathbb{Z}[x,y,z]$ that appeared in [75]. As an example, we consider the venerable identity of Girard-Waring.

$$\sum_{0\leq k\leq n/2}(-1)^k\binom{n-k}{k}(xy)^k(x+y)^{n-2k}=\frac{x^{n+1}-y^{n+1}}{x-y},\tag{4.266}$$

where $xy\neq 0$ and $x\neq y$. This can be put into standard triangular form by reversing and aerating the rows by replacing $\binom{n-k}{k}$ by $\binom{\frac{n+k}{2}}{\frac{n-k}{2}}$ when $n\equiv k\,(mod\ 2)$ and 0 otherwise.

We now look at the Riordan arrays $(1/(1+at^2), bt/(1+at^2))$ with $b^2\geq 4a>0$. Then, its entries

$$\begin{aligned} d_{n,k} &= [t^n]\frac{b^kt^k}{(1+at^2)^{k+1}}\\ &= [t^n]\sum_{j\geq 0}\binom{j+k}{j}(-at^2)^jb^kt^k\\ &= [t^n]\sum_{j\geq 0}\binom{j+k}{j}(-a)^jb^kt^{2j+k}\\ &= \begin{cases}\binom{\frac{n+k}{2}}{\frac{n-k}{2}}(-1)^{(n-k)/2}a^{(n-k)/2}b^k, & when\ n\equiv k\ (mod\ 2),\\ 0, & when\ n\not\equiv k\ (mod\ 2),\end{cases}\end{aligned}\tag{4.267}$$

where the last step is from that $j=(n-k)/2$ when $n\equiv k(mod\ 2)$. Hence, the row sum of $(1/(1+at^2), bt/(1+at^2))=(d_{n,k})_{n,k\geq 0}$ is

$$\begin{aligned}\sum_{k=0}^n d_{n,k} &= \sum_{\substack{0\leq k\leq n\\ n\equiv k\ (mod\ 2)}}\binom{\frac{n+k}{2}}{\frac{n-k}{2}}(-1)^{(n-k)/2}a^{(n-k)/2}b^k\\ &= \sum_{k=0}^{n/2}(-1)^k\binom{n-k}{k}a^kb^{n-2k},\end{aligned}\tag{4.268}$$

where the last step is from the transform $(n-k)/2\to k$ and the assumption of $n\equiv k\ (mod\ 2)$.

The generating function of the row sums of $(1/(1+at^2), bt/(1+at^2))$ can be represented as

$$\left(\frac{1}{1+at^2},\frac{bt}{1+at^2}\right)\frac{1}{1-t}=\frac{\frac{1}{1+at^2}}{1-\frac{bt}{1+at^2}}=\frac{1}{at^2-bt+1}.\tag{4.269}$$

Since $b^2\geq 4a>0$, we have

$$at^2-bt+1=a(t-t_1)(t-t_2),$$

where

$$t_1=\frac{b+\sqrt{b^2-4a}}{2a}\quad\text{and}\quad t_2=\frac{b-\sqrt{b^2-4a}}{2a}.\tag{4.270}$$

Hence, (4.269) can be written as

$$\begin{aligned}
&\left(\frac{1}{1+at^2}, \frac{bt}{1+at^2}\right)\frac{1}{1-t} = \frac{1}{a(t-t_1)(t-t_2)} \\
= &\ \frac{1}{a(t_1-t_2)}\left(\frac{1}{t-t_1} - \frac{1}{t-t_2}\right) \\
= &\ \frac{1}{at_2(t_1-t_2)}\frac{1}{1-\frac{t}{t_2}} - \frac{1}{at_1(t_1-t_2)}\frac{1}{1-\frac{t}{t_1}} \\
= &\ \frac{1}{at_2(t_1-t_2)}\sum_{n\geq 0}\left(\frac{t}{t_2}\right)^n - \frac{1}{at_1(t_1-t_2)}\sum_{n\geq 0}\left(\frac{t}{t_1}\right)^n \\
= &\ \sum_{n\geq 0}\frac{1}{a(t_1-t_2)}\left(\frac{1}{t_2^{n+1}} - \frac{1}{t_1^{n+1}}\right)t^n. \qquad (4.271)
\end{aligned}$$

From the last expression of the sum function and the row sums given in (4.268), we obtain the identity

$$\sum_{k=0}^{n/2}(-1)^k\binom{n-k}{k}a^k b^{n-2k} = \frac{1}{a(t_1-t_2)}\left(\frac{1}{t_2^{n+1}} - \frac{1}{t_1^{n+1}}\right), \qquad (4.272)$$

where t_1 and t_2 are given in (4.270).

Specifically, we consider the case of $a = xy$ and $b = x+y$, where $b^2 - 4a = (x-y)^2 \geq 0$. Then the corresponding Riordan array is

$$\left(\frac{1}{1+xyt^2}, \frac{(x+y)\,t}{1+xyt^2}\right) =$$

$$\begin{bmatrix}
1 & 0 & 0 & 0 & 0 & 0 \\
0 & 1\,(x+y) & 0 & 0 & 0 & 0 \\
-1\,(xy) & 0 & 1\,(x+y)^2 & 0 & 0 & 0 \\
0 & -2\,(xy)\,(x+y) & 0 & 1\,(x+y)^3 & 0 & 0 \\
(xy)^2 & 0 & -3\,(xy)\,(x+y)^2 & 0 & 1\,(x+y)^4 & 0 \\
0 & 3\,(xy)^2\,(x+y) & 0 & -4\,(xy)\,(x+y)^3 & 0 & 1\,(x+y)^5 \\
 & & & \cdots & &
\end{bmatrix}$$

Thus, Girard-Waring identity (4.266) holds in the ring of polynomials $\mathbb{Z}\,[x, y,]$ since it is a special case of (4.272) for $t_1 = 1/y$ and $t_2 = 1/x$.

Another special case is setting $a = 1$ and $b = 2x$, then, using (4.269), the row sum generating function of $(1/(1+t^2), 2xt/(1+t^2))$ is

$$\left(\frac{1}{1+t^2}, \frac{2xt}{1+t^2}\right)\frac{1}{1-t} = \frac{\frac{1}{1+t^2}}{1-\frac{2xt}{1+t^2}} = \frac{1}{t^2-2xt+1}.$$

Thus, the row sums of $(1/(1+t^2), 2xt/(1+t^2))$ are the Chebyshev polynomials of the second kind, $(U_n(x)) = (1, 2x, 4x^2-1, 8x^3-4x, \ldots)$. Let $x \geq 1$ or $x \leq -1$. Then the roots of $t^2 - 2xt + 1$ are

$$t_1 = x + \sqrt{x^2-1} \quad \text{and} \quad t_2 = x - \sqrt{x^2-1}.$$

Thus, $t_1^{-1} = t_2$, $t_2^{-1} = t_1$, and

$$U_n(x) = \frac{(x+\sqrt{x^2-1})^{n+1} - (x-\sqrt{x^2-1})^{n+1}}{2\sqrt{x^2-1}}.$$

Similarly, if we set $b = -2x$, $a = -1$ and 2, then the corresponding row sum sequences are the Pell polynomial sequence, $(P_{n+1}(x))_{n\geq 0}$, and the Fermat polynomial sequence of the first kind, $(F_{n+1}(x))_{n\geq 0}$, respectively:

$$P_{n+1}(x) = \frac{(x+\sqrt{x^2+1})^{n+1} - (x-\sqrt{x^2+1})^{n+1}}{2\sqrt{x^2+1}},$$
$$F_{n+1}(x) = \frac{(x+\sqrt{x^2-2})^{n+1} - (x-\sqrt{x^2-2})^{n+1}}{2\sqrt{x^2-2}}.$$

The above polynomials can be sorted into the class of the generalized Gegenbauer-Humbert polynomials (cf. for example, [72, 76], and [106]). Their expressions can be constructed by using the Binet formula (cf. [111]).

Identity (4.272) can be extended to the case of the ring of polynomials $\mathbb{Z}[x, y, z]$ or higher dimensional cases by using a similar argument.

Let polynomial $at^3 + ct^2 - bt + 1$ has three distinct roots t_1, t_2, and t_3. Then

$$\begin{aligned} & \frac{1}{at^3+ct^2-bt+1} \\ = & \frac{1}{a}\left(\frac{1}{(t-t_1)(t_1-t_2)(t_1-t_3)} + \frac{1}{(t-t_2)(t_2-t_1)(t_2-t_3)}\right. \\ & \left. + \frac{1}{(t-t_3)(t_3-t_1)(t_3-t_2)}\right). \end{aligned} \tag{4.273}$$

Considering Riordan array $(1/(1+ct^2+at^3), bt/(1+ct^2+at^3))$, if $c \neq 0$, its entries may be written as

$$\begin{aligned} d_{n,k} &= [t^n]\frac{b^k t^k}{(1+ct^2+at^3)^{k+1}} \\ &= [t^n]\sum_{j\geq 0}\binom{j+k}{j}(-1)^j(ct^2+at^3)^j b^k t^k \\ &= [t^n]\sum_{j\geq 0}\sum_{\ell=0}^{j}\binom{j+k}{j}\binom{j}{\ell}(-1)^j a^\ell b^k c^{j-\ell} t^{2j+k+\ell} \\ &= \begin{cases} \displaystyle\sum_{\substack{\ell\geq 0,\\ \ell\equiv n-k\ (mod\ 2)}} \binom{n+k-\ell}{\frac{n-k-\ell}{2}}\binom{\frac{n-k-\ell}{2}}{\ell}(-1)^{\frac{n-k-\ell}{2}} a^\ell b^k c^{\frac{n-k-3\ell}{2}}, & \\ \qquad\qquad when\ n\equiv k+\ell\ (mod\ 2), & \\ 0, & otherwise. \end{cases} \end{aligned}$$

Therefore, the row sums of the Riordan array $(1/(1+ct^2+at^3), bt/(1+ct^2+at^3))$ are

$$\sum_{k=0}^{n} d_{n,k} = \sum_{0\leq k\leq n} \sum_{\substack{\ell\geq 0,\\ \ell\equiv n-k\ (mod\ 2)}} \binom{n+k-\ell}{\frac{n-k-\ell}{2}}\binom{\frac{n-k-\ell}{2}}{\ell}(-1)^{\frac{n-k-\ell}{2}} a^\ell b^k c^{\frac{n-k-3\ell}{2}}. \tag{4.274}$$

If $c=0$, i.e., $t_1+t_2+t_3=0$, the entires of the Riordan array $(1/(1+at^3), bt/(1+at^3))$ can be represented as

$$\begin{aligned} d_{n,k} &= [t^n]\frac{b^k t^k}{(1+at^3)^{k+1}} \\ &= [t^n]\sum_{j\geq 0}\binom{j+k}{j}(-1)^j a^j b^k t^{3j+k} \\ &= \begin{cases} \binom{\frac{n+2k}{3}}{\frac{n-k}{3}}(-a)^{\frac{n-k}{3}} b^k, & when\ n\equiv k\ (mod\ 3), \\ 0, & otherwise. \end{cases} \end{aligned}$$

Therefore, the row sums of the Riordan array $(1/(1+at^3), bt/(1+at^3))$ are

$$\begin{aligned} \sum_{\substack{0\leq k\leq n,\\ k\equiv n\ (mod\ 3)}} d_{n,k} &= \sum_{\substack{0\leq k\leq n,\\ k\equiv n\ (mod\ 3)}} \binom{\frac{n+2k}{3}}{\frac{n-k}{3}}(-a)^{\frac{n-k}{3}} b^k \\ &= \sum_{0\leq k\leq n/3}\binom{n-2k}{k}(-a)^k b^{n-3k}. \end{aligned} \tag{4.275}$$

Using the FTRA, the generating function of the row sums of $(1/(1+at^3), bt/(1+at^3))$ can be written as

$$\left(\frac{1}{1+at^3}, \frac{bt}{1+at^3}\right)\frac{1}{1-t} = \frac{\frac{1}{1+at^3}}{1-\frac{bt}{1+at^3}} = \frac{1}{at^3-bt+1}.$$

If at^3-bt+1 has distinct roots t_1, t_2, and t_3 with $t_1+t_2+t_3=0$, then from the above equations and (4.273), there holds

$$\begin{aligned} &\left(\frac{1}{1+at^3}, \frac{bt}{1+at^3}\right)\frac{1}{1-t} \\ &= \frac{1}{a}\left(\frac{1}{(t-t_1)(t_1-t_2)(t_1-t_3)} + \frac{1}{(t-t_2)(t_2-t_1)(t_2-t_3)} + \frac{1}{(t-t_3)(t_3-t_1)(t_3-t_2)}\right) \\ &= \frac{1}{a}\sum_{n\geq 0}\left(\frac{-1}{(t_1-t_2)(t_1-t_3)t_1^{n+1}} + \frac{-1}{(t_2-t_1)(t_2-t_3)t_2^{n+1}} + \frac{-1}{(t_3-t_1)(t_3-t_2)t_3^{n+1}}\right)t^n. \end{aligned}$$

From the rightmost expression of the generating function of the row sums of $(1/(1+at^3), bt/(1+at^3))$ and the formula (4.275), we obtain the identity

$$\sum_{0\leq k\leq n/3}\binom{n-2k}{k}(-a)^k b^{n-3k}$$
$$=\frac{1}{a}\left(\frac{-1}{(t_1-t_2)(t_1-t_3)t_1^{n+1}}+\frac{-1}{(t_2-t_1)(t_2-t_3)t_2^{n+1}}+\frac{-1}{(t_3-t_1)(t_3-t_2)t_3^{n+1}}\right) \tag{4.276}$$

for all $at^3-bt+1=a(t-t_1)(t-t_2)(t-t_3)$ with $t_1+t_2+t_3=0$. Denote

$$t_1=\frac{1}{x},\quad t_2=\frac{1}{y},\quad \textit{and}\quad t_3=\frac{1}{z}$$

with $xyz\neq 0$ and $xy+yz+zx=0$. Then

$$at^3-bt+1=a(t-t_1)(t-t_2)(t-t_3)=-xyz\left(t-\frac{1}{x}\right)\left(t-\frac{1}{y}\right)\left(t-\frac{1}{z}\right),$$

where $a=-xyz$ and $b=x+y+z$. Additionally, (4.276) leads the Girard-Waring identity in the ring of polynomials $\mathbb{Z}[x,y,z]$ as follows:

$$\sum_{0\leq k\leq n/3}\binom{n-2k}{k}(xyz)^k(x+y+z)^{n-3k}$$
$$=\frac{x^{n+2}}{(x-y)(x-z)}+\frac{y^{n+2}}{(y-x)(y-z)}+\frac{z^{n+2}}{(z-x)(z-y)}, \tag{4.277}$$

where x,y,z are distinct with $xyz\neq 0$ and $xy+yz+zx=0$.

Similarly, we may have the Girard-Waring identity in the ring of polynomials $\mathbb{Z}[x_1,x_2,x_3,x_4]$:

$$\sum_{0\leq k\leq n/4}\binom{n-3k}{k}(-x_1x_2x_3x_4)^k(x_1+x_2+x_3+x_4)^{n-4k}$$
$$=\frac{x_1^{n+3}}{(x_1-x_2)(x_1-x_3)(x_1-x_4)}+\frac{x_2^{n+3}}{(x_2-x_1)(x_2-x_3)(x_2-x_4)}$$
$$+\frac{x_3^{n+3}}{(x_3-x_1)(x_3-x_2)(x_3-x_4)}+\frac{x_4^{n+3}}{(x_4-x_1)(x_4-x_2)(x_4-x_3)}, \tag{4.278}$$

where x_1,x_2,x_3, and x_4 are distinct real numbers satisfying

$$x_1x_2x_3x_4\neq 0,\quad \sum_{0\leq i<j<k\leq 4}x_ix_jx_k=0,\ \textit{and}\ \sum_{0\leq i<j\leq 4}x_ix_j=0.$$

Remark 4.3.25 *From the above condition for the identity (4.277), $z=-(xy)/(x+y)$, we may know that the identity holds when $x,y,z\neq 0$ and*

$x/y, y/z, z/x \neq -2$ and $-1/2$. From the above conditions for the identity (4.278), we also know that the identity holds when $x_1x_2x_3x_4 \neq 0$ and

$$
\begin{aligned}
x_3 &= \frac{-x_1^2x_2 - x_1x_2^2 - \sqrt{-3x_1^4x_2^2 - 2x_1^3x_2^3 - 3x_1^2x_2^4}}{2(x_1^2 + x_1x_2 + x_2^2)}, \\
x_4 &= \frac{1}{2(x_1^2 + x_1x_2 + x_2^2)^2}\left(-x_1x_2(x_1^3 + 2x_1^2x_2 + 2x_1x_2^2 + x_2^3)\right. \\
&\quad \left. +(x_1^2 + x_1x_2 + x_2^2)\sqrt{-3x_1^4x_2^2 - 2x_1^3x_2^3 - 3x_1^2x_2^4}\right).
\end{aligned}
$$

In general, we have the following result.

Theorem 4.3.26 *For integer $\ell \geq 2$, we have identities*

$$
\begin{aligned}
&\sum_{0 \leq k \leq n/\ell} \binom{n - (\ell - 1)k}{k} (-1)^{\ell+1} \left(\Pi_{i=1}^{\ell} x_i\right)^k \left(\sum_{i=1}^{\ell} x_i\right)^{n-\ell k} \\
&= \sum_{i=1}^{\ell} \frac{x_i^{n+\ell-1}}{\Pi_{j=1, j\neq i}^{\ell}(x_i - x_j)},
\end{aligned} \tag{4.279}
$$

where x_i $(1, 2, \ldots, \ell)$ are distinct and satisfy $\Pi_{i=1}^{\ell} x_i \neq 0$ and

$$
\sum_{0 \leq i_1 < i_2 < \cdots < i_{\ell-j} \leq \ell} x_{i_1} x_{i_2} \cdots x_{i_{\ell-j}} = 0
$$

for $j = 1, 2, \ldots, \ell - 2$. Obviously, when $\ell = 2, 3$, and 4, we obtain identities (4.266), (4.277), and (4.278), respectively.

4.3.6 Summation formulas related to Fuss-Catalan numbers

A *Dyck path* of length $2n$ is a path in the plane lattice $\mathbb{Z} \times \mathbb{Z}$ from the origin $(0, 0)$ to $(2n, 0)$ using the steps $(1, 1)$ and $(1, -1)$. The other requirement is that a path can never go below the x-axis. We refer to n as the semilength of the path. It is well-known that the number of Dyck paths of semilength n is the n-th Catalan number C_n. A partial Dyck path is a Dyck path without requiring that end point is on the x-axis. Hence, $\frac{k}{2n+k}\binom{2n+k}{n}$ is the number of the partial Dyck paths from $(0, 0)$ to $(2n + k, k)$.

For a positive integer m, an *m-Dyck path* is a path from the origin to $(mn, 0)$ using the steps $(1, 1)$ and $(1, 1 - m)$ and again not going below the x-axis. We refer to mn as the length of the path. A partial m-Dyck path is defined as an m-partial Dyck path. It is well-known that the number of m-Dyck paths of length mn is (cf. for example, [79, 109])

$$
F_m(n, 1) = \frac{1}{mn + 1}\binom{mn + 1}{n},
$$

which are the *Fuss-Catalan numbers*. For $m = 2$, the Fuss-Catalan numbers are the Catalan numbers $F_2(n,1) = c_n$. More generally, the Fuss-Catalan numbers are

$$F_m(n,r) := \frac{r}{mn+r}\binom{mn+r}{n},$$

which are named after N. I. Fuss and E. C. Catalan (cf. [64, 79, 149, 166, 176]). The Fuss-Catalan numbers have many combinatorial applications (cf. for example, [109]).

The generating function $F_m(t)$ of the mth order Fuss-Catalan numbers $(F_m(n,\ 1))_{n\geq 0}$ is called the generalized binomial series in [79]. Hence from Lambert's formula for the Taylor expansion of the powers of $F_m(t)$ (cf. P. 201 of [79]), we have

$$F_m^r \equiv F_m(t)^r = \sum_{n\geq 0} \frac{r}{mn+r}\binom{mn+r}{n} t^n \tag{4.280}$$

for all $r \in \mathbb{R}$. A closed form of F_m^r for $r \in \mathbb{N}$ is given by (4.182) in Example 4.3.3. For $r = 1$, $F_m(t)$ has the expression

$$F_m(t) = \sum_{k\geq 0} \frac{(mk)!}{(m-1)k+1)!}\frac{t^k}{k!} = \sum_{k\geq 0} \frac{1}{(m-1)k+1}\binom{mk}{k} t^k. \tag{4.281}$$

Example 4.3.3 gives

$$\begin{aligned} F_0(t) &= 1+t, \\ F_1(t) &= \sum_{k\geq 0} t^k = \frac{1}{1-t}, \\ F_2(t) &= \sum_{k\geq 0} \frac{1}{k+1}\binom{2k}{k} t^k = C(t). \end{aligned}$$

The key case (4.280) leads the following formula for $F_m(t)$, which has been established in Example 4.3.3:

$$F_m(t) = 1 + tF_m^m(t). \tag{4.282}$$

Equation (4.282) can also be used to define the function $F_m(t)$. Actually, formula (4.282) can be proved directly by using the definition (4.280) as follows.

$$\begin{aligned} 1 + tF_m^m(t) &= 1 + \sum_{n\geq 0} \frac{m}{mn+m}\binom{mn+m}{n} t^{n+1} \\ &= 1 + \sum_{n\geq 1} \frac{m}{mn}\binom{mn}{n-1} t^n \\ &= \sum_{n\geq 0} \frac{1}{mn+1}\binom{mn+1}{n} t^n = F_m(t). \end{aligned}$$

For the cases $m = 1$ and 2, $F_1 = 1/(1-t)$ and $F_2 = C(t)$ satisfy $1 + t/(1-t) = 1/(1-t)$ and $1 + tC(t)^2 = C(t)$, respectively. When $m = 3$, the Fuss-Catalan number sequence $(F_3(n,1))_n$ is the sequence $A001764$ (cf. [197]), $(1, 1, 3, 12, 55, 273, 1428, \ldots)$, which is called the *ternary number sequence*. The ternary numbers count the number of 3-Dyck paths or ternary paths. The generating function $T(t)$ of the ternary number sequence (T_n) is $T(t) = \sum_{n=0}^{\infty} T_n t^n$, where $T_n = \frac{1}{3n+1}\binom{3n+1}{n}$. It can be seen that the function $T(t)$ satisfies the equation $T(t) = 1 + tT(t)^3$.

Now, we assume that m is arbitrarily fixed. Unless there is a special need to indicate its subscript, we will represent F_m as F. We say a Riordan array (g, f) is an involution if $(g, f)^{-1} = (g, f)$, or (g, f) has order 2. Several important Riordan arrays such as the Pascal triangle, the RNA array, and the directed animal array are almost involutions in the sense that $(g(t), f(t))^{-1} = (g(-t), f(-t))$ or equivalently $(g(t), f(t))(1, -t)$ is an involution. We call such arrays pseudo involutions. The pseudo involution can be more "combinatorial" in the sense of having non-negative terms.

Theorem 4.3.27 *Let $f(t) = -tF(t)^r$. Then f is an involution, i.e., the compositional inverse of $f(t)$ satisfies $\bar{f}(t) = f(t)$ if and only if $r = 2m-1$. Therefore, (F^l, tF^r) is a pseudo (Riordan) involution, if and only if $r = 2m-1$.*

Proof. Let $f(t) = -tF(t)^r$, and let $\bar{f}$ be its compositional inverse. Using the Lagrange inversion formula(4.175), from (4.280) we have

$$[t^{n+1}]\bar{f}(t) = \frac{1}{n+1}[t^n]\left(\frac{t}{f(t)}\right)^{n+1}$$

$$\begin{aligned}
&= \frac{1}{n+1}[t^n]\left(-F(t)^{-r}\right)^{n+1} \\
&= (-1)^{n+1}\frac{-r}{mn - r(n+1)}\binom{mn - r(n+1)}{n} \\
&= (-1)^{n+1}\frac{-r}{-n(r-m) - r}\binom{-n(r-m) - r}{n} \\
&= \frac{-r}{n(r-m)+r}\binom{n(r-m+1)+r-1}{n} \\
&= (-r)\frac{(n(r-m+1)+r-1)!}{n!(n(r-m)+r)!}.
\end{aligned}$$

On the other hand, from (4.280)

$$\begin{aligned} & [t^{n+1}](-t)F(t)^r = -[t^n]F(t)^r \\ = & \frac{-r}{mn+r}\binom{mn+r}{n} \\ = & (-r)\frac{(mn+r-1)!}{n!(n(m-1)+r)!}. \end{aligned}$$

Hence, if $r = 2m-1$, then

$$[t^{n+1}]\bar{f}(t) = (-r)\frac{(mn+2m)!}{n!(n(m-1)+2m-1)!} = [t^{n+1}](-t)F(t)^{2m-1},$$

i.e.,

$$\bar{f}(t) = -tF(t)^{2m-1}.$$

Conversely, if the above result holds, then we must have

$$(-r)\frac{(n(r-m+1)+r-1)!}{n!(n(r-m)+r)!} = (-r)\frac{(mn+r-1)!}{n!(n(m-1)+r)!}$$

for all integers $n \geq 0$. When $n = 2$, the left-hand side and the right-hand side of the above equation are $3r-2m+1$ and $r+2m-1$, respectively, which implies that $3r-2m+1 = r-2m-1$, i.e., $r = 2m-1$.

Since

$$\begin{aligned} & \left(F^l, -tF^r\right)\left(F^l, -tF^r\right) \\ = & \left(F(t)^l\left(F\left(-tF(t)^r\right)\right)^l, tF(t)^r\left(F\left(-tF(t)^r\right)\right)^r\right), \end{aligned}$$

and

$$tF(t)^r\left(F\left(-tF(t)^r\right)\right)^r = t$$

if and only if $r = 2m-1$, we have

$$F(t)F\left(-tF(t)^r\right) = 1$$

and

$$\left(F^l, -tF^r\right)\left(F^l, -tF^r\right) = I$$

if and only if $r = 2m-1$. This completes the proof of the theorem.

■

From Theorems 4.3.1 and 4.3.4, we immediately know that the generating functions of the A-sequence and the Z-sequence of $(F^\ell, -tF^{2m-1})$ are (cf. Exercise 4.13)

$$A = -F^{1-2m}(t) \tag{4.283}$$

and

$$Z = \frac{1-F(t)^\ell}{-tF^{2m-1}(t)}. \tag{4.284}$$

Theorem 4.3.28 *The* (n,k) *entry* $d_{n,k}$ *of* (F^ℓ, tF^s) *is given by*

$$d_{n,k} = \frac{sk+\ell}{m(n-k)+sk+\ell}\binom{m(n-k)+sk+\ell}{n-k}, \tag{4.285}$$

and $d_{n,k}$ *counts the number of partial* m*-Dyck paths from the origin to* $(mn - mk + sk + \ell - 1, sk + \ell - 1)$.

Particularly, if $s = m$, *then the* (n,k) *entry of* (F^ℓ, tF^m) *is*

$$d_{n,k} = \frac{mk+\ell}{mn+\ell}\binom{mn+\ell}{n-k}. \tag{4.286}$$

Proof. From (4.280) we have that the (n,k) entry $d_{n,k}$ of (F^ℓ, tF^s) is

$$\begin{aligned} d_{n,k} =&[t^n]F^\ell(tF^s)^k = [t^{n-k}]F^{sk+\ell} \\ =&[t^{n-k}]\sum_{j=0}^{\infty}\frac{sk+\ell}{mj+sk+\ell}\binom{mj+sk+\ell}{j}t^j \\ =&\frac{sk+\ell}{m(n-k)+sk+\ell}\binom{m(n-k)+sk+\ell}{n-k}. \end{aligned}$$

The combinatorial interpretation follows.

■

We now come to a collection of odd and interesting facts. For instance $(C, tC^2)\frac{1+t}{1-3t+t^2} = 1/(1-5t)$ which links the Catalan numbers, the alternate Lucas numbers and the powers of 5. The common theme is that we use arrays in the Fuss-Catalan family, input sequences given by second order linear recursions and end up with geometric sequences or other second order sequences. In other words, we may consider the sequence transformations by using Fuss-Catalan matrices or other Riordan arrays. First we establish the following result.

Theorem 4.3.29 *Let* $F_m(t)$ *be defined by* (4.282), *and let* $f_1(t) = tF_m^r(t)$ *and* $f_2(t) = -tF_m^r(t)$, $r \in \mathbb{Z}$. *Denote by* $\bar{f}_1(t)$ *and* $\bar{f}_2(t)$, *the compositional inverse of* f_1 *and* f_2, *respectively. Then*

$$\bar{f}_1(t) = tF_{m-r}(t)^{-r}, \tag{4.287}$$

$$\bar{f}_2(t) = -tF_{m-r}(-t)^{-r} = \bar{f}_1(-t), \tag{4.288}$$

$$F_{m-r}(tF_m(t)^r) = F_m(t), \tag{4.289}$$

$$(F_m(t)^\ell, tF_m(t)^r)(F_{m-r}(t)^p, tF_{m-r}(t)^q) = (F_m(t)^{\ell+p}, tF_m(t)^{r+q}), \tag{4.290}$$

$$(F_m(t)^\ell, tF_m(t)^r)^{-1} = (F_{m-r}(t)^{-\ell}, tF_{m-r}(t)^{-r}), \tag{4.291}$$

$$(F_m(t)^\ell, tF_m(t)^m)^{-1} = \left(\frac{1}{(1+t)^\ell}, \frac{t}{(1+t)^m}\right). \tag{4.292}$$

Proof. Similar to the initial steps of the proof of Theorem 4.3.27, we use the Lagrange inversion formula(4.178) and expression (4.280) to obtain

$$
\begin{aligned}
[t^{n+1}]\bar{f}_1(t) &= \frac{1}{n+1}[t^n]\left(\frac{t}{f_1(t)}\right)^{n+1} \\
&= \frac{1}{n+1}[t^n]\left(F_m(t)^{-r}\right)^{n+1} \\
&= \frac{-r}{mn-r(n+1)}\binom{mn-r(n+1)}{n} \\
&= \frac{-r}{(m-r)n-r}\binom{(m-r)n-r}{n} = [t^n]F_{m-r}(t)^{-r},
\end{aligned}
$$

which implies (4.287). Noting $f_2(t) = -tF_m(t)^r = -f_1(t)$, we have

$$t = \bar{f}_2(f_2(t)) = \bar{f}_2(-f_1(t)),$$

which implies $\bar{f}_2(t) = \bar{f}_1(-t)$. Hence, we get (4.288). Since

$$t = \bar{f}_1(f_1(t)) = tF_m(t)^r F_{m-r}(tF_m(t)^r)^{-r},$$

we have

$$F_m(t)^r = F_{m-r}(tF_m(t)^r)^r,$$

which is equivalent to (4.289). Hence,

$$
\begin{aligned}
&(F_m(t)^\ell, tF_m(t)^r)(F_{m-r}(t)^p, tF_{m-r}(t)^q) \\
=&(F_m(t)^\ell F_{m-r}(tF_m(t)^r)^p, tF_m(t)^r F_{m-r}(tF_m(t)^r)^q) \\
=&(F_m^\ell F_m^p, tF_m^r F_m^q),
\end{aligned}
$$

which gives (4.290). Since

$$(F_m(t)^\ell, tF_m(t)^r)(F_{m-r}(t)^{-\ell}, tF_{m-r}(t)^{-r}) = (F_m^{\ell-\ell}, tF_m^{r-r}) = (1,t),$$

we obtain (4.291). For $F_0(t) = 1/(1+t)$, (4.292) is obtained because it is a special case of (4.291) for $r = m$.

■

Now we can take any target sequence such as $(1, k, k^2, k^3, \cdots)$ with its generating function $\frac{1}{1-kt}$, compute $\left(\frac{1}{(1+t)^l}, \frac{t}{(1+t)^m}\right)\frac{1}{1-kt}$, and have the generating function for the sequence to input.

Example 4.3.30 *Here are a few examples. If* $m = 2$ *then*

$$\left(C^2, tC^2\right) M(t) = \frac{1}{1-5t},$$

i.e.,

$$\begin{bmatrix} 1 & 0 & 0 & 0 & 0 & \\ 2 & 1 & 0 & 0 & 0 & \\ 5 & 4 & 1 & 0 & 0 & \cdots \\ 14 & 14 & 6 & 1 & 0 & \\ 42 & 48 & 27 & 8 & 1 & \\ & & \cdots & & & \end{bmatrix} \begin{bmatrix} 1 \\ 3 \\ 8 \\ 21 \\ 55 \\ \vdots \end{bmatrix} = \begin{bmatrix} 1 \\ 5 \\ 25 \\ 125 \\ 625 \\ \vdots \end{bmatrix}.$$

We suspect that the mystery sequence and generating function, $M(t)$, represents the even Fibonacci numbers, viz.

$$M(t) = \frac{1}{1-3t+t^2} = \sum_{n=0}^{\infty} F_{2n+2} t^n \tag{4.293}$$

while

$$\frac{1-t}{1-3t+t^2} = \sum_{n=0}^{\infty} F_{2n+1} t^n \tag{4.294}$$

represents the odd Fibonacci numbers.

Since (4.292) *shows that*

$$(C^2, tC^2)^{-1} = \left(\frac{1}{(1+t)^2}, \frac{t}{(1+t)^2} \right),$$

which derives the following from $(C^2, tC^2)M(t) = 1/(1-5t)$

$$M(t) = \left(\frac{1}{(1+t)^2}, \frac{t}{(1+t)^2} \right) \frac{1}{1-5t} = \frac{1}{(1+t)^2} \cdot \frac{1}{1-5\left(\frac{t}{(1+t)^2}\right)} = \frac{1}{1-3t+t^2}.$$

By using (4.285), $(C^2, tC^2)M(t) = 1/(1-5t)$ *and* $(1/(1+t)^2, t/(1+t)^2)/(1-5t) = M(t)$ *can be written as the pair of inversion identities as follows:*

$$\sum_{k=0}^{n} \frac{k+1}{n+1} \binom{2n+2}{n-k} F_{2k+2} = 5^n, \tag{4.295}$$

$$\sum_{k=0}^{n} (-1)^{n-k} 5^k \binom{n+k+1}{n-k} = F_{2n+2}. \tag{4.296}$$

Similarly

$$\left(C, tC^2\right) M(t) = \frac{1}{1-4t},$$

i.e.,

$$\begin{bmatrix} 1 & 0 & 0 & 0 & 0 & \\ 1 & 1 & 0 & 0 & 0 & \\ 2 & 3 & 1 & 0 & 0 & \cdots \\ 5 & 9 & 5 & 1 & 0 & \\ 14 & 28 & 20 & 7 & 1 & \\ & & \cdots & & & \end{bmatrix} \begin{bmatrix} 1 \\ 3 \\ 5 \\ 7 \\ 9 \\ \vdots \end{bmatrix} = \begin{bmatrix} 1 \\ 4 \\ 16 \\ 64 \\ 256 \\ \vdots \end{bmatrix}.$$

The inverting gives

$$\left(\frac{1}{1+t}, \frac{t}{(1+t)^2}\right)\frac{1}{1-4t} = \frac{1}{1+t}\cdot\frac{1}{1-4\left(\frac{t}{(1+t)^2}\right)} = \frac{1+t}{(1-t)^2},$$

which is the generating function for the odd numbers. Hence, we obtain the following pair of inversion identities.

$$\sum_{k=0}^{n}\frac{(2k+1)^2}{2n+1}\binom{2n+1}{n=k} = 4^n, \tag{4.297}$$

$$\sum_{k=0}^{n}(-1)^{n-k}4^k\binom{n+k}{n-k} = 2n+1. \tag{4.298}$$

Using Theorem 4.3.29, we have the following result.

Theorem 4.3.31 *Let $F(t)$ be defined by (4.282). Then*

$$(F^{m-1}, tF^m)\frac{1+t}{(1+t)^m-(2m+1)t} = \frac{1}{1-(2m+1)t}, \tag{4.299}$$

Particularly, for $m = 1$, and 2, we have

$$\left(1, \frac{t}{1-t}\right)\frac{1+t}{1-2t} = \frac{1}{1-3t}, \tag{4.300}$$

and

$$(C, tC^2)\frac{1+t}{1-3t+t^2} = \frac{1}{1-5t}, \tag{4.301}$$

respectively.

The coefficient set (a_k) of the denominator polynomial $(1+t)^m-(2m+1)t$ shown in Theorem 4.3.31 satisfies the relation

$$\sum_{k=1}^{m}(-1)^k a_k = 2m.$$

For instance, for $m = 2$, $(a_k)_{k\geq 0}$ is

$$(1, 4, 11, 29, 76, 199, 521, 1364, 3571, 9349, \ldots),$$

which is the bisection of the Lucas sequence L_{2n+1}, A002878 (cf. [197]).

For $m = 3$, we have the sequence

$$(1, 5, 17, 52, 152, 435, 1232, 3471, 9753, \ldots),$$

which is a new sequence not be found in the OEIS [197].

Many identities involving Catalan numbers can be obtained from Theorem 4.3.31.

We have established (4.301) and

$$(C^2, tC^2)\frac{1}{1-3t+t^2} = \frac{1}{1-5t}. \tag{4.302}$$

Denote by $d_{n,k}$ and $e_{n,k}$ the (n,k) entries of (C, tC^2) and (C^2, tC^2), respectively.

The first five rows of these matrices are

$$(C, tC^2) = (d_{n,k}) = \begin{bmatrix} 1 & 0 & 0 & 0 & 0 \\ 1 & 1 & 0 & 0 & 0 \\ 2 & 3 & 1 & 0 & 0 \\ 5 & 9 & 5 & 1 & 0 \\ 14 & 28 & 20 & 7 & 1 \end{bmatrix} \quad \text{and}$$

$$(C^2, tC^2) = (e_{n,k}) = \begin{bmatrix} 1 & 0 & 0 & 0 & 0 \\ 2 & 1 & 0 & 0 & 0 \\ 5 & 4 & 1 & 0 & 0 \\ 14 & 14 & 6 & 1 & 0 \\ 42 & 48 & 27 & 8 & 1 \end{bmatrix}.$$

Then (4.301) and (4.295) imply the following corollary.

Corollary 4.3.32

$$\sum_{k=0}^{n} d_{n,k}L_{2k+1} = 5^n = \sum_{k=0}^{n} e_{n,k}F_{2k+2}, \tag{4.303}$$

or

$$\sum_{k=0}^{n} \frac{2k+1}{2n+1}\binom{2n+1}{n+k+1}L_{2k+1} = 5^n = \sum_{k=0}^{n} \frac{k+1}{n+1}\binom{2n+2}{n+k+2}F_{2k+2}. \tag{4.304}$$

where (L_n) and (F_n) are the Lucas and Fibonacci numbers, respectively. Furthermore, we have

$$(C, tC^2)\frac{1-t}{1-3t+t^2} = \frac{\sqrt{1-4t}}{1-5t} \quad \textit{and}$$
$$(C, tC^2)\frac{1}{1-3t+t^2} = \frac{1}{(1-5t)C},$$

or equivalently,

$$\sum_{k=0}^{n} d_{n,k}F_{2k+1} = \sum_{k=0}^{n} \frac{2k+1}{2n+1}\binom{2n+1}{n+k+1}F_{2k+1} = u_n \quad \textit{and}$$
$$\sum_{k=0}^{n} d_{n,k}F_{2k+2} = \sum_{k=0}^{n} \frac{2k+1}{2n+1}\binom{2n+1}{n+k+1}F_{2k+2} = 5^n - \sum_{k=1}^{n} 5^{n-k}C_{k-1},$$

where $(u_n) = (1, 3, 13, 61, 295, ...)$*, the row sum sequence of* $\left(\frac{1}{\sqrt{1-4t}}, \frac{t}{1-4t}\right)$*, which is the integer sequence A046748. Similarly,*

$$(C^2, tC^2)\frac{1-t}{1-3t+t^2} = \frac{2-C}{1-5t} \quad \textit{and}$$
$$(C^2, tC^2)\frac{1+t}{1-3t+t^2} = \frac{C}{1-5t}.$$

We now discuss transformations via Fuss-Catalan matrices to produce identities connecting some well-known sequences. We omit the similar proofs for the sake of brevity. We note that the generating functions shown above such as

$$\frac{1}{1-3t+t^2}, \frac{1+t}{1-3t+t^2}, \frac{1-t}{1-3t+t^2}, \textit{ and } \frac{1}{1-5t}$$

are generating functions of recursive sequences. More precisely, the first three of the above generating functions yield recursive sequences of order 2 generated by the recursive relation

$$a_n = 3a_{n-1} - a_{n-2} \tag{4.305}$$

for $n \geq 2$ with the initial pairs $(a_0, a_1) = (1, 3), (1, 4)$, and $(1, 2)$, respectively. The last function generates a recursive sequence of order 1, $a_n = 5a_{n-1}$ with the initial $a_0 = 1$. Therefore, Corollary 4.3.32 can be viewed as a transformation of recursive sequences. Two questions arise naturally: (1) Can we find a recursive sequence such that (C, tC^2) transfers it to the generating function $1/(1-kt)$? (2) Given two recursive sequences (a_n) and (b_n) of order 2, can we always find a Riordan array to transfer one from another one? The answers for both questions are "yes". In addition, we will see that the results can be extended to higher order cases.

Theorem 4.3.33 *Let* $(a_n)_{n\geq 0}$ *be a recursive sequence of order 2 defined by*

$$\sum_{n\geq 0} a_n t^n = \frac{1+t}{1-(k-2)t+t^2}.$$

Then $a_{n+1} = (k-2)a_n - a_{n-1}$, $n \geq 1$ *with the initial pair* $(a_0, a_1) = (1, k-1)$*. Thus we have*

$$(C, tC^2)\frac{1+t}{1-(k-2)t+t^2} = \frac{1}{1-kt}. \tag{4.306}$$

Theorem 4.3.34 *Let* (a_n) *be a recursive sequence of order 3 defined by*

$$\sum_{n\geq 0} a_n t^n = \frac{1+t}{1-(k-3)t+3t^2+t^3}.$$

Then $a_{n+1} = (k-3)a_n - 3a_{n-1} + a_{n-2}$ *with the initial triple* $(a_0, a_1, a_2) = \left(1, k-2, k^2-5k+3\right)$*. Thus, we have*

$$(T^2, tT^3)\frac{1+t}{1-(k-3)t+3t^2+t^3} = \frac{1}{1-kt}$$

Theorem 4.3.35 *Let (a_n) be a recursive sequence of order m satisfying*

$$a_n = \sum_{k=1}^{m} p_k a_{n-k} \tag{4.307}$$

for $n \geq m$ with initial terms $a_0, a_1, \ldots, a_m$, and let $m \geq \ell \geq 1$ be integers. Then (a_n) can be transferred by Riordan array operator (F^{m-1}, tT^m) to $1/(1-kt)$ if and only if the characteristic polynomial of (4.307) is $p_m(t) = (1+t)^m - kt$, and the initial conditions are $a_0 = 1$ and $\sum_{n=1}^{r-1}\left(a_n - \sum_{j=1}^{n} p_j a_{n-j}\right) = \delta_{1,n}$. Thus we have

$$(F^{m-1}, tF^m)\frac{1+t}{(1+t)^m - kt} = \frac{1}{1-kt}. \tag{4.308}$$

The above results on the transformations via Fuss-Catalan matrices (F^{m-1}, tF^m) can be extended to the transformations via general Fuss-Catalan matrices (F^ℓ, tF^m) by using similar arguments.

4.3.7 One-pth Riordan arrays and Andrews' identities

We now apply the procedure described before and Theorem 4.3.13 on the one-pth Riordan arrays to provide a new proof of some identities obtained by Andrews in [7], namely

$$F_n = \sum_{k=-\infty}^{\infty} (-1)^k \binom{n-1}{\lfloor \frac{1}{2}(n-1-5k) \rfloor} \tag{4.309}$$

and

$$F_n = \sum_{k=-\infty}^{\infty} (-1)^k \binom{n}{\lfloor \frac{1}{2}(n-1-5k) \rfloor}, \tag{4.310}$$

where (F_n) is the sequence of Fibonacci numbers, defined by $F_0 = 0$, $F_1 = 1$, and $F_{n+2} = F_{n+1} + F_n$. Different proofs of (4.309) and (4.310) were given by Gupta in [81], Hirschhorn in [115, 117], and Brietzke in [25], respectively. The first three papers are rather involved, though elementary, and they are specifically designed to deal with the case of Pascal's triangle. The last paper makes use of Riordan array, which will be introduced in this subsection. As indicated in [81, 115], identities (4.309) and (4.310) are equivalent to

$$F_{2n+2} = \sum_{j=-\infty}^{\infty} \left[\binom{2n+1}{n-5j} - \binom{2n+1}{n-5j-2}\right], \tag{4.311}$$

$$F_{2n+1} = \sum_{j=-\infty}^{\infty} \left[\binom{2n}{n-5j} - \binom{2n}{n-5j-2}\right], \tag{4.312}$$

and

$$F_{2n+1} = \sum_{j=-\infty}^{\infty} \left[\binom{2n+1}{n-5j} - \binom{2n+1}{n-5j-1} \right], \quad (4.313)$$

$$F_{2n+2} = \sum_{j=-\infty}^{\infty} \left[\binom{2n+2}{n-5j} - \binom{2n+2}{n-5j-1} \right], \quad (4.314)$$

respectively. In [8] Andrews proves these identities in the context of identities of the Rogers-Ramanujan type. In [47], identities (4.311) through (4.314), as well as several other similar identities for trinomial coefficients and Catalan's triangle, have been proved in an elementary and direct way.

Brietzke proved (4.314) by using a Riordan array approach, which illustrates how identities (4.311) through (4.313) can be obtained similarly by this Riordan array technique. This subsection gives a modification of his proof by using the one-pth Riordan arrays. Replacing n by $n-1$ in (4.314), we may write it as

$$F_{2n} = \sum_{j=-\infty}^{\infty} \left[\binom{2n}{n-5j-1} - \binom{2n}{n-5j-2} \right]. \quad (4.315)$$

Identity (4.315) corresponds to adding in each row the elements marked with a plus sign and subtracting the ones marked with a minus. By symmetry, we can represent this sum using the following Riordan array $(\tilde{d}_{n,k} = \binom{2n}{n+k})_{n,k\geq 0}$ with marked plus and minus entries, where $\tilde{d}_{n,k}$ are found by using Theorem 4.3.13 when $p=2$ and $r=0$.

$$\begin{bmatrix} 1 & 0 & 0 & 0 & 0 \\ 2 & 1 & 0 & 0 & 0 \\ 6 & 4 & -1 & 0 & 0 \\ 20 & 15 & -6 & -1 & 0 \\ 70 & 56 & -28 & -8 & 1 \end{bmatrix}$$

In order to prove (4.315), we wish to evaluate the sum (4.315), i.e.,

$$S_n = \sum_{k=0}^{\infty} \left[\binom{2n}{n+5k+1} - \binom{2n}{n+5k+2} - \binom{2n}{n+5k+3} + \binom{2n}{n+5k+4} \right].$$

In terms of the Riordan array $(\tilde{d}_{n,k})_{n,k\geq 0} = (g, f)$, we have

$$S_n = \sum_{k=0}^{\infty} h_k \tilde{d}_{n,k},$$

where

$$h(t) = \sum_{k=0}^{\infty} h_k t^k = t - t^2 - t^3 + t^4 + t^6 - t^7 - t^8 + t^9 + \cdots$$

$$=\frac{t-t^2-t^3+t^4}{1-t^5}=\frac{t(1-t)(1-t^2)}{(1-t)(1+t+t^2+t^3+t^4)}$$
$$=\frac{t^{-1}-t}{t^{-2}+t^{-1}+1+t+t^2}.$$

It is obvious that $g(t)=\sum_{j\geq 0}\binom{2n}{n}t^n=1/\sqrt{1-4t}$, i.e., the generating function for the central binomial coefficients, which is usually denoted by $B(t)$. From Theorem 4.3.13 and noting $\phi=\overline{t^2/(t/(1-t))}=\overline{t(1-t)}=tC(t)$, we have

$$f(t)=\frac{\phi}{1-\phi}=\frac{tC(t)}{1-tC(t)}=tC(t)^2=\frac{1-2t-\sqrt{1-4t}}{2t}.$$

Hence, $(\tilde{d}_{n,k})_{n,k\geq 0}=(1/\sqrt{1-4t},(1-2t-\sqrt{1-4t})/(2t))$. By using FTRA to the Riordan array, we get

$$S_n=[t^n]d(t)h(f(t))=[t^n]d\frac{f^{-1}+f}{(f^{-2}+f^2)+(f^{-1}+f)+1}.$$

Since

$$f^{-1}-f=\frac{1-2t+\sqrt{1-4t}}{2t}-\frac{1-2t-\sqrt{1-4t}}{2t}=\frac{\sqrt{1-4t}}{t},$$
$$f^{-1}+f=\frac{1-2t+\sqrt{1-4t}}{2t}+\frac{1-2t-\sqrt{1-4t}}{2t}=\frac{1-2t}{t},\quad \text{and}$$
$$f^{-2}+f^2=(f^{-1}+f)^2-2=\frac{1-4t+2t^2}{t^2},$$

we may write S_n as

$$S_n=[t^n]\frac{1}{\sqrt{1-4t}}\frac{\frac{\sqrt{1-4t}}{t}}{\frac{1-4t+2t^2}{t^2}+\frac{1-2t}{t}+1}=[t^n]\frac{t}{1-3t+t^2}=F_{2n},$$

where the last step holds because of (4.293) in Example 4.3.30. From the last expression, (4.315) and (4.314) follow.

Identities (4.311)–(4.313) can be obtained in a similar way, which are left as Exercise 4.14.

In Subsection 4.3.1, we have presented some identities constructed by using Theorem 4.3.13. We now give more identities by using this Riordan approach. Some of these identities are well known, while others are new. We need one more property of Riordan arrays, which generalizes a well-known property of Pascal's triangle.

Theorem 4.3.36 *Let $(d_{n,k})_{n,k\geq 0}=(g,f)$ be a Riordan array. Then for any integers $k\geq s\geq 1$ we have*

$$d_{n,k}=\sum_{j=s}^{n}d_{n-j,k-s}[t^j](f(t))^s. \tag{4.316}$$

Particularly, for $s=1$, $d_{n,k}=\sum_{j=1}^{n}f_jd_{n-j,k-1}$, where $f_j=[t^j]f(t)$.

Proof. The (n,k) entry of the Riordan array (g,f) can be written as

$$\begin{aligned} d_{n,k} =& [t^n]g(t)(f(t))^k = [t^n]g(t)(f(t))^{k-s}((f(t))^s \\ =& \sum_{j=s}^{n} \left([t^{n-j}]g(t)(f(t))^{k-s}\right)\left([t^j](f(t))^s\right) \\ =& \sum_{j=s}^{n} d_{n-j,k-s}[t^j]((f(t))^s. \end{aligned}$$

Example 4.3.37 *If* $(g,f) = (1/(1-t), t/(1-t))$, *then* $f_j = [t^j](t/(1-t)) = 1$ *for all* $j \geq 1$. *We have the well-known identity*

$$\sum_{j=1}^{n} \binom{n-j}{k-1} = \binom{n}{k}. \tag{4.317}$$

More generally, for the Pascal triangle $(g,f) = (1/(1-t), t/(1-t))$, *we have*

$$[t^j](f(t))^s = [t^j]\frac{t^s}{(1-t)^s} = [t^{j-s}](1-t)^{-s} = [t^{j-s}]\sum_{i\geq 0}\binom{s+i-1}{i}t^i = \binom{j-1}{s-1}.$$

Consequently, (4.316) *becomes the Chu-Vandermonde identity* (4.79),

$$\sum_{j=s}^{n} \binom{n-j}{k-s}\binom{j-1}{s-1} = \binom{n}{k},$$

which contains (4.317) *as a special case.*

We now give more examples of Theorem 4.3.13 related to Fuss-Catalan numbers. First, we establish the relation between the Fuss Catalan numbers and the Riordan array $(\tilde{g}, \tilde{f}) = (\tilde{d}_{n,k})_{n,k\geq 0}$, where $\tilde{d}_{n,k} = d_{pn+r,(p-1)n+r+k}$ and $d_{n,k}$ is the (n,k) entry of the Pascal's triangle $(g,f) = (1/(1-t), t/(1-t))$.

Theorem 4.3.38 *Let* $(d_{n,k})_{n,k\geq 0} = (1/(1-t), t/(1-t))$ *be the Pascal triangle, for any integers* $p \geq 2$ *and* $r \geq 0$ *and a given Riordan array* (g,f) *let* $(\tilde{d}_{n,k} = d_{pn+r,(p-1)n+r+k})_{n,k\geq 0} = (\tilde{g}, \tilde{f})$ *be the one-pth Riordan array of* (g,f) *with respect to* r. *Then*

$$\tilde{g}(t) = \sum_{n\geq 0}\binom{pn+r}{n}t^n = \left.\frac{(1+w)^{r+1}}{1-(p-1)w}\right|_{w=t(1+w)^p} \tag{4.318}$$

$$\tilde{f}(t) = \sum_{n=1}^{\infty}\frac{1}{pn+1}\binom{pn+1}{n}t^n = F_p(t) - 1 = tF_p^p(t), \tag{4.319}$$

where $F_p(t)$ *is the pth order Fuss-Catalan function satisfying*

$$F_p\left(t(1-t)^{p-1}\right) = \frac{1}{1-t}. \tag{4.320}$$

Proof. For expression (4.318), we find

$$\begin{aligned}[t^n]\tilde{g} =&\tilde{d}_{n,0} = d_{pn+r,(p-1)n+r} = \binom{pn+r}{n}\\ =&[t^n](1+t)^{pn+r} = [t^n](1+t)^r((1+t)^p)^n\\ =&[t^n]\frac{(1+w)^r}{1-t(d/dw)((1+w)^p)}\bigg|_{w=t(1+w)^p},\end{aligned}$$

which implies (4.318).

From (4.192) of Theorem 4.3.13 we know that

$$(\tilde{g},\tilde{f}) = \left(\frac{t\phi'(t)g(\phi)f(\phi)^r}{\phi^{r+1}}, f(\phi)\right), \tag{4.321}$$

where $\phi(t) = \overline{\frac{t^p}{(f(t))^{p-1}}}$, the compositional inverse of $t^p/(f(t))^{p-1}$. Moreover, the generating function of the A-sequence of the new array $(\tilde{g},\tilde{f})$ is $(A(t))^p$, where $A(t)$ is the generating function of the A-sequence of the original Riordan array (g,f). By using the Lagrange Inversion Formula (4.174), we have

$$[t^n]\tilde{f} = \frac{1}{n}[t^{n-1}](A(t))^{pn} = \frac{1}{n}[t^{n-1}](1+t)^{pn} = \frac{1}{n}\binom{pn}{n-1}.$$

Therefore,

$$\tilde{f} = \sum_{n=1}^{\infty}\frac{(pn)!}{((p-1)n+1)!n!}t^n = \sum_{n=1}^{\infty}\frac{1}{pn+1}\binom{pn+1}{n}t^n = F_p(t)-1.$$

Since $F_p = 1 + tF_p^p$ (cf. (4.282)), we obtain (4.319). From (4.321),

$$f(\phi) = \tilde{f}(t) = tF_p^p(t).$$

Therefore, noting $f(t) = t/(1-t)$ and the last equation, we get

$$\begin{aligned}\frac{t}{1-t} = f(t) = \overline{\phi}F_p^p\left(\overline{\phi}\right) &= \frac{t^p}{(f(t))^{p-1}}F_p^p\left(\frac{t^p}{(f(t))^{p-1}}\right)\\ &= t(1-t)^{p-1}F_p^p\left(t(1-t)^{p-1}\right).\end{aligned}$$

Consequently, (4.320) follows from the comparison of the leftmost side and the rightmost side of the last equation.

■

For example, if $p=2$ and $r\geq 0$, then

$$\tilde{f} = tF_2^2(t) = t(C(t))^2.$$

Since $w = t(1+w)^2$ has a solution

$$w = \frac{1-2t-\sqrt{1-4t}}{2t} = C(t)-1,$$

we have

$$\tilde{g} = \left.\frac{(1+w)^{r+1}}{1-w}\right|_{w=t(1+w)^2} = \frac{(C(t))^{r+1}}{2-C(t)} = \frac{(C(t))^r}{\sqrt{1-4t}} = B(t)(C(t))^r,$$

where $B(t)$ is the generating function for the central binomial coefficients. Thus, $(\tilde{d}_{n,k})_{n,k\geq 0} = (d_{2n+r,n+r+k})_{n,k\geq 0}$ is the Riordan array

$$(\tilde{g},\tilde{f}) = \left(B(t)C^r, t(C(t))^2\right).$$

Example 4.3.39 *For fixed integers $p \geq 2$ and $r \geq 0$, starting with the Pascal triangle and using Theorems 4.3.13 and 4.3.38, we obtain the Riordan array $(\tilde{g},\tilde{f})$ with its (n,k) entry as*

$$\tilde{d}_{n,k} = \binom{pn+r}{(p-1)n+r+k} = \binom{pn+r}{n-k}$$

possesses the formal power series $\tilde{f}(t) = tF_p^p(t)$. Thus,

$$\begin{aligned}[t^j](\tilde{f}(t))^s =& [t^{j-s}]F_p^{ps}(t) = [t^{j-s}]\frac{ps}{pn+ps}\binom{pn+ps}{n}\\ =& \frac{ps}{p(j-s)+ps}\binom{p(j-s)+ps}{j-s} = \frac{s}{j}\binom{pj}{j-s}.\end{aligned}$$

From the expression (4.316) *in Theorem 4.3.36, we obtain the identity*

$$\sum_{j=s}^{n}\frac{s}{j}\binom{pj}{j-s}\binom{p(n-j)+r}{n-j-k+s} = \binom{pn+r}{n-k}. \tag{4.322}$$

Particularly, if $s=1$, then (4.322) *becomes*

$$\sum_{j=1}^{n}\frac{1}{pj+1}\binom{pj+1}{j}\binom{p(n-j)+r}{n-j-k+1} = \binom{pn+r}{n-k}$$

and, finally, adding to the both sides $\binom{pn+r}{n-k+1}$, we have

$$\sum_{j=0}^{n}\frac{1}{pj+1}\binom{pj+1}{j}\binom{p(n-j)+r}{n-j-k+1} = \binom{pn+r+1}{n-k+1}.$$

Setting $j=i+s$, $x=ps$, $y=pk-ps+r$, and replacing n by $n+k$, we find that identity (4.322) *becomes formula (5.62) of [79]:*

$$\sum_{i=0}^{n}\frac{x}{x+pi}\binom{x+pi}{i}\binom{y+p(n-i)}{n-i} = \binom{x+y+pn}{n}.$$

Substituting $p = -q$, $x = r$, *and* $y + pn = p$, *we may know that the above identity is equivalent to Gould identity:*

$$\sum_{i=0}^{n} \frac{r}{r - qi} \binom{r - qi}{i} \binom{p + qi}{n - i} = \binom{r + p}{n}.$$

More identities can be found from the source formula (4.322).

Exercises

4.1. Use the expansion formula (4.1) or (4.2) to present the summation (4.7).

Hint: See Hsu and Shiue [133].

4.2. Prove the summation formula (4.8).

Hint: See Wang and Hsu [210].

4.3. Prove the summation formulas (4.31)–(4.55).

4.4. Prove Corollaries 4.2.15–4.2.17

4.5. Prove the implication (4.85) ⇒ (4.86) in Theorem 4.2.6.

4.6. Show that the following six subsets of the Riordan group are subgroups of the Riordan group.

- the *Appell subgroup* $\{(g(t), t)\}$.
- the *Lagrange (associated) subgroup* $\{(1, f(t))\}$.
- the *k-Bell subgroup* $\{(g(t), t(g(t))^k)\}$, where k is a fixed positive integer.
- the *hitting-time subgroup* $\{(tf'(t)/f(t), f(t))\}$.
- the *derivative subgroup* $\{(f'(t), f(t))\}$.
- the *checkerboard subgroup* $\{(g(t), f(t))\}$, where g is an even function and f is an odd function,

where $g \in \mathcal{F}_0$ and $f \in \mathcal{F}_1$.

4.7. Prove Theorem 4.3.4. Hint: cf. [158]

4.8. Prove Theorem 4.3.8. Hint: cf. [113].

4.9. Prove Theorem 4.3.9. Hint: cf. [113].

4.10. Prove Theorem 4.3.10. Hint: cf. [113].

4.11 Prove Theorem 4.3.16.

Hint: Since $(f(t))^p$ is the generating function of the A-sequence of $(\tilde{d}_{n,k})$, where $\tilde{d}_{n,k} = d_{pn+r,(p-1)n+r+k}$, we may obtain (4.216) from the definition of A-sequence, where β_j can be found from (4.204) and (4.212).

4.12. For $g \in \mathcal{F}_0$ and $f \in \mathcal{F}_1$, show that the following statements are true.

- The *Appell subgroup*, a normal subgroup of the Riordan group, defined by $A = \{(g(t), t)\}$, has the characterization:

$$R^+ = \frac{1-t}{t}R^E, \ R^E = \frac{t}{1-t}R^+ \text{ and } g = R^+(1-t) = R^E\frac{(1-t)^2}{t}. \quad (4.323)$$

- The *Associate subgroup (or Lagrange subgroup)*, defined by $L = \{(1, f(t))\}$, has the characterization:

$$(R^+)^2 - R^+ - R^E = 0, \quad \text{and } f(t) = \frac{R^+ - 1}{R^+} = \frac{R^E}{(R^+)^2} \quad (4.324)$$

- The *Bell subgroup*, defined by $B = \{(g(t),\, tg(t))\}$, has the characterization:

$$t(R^+)^2 - R^E = 0 \quad \text{and} \quad g = \frac{R^+}{1+tR^+} = \frac{R^E}{t(R^E + R^+)} \tag{4.325}$$

4.13. Prove the generating functions of the A-sequence and the Z-sequence of $(F^\ell, -tF^{2m-1})$ are (4.283) and (4.284), respectively.

4.14. Prove the formulas (4.311)–(4.313).

Hint: One may use the same Riordan array $(\tilde{d}_{n,k})_{n,k\geq 0} = (1/\sqrt{1-4t}, (1-2t-\sqrt{1-4t})/(2t))$ in the proof of (4.314) to prove (4.313) while use the Riordan array

$$(\tilde{d}_{n,k})_{n,k\geq 0} = \left(\frac{1}{2t}\left(\frac{1}{\sqrt{1-4t}} - 1\right), \frac{1-2t-\sqrt{1-4t}}{2t}\right)$$

to prove the formulas (4.311) and (4.312). Both Riordan arrays are from Theorem 4.3.13.

5

Extension Methods

CONTENTS

In this chapter, we shall present the methods extended from the previous chapters for constructing various identities and inversion relations. The identities constructed in this chapter include the identities of high dimensions, convolution-type, Abel-type, and those related Bernoulli polynomials and numbers, Euler polynomials and numbers, etc. The methods represented in this chapter are related to generalized Riordan arrays and groups with different bases and their analogs, Sheffer polynomial sequences and the Sheffer group. Some identities and inversion relations are constructed by using dual sequences with Riordan array representation, pseudo-Riordan involutions. Finally, an extensions of W-Z algorithm and Zeilberger's creative telescoping

DOI: 10.1201/9781003051305-5

algorithm is represented and used to construct and prove the identities for Bernoulli polynomials and numbers.

5.1 Identities and Inverse Relations Related to Generalized Riordan Arrays and Sheffer Polynomial Sequences

The concepts of Riordan arrays and the Riordan group have been introduced in Section 4.3. In this section, we will study their extensions, generalized Riordan arrays and the generalized Riordan group, as well as their analogs, Sheffer polynomial sequences and the Sheffer group. Several recurrence relations of the entries of the generalized Riordan arrays will be represented in the first subsection, which make important roles in the construction of identities.

A composition operator performed on the set of all Sheffer polynomial sequences is introduced so it forms a group called the Sheffer group, which gives a general pattern consisting of various special Sheffer-type polynomial sequences as elements. We will show that every element of the group and its inverse are the potential polynomials of a pair of generalized Stirling numbers (GSN's) (cf. Corollary 5.1.35), and we also show the isomorphism between the Sheffer group and the Riordan group (cf. Theorem 5.1.20). Hence, the established results of the Sheffer group connect the Riordan group, GSN pairs, and Riordan arrays so that all those subjects can be fully studied comprehensively. For instance, the Sheffer group and the related GSN-pairs and their inverse relationships can be used to derive combinatorial identities and algebraic identities involving the Sheffer-type polynomials.

The problem of finding inverse relationships is one of the most interesting subjects in combinatorics, and there is a vast literature on it including various methods for constructing inversion formulas. The inversion formulas for sequences appear in pairs, which two formulas give the mutual expressions of two sequences. More preciously, by the so-called inverse series transform we mean a pair of inverse formulas of the form

$$f_n = \sum_{k}^{n} r_{n,k} g_k \Longleftrightarrow g_n = \sum_{k}^{n} r^*_{n,k} f_k, \tag{5.1}$$

where $r_{n,k}$ and $r^*_{n,k}$ are called the kernels of series transformation, and (f_k) and (g_k) are any two sequences of real or complex numbers connected by the inverse relations. If $r_{n,k} = r^*_{n,k}$, the transformation is said to be a self-inverse relation. For a fundamental discussion, the reader is referred to the Riordan's books [182, 183]. In 1958, Riordan [182] had been proposed the general problem of inverting combinatorial sums shown in (5.1).

The following known as *Gould-Hsu inverse relation* (cf. [77]) has sparked numerous follow-up research work and was used to generalize many special combinatorial inversion formulas.

Theorem 5.1.1 *Let (a_j) and (b_j) be two sequences of numbers such that*

$$\psi(x,n)=\Pi_{j=1}^{n}(a_j+b_jx)\neq 0 \tag{5.2}$$

for all non-negative integers x and n with $\psi(x,0)=1$. Then we have the following inversion formulas

$$f(n)=\sum_{k=0}^{n}(-1)^k\binom{n}{k}\psi(k,n)g(k) \quad \textit{and} \tag{5.3}$$

$$g(n)=\sum_{k=0}^{n}(-1)^k\binom{n}{k}(a_{k+1}+kb_{k+1})\psi(n,k+1)^{-1}f(k), \tag{5.4}$$

which is a generalization of Gould inverse relation

$$f(n)=\sum_{k=0}^{n}(-1)^k\binom{n}{k}\binom{a+bk}{n}g(k) \quad \textit{and} \tag{5.5}$$

$$g(n)=\sum_{k=0}^{n}(-1)^k\frac{a+bk-k}{a+bn-k}\binom{a+bn-k}{n-k}\binom{a+bn}{n}^{-1}f(k). \tag{5.6}$$

Gould-Hsu inverse relation can be proved by using difference of polynomials. Substitution of (5.4) into (5.3) gives

$$\sum_{j=0}^{n}f(j)(a_{j+1}+jb_{j+1})\binom{n}{j}\sum_{k=0}^{n-j}(-1)^k\binom{n-j}{k}\psi(k+j,n)\psi(k+j,j+1)^{-1},$$

which will reduce to $f(n)$ if we ca show that

$$\sum_{k=0}^{n-j}(-1)^k\binom{n-j}{k}\frac{\psi(k+j,n)}{\psi(k+j,j+1)}=\begin{cases}0, & n>j\geq 0,\\ \frac{1}{a_{n+1}+nb_{n+1}}, & n=j.\end{cases} \tag{5.7}$$

What happens for $n=j$ is obvious. We note $\psi(k+j,n)\psi(k+j,j+1)^{-1}=\Pi_{\ell=j+2}^{n}(a_\ell+(k+j)b_\ell)$ is a polynomial of degree $n-j-1$ and (5.7) is an $(n-j)$-th difference of the polynomial, which is therefore zero for $n>j\geq 0$. The Gould inverse relation follows at once when we choose $a_j=a-j+1$ and $b_j=b$, i.e., $\psi(k,n)=n!\binom{a+bk}{n}$.

Taking $a_j=1$ and $b_j=0$, then $\psi(x,n)=1$, the inverse relation in Theorem 5.1.1 reduces to a self-inverse relation, and $(f_n)=(g_n)$ is a self-dual sequence that will be studied in Subsection 5.1.4. If $\psi(k,n)=(a+bk)^n$, $\psi(k,n)=n!\binom{a+n+bk}{n}$, etc. the corresponding Gould inverse relations can be found in [70, 71].

In this section, we will focus on a special matrix method, Riordan array method, in the construction of inverse relationships for number sequences and polynomial sequences. The Riordan array method will also be used to find the series summations of number series and function series. A matrix representation $f = Rg$ where $f = (f_0, f_1, ...)^T$, $g = (g_0, g_1, ...)^T$ and $R = (r_{n,k})_{n,k\geq 0}$ of these combinatorial sums is useful for inverting the sums. For example, the combinatorial sums involving binomial coefficients can be represented by Riordan arrays. Sprugnoli [201] showed that the sums involving the rows of a Riordan array can be performed by operating a suitable transformation on a generating function and then by extracting a coefficient from the resulting function. In the view of the isomorphism between the Riordan group and the Sheffer group, the inverse relations can be constructed for Sheffer-type polynomials. All of these results will be shown in the second subsection.

The results of subsection 5.1.2 will be extended to high dimensional case in the subsection5.1.3. The final subsection of this section diverts to dual sequences.

5.1.1 Generalized Riordan arrays and the recurrence relations of their entries

We start from the definition of generalized Riordan arrays.

Definition 5.1.2 *Let $g(t)$ and $f(t)$ be any given formal power series over the real number field $\mathbb{R}$ or complex number field $\mathbb{C}$ with $g(0) = 1$, $f(0) = 0$ and $f'(0) \neq 0$, and let $(c_n)_{n\in\mathbb{N}_0}$ be a non-zero number sequence with $c_0 = 1$. We call the infinite matrix $D = (d_{n,k})_{n,k\geq 0}$ with real entries or complex entries a generalized Riordan array with respect to the sequence $(c_n)_{n\in\mathbb{N}_0}$ if its kth column satisfies*

$$\sum_{n\geq 0} d_{n,k}\frac{t^n}{c_n} = g(t)\frac{(f(t))^k}{c_k}; \tag{5.8}$$

that is,

$$d_{n,k} = \left[\frac{t^n}{c_n}\right] g(t)\frac{(f(t))^k}{c_k}. \tag{5.9}$$

The generalized Riordan array is still denoted by $(d_{n,k})$ or $(g(t), f(t))$ without causing confusion.

As what mentioned in [171] (see also in [21]), “The concept of representing columns of infinite matrices by formal power series is not new and goes back to Schur's paper on Faber polynomials in 1945 (cf. [192])”. For a given sequence $(b_n)_{n\geq 0}$, a formal power series in auxiliary variable t of the form with a sequence $(c_n \neq 0)$ $(n \in \mathbb{N}_0)$ with $c_0 = 1$ represented as

$$b(t) = b_0\frac{1}{c_0} + b_1\frac{t}{c_1} + b_2\frac{t^2}{c_2} + \cdots = \sum_{n\geq 0} b_n\frac{t^n}{c_n}$$

is called (c)- generating function of the sequence (b_n). If $c_n = 1$ for $n \geq 0$, then the generating function $b(t) = \sum_{n\geq 0} b_n t^n$ is called the ordinary generating function of the sequence (b_n), while a formal power series of the form $b(t) = \sum_{n\geq 0} b_n t^n/n!$ with $c_n = n!$ $(n \in \mathbb{N}_0)$ is called the exponential generating function of the sequence (b_n). As usual, the notation $[t^n]$ stands for the "coefficient of" operator, and if $f(t) = \sum_{k=0}^{\infty} f_k t^k$, then $[t^n]f(t) = f_n$. Similarly, if $f(t) = \sum_{k=0}^{\infty} f_k t^k/c_k$, then $[t^n/c_n]f(t) = f_n$. It is easy to see that $[t^n/c_n]f(t) = c_n[t^n]f(t)$.

Hence, by the definition, the classical Riordan arrays introduced in Section 4.3 correspond to the case of $c_n = 1$, and the exponential Riordan arrays represented in Wang and Wang [209] and Zhao and Wang [227] correspond to the case of $c_n = n!$. In Gould and He's paper [76], the generalized Riordan arrays are referred to as (c)-Riordan arrays including the special case of $c_n = n!$ $(n \geq 0)$.

As we have seen in Section 4.3, one of the most important applications of the theory of Riordan arrays is to deal with summations of the form $\sum_{k=0}^{n} d_{n,k}h_k$. In the context of the generalized Riordan arrays, we have the following theorem on the summation.

Theorem 5.1.3 *Let $D = (g(t), f(t)) = (d_{n,k})_{n,k\in\mathbb{N}_0}$ be a Riordan array with respect to $(c_n)_{n\in\mathbb{N}_0}$ and let $h(t) = \sum_{k=0}^{\infty} h_k t^k/c_k$ be the generating function of the sequence $(h_n)_{n\in\mathbb{N}_0}$. Then we have*

$$\sum_{k=0}^{n} d_{n,k}h_k = \left[\frac{t^n}{c_n}\right] g(t)h(f(t)), \tag{5.10}$$

or equivalently, $(g(t), f(t))h(t) = g(t)h(f(t))$.

Proof. Based on the definition, we have

$$\begin{aligned}\sum_{k=0}^{n} d_{n,k}h_k &= \sum_{k=0}^{\infty} \left[\frac{t^n}{c_n}\right] g(t)\frac{(f(t))^k}{c_k}h_k = \left[\frac{t^n}{c_n}\right] g(t)\sum_{k=0}^{\infty} h_k\frac{(f(t))^k}{c_k} \\ &= \left[\frac{t^n}{c_n}\right] g(t)h(f(t)).\end{aligned}$$

This completes the proof.

■

Similarly to the classical case [201, Theorem 1.2], we can prove the converse of Theorem 5.1.3: For an infinite triangle $D = (d_{n,k})_{0\leq k\leq n<\infty}$ such that for every sequence $(h_k)_{k\in\mathbb{N}_0}$, we have $\sum_{k=0}^{n} d_{n,k}h_k = [t^n/c_n]g(t)h(f(t))$, where $h(t) = \sum_{k=0}^{\infty} h_k t^k/c_k$, and $g(t), f(t)$ are two formal power series not depending on $h(t)$. Then the triangle defined by the Riordan array $(g(t), f(t))$ coincides with $(d_{n,k})_{n,k\in\mathbb{N}_0}$.

With Theorem 5.1.3, we can further compute the product of two Riordan arrays $(g(t), f(t))(h(t), l(t))$. In fact, the column generating function of

$(h(t), l(t))$ is $h(t)(l(t))^k/c_k$. Thus, by matrix multiplication, the column generating function of the product $(g(t), f(t))(d(t), h(t))$ is

$$g(t)d(f(t))\frac{(h(f(t)))^k}{c_k},$$

which means that the product is also a Riordan array, i.e.,

$$(g(t), f(t))(d(t), h(t)) = (g(t)d(f(t)), h(f(t))). \tag{5.11}$$

Similarly to the classical case, the next theorem holds.

Theorem 5.1.4 *For any fixed sequence $(c_n)_{n\in\mathbb{N}_0}$ defined above, the set of all Riordan arrays $(g(t), f(t))$, with $g(t)$ an invertible series and $f(t)$ a delta series, is a group under matrix multiplication. Moreover, the identity of this group is $(1, t)$ and the inverse of the array $(g(t), f(t))$ is $(1/g(\bar{f}(t)), \bar{f}(t))$, where $\bar{f}(t)$ is the compositional inverse of $f(t)$.*

Proof. Denote the set by $\mathscr{R}$. Then $\mathscr{R}$ is closed under matrix multiplication according to (5.11), and the multiplication is associative. The array $(1, t)$ is an element of $\mathscr{R}$, and for each array $(g(t), f(t)) \in \mathscr{R}$, there exists an array $(1/g(\bar{f}(t)), \bar{f}(t)) \in \mathscr{R}$, for which we have

$$\begin{aligned}&(g(t), f(t))(1, t) = (g(t), f(t)) = (1, t)(g(t), f(t)),\\&(g(t), f(t))\left(\frac{1}{g(\bar{f}(t))}, \bar{f}(t)\right) = (1, t) = \left(\frac{1}{g(\bar{f}(t))}, \bar{f}(t)\right)(g(t), f(t)).\end{aligned}$$

Thus, $\mathscr{R}$ is a group and the proof is complete. ∎

The group introduced in Theorem 5.1.4 is called the *Riordan group* with respect to $(c_n)_{n\in\mathbb{N}_0}$. It should be noticed that, for a sequence $(c_n)_{n\in\mathbb{N}_0}$ defined above, the identity $(1, t)$ of the Riordan group R is the usual infinite identity matrix I. Actually, by Eq. (5.9), the (n, k) entry of $(1, t)$ is

$$d_{n,k} = \left[\frac{t^n}{c_n}\right]\frac{t^k}{c_k} = \frac{c_n}{c_k}[t^{n-k}]1 = \delta_{n,k},$$

where $\delta_{n,k}$ is the Kronecker delta defined by $\delta_{n,n} = 1$ and $\delta_{n,k} = 0$ for $n \neq k$.

Theorem 5.1.5 *The quantity $d_{n,k}$ is the (n, k) entry of the generalized Riordan array $(g(t), f(t))$ with respect to $(c_n)_{n\in\mathbb{N}_0}$ if and only if $c_k d_{n,k}/c_n$ is the (n, k) entry of the classical Riordan array $(g(t), f(t))$.*

Proof. By Definition 5.1.2, we have

$$d_{n,k} = \left[\frac{t^n}{c_n}\right]g(t)\frac{(f(t))^k}{c_k} = \frac{c_n}{c_k}[t^n]g(t)(f(t))^k,$$

which is equivalent to the fact that $c_k d_{n,k}/c_n = [t^n]g(t)(f(t))^k$. This completes the proof.

■

Despite its simple proof, Theorem 5.1.5 is an important result, because it shows that the generalized Riordan arrays can always be reduced to the classical case. By Theorem 5.1.5, the next two theorems can be obtained without difficulty.

Theorem 5.1.6 *For any generalized Riordan array $(g(t), f(t)) = (d_{n,k})_{n,k\in\mathbb{N}_0}$, every element $d_{n+1,k+1}$, where $n, k \in \mathbb{N}_0$, can be expressed as follows:*

$$d_{n+1,k+1} = \sum_{j=0}^{\infty} \frac{c_{n+1}c_{k+j}}{c_{k+1}c_n} a_j d_{n,k+j}\,, \tag{5.12}$$

where the sum is actually finite and the sequence $A = (a_k)_{k\in\mathbb{N}_0}$ is fixed. It is called the A-sequence of the generalized Riordan array and it only depends on $f(t)$. Particularly, let $A(t) = \sum_{k=0}^{\infty} a_k t^k$, then

$$f(t) = tA(f(t)) \quad \textit{and} \quad A(t) = \frac{t}{\bar{f}(t)}\,. \tag{5.13}$$

Proof. We have shown that for the classical Riordan array $(g(t), f(t)) = (d^*_{n,k})_{n,k\in\mathbb{N}_0}$, there exists a unique sequence $A = (a_k)_{k\in\mathbb{N}_0}$ satisfying the statements of the theorem. Then

$$d^*_{n+1,k+1} = \sum_{j=0}^{\infty} a_j d^*_{n,k+j}$$

and $f(t) = tA(f(t))$. By Theorem 5.1.5, $d^*_{n,k} = c_k d_{n,k}/c_n$, so we have

$$\frac{c_{k+1}}{c_{n+1}} d_{n+1,k+1} = \sum_{j=0}^{\infty} a_j \frac{c_{k+j}}{c_n} d_{n,k+j}\,,$$

which yields the recurrence relation (5.12) at once.

■

Theorem 5.1.7 *Let $(c_n)_{n\in\mathbb{N}_0}$ be a sequence of non-zero constants with $c_0 = 1$, and let $D := (d_{n,k})_{0\le k\le n<\infty}$ be an infinite triangle such that $d_{n,n} \neq 0$, $\forall n \in \mathbb{N}_0$, and for which the relation (5.12) holds for some sequence $A = (a_k)_{k\in\mathbb{N}_0}$, $a_0 \neq 0$. Then D is a generalized Riordan array $(g(t), f(t))$ with respect to (c_n), where $g(t) = \sum_{k=0}^{\infty} d_{k,0}t^k/c_k$ and $f(t)$ is the unique solution of $f(t) = tA(f(t))$ with $A(t) = \sum_{k=0}^{\infty} a_k t^k$.*

Proof. Define $d^*_{n,k} = c_k d_{n,k}/c_n$, then $d^*_{n,n} \neq 0$, $\forall n \in \mathbb{N}_0$ and $d^*_{n+1,k+1} = \sum_{j=0}^{\infty} a_j d^*_{n,k+j}$. In view of [184, 201], the infinite triangle $D^* = (d^*_{n,k})_{n,k\in\mathbb{N}_0}$ is the classical Riordan array $(g(t), f(t))$, where $g(t) = \sum_{k=0}^{\infty} d^*_{k,0} t^k$ and $f(t)$ is the unique solution of $f(t) = tA(f(t))$. Thus, $g(t) = \sum_{k=0}^{\infty} c_0 d_{k,0} t^k/c_k = \sum_{k=0}^{\infty} d_{k,0} t^k/c_k$, and by Theorem 5.1.5, $D = (d_{n,k})_{n,k\in\mathbb{N}_0}$ is the generalized Riordan array $(g(t), f(t))$ with respect to c_n.

■

Next, we demonstrate another three recurrences related to the entries of the generalized Riordan arrays.

Theorem 5.1.8 *For any generalized Riordan array* $(g(t), f(t)) = (d_{n,k})_{n,k\in\mathbb{N}_0}$, *we have*

$$d_{n,k} - \frac{c_n}{nc_{n-1}} \tilde{d}_{n-1,k} = \sum_{l=k}^{n} \frac{c_n}{c_{l-1}c_{n-l+1}} \frac{n-l+1}{n} f_{n-l+1} \frac{kc_{k-1}}{c_k} d_{l-1,k-1}, \; n, k \geq 1, \tag{5.14}$$

where $\tilde{d}_{n,k}$ *is the* (n, k) *entry of the Riordan array* $(g'(t), f(t))$, $g'(t)$ *is the derivative of* $g(t)$, *and* f_k *are the coefficients of the delta series* $f(t) = \sum_{k=1}^{\infty} f_k t^k/c_k$, *where a delta series is a series in* $\mathcal{F}_1$.

Proof. The column generating function is

$$\sum_{n=k}^{\infty} d_{n,k} \frac{t^n}{c_n} = g(t) \frac{(f(t))^k}{c_k}. \tag{5.15}$$

Differentiating (5.15) with respect to t gives

$$\begin{aligned}
\sum_{n=k}^{\infty} d_{n,k} \frac{nt^{n-1}}{c_n} &= g'(t) \frac{(f(t))^k}{c_k} + \frac{kc_{k-1}}{c_k} g(t) f'(t) \frac{(f(t))^{k-1}}{c_{k-1}} \\
&= \sum_{n=k}^{\infty} \tilde{d}_{n,k} \frac{t^n}{c_n} + \frac{kc_{k-1}}{c_k} \sum_{i=k-1}^{\infty} d_{i,k-1} \frac{t^i}{c_i} \sum_{j=0}^{\infty} (j+1) f_{j+1} \frac{t^j}{c_{j+1}} \\
&= \sum_{n=k}^{\infty} \tilde{d}_{n,k} \frac{t^n}{c_n} + \frac{kc_{k-1}}{c_k} \sum_{n=k-1}^{\infty} \sum_{i=k-1}^{n} \frac{c_n}{c_i c_{n-i+1}} (n-i+1) f_{n-i+1} d_{i,k-1} \frac{t^n}{c_n}.
\end{aligned}$$

Identifying the coefficients of t^{n-1}/c_{n-1} in the last equation yields

$$\frac{nc_{n-1}}{c_n} d_{n,k} = \tilde{d}_{n-1,k} + \sum_{i=k-1}^{n-1} \frac{c_{n-1}}{c_i c_{n-i}} (n-i) f_{n-i} \frac{kc_{k-1}}{c_k} d_{i,k-1},$$

which, after some transformations, leads (5.14) finally.

■

For the iteration matrix with respect to (c_n), we have $g'(t) = 0$. This fact indicates that $\tilde{d}_{n-1,k} = 0$. Thus, (5.14) reduces to

$$d_{n,k} = \sum_{l=k}^{n} \frac{c_n}{c_{l-1}c_{n-l+1}} \frac{n-l+1}{n} f_{n-l+1} \frac{kc_{k-1}}{c_k} d_{l-1,k-1}\,, \tag{5.16}$$

which has been given in [209, Lemma 3.1].

Theorem 5.1.9 *For any generalized Riordan array* $(g(t), f(t)) = (d_{n,k})_{n,k\in\mathbb{N}_0}$, *we have*

$$\frac{c_k}{c_{k-1}} d_{n,k} = \sum_{l=k}^{n} \frac{c_n}{c_{l-1}c_{n-l+1}} f_{n-l+1} d_{l-1,k-1}\,, \quad n, k \geq 1\,, \tag{5.17}$$

where f_k *are the coefficients of the delta series* $f(t) = \sum_{k=1}^{\infty} f_k t^k / c_k$.

Proof. From (5.15), we have

$$\begin{aligned}\sum_{n=k}^{\infty} \frac{c_k}{c_{k-1}} d_{n,k} \frac{t^n}{c_n} &= f(t)g(t)\frac{(f(t))^{k-1}}{c_{k-1}} = \sum_{i=1}^{\infty} f_i \frac{t^i}{c_i} \sum_{j=k-1}^{\infty} d_{j,k-1} \frac{t^j}{c_j} \\ &= \sum_{n=k}^{\infty} \sum_{j=k-1}^{n-1} f_{n-j} d_{j,k-1} \frac{c_n}{c_j c_{n-j}} \frac{t^n}{c_n}\,.\end{aligned}$$

By equating the coefficients of t^n/c_n, we obtain the desired result.

■

Theorem 5.1.10 *For any generalized Riordan array* $(g(t), f(t)) = (d_{n,k})_{n,k\in\mathbb{N}_0}$, *we have*

$$\frac{c_{n+1}}{c_n} d_{n,k} = \sum_{l=k}^{n} \frac{c_{l+1}}{c_{l+1-k}c_k} \bar{f}_{l+1-k} d_{n+1,l+1}\,, \tag{5.18}$$

where $\bar{f}_j$ *are the coefficients of the delta series* $\bar{f}(t) = \sum_{j=1}^{\infty} \bar{f}_j t^j / c_j$, *and* $\bar{f}(t)$ *is the compositional inverse of* $f(t)$.

Proof. Based on (5.9), we have

$$\begin{aligned}&\sum_{n=k}^{\infty} \left(\sum_{l=k}^{n} \frac{c_{l+1}}{c_{l+1-k}c_k} \bar{f}_{l+1-k} \frac{d_{n+1,l+1}}{c_{n+1}} \right) t^n = \frac{1}{c_k} \sum_{l=k}^{\infty} \frac{\bar{f}_{l+1-k} c_{l+1}}{c_{l+1-k}} \sum_{n=l}^{\infty} d_{n+1,l+1} \frac{t^n}{c_{n+1}} \\ &= \frac{g(t)(f(t))^k}{c_k t} \sum_{l=k}^{\infty} \frac{\bar{f}_{l+1-k}}{c_{l+1-k}} (f(t))^{l+1-k} = \frac{g(t)(f(t))^k}{c_k t} \sum_{j=1}^{\infty} \frac{\bar{f}_j}{c_j} (f(t))^j \\ &= g(t) \frac{(f(t))^k}{c_k}\,,\end{aligned}$$

which is coincident with the generating function of $d_{n,k}/c_n$. Therefore, the recurrence relation (5.18) is established.

■

For convenience, we give the specializations of the recurrence relations (5.12), (5.14), (5.17), and (5.18) for the cases $c_n = 1$ and $c_n = n!$, respectively.

Corollary 5.1.11 *For any classical Riordan array* $(g(t), f(t)) = (d_{n,k})_{n,k\in\mathbb{N}_0}$, *we have*

$$d_{n+1,k+1} = \sum_{j=0}^{\infty} a_j d_{n,k+j}\,, \tag{5.19}$$

$$d_{n,k} = \sum_{l=k}^{n} \frac{k}{n}(n-l+1) f_{n-l+1} d_{l-1,k-1} + \frac{1}{n}\tilde{d}_{n-1,k}\,, \tag{5.20}$$

$$d_{n,k} = \sum_{l=k}^{n} f_{n-l+1} d_{l-1,k-1}\,, \tag{5.21}$$

$$d_{n,k} = \sum_{l=k}^{n} \bar{f}_{l+1-k} d_{n+1,l+1}, \tag{5.22}$$

where (5.21) *is the special case of* (4.316) *of Theorem 4.3.36 for* $j = n - l + 1$ *and* $s = 1$.

Corollary 5.1.12 *For any exponential Riordan array* $(g(t), f(t)) = (d_{n,k})_{n,k\in\mathbb{N}_0}$, *we have*

$$d_{n+1,k+1} = \sum_{j=0}^{\infty} \frac{n+1}{k+1}\binom{k+j}{j} j! a_j d_{n,k+j}\,, \tag{5.23}$$

$$d_{n,k} = \sum_{l=k}^{n} \binom{n-1}{l-1} f_{n-l+1} d_{l-1,k-1} + \tilde{d}_{n-1,k}\,, \tag{5.24}$$

$$kd_{n,k} = \sum_{l=k}^{n} \binom{n}{l-1} f_{n-l+1} d_{l-1,k-1}\,, \tag{5.25}$$

$$(n+1)d_{n,k} = \sum_{l=k}^{n} \binom{l+1}{k} \bar{f}_{l+1-k} d_{n+1,l+1}\,. \tag{5.26}$$

According to Theorem 5.1.5, the recurrence relations represented in Corollaries 5.1.11 and 5.1.12 are in fact equivalent to the general recurrence relations (5.12), (5.14), (5.17) and (5.18).

Theorem 5.1.13 *[56] Let* $(d_{n,k})_{n\geq k\geq 0} = (g(t), f(t))$ *be an exponential Riordan array and let*

$$c(x) = c_0 + c_1 t + c_2 t^2 + \cdots, \quad r(x) = r_0 + r_1 t + r_2 t^2 + \cdots \tag{5.27}$$

be two formal power series such that

$$c(f(t)) = g'(t)/g(t), \qquad r(f(t)) = f'(t).$$

Then

$$d_{n+1,0} = \sum_{i\geq 0} i! c_i d_{n,i}, \tag{5.28}$$

$$d_{n+1,k} = r_0 d_{n,k-1} + \frac{1}{k!}\sum_{i\geq k} i!(c_{i-k} + k r_{i-k+1}) d_{n,i}, \tag{5.29}$$

or, defining $c_{-1} = 0$,

$$d_{n+1,k} = \frac{1}{k!}\sum_{i\geq k-1} i!(c_{i-k} + k r_{i-k+1}) d_{n,i} \tag{5.30}$$

for all $k \geq 0$.

Conversely, starting from the sequence defined by (5.27), the infinite array $(d_{n,k})_{n\geq k\geq 0}$ *defined by (5.30) is an exponential Riordan array.*

Proof. Let $(g(t), f(t))$ be an exponential Riordan array, and let $g(t) = \sum_{n\geq 0} g_n t^n/n!$ and $f(t) = \sum_{n\geq 1} f_n t^n/n!$. Then, we may write the LHS of (5.28) as

$$d_{n+1,0} = \left[\frac{t^{n+1}}{(n+1)!}\right] g(t) = g_{n+1}$$

and simplify the RHS of (5.28) as

$$\begin{aligned}\sum_{i\geq 0} i! c_i d_{n,i} &= \sum_{i\geq 0} i! c_i \left[\frac{t^n}{n!}\right] g(t)\frac{f(t)^i}{i!} = \left[\frac{t^n}{n!}\right] g(t)\sum_{i\geq 0} c_i f(t)^i = \left[\frac{t^n}{n!}\right] g(t) c(f(t))\\ &= \left[\frac{t^n}{n!}\right] g(t)\frac{g'(t)}{g(t)} = \left[\frac{t^n}{n!}\right] g'(t) = \left[\frac{t^n}{n!}\right]\sum_{n\geq 1} g_n \frac{n t^{n-1}}{n!} = g_{n+1},\end{aligned}$$

which presents the equality of the LHS and RHS of (5.28).

Similarly, the RHS of (5.29) becomes to

$$\begin{aligned}&\frac{1}{k!}\sum_{i\geq k} i! c_{i-k} d_{n,i} + \frac{1}{k!}\sum_{i\geq k-1} k r_{i-k+1} d_{n,i}\\ =&\frac{1}{k!}\left[\frac{t^n}{n!}\right] g(t)\sum_{i\geq k} i! c_{i-k}\frac{f(t)^i}{i!} + \frac{1}{(k-1)!}\left[\frac{t^n}{n!}\right] g(t)\sum_{i\geq k-1} i! r_{i-k+1}\frac{f(t)^i}{i!}\\ =&\frac{1}{k!}\left[\frac{t^n}{n!}\right] g(t) f(t)^k c(f(t)) + \frac{1}{(k-1)!}\left[\frac{t^n}{n!}\right] g(t) f(t)^{k-1} r(f(t))\\ =&\frac{1}{k!}\left[\frac{t^n}{n!}\right] g(t) f(t)^k \frac{g'(t)}{g(t)} + \frac{1}{(k-1)!}\left[\frac{t^n}{n!}\right] g(t) f(t)^{k-1} f'(t)\end{aligned}$$

$$
\begin{aligned}
&= \left[\frac{t^n}{n!}\right] g'(t)\frac{f(t)^k}{k!} + \frac{1}{(k-1)!}\left[\frac{t^n}{n!}\right] g(t)f(t)^{k-1}f'(t) \\
&= \tilde{d}_{n,k} + \frac{1}{(k-1)!}\sum_{\ell=1}^{n+1} n![t^{\ell-1}]g(t)f(t)^{k-1}[t^{n-\ell+1}]f'(t),
\end{aligned}
$$

where $\tilde{d}_{n,k}$ is the (n,k) entry of the exponential Riordan array $(g'(t), f(t))$ and $g'(t)$ is the derivative of $g(t)$. Since $\ell \le k$, the second term of the rightmost side of the last equation can be written as

$$
\begin{aligned}
&\frac{1}{(k-1)!}\sum_{\ell=k}^{n+1} n![t^{\ell-1}]g(t)f(t)^{k-1}[t^{n-\ell+1}]f'(t) \\
&= \sum_{\ell=k}^{n+1}\binom{n}{\ell-1}\left[\frac{t^{\ell-1}}{(\ell-1)!}\right] g(t)\frac{f(t)^{k-1}}{(k-1)!}\left[\frac{t^{n-\ell+1}}{(n-\ell+1)!}\right]\frac{d}{dt}\left(\sum_{j\ge 1} f_j\frac{t^j}{j!}\right) \\
&= \sum_{\ell=k}^{n+1}\binom{n}{\ell-1} f_{n-\ell+2}d_{\ell-1,k-1}.
\end{aligned}
$$

Thus the RHS of (5.29) can be written as

$$
\tilde{d}_{n,k} + \sum_{\ell=k}^{n+1}\binom{n}{\ell-1} f_{n-\ell+2}d_{\ell-1,k-1}.
$$

Using (5.24) and replacing its n by $n+1$, the last expression yields $d_{n+1,k}$ and (5.29) is proved.

■

Example 5.1.14 *Consider the classical Riordan array* $(\frac{1}{1-t}, \frac{t}{1-t})$*, where its* (n,k) *entry is*

$$
[t^n]\frac{1}{1-t}\left(\frac{t}{1-t}\right)^k = \binom{n}{k},
$$

then $(\frac{1}{1-t}, \frac{t}{1-t})$ *is the well-known Pascal matrix. The corresponding row generating functions are* $(x+1)^n$*, which form the Sheffer sequence for* $(\frac{1}{1+t}, \frac{t}{1+t})$*.*

Since $\bar{f}(t) = t/(1+t)$*, then* $A(t) = t/\bar{f}(t) = 1+t$*, and (5.19) gives the relation*

$$
\binom{n+1}{k+1} = \binom{n}{k} + \binom{n}{k+1}.
$$

Next, because $g'(t) = 1/(1-t)^2$*, we have*

$$
\tilde{d}_{n-1,k} = [t^{n-1}]g'(t)(f(t))^k = [t^{n-1-k}](1-t)^{-k-2} = \binom{n}{k+1}.
$$

Thus, in view of $f_k = 1$, we deduce from (5.20) and (5.21) that

$$\binom{n+1}{k+1} = \sum_{l=k}^{n} (n-l+1)\binom{l-1}{k-1},$$
$$\binom{n}{k} = \sum_{l=k}^{n} \binom{l-1}{k-1}.$$

Finally, since the coefficient of t^j in $\bar{f}(t)$ is $\bar{f}_j = (-1)^{j-1}$, then by Eq. (5.22), the following recurrence relation holds:

$$\binom{n}{k} = \sum_{l=k}^{n} (-1)^{l-k}\binom{n+1}{l+1}. \tag{5.31}$$

Example 5.1.15 *As we present in Chapter 4, the (n,k) entry of the exponential Riordan array $(1, \log(1+t))$, the Stirling matrix of the first kind, is*

$$\left[\frac{t^n}{n!}\right]\frac{1}{k!}(\log(1+t))^k = (-1)^k \begin{bmatrix} n \\ k \end{bmatrix},$$

where $((-1)^{n-k}\begin{bmatrix} n \\ k \end{bmatrix})$ are the Stirling numbers of the first kind. The row generating functions are $\sum_{k=0}^{n}(-1)^{n-k}\begin{bmatrix} n \\ k \end{bmatrix}x^k = (x)_n$, which are the falling factorials defined by $(x)_0 = 1$ and $(x)_n = x(x-1)\cdots(x-n+1)$ for $n = 1, 2, \ldots$. The sequence $(\sum_{k=0}^{n}(-1)^{n-k}\begin{bmatrix} n \\ k \end{bmatrix}x^k)_{n\geq 0}$ is associated to $\mathrm{e}^t - 1$ (cf. Section 4.1.2, [186]).

By (5.13), the generating function of the A-sequence of $(1, \log(1+t))$ is

$$A(t) = \frac{t}{\bar{f}(t)} = \frac{t}{\mathrm{e}^t - 1} = \sum_{j=0}^{\infty} B_j \frac{t^j}{j!},$$

where B_j are the Bernoulli numbers (cf. for example, P. 48, [44]). Then Eq. (5.23) reduces to

$$\begin{bmatrix} n+1 \\ k+1 \end{bmatrix} = \sum_{j=0}^{\infty} (-1)^j \frac{n+1}{k+1}\binom{k+j}{j} B_j \begin{bmatrix} n \\ k+j \end{bmatrix}.$$

Additionally, using $f_k = (-1)^{k-1}(k-1)!$, we obtain from (5.24) and (5.25) that

$$\begin{bmatrix} n \\ k \end{bmatrix} = \sum_{l=k}^{n} \binom{n-1}{l-1}(n-l)! \begin{bmatrix} l-1 \\ k-1 \end{bmatrix},$$
$$k\begin{bmatrix} n \\ k \end{bmatrix} = \sum_{l=k}^{n} \binom{n}{l-1}(n-l)! \begin{bmatrix} l-1 \\ k-1 \end{bmatrix}.$$

Finally, since $\bar{f}(t) = \mathrm{e}^t - 1$, *then* $\bar{f}_j = 1$, *and by Eq. (5.26) we have*

$$(n+1)\begin{bmatrix} n \\ k \end{bmatrix} = \sum_{l=k}^{n} (-1)^{k-\ell} \binom{l+1}{k} \begin{bmatrix} n+1 \\ \ell+1 \end{bmatrix}.$$

The inverse of the array $(1, \log(1+t))$ *is* $(1, \mathrm{e}^t - 1)$, *which is called the Stirling matrix of the second kind, and its* (n,k) *entry is*

$$\left[\frac{t^n}{n!}\right] (\mathrm{e}^t - 1)^k / k! = \left\{ \begin{matrix} n \\ k \end{matrix} \right\},$$

which is the Stirling number of the second kind. The row generating functions are $\sum_{k=0}^{n} \left\{ \begin{smallmatrix} n \\ k \end{smallmatrix} \right\} x^k$, *which are called the exponential polynomials and denoted by* $\phi_n(x)$. *The sequence* $(\phi_n(x))_{n \in \mathbb{N}_0}$ *is associated to* $\log(1+t)$ *(cf. Section 4.1.3, [186]).*

The generating function of the A-sequence of the array $(1, \mathrm{e}^t - 1)$ *is*

$$A(t) = \frac{t}{\bar{f}(t)} = \frac{t}{\log(1+t)} = \sum_{j=0}^{\infty} b_j(0) \frac{t^j}{j!},$$

where $b_j(0)$ *are the Bernoulli numbers of the second kind [186, p. 114], and they are also called the Cauchy numbers of the first kind (cf. P. 294, [44] and [159]). Thus, we have*

$$\left\{ \begin{matrix} n+1 \\ k+1 \end{matrix} \right\} = \sum_{j=0}^{\infty} \frac{n+1}{k+1} \binom{k+j}{j} b_j(0) \left\{ \begin{matrix} n \\ k+j \end{matrix} \right\}.$$

Next, because $f_k = 1$, *equations (5.24) and (5.25) yields at once*

$$\left\{ \begin{matrix} n \\ k \end{matrix} \right\} = \sum_{l=k}^{n} \binom{n-1}{l-1} \left\{ \begin{matrix} \ell-1 \\ k-1 \end{matrix} \right\},$$

$$k \left\{ \begin{matrix} n \\ k \end{matrix} \right\} = \sum_{l=k}^{n} \binom{n}{l-1} \left\{ \begin{matrix} \ell-1 \\ k-1 \end{matrix} \right\}.$$

Now, $\bar{f}(t) = \log(1+t)$, *so* $\bar{f}_j = (-1)^{j-1}(j-1)!$, *and (5.26) gives the recurrence relation*

$$(n+1)\left\{ \begin{matrix} n \\ k \end{matrix} \right\} = \sum_{l=k}^{n} (-1)^{l-k} (l-k)! \binom{l+1}{k} \left\{ \begin{matrix} n+1 \\ \ell+1 \end{matrix} \right\}.$$

5.1.2 The Sheffer group and the Riordan group

The generalized Riordan arrays and the Riordan group are tightly related to the Sheffer polynomial sequences and the Sheffer group. We first give the definition of the *Sheffer-type polynomials.*

Definition 5.1.16 *Let $g(t)$ and $f(t)$ be any given formal power series over the real number field $\mathbb{R}$ or complex number field $\mathbb{C}$ with $g \in \mathcal{F}_0$ and $f \in \mathcal{F}_1$. Then the polynomials $p_n(x)$ $(n = 0, 1, 2, \cdots)$ defined by the generating function (GF)*

$$g(t)e^{xf(t)} = \sum_{n\geq 0} p_n(x)t^n \tag{5.32}$$

are called Sheffer-type polynomials with $p_0(x) = 1$. Accordingly, $p_n(D)$ with $D = d/dt$ is called Sheffer-type differential operator of degree n associated with $g(t)$ and $f(t)$. In particular, $p_0(D) = I$ is the identity operator.

The set of all Sheffer-type polynomial sequences $(p_n(x) = [t^n]g(t)e^{xf(t)})$ with an operation, "umbral composition" (cf. [186] and [187]), shown later forms a group called the Sheffer group. We will also see that the Riordan group and the Sheffer group are isomorphic.

In Roman's book [186], $(S_n = n!p_n(x))$ is called Sheffer sequence (also cf. [187]-[188]). Certain recurrence relations of $p_n(x)$ can be found in Hsu-Shiue's paper [134] . There are two special kinds of weighted Stirling numbers defined by Carlitz [32] (see also [26, 119]). We now give the following definition of the generalized Stirling numbers.

Definition 5.1.17 *Let $g(t)$ and $f(t)$ be defined as 5.1.16, and let*

$$\frac{1}{k!}g(t)(f(t))^k = \sum_{n\geq k} \sigma(n,k)\frac{t^n}{n!}. \tag{5.33}$$

Then $\sigma(n,k)$ is called the generalized Stirling number with respect to $g(t)$ and $f(t)$.

The special case of $g(t) = 1$ was studied in [126], and a more generalized case was studied in [87, 88].

As having been commonly employed in the Calculus of Finite Differences as well as in Combinatorial Analysis, the operators E, Δ, D are defined in Chapter 2. For the sake of readers' convenience, we recall them briefly below:

$$Ef(t) = f(t+1), \quad \Delta f(t) = f(t+1) - f(t), \quad Df(t) = \frac{d}{dt}f(t).$$

Powers of these operators are defined in the usual way. In particular, for any real numbers x, one may define $E^x f(t) = f(t+x)$. Also, the number 1 may be used as an identity operator, viz. $1f(t) \equiv f(t)$. It is easy to verify that these operators satisfy the formal relations (cf. [141])

$$E = 1 + \Delta = e^D, \quad \Delta = E - 1 = e^D - 1, \quad D = \log(1+\Delta).$$

From Definitions 5.1.16 and 5.1.17 we have

$$\begin{aligned} p_n(x) &= [t^n]g(t)e^{xf(t)} = [t^n]\sum_{k\geq 0}\frac{1}{k!}g(t)(f(t))^k x^k \\ &= \sum_{k=0}^{n} d_{n,k}\frac{x^k}{k!} = \frac{1}{n!}\sum_{k=0}^{n}\sigma(n,k)x^k, \end{aligned} \tag{5.34}$$

where we use $d_{n,k} = \sigma(n,k) = 0$ for all $k > n$. Therefore, with a constant multiple, $1/(k!)$, of the kth column, the rows of the Riordan array represent the coefficients of the Sheffer-type polynomial sequences. As an example, the rows of the Riordan array $(1/(1-t), t/(t-1)) = [(-1)^k\binom{n}{k}]_{0\leq k\leq n}$ give the coefficients of the Laguerre polynomial sequences $(p_n(x) = \sum_{k=0}^{n}(-1)^k\binom{n}{k}x^k/k!)_{0\leq n}$.

Let $(p_n(x) = \sum_{k=0}^{n} p_{n,k}x^k)$ and $(q_n(x) = \sum_{k=0}^{n} q_{n,k}x^k)$ be two Sheffer-type polynomial sequences. Then we define an operation, #, of $(p_n(x))$ and $(q_n(x))$, called the (polynomial) sequence multiplication (or the "umbral composition" see [18,19]), as

$$(p_n(x))\#(q_n(x)) := (r_n(x) = \sum_{k=0}^{n} r_{n,k}x^k), \tag{5.35}$$

where

$$r_{n,k} = \sum_{\ell=k}^{n} \ell! p_{n\ell}q_{\ell k}, \quad n\geq \ell\geq k. \tag{5.36}$$

It is clear that the defined operation is not commutative. Sheffer group under the "umbral composition" was defined with the $n!$-umbral calculus in Roman [186]. We now give a formulation with the matrix form of the 1-umbral calculus (cf. Sprugnoli [201]).

Theorem 5.1.18 *The set of all Sheffer-type polynomial sequences defined by Definition 5.1.16 with the operation # defined by (5.36) forms a group called the Sheffer group and denoted by* $\{(p_n(x)), \#\}$. *The identity of the group is* $(\frac{x^n}{n!})$. *The inverse of* $(p_n(x))$ *in the group, denoted by* $(p_n(x))^{(-1)}$, *is the Sheffer-type polynomial sequence generated by* $1/g(\bar{f}(t))exp(x\bar{f}(t))$, *where* $\bar{f}$ *is the compositional inverse of* f.

Proof. We now give a sketch of the proof. Some obvious details are omitted. Let $(p_n(x) = \sum_{k=0}^{n} p_{n,k}x^k)$, $(q_n(x) = \sum_{k=0}^{n} q_{n,k}x^k)$, and $(r_n(x) = \sum_{k=0}^{n} r_{n,k}x^k)$ be three Sheffer-type polynomial sequences. It can also be found the operation of the sequence multiplication satisfies the associative law, namely,

$$\begin{aligned} (p_n(x))\#((q_n(x))\#(r_n(x))) &= \left\{\sum_{k=0}^{n}\left(\sum_{u=k}^{n}\sum_{\ell=u}^{n}\ell!u!p_{n,\ell}q_{\ell,u}r_{u,k}\right)x^k\right\} \\ &= ((p_n(x))\#(q_n(x)))\#(r_n(x)). \end{aligned}$$

It is clear that $(p_n(x))\#(x^n/n!) = (x^n/n!)\#(p_n(x)) = (p_n(x))$. Hence, the set of all Sheffer-type polynomial sequences forms a group.

■

From (5.34) we can establish the mapping $\theta : (d_{n,k}) \mapsto (p_n(x))$ or $\theta : (g(t), f(t)) \mapsto (p_n(x))$ as follows.

$$\theta((d_{n,k})_{n\geq k\geq 0}) := \sum_{j=0}^{n} d_{n,j}x^j/j! = (d_{n,k})_{n\geq k\geq 0}X \tag{5.37}$$

for fixed n, where $X = (1, x, x^2/2!, \ldots)^T$, or equivalently,

$$\theta((g(t), f(t)) := [t^n]g(t)e^{xf(t)}. \tag{5.38}$$

It is clear that $(1, t)$, the identity Riordan array, maps to the identity Sheffer-type polynomial sequence $(p_n(x) \equiv x^n/(n!))_{n\geq 0}$. From the definitions 5.1.16 we immediately know that

$$p_n(x) = [t^n]g(t)e^{xf(t)} \text{ if and only if } d_{n,k} = [t^n]g(t)\left(f(t)\right)^k. \tag{5.39}$$

Hence, the mapping θ is one-to-one and onto. From the mapping defined by (5.37), we understand that the operation # defined in the Sheffer group is equivalent to the matrix multiplication of two Riordan arrays in the Riordan group. In [170], the connection between usual matrix multiplication and Riordan array multiplication is given. Hence, a connection between usual matrix multiplication and the Sheffer-type sequence multiplication can be established similarly. Using symbolic calculus with operators D and E, we find via (5.38) or Definition 5.1.16

$$g(t)h(f(t)) = g(t)E^{f(t)}h(0) = g(t)e^{f(t)D}h(0) = \sum_{k\geq 0} t^k p_k(D)h(0). \tag{5.40}$$

This is the desired expression given in [103] to expand the composite function $g(t)h(f(t))$. Hence, we have $p_n(D)h(0) = [t^n]g(t)h(f(t))$, which, from (5.39) and the multiplication in the Riordan group, is equivalent to

$$\sum_{k=0}^{n} d_{n,k}h_k = [t^n]g(t)h(f(t)), \tag{5.41}$$

where $(h_0, h_1, \ldots)$ has the generating function $h(t)$. The last expression was used to find the Riordan subgroup by Shapiro (cf. Section 4.3 or [195]). Equation (5.41) can also be considered as a linear transform to $h(t)$ or $(h_0, h_1, \ldots)$ represented by Riordan array $(g(t), f(t))$. Thus, (5.38) is the linear transform of e^{xt}. With the aid of (5.41), we may transfer a property of the Riordan group to the Sheffer group.

Example 5.1.19 *We now consider* $\left(\frac{1}{1-t}, \frac{t}{t-1}\right)$, *an involution in the Riordan group (cf. [28]), that possesses the matrix form*

$$(d_{n,k})_{n\geq k\geq 0} = \begin{bmatrix} 1 & & & & & \\ 1 & -1 & & & & \\ 1 & -2 & 1 & & & \\ 1 & -3 & 3 & -1 & & \\ 1 & -4 & 6 & -4 & 1 & \\ & \cdots & & \cdots & & \ddots \end{bmatrix}. \tag{5.42}$$

It is easy to find that

$$d_{n,k} = (-1)^k \binom{n}{k}.$$

Consequently,

$$p_n(x) = \theta(d_{n,k}) = \sum_{k=0}^{n} (-1)^k \binom{n}{k} \frac{x^k}{k!}, \tag{5.43}$$

which is the Laguerre polynomial of order zero. Conversely, for the given polynomials (5.43), we obtain the matrix (5.42), which entries satisfy

$$-d_{n,k-1} + d_{n,k} = d_{n+1,k}.$$

Hence, its generating functions satisfy

$$-tg(t)(f(t))^{k-1} + tg(t)(f(t))^k = g(t)(f(t))^k.$$

It follows that $f(t) = t/(t-1)$. *From the first column of matrix (5.42) we also obtain* $g(t) = 1/(1-t)$.

If the sequences $(p_n(x))$ and $(q_n(x))$ are mapped from the Riordan arrays $(d_{n,k})$ and $(c_{n,k})$, respectively, then from the defined operation of the polynomial sequence multiplication in (5.36), the coefficients of the polynomials $p_n(x)$ and $q_n(x)$ are respectively $p_{n,k} = d_{n,k}/(k!)$ and $q_{n,k} = c_{n,k}/(k!)$, and hence, we have

$$(k! r_{n,k})_{n\geq k\geq 0} = (d_{n,k})(c_{n,k}),$$

where $r_{n,k}$ are obtained in (5.36). Consequently, the Sheffer-type polynomial sequence $(r_n(x))$ is mapped from the Riordan array

$$(e_{n,k}) := (d_{n,k})(c_{n,k}),$$

where $e_{n,k} = \sum_{\ell=0}^{n} d_{n,\ell} c_{\ell,k}$ $(n \geq \ell \geq k)$.

Similarly, $(L_n^{(p-1)}(x))^{-1} = (L_n^{(p-1)}(x))$ because of

$$\left(\frac{1}{(1-t)^p}, \frac{t}{t-1}\right)^{-1} = \left(\frac{1}{(1-t)^p}, \frac{t}{t-1}\right).$$

Theorem 5.1.20 *The Sheffer group and the Riordan group are isomorphic.*

Proof. Let $(p_n(x) = \sum_{k=0}^{n} p_{n,k}x^k)$ and $(q_n(x) = \sum_{k=0}^{n} q_{n,k}x^k)$ be two Sheffer-type polynomial sequences mapped by θ from $(g(t), f(t))$ and $(d(t), h(t))$, respectively, i.e., $\theta(g(t), f(t)) = (p_n(x))$ and $\theta(d(t), h(t)) = (q_n(x))$ we have

$$\begin{aligned} & (p_n(x))\#(q_n(x)) = \theta\left((g(t), f(t))\right)\#\theta\left((d(t), h(t)\right) \\ = \; & \theta\left((g(t), f(t))(d(t), h(t)\right). \end{aligned} \tag{5.44}$$

Since mapping $\theta : (d_{n,k}) \mapsto (p_n(x))$ or equivalently, $\theta : (g(t), f(t)) \mapsto (p_n(x))$ is one-to-one and onto and satisfies (5.44), we obtain the theorem.

■

It is clear that the identity Sheffer polynomial sequence, $(x^n/n!)$, is the mapping from the Riordan array $(1, t)$. Hence, the inverse of a Sheffer-type polynomial sequence $(p_n(x))$, denoted by $(p_n(x))^{-1}$, is defined as $\theta\left(1/(g(\bar{f}(t))), \bar{f}(t)\right)$, where $(p_n(x)) = \theta\left(g(t), f(t)\right)$ and $\bar{f}(t)$ is the compositional inverse of $f(t)$.

Example 5.1.21 *We now consider an exponential Riordan array* $(\frac{t}{e^t-1}, t)$. *Since* $\theta\left(\frac{t}{e^t-1}, t\right) = \left(\frac{1}{n!}B_n(x)\right)$ *and* $\left(\frac{t}{e^t-1}, t\right)^{-1} = \left(\frac{e^t-1}{t}, t\right)$, *we have*

$$\begin{aligned} \left(\frac{1}{n!}B_n(x)\right)^{-1} &= \left(\frac{e^t-1}{t}, t\right) \\ &= \left(\frac{1}{n!}\sum_{k=0}^{n}\binom{n}{k}\frac{x^{n-k}}{k+1}\right). \end{aligned}$$

Similarly, $(L_n^{(p-1)}(x))^{-1} = (L_n^{(p-1)}(x))$ *because of*

$$\left(\frac{1}{(1-t)^p}, \frac{t}{t-1}\right)^{-1} = \left(\frac{1}{(1-t)^p}, \frac{t}{t-1}\right).$$

Example 5.1.22 *We now consider the sequence multiplication of the* $(p-1)$*st order Laguerre polynomial sequences generated by involution* $(\frac{1}{(1-t)^p}, \frac{t}{t-1})$ *in the Riordan group (cf. [28]),*

$$(L_n^{p-1}(x))\#(L_n^{p-1}(x)) = \left(\frac{1}{(1-t)^p}, \frac{t}{t-1}\right)\left(\frac{1}{(1-t)^p}, \frac{t}{t-1}\right) = (1, t),$$

or equivalently

$$(L_n^{p-1}(x))\#(L_n^{p-1}(x)) = \left(\frac{x^n}{n!}\right),$$

which implies that $(L_n^{p-1}(x))$ *is self-invertible. Thus, the following identity holds:*

$$\sum_{\ell=0}^{n}(-1)^{k+\ell}\frac{(n+p-1)!}{(p+k-1)!(n-\ell)!(\ell-k)!} = \delta_{n,k}, \tag{5.45}$$

where $\delta_{n,k}$ *is the Kronecker delta.*

Example 5.1.23 *Since the multiplication of two exponential Riordan arrays*

$$\left(\frac{t}{e^t-1}, t\right)\left(\frac{2}{e^t+1}, t\right) = \left(\frac{2t}{e^{2t}-1}, t\right),$$

we can present the result of the sequence operation of the Bernoulli polynomial sequence and the Euler polynomial sequence as

$$\left(\frac{1}{n!}B_n(x)\right) \# \left(\frac{1}{n!}E_n(x)\right) = \left(\frac{1}{n!}B_n\left(\frac{x}{2}\right)\right).$$

Remark 5.1.24 *Let* $p_n(x) = a_0 + a_1x + \cdots a_nx^n$, $a_i \in \mathbb{R}$, $n = 0, 1, \ldots$, *with the corresponding lower triangular array*

$$A = \begin{bmatrix} d_{0,0} & & & & \\ d_{1,0} & d_{1,1} & & & \\ d_{2,0} & d_{2,1} & d_{2,2} & & \\ \vdots & \vdots & \vdots & \ddots & \\ d_{n,0} & d_{n,1} & d_{n,2} & \cdots & d_{n,n} \\ \vdots & \vdots & \vdots & \vdots & \vdots \end{bmatrix}.$$

Then the Sheffer polynomial sequence $(p_n(x))$ *can be regarded as the matrix transformation* θ *defined by*

$$\theta : A \mapsto A \begin{bmatrix} 1 \\ x \\ \frac{x^2}{2!} \\ \vdots \\ \frac{x^n}{n!} \\ \vdots \end{bmatrix}. \tag{5.46}$$

By using (5.46), we may find subgroups of the Sheffer group from the corresponding subgroups of the Riordan group. In addition, the above consideration can be extended to the higher dimensional setting.

Furthermore, let $\theta_A X = (p_n(x))$ and $\theta_B X = (q_n(x))$, and $X = \left[1, x, \frac{x^2}{2!}, \cdots, \frac{x^n}{n!}, \ldots\right]^T$, where $(p_n(x))$ and $(q_n(x))$ are two sequence elements in the Sheffer group with respect to the corresponding Riordan arrays A and B, respectively. Then the operation defined by (5.35) and (5.36) can be written as

$$(p_n(x))\#(q_n(x)) := (\theta_{AB}) \begin{bmatrix} 1 \\ x \\ \frac{x^2}{2!} \\ \vdots \\ \frac{x^n}{n!} \\ \vdots \end{bmatrix} = (AB) \begin{bmatrix} 1 \\ x \\ \frac{x^2}{2!} \\ \vdots \\ \frac{x^n}{n!} \\ \vdots \end{bmatrix},$$

where AB is the regular matrix multiplication of A and B. Based on this point of view, $((p_n(x)), \#, +)$ can be considered as a ring, where $+$ is a used addition of matrices.

Definition 5.1.25 *Let $(p_n(x))$ and $(q_n(x))$ be two Sheffer polynomial sequences. We say $(p_n(x))$ and $(q_n(x))$ are combinatorial orthogonal if they satisfy*

$$(p_n(x))\#(q_n(x)) = (q_n(x))\#(p_n(x)) = \left(\frac{x^n}{n!}\right),$$

and we denote $(p_n(x)) \perp_{com} (q_n(x))$.

Example 5.1.26 *Laguerre polynomial sequence is combinatorial orthogonal, i.e.,*

$$(L_n^{(p-1)}(x)) \perp_{com} (L_n^{(p-1)}(x)).$$

Although Laguerre polynomials are also analytic orthogonal, i.e., orthogonal in an inner product sense, it is not necessary that the analytic orthogonality implies the combinatorial orthogonality and vice versa. For instance, $\theta(1/(1-t), t) = (1, 1+x, 1+x+x^2/2, \ldots)$ and $\theta(1-t, t) = (1, -1+x, -x+x^2/2, \ldots)$ are combinatoric orthogonal, but not analytic orthogonal.

$g(t)$	$f(t)$	$p_n(x)$	*Name of polynomials*
$t/(e^t-1)$	t	$\frac{1}{n!}B_n(x)$	Bernoulli
$2/(e^t+1)$	t	$\frac{1}{n!}E_n(x)$	Euler
e^t	$\log(1+t)$	$(PC)_n(x)$	Poisson-Charlier
$e^{-\alpha t}(\alpha \neq 0)$	$\log(1+t)$	$\hat{C}_n^{(\alpha)}(x)$	Charlier
1	$\log(1+t)\ /\ (1-t)$	$(ML)_n(x)$	Mittag-Leffler
$(1-t)^{-1}$	$\log(1+t)\ /\ (1-t)$	$p_n(x)$	Pidduck
$(1-t)^{(-p)}(p>0)$	$t/(t-1)$	$L_n^{(p-1)}(x)$	Laguerre
$e^{\lambda t}(\lambda \neq 0)$	$1-e^t$	$(Tos)_n^{(\lambda)}(x)$	Toscano
1	e^t-1	$\tau_n(x)$	Touchard
$1/(1+t)$	$t/(t-1)$	$A_n(x)$	Angelescu
$(1-t)/(1+t)^2$	$t/(t-1)$	$(De)_n(x)$	Denisyuk
$(1-t)^{-p}(p>0)$	e^t-1	$T_n^{(p)}(x)$	Weighted-Touchard

At the end of this subsection, we give the above list of some Sheffer polynomials for the interested readers. In the table, we can see many array components are exponential Riordan array components.

We now consider the definition of the Riordan pairs.

Definition 5.1.27 *Let $g(t)$ and $f(t)$ be given as in Definition 5.1.16. Then we have a Riordan pair $\{d_{n,k}, \bar{d}_{n,k}\}$ defined by*

$$\left\{ \begin{array}{l} g(t)(f(t))^k = \sum_{n=k}^{\infty} d_{n,k} t^n, \\ g(\bar{f}(t))^{-1}(\bar{f}(t))^k = \sum_{n=k}^{\infty} \bar{d}_{n,k} t^n, \end{array} \right. \tag{5.47}$$

where $\bar{f}$ is the compositional inverse of f with $\bar{f}(0) = 0$, $[t]\bar{f}(t) \neq 0$ (i.e., $f \in \mathcal{F}_1$), and $d_{0,0} = \bar{d}_{0,0} = 1$.

We also need the following definition of generalized Stirling number pairs (cf. [126] for the case of $A \equiv 1$, and a special example has been also studied in [201]))

Definition 5.1.28 *Let $g(t)$ and $f(t)$ be given as in Definition 5.1.16. Then we have a generalized Stirling number pair $\{\sigma(n,k), \bar{\sigma}(n,k)\}$ as defined by*

$$\left\{ \begin{array}{l} \frac{1}{k!} g(t)(f(t))^k = \sum_{n=k}^{\infty} \sigma(n,k) \frac{t^n}{n!}, \\ \frac{1}{k!} g(\bar{f}(t))^{-1}(\bar{f}(t))^k = \sum_{n=k}^{\infty} \bar{\sigma}(n,k) \frac{t^n}{n!}, \end{array} \right. \tag{5.48}$$

where $\bar{g} \equiv g^{\langle -1 \rangle}$ is the compositional inverse of g with $\bar{f}(0) = 0$, $[t]\bar{f}(t) \neq 0$ (i.e., $f \in \mathcal{F}_1$), and $\sigma(0,0) = \bar{\sigma}(0,0) = 1$.

Remark 5.1.29 *A closed connection between (5.47) and (5.48) is apparent. A special case of pair (5.48) for $g(t) = 1$ was established in [125], which were later applied to derive some combinatorial identities (cf. [227]).*

Remark 5.1.30 *If in (5.48) let $g(t)$ and $f(t)$ be defined by*

$$g(t) = (1 + \alpha t)^{\gamma/\alpha}, \quad f(t) = \left((1 + \alpha t)^{\beta/\alpha} - 1\right)/\beta,$$

where α, β, and γ are real or complex numbers with $\alpha\beta \neq 0$, then

$$g(\bar{f}(t)) = (1 + \beta t)^{\gamma/\beta}, \quad \bar{f}(t) = \left((1 + \beta t)^{\alpha/\beta} - 1\right)/\alpha,$$

so that $\sigma(n,k) = S(n,k;\alpha,\beta,\gamma)$ and $\bar{\sigma}(n,k) = S(n,k;\beta,\alpha,-\gamma)$ just form a pair of GSN's with three parameters α, β, and γ. Note that such a class of GSN-pairs includes various useful special number-pairs. A detailed investigation of GSNs was given in [132] in 1998. For a very recent development relating to this subject, see [181]. Many of these results have been surveyed in Chapters 2 and 3.

Note that (5.47) and (5.48) imply the orthogonality relations

$$\sum_{k\le n\le m} d_{m,n}\bar{d}_{n,k} = \sum_{k\le n\le m} \bar{d}_{m,n}d_{n,k} = \delta_{m,k}$$

and

$$\sum_{k\le n\le m} \sigma(m,n)\bar{\sigma}(n,k) = \sum_{k\le n\le m} \bar{\sigma}(m,n)\sigma(n,k) = \delta_{m,k}$$

with $\delta_{m,k}$ denoting the Kronecker delta, and it follows that there hold the inverse relations

$$\frac{f_n}{n!} = \sum_{k=0}^{n} d_{n,k}\frac{g_k}{k!} \Longleftrightarrow \frac{g_n}{n!} = \sum_{k=0}^{n} \bar{d}_{n,k}\frac{f_k}{k!}. \tag{5.49}$$

and

$$f_n = \sum_{k=0}^{n} \sigma(n,k)g_k \Longleftrightarrow g_n = \sum_{k=0}^{n} \bar{\sigma}(n,k)f_k. \tag{5.50}$$

For an element $(p_n(x))$ in the Sheffer group $\{(p_n(x)), \#\}$, it is easy to write its inverse $(\bar{p}_n(x)) = (p_n(x))^{-1}$ as

$$\bar{p}_n(x) = \sum_{k=0}^{n} \bar{d}_{n,k}\frac{x^k}{k!} = \frac{1}{n!}\sum_{k=0}^{n} \bar{\sigma}(n,k)x^k,$$

which are generated by

$$g(\bar{f}(t))^{-1}e^{x\bar{f}(t)} = \sum_{n\ge 0} \bar{p}_n(x)t^n,$$

with $\bar{p}_0(x) = 1$.

We shall give an application of the inverse relations (5.49) and (5.50) based on the following result.

Theorem 5.1.31 *The Sheffe-type operator $p_n(D)$ has an expression of the form*

$$p_n(D) = \frac{1}{n!}\sum_{k=0}^{n} \sigma(n,k)D^k, \tag{5.51}$$

where $\sigma(n, k)$ (associated with $g(t)$ and $f(t)$) may be written in the form

$$\sigma(n,k) = \sum_{r=k}^{n} \binom{n}{r} \alpha_{n-r}B_{rk}(a_1, a_2, \cdots), \tag{5.52}$$

where B_{rk} are incomplete exponential Bell polynomials (cf. (4.198) in Subsection 4.3.2), provided that $g(t) = \sum_{m\ge 0} \alpha_m t^m/m!$ and $f(t) = \sum_{m\ge 1} a_m t^m/m!$ with $\alpha_0 = 1, a_1 \ne 0$.

Proof. Note that (5.52) follows from (5.34). Moreover, recall a known expression for potential polynomials (cf. e.g., Comtet [44, § 3.5, Theorem B, etc.]). We have

$$\frac{1}{k!}(f(t))^k = \sum_{r\geq k}^{\infty} \frac{t^r}{r!} B_{rk}(a_1, a_2, \cdots).$$

Substituting this into (5.48) and comparing the resulting expression with the RHS of (5.48), we see that (5.52) is true.

■

Corollary 5.1.32 *Formula (5.39) may be rewritten in the form*

$$\begin{aligned} g(t)h(f(t)) &= \sum_{n=0}^{\infty} \frac{t^n}{n!} (\sum_{k=0}^{n} \sigma(n,k) f^{(k)}(0)) \\ &= \sum_{n=0}^{\infty} t^n (\sum_{k=0}^{n} \frac{d_{n,k}}{k!} f^{(k)}(0)), \end{aligned} \tag{5.53}$$

where $\sigma(n,k)$'s are defined by (5.47) and represented by (5.52).

Corollary 5.1.33 *For the case $g(t) = 1$, equation (5.52) gives*

$$\sigma(n,k) = B_{n,k}(a_1, a_2, \cdots)$$

and

$$d_{n,k} = \frac{k!}{n!} B_{n,k}(a_1, a_2, \cdots),$$

where the incomplete Bell polynomial $B_{n,k}(a_1, a_2, \cdots)$ has an explicit expression (cf. (4.198) in Subsection 4.3.2 or Comtet [44])

$$B_{n,k}(a_1, a_2, \cdots) = \sum_{(c)} \frac{n!}{c_1! c_2! \cdots} \left(\frac{a_1}{1!}\right)^{c_1} \left(\frac{a_2}{2!}\right)^{c_2} \cdots,$$

where the summation extends over all integers $c_1, c_2, \cdots \geq 0$, such that $c_1 + 2c_2 + 3c_3 + \cdots = n, c_1 + c_2 + \cdots = k$.

Corollary 5.1.34 *The generalized exponential polynomials related to the generalized Stirling numbers $\sigma(n,k)$ and $\bar{\sigma}(n,k)$ are given respectively by the following*

$$n! p_n(x) = \sum_{k=0}^{n} \sigma(n,k) x^k \tag{5.54}$$

and

$$n! \bar{p}_n(x) = \sum_{k=0}^{n} \bar{\sigma}(n,k) x^k, \tag{5.55}$$

where $p_n(x)$ and $\bar{p}_n(x)$ are Sheffer-type polynomials associated with $\{g(t), f(t)\}$ and $\{g(\bar{f}(t))^{-1}, \bar{f}(t)\}$, respectively

Applying the reciprocal relations (5.50) to (5.54) and (5.55) we get

Corollary 5.1.35 *There hold the relations*

$$\sum_{k=0}^{n} \bar{\sigma}(n,k)k!p_k(x) = x^n \tag{5.56}$$

and

$$\sum_{k=0}^{n} \sigma(n,k)k!\bar{p}_k(x) = x^n. \tag{5.57}$$

These may be used as the recurrence relations for $p_n(x)$ *and* $p_n^*(x)$*, respectively.*

Equations (5.56) and (5.57) are equivalently

$$\begin{cases} \sum_{k=0}^{n} \bar{d}_{n,k}p_k(x) = \frac{x^n}{n!}, \\ \sum_{k=0}^{n} d_{n,k}\bar{p}_k(x) = \frac{x^n}{n!}. \end{cases} \tag{5.58}$$

Evidently (5.50) and (5.53) imply a higher derivative formula for $g(t)h(f(t))$ at $t = 0$, namely

$$(\frac{d^n}{dt^n})\,(g(t)h(f(t)))|_{t=0} = \sum_{k=0}^{n} \sigma(n,k)f^{(k)}(0) = n!p_n(D)h(0).$$

Certainly, this will reduce to the Faà di Bruno formula when $g(t) = 1$.

Example 5.1.36 *As a simple instance, take* $\{\sigma(n,k), \bar{\sigma}(n,k)\}$ *to be the ordinary Stirling numbers* $\{(-1)^{n-k}\left[{n \atop k}\right], \left\{{n \atop k}\right\}\}$ *of the 1st and 2nd kinds. Then (5.55) yields the Bell number* $W(n)$ *at* $x = 1$*, namely,*

$$W(n) = n!\bar{p}_n(1) = \sum_{k=0}^{n} \left\{{n \atop k}\right\}.$$

Consequently, (5.57) gives the simple identity

$$\sum_{k=0}^{n} (-1)^{n-k} \left[{n \atop k}\right] W(k) = 1.$$

More examples could be constructed using Sheffer polynomials listed in the table shown above.

5.1.3 Higher dimensional extension of the Riordan group

We now extend the Riordan group to the higher dimensional setting. In what follows we shall adopt the multi-index notational system. Denote

$$\begin{aligned} &\hat{t} = (t_1, \cdots, t_r),\ \hat{x} = (x_1, \cdots, x_r), \\ &\hat{t} + \hat{x} = (t_1 + x_1, \cdots, t_r + x_r), \end{aligned}$$

$$\hat{0} = (0, \cdots, 0),\ \widehat{D} = (D_1, \cdots, D_r),$$
$$\widehat{g(t)} = (g_1(t_1), \cdots, g_r(t_r)),$$
$$\hat{x} \cdot \widehat{D} = \sum_{i=1}^{r} x_i D_i,\ \hat{x} \cdot \widehat{g(t)} = \sum_{i=1}^{r} x_i g_i(t_i),$$

where we define $D_i = \partial/\partial t_i$. Also, E_i means the shift operator acting on t_i, namely for $1 \le i \le r$,

$$\begin{aligned} E_i f(\cdots, t_i, \cdots) &= f(\cdots, t_i + 1, \ldots), \\ E_i^{x_i} f(\cdots, t_i, \cdots) &= f(\cdots, t_i + x_i, \cdots). \end{aligned}$$

Formally we may denote $E_i = e^{D_i} = \exp(\partial/\partial t_i)$. Moreover, we write $t^\lambda := t_1^{\lambda_1} \cdots t_r^{\lambda_r}$ with $\lambda := \hat{\lambda} = (\lambda_1, \cdots, \lambda_r), r$ being positive integer. Also, $\lambda \ge \hat{0}$ means $\lambda_i \ge 0$ $(i = 1, \cdots, r)$, and $\lambda \ge \mu$ means $\lambda_i \ge \mu_i$ for all $i = 1, \cdots, r$. Since we assume that all series expansions are formal so that the symbolic calculus with formal differentiation operator $\widehat{D}$ and shift operator $E = (E_1, \cdots, E_r)$ can be applied to all formal series, where $E^{\hat{x}}$ $(\hat{x} \in \mathbb{C}^r)$ is defined by

$$E^{\hat{x}} f(\hat{t}) = E_1^{x_1} \cdots E_r^{x_r} f(\hat{t}) = f(\hat{t} + \hat{x}) \quad \hat{x} \in \mathbb{C}^r,$$

and satisfies the formal relations

$$E^{\hat{x}} = e^{\hat{x} \cdot \widehat{D}}, \quad \hat{x} \in \mathbb{C}^r, \tag{5.59}$$

because of the following formal process:

$$E^{\hat{x}} f(\hat{0}) = f(\hat{x}) = \sum_{k=0}^{\infty} \frac{1}{k!} (\hat{x} \cdot \widehat{D})^k f(\hat{0}) = e^{\hat{x} \cdot \widehat{D}} f(\hat{0}).$$

We first give an analog of Definition 5.1.16

Definition 5.1.37 *Let $\hat{t} = (t_1, t_2, \cdots, t_r)$, $g(\hat{t})$, $\widehat{f(t)} = (f_1(t_1), f_2(t_2), \cdots, f_r(t_r))$ and $h(\hat{t})$ be any given formal power series over the complex number field $\mathbb{C}^r$ with $g(\hat{0}) = 1$, $f_i(0) = 0$ and $f_i'(0) \neq 0$ $(i = 1, 2, \cdots, r)$. Then the polynomials $p_{\hat{n}}(\hat{x})$ $(\hat{n} \in \mathbb{N}^r \cup \hat{0})$ as defined by the generating function*

$$g(\hat{t}) e^{\hat{x} \cdot \widehat{f(t)}} = \sum_{\hat{n} \ge \hat{0}} p_{\hat{n}}(\hat{x}) t^{\hat{n}} \tag{5.60}$$

are called Sheffer-type polynomials with $p_{\hat{0}}(\hat{x}) = 1$. Accordingly, $p_{\hat{n}}(\hat{D})$ with $\hat{D} \equiv (D_1, D_2 \cdots, D_r)$ is called Sheffer-type differential operator of degree $\hat{n}$ associated with $g(\hat{t})$ and $\widehat{f(t)}$. In particular, $p_{\hat{0}}(\hat{D}) = I$ is the identity operator.

For the formal power series $h(\hat{t})$, the coefficient of $t^\lambda = (t_1^{\lambda_1}, t_2^{\lambda_2}, \cdots, t_r^{\lambda_r})$ is usually denoted by $[t^\lambda]h(\hat{t})$. Accordingly, (5.60) is equivalent to the expression $p_\lambda(\hat{x}) = [t^\lambda]g(\hat{t})e^{\hat{x}\cdot\widehat{f(t)}}$. Also, we shall frequently use the notation

$$p_\lambda(\hat{D})h(\hat{0}) = [p_\lambda(\hat{D})h(\hat{t})]_{\hat{t}=\hat{0}} \tag{5.61}$$

and $\lambda! \equiv \hat{\lambda}! = \lambda_1!\lambda_2!\cdots\lambda_r!$.

Definition 5.1.38 *Let $g(\hat{t})$ and $\widehat{f(t)}$ be any formal power series defined on $\mathbb{C}^r$, with $g(\hat{0}) = 1, f_i(0) = 0$ and $f_i'(0) \neq 0$ $(i = 1, 2, \cdots, r)$. Then we have a multivariate weighted Stirling-type pair $\{\sigma(\hat{n}, \hat{k}), \sigma^*(\hat{n}, \hat{k})\}$ as defined by*

$$\frac{1}{\hat{k}!}g(\hat{t})\Pi_{i=1}^r \left(f_i(t_i)\right)^{k_i} = \sum_{\hat{n}\geq\hat{k}} \sigma(\hat{n}, \hat{k})\frac{t^{\hat{n}}}{\hat{n}!} \tag{5.62}$$

$$\frac{1}{\hat{k}!}g(\widehat{\bar{f}(t)})^{-1}\Pi_{i=1}^r \left(\bar{f}_i(t_i)\right)^{k_i} = \sum_{\hat{n}\geq\hat{k}} \sigma^*(\hat{n}, \hat{k})\frac{t^{\hat{n}}}{\hat{n}!}, \tag{5.63}$$

where $\widehat{\bar{f}(t)} = (\bar{f}_1(t_1), \bar{f}_2(t_2), \cdots, \bar{f}_r(t_r))$, $\bar{f}_i$ is the compositional inverse of f_i $(i = 1, 2, \cdots, r)$ with $\bar{f}_i(0) = 0$, $[t_i]\bar{f}_i(t_i) \neq 0$, and $\sigma(\hat{0}, \hat{0}) = \sigma^*(\hat{0}, \hat{0}) = 1$. We call $\sigma(\hat{n}, \hat{k})$ the dual of $\sigma^*(\hat{n}, \hat{k})$ and vice versa. We will also call

$$d_{\hat{n},\hat{k}} := \frac{\hat{k}}{\hat{n}}\sigma(\hat{n}, \hat{k})$$
$$d^*_{\hat{n},\hat{k}} := \frac{\hat{k}}{\hat{n}}\sigma^*(\hat{n}, \hat{k})$$

the multivariate Riordan arrays and denote them by $\left(g(\hat{t}), \widehat{f(t)}\right)$ and $\left(1/g(\widehat{\bar{f}(t)}), \widehat{\bar{f}(t)}\right)$, respectively.

Example 5.1.39 *As an example, considering $g(\hat{t}) = 1$ and $\widehat{f(t)} = \hat{t}$, we obtain the $p_\lambda(\hat{x})$ defined by (5.60), namely,*

$$p_\lambda(\hat{x}) = \frac{x^\lambda}{\lambda!}.$$

Thus, the multivariate weighted Stirling-type pair $\{\sigma(\hat{n}, \hat{k}), \sigma^*(\hat{n}, \hat{k})\}$ defined as (5.62) and (5.63) is $(\sigma(\hat{n}, \hat{k}), \sigma^*(\hat{n}, \hat{k}))$, where

$$\sigma(\hat{n}, \hat{k}) = \sigma^*(\hat{n}, \hat{k}) = \delta_{\hat{n},\hat{k}}.$$

The corresponding multivariate Riordan array pair is $(d_{\hat{n},\hat{k}}, d^*_{\hat{n},\hat{k}})$, where

$$d_{\hat{n},\hat{k}} = d^*_{\hat{n},\hat{k}} = \frac{\hat{k}}{\hat{n}}\delta_{\hat{n},\hat{k}}.$$

A similar argument as (5.37) and (5.38) can be established as follows.

Theorem 5.1.40 *The equations (5.62) and (5.63) imply the biorthogonality relations*

$$\sum_{\hat{m}\geq\hat{n}\geq\hat{k}} \sigma(\hat{m},\hat{n})\sigma^*(\hat{n},\hat{k}) = \sum_{\hat{m}\geq\hat{n}\geq\hat{k}} \sigma^*(\hat{m},\hat{n})\sigma(\hat{n},\hat{k}) = \delta_{\hat{m},\hat{k}} \tag{5.64}$$

with $\delta_{\hat{m},\hat{k}}$ denoting the Kronecker delta, i.e., $\delta_{\hat{m},\hat{k}} = 1$ if $\hat{m} = \hat{k}$ and 0 otherwise, and it follows that there hold the inverse relations

$$f_{\hat{n}} = \sum_{\hat{n}\leq\hat{k}\leq\hat{0}} \sigma(\hat{n},\hat{k})g_{\hat{k}} \Longleftrightarrow g_{\hat{n}} = \sum_{\hat{n}\leq\hat{k}\leq\hat{0}} \sigma^*(\hat{n},\hat{k})f_{\hat{k}}. \tag{5.65}$$

Proof. Transforming t_i by $\bar{f}_i(t_i)$ in (5.62) and multiplying $A\left(\widehat{\bar{f}(t)}\right)^{-1}(\hat{k}!)$ on the both sides of the resulting equation yields

$$t^{\hat{k}} = \sum_{\hat{n}\geq\hat{k}} \sigma(\hat{n},\hat{k})\frac{\hat{k}!}{\hat{n}!}g(\widehat{\bar{f}(t)})^{-1}\Pi_{i=1}^r\left(\bar{f}_i(t_i)\right)^{n_i}. \tag{5.66}$$

By substituting (5.63) into the above equation, we obtain

$$\begin{aligned} t^{\hat{k}} &= \sum_{\hat{n}\geq\hat{k}} \sigma(\hat{n},\hat{k}) \sum_{\hat{m}\geq\hat{n}} \sigma^*(\hat{m},\hat{n})\frac{\hat{k}!}{\hat{m}!}t^{\hat{m}} \\ &= \sum_{\hat{m}\geq\hat{k}} \frac{\hat{k}!}{\hat{m}!}t^{\hat{m}} \sum_{\hat{m}\geq\hat{n}\geq\hat{k}} \sigma^*(\hat{m},\hat{n})\sigma(\hat{n},\hat{k}). \end{aligned}$$

Equating the coefficients of the terms $t^{\hat{m}}$ on the leftmost side and the rightmost side of the above equation leads (5.64) and (5.65). This completes the proof. ■

Remark 5.1.41 *$[\sigma(\hat{m},\hat{n})]$ and $[\sigma^*(\hat{n},\hat{k})]$ are a pair of inverse r-dimensional matrices, which may be useful in the higher dimensional matrix theory.*

From (5.64), we can see that the $2r$ dimensional infinite matrices $\sigma(\hat{n},\hat{k})$ and $\sigma^*(\hat{n},\hat{k})$ are invertible for each other, i.e., their product is the identity matrix $\left[\delta_{\hat{n},\hat{k}}\right]_{\hat{n}\geq\hat{k}\geq\hat{0}}$.

By introducing group multiplication

$$\left(g(\hat{t}),\widehat{f(t)}\right) * \left(d(\hat{t}),\widehat{h(t)}\right) = \left(g(\hat{t})d(\widehat{f(t)}),\widehat{h(f(t))}\right), \tag{5.67}$$

where $\widehat{h(f(t))} = (h_1(g_1(t_1)),\cdots,h_r(g_r(t_r)))$, from Theorem 5.1.40, we immediately see that the inverse of $\left(g(\hat{t}),\widehat{f(t)}\right)$ is $\left(1/g(\widehat{\bar{f}(t)}),\widehat{\bar{f}(t)}\right)$ because their multiplication result is the identity $I = (1,\hat{t})$. Hence, similar to [196], we obtain the following corollary.

Corollary 5.1.42 *Let $g(\hat{t})$ and $\widehat{f(t)}$ be any formal power series defined on $\mathbb{C}^r$, with $g(\hat{0}) = 1, g_i(0) = 0$ and $g_i'(0) \neq 0$ ($i = 1, 2, \cdots, r$). Then with respect to the multiplication defined by (5.67), $\{\left(g(\hat{t}), \widehat{f(t)}\right)\}$ forms a group with the identity $I = (1, \hat{t})$ and for any element $\left(g(\hat{t}), \widehat{f(t)}\right)$ in the group, its inverse is $\left(1/g(\widehat{\bar{f}(t)}), \widehat{\bar{f}(t)}\right)$, where $\widehat{\bar{f}(t)} = (\bar{f}_1(t_1), \bar{f}_2(t_2), \cdots, \bar{f}_r(t_r))$, $\bar{f}_i$ is the compositional inverse of f_i ($i = 1, 2, \cdots, r$) with $\bar{f}_i(0) = 0$, $[t_i]\bar{f}_i(0) \neq 0$.*

Proof. This proof is an analog of the proof on the one variable Riordan group (cf. [196]). Indeed, from (5.67) we have

$$\left(g(\hat{t}), \widehat{f(t)}\right) I = \left(g(\hat{t}), \widehat{f(t)}\right),$$

$$\begin{aligned} & \left(\left(g(\hat{t}), \widehat{f(t)}\right)\left(d(\hat{t}), \widehat{h(t)}\right)\right)\left(u(\hat{t}), \widehat{v(t)}\right) \\ = & \left(g(\hat{t})d\left(\widehat{f(t)}\right)u\left(\widehat{h(f(t))}\right), v(\widehat{h(f(t)))}\right) \\ = & \left(g(\hat{t}), \widehat{f(t)}\right)\left(\left(d(\hat{t}), \widehat{h(t)}\right)\left(u(\hat{t}), \widehat{v(t)}\right)\right) \end{aligned}$$

and

$$\begin{aligned} & \left(g(\hat{t}), \widehat{f(t)}\right)\left(\frac{1}{g\left(\widehat{\bar{f}(t)}\right)}, \widehat{\bar{f}(t)}\right) = \left(g(\hat{t})\frac{1}{g\left(\widehat{\bar{f}(f(t))}\right)}, \widehat{\bar{f}(f(t))}\right) \\ = & (1, \hat{t}) = I. \end{aligned}$$

This completes the proof of the corollary. ■

From Definition 5.1.37, we have

$$p_\lambda(\hat{x}) = [t^\lambda]g(\hat{t})e^{\hat{x}\cdot\widehat{f(t)}} = [t^\lambda]\sum_{\hat{k}\geq\hat{0}}\frac{1}{\hat{k}!}g(\hat{t})\Pi_{i=1}^r\left(g_i(t_i)\right)^{\lambda_i}x_i^{\lambda_i} = \sum_{\lambda\geq k\geq\hat{0}}d_{\lambda,\hat{k}}\frac{x^\lambda}{\hat{k}!}.$$

Therefore, we establish a one-to-one and onto mapping θ^r from $(d_{\hat{n},\hat{k}})$ to $p_{\hat{n}}(\hat{x})$, where $\theta^1 = \theta$ shown in (5.43). By defining the operation, denoted as #, to two higher dimensional Sheffer type polynomial sequences, $(p_{\hat{n}}(\hat{x}) = \sum_{\hat{n}\geq\lambda\geq\hat{0}}p_{\hat{n},\lambda}x^\lambda)$ and $(q_{\hat{n}}(\hat{x}) = \sum_{\hat{n}\geq\lambda\geq\hat{0}}q_{\hat{n},\lambda}x^\lambda)$, as follows, the set $\{(p_{\hat{n}}), \#\}$ forms a group, called the higher dimensional Sheffer group.

$$(p_{\hat{n}})\#(q_{\hat{n}}) = (\sum_{\hat{n}\geq\lambda\geq\hat{0}}r_{\hat{n},\lambda}x^\lambda),$$

where

$$r_{\hat{n},\lambda} = \sum_{\hat{n}\geq\hat{\ell}\geq\lambda}\hat{\ell}!p_{\hat{n},\hat{\ell}}q_{\hat{\ell},\lambda}.$$

Similar to Theorem 5.1.20, we can establish the following result.

Theorem 5.1.43 *The set $\{(p_{\hat{n}}), \#\}$ with the operation $\#$ is a group, called the higher dimensional Sheffer group that is isomorphic to the higher dimensional Riordan group defined in Corollary 5.1.42.*

Example 5.1.44 *As an example of (5.60), we set $g(\hat{t}) = 1$ and $exp(\hat{x}\cdot\widehat{f(t)}) = exp(x_1(e^{t_1} - 1) + x_2(e^{t_2} - 1) + \cdots + x_r(e^{t_r} - 1))$ in (5.60) and obtain*

$$exp(\hat{x}\cdot\widehat{f(t)}) = \sum_{\lambda\geq\hat{0}} \hat{\tau}_\lambda(\hat{x})t^\lambda, \tag{5.68}$$

where

$$\hat{\tau}_\lambda(\hat{x}) = \Pi_{j=1}^r \tau_{\lambda_j}(x_j)$$

and $\tau_u(t)$ is the Touchard polynomial of degree u. Hence, we may call $\hat{\tau}_\lambda(\hat{x})$ the higher dimensional Touchard polynomial of order λ.

Example 5.1.45 *Sheffer-type expansion (5.60) also includes the following two special cases as shown in [148]. Let $g(\hat{t}) = 2^m/(exp\sum_{i=1}^r t_i + 1)^m$ and $exp\left(\hat{x}\cdot\widehat{f(t)}\right) = exp\left(\sum_{i=1}^r x_i t_i\right)$. Then the corresponding Sheffer-type expansion of (5.60) shown in [148] has the form*

$$g(\hat{t})exp\left(\hat{x}\cdot\widehat{f(t)}\right) = \sum_{\lambda\geq\hat{0}} \frac{E_\lambda^{(m)}(\hat{x})}{\lambda!}t^\lambda,$$

where $E_\lambda^{(m)}(\hat{x})$ $(\lambda \geq \hat{0})$ is the mth order r-variable Euler's polynomial defined in [148].

Similarly, substituting $g(\hat{t}) = \left(\sum_{i=1}^r t_i\right)^m/(exp\sum_{i=1}^r t_i - 1)^m$ and $exp\left(\hat{x}\cdot\widehat{f(t)}\right) = exp\left(\sum_{i=1}^r x_i t_i\right)$ into (5.60) yields

$$g(\hat{t})exp\left(\hat{x}\cdot\widehat{f(t)}\right) = \sum_{\lambda\geq\hat{0}} \frac{B_\lambda^{(m)}(\hat{x})}{\lambda!}t^\lambda,$$

where $B_\lambda^{(m)}(\hat{x})$ $(\lambda \geq \hat{0})$ is called in [148] the mth order r-variable Bernoulli polynomial. Some basic properties of $E_\lambda^{(m)}(\hat{x})$ and $B_\lambda^{(m)}(\hat{x})$ were studied in [148].

Since

$$\begin{aligned}&\left(\frac{2^m}{\left(exp\sum_{i=1}^r t_i + 1\right)^m}, \hat{t}\right)\left(\frac{\left(\sum_{i=1}^r t_i\right)^m}{\left(exp\sum_{i=1}^r t_i + 1\right)^m}, \hat{t}\right)\\ = &\left(\frac{\left(\sum_{i=1}^r 2t_i\right)^m}{\left(exp\sum_{i=1}^r 2t_i + 1\right)^m}, \hat{t}\right),\end{aligned}$$

we have

$$\left(E_\lambda^{(m)}(\hat{x})\right)\#\left(B_\lambda^{(m)}(\hat{x})\right) = \left(B_\lambda^{(m)}(2\hat{x})\right).$$

Example 5.1.46 *For the case* $g(\hat{t}) = 1$ *and* $\widehat{f(t)} = \sum_{\hat{m}\geq(1,\cdots,1)} a_m\ t^m/(\hat{m})!$, *where* $a_m = a^{(1)}{}_{m_1}\cdots a^{(r)}{}_{m_r}$, *it follows that* $e^{\hat{x}\cdot\widehat{f(t)}}$ *can be written in the form*

$$\exp(\hat{x}\cdot\widehat{f(t)}) = \Pi_{\ell=1}^{r} exp\ \{x_\ell \sum_{m_\ell\geq 1} a^{(\ell)}{}_{m_\ell}\frac{t_\ell{}^{m_\ell}}{m_\ell!}\}$$

$$= \Pi_{\ell=1}^{r}\left(1+\sum_{k_\ell\geq 1}\frac{t_\ell{}^{k_\ell}}{k_\ell!}\{\sum_{j_\ell=1}^{k_\ell} x_\ell{}^{j_\ell}B_{k_\ell j_\ell}(a^{(\ell)}{}_1, a^{(\ell)}{}_2,\cdots)\}\right) \quad (5.69)$$

so that

$$p_\lambda(\hat{x}) = [t^\lambda]e^{\hat{x}\cdot\widehat{f(t)}} = \Pi_{\ell=1}^{r}\frac{1}{\lambda_\ell!}\sum_{j_\ell=1}^{\lambda_\ell} x_\ell{}^{j_\ell}B_{\lambda_\ell j_\ell}(a^{(\ell)}{}_1, a^{(\ell)}{}_2,\cdots),$$

where $B_{\lambda_\ell j_\ell}(a^{(\ell)}{}_1, a^{(\ell)}{}_2,\cdots)$ *is the Bell polynomial. Consequently, we have*

$$[t^\lambda]h(\widehat{f(t)}) = \Pi_{\ell=1}^{r}\frac{1}{\lambda_\ell!}\sum_{j_\ell=1}^{\lambda_\ell} B_{\lambda_\ell j_\ell}(a^{(\ell)}{}_1, a^{(\ell)}{}_2,\cdots)D_\ell{}^{j_\ell}h(\hat{0}).$$

This is precisely the multivariate extension of the univariate Faà di Bruno formula (cf. [46] for another type extension.)

$$[(d/dt)^k h(f(t))]_{t=0} = \sum_{j=1}^{k} B_{kj}(g'(0), g''(0),\cdots)f^{(j)}(0). \quad (5.70)$$

In this subsection, examples seems not so enough to illustrate the merit of the theoretical results obtained. Interested reader might do something including more applications subsequently.

Remark 5.1.47 *The properties of the higher dimensional Sheffer group such as construction of the subgroup with certain orders, application to the multivariate expansions, combinatorial identities, etc., remain much to be investigated while some application results in this topic can be referred to [103].*

The last part of this subsection diverts to the following expansion problem.

Problem. Let $\hat{t} = (t_1, t_2, \cdots, t_r)$, $g(\hat{t})$, $\widehat{f(t)} = (f_1(t_1), f_2(t_2), \cdots, f_r(t_r))$, and $h(\hat{t})$ be any given formal power series over the complex number field $\mathbb{C}^r$ with $g(\hat{0}) = 1$, $f_i(0) = 0$, and $f_i'(0) \neq 0$ $(i = 1, 2, \cdots, r)$. We wish to find the power series expansion in $\hat{t}$ of the composite function $g(\hat{t})h(\widehat{f(t)})$.

For this problem, there is a significant body of relevant work in terms of the choices of univariate functions $g(t)$ and $f(t)$ (cf. for example, Comtet [44]). Certainly, such the problem is of fundamental importance in combinatorial analysis as well as in the special function theory, inasmuch as various generating functions often used or required of the form $g(\hat{t})h(\widehat{f(t)})$. In the case of $r = 1$, it is known that [44] has dealt with various explicit expansions of

$h(f(t))$ using either Faà di Bruno's formula or Bell polynomials. In addition, for $r = 1$, such a problem also gives a general extension of the row sums of Riordan arrays. In the following, we will show that a power series expansion of $g(\hat{t})h(\widehat{f(t)})$ could quite readily be obtained via the use of Sheffer-type differential operators. Also it will be shown that some generalized weighted Stirling numbers would be naturally entering into the coefficients of the general expansion formula developed. We shall frequently use the notation (5.61),

$$p_\lambda(\widehat{D})f(\hat{0}) = \left[p_\lambda(\widehat{D})f(\hat{t})\right]_{\hat{t}=\hat{0}},$$

where $p_\lambda(\hat{t})$ is defined in Definition 5.1.37.

As may be observed, the following theorem contains a constructive solution to the problem shown before.

Theorem 5.1.48 *(First Expansion Theorem) Let $g(\hat{t})$, $\widehat{f(t)}$, and components of $\widehat{h(t)}$ each be a formal power series over $\mathbb{C}^r$, which satisfy $g(\hat{0}) = 1$, $f_i(0) = 0$, and $f_i'(0) \neq 0$ $(i = 1, \cdots, r)$. Then there holds an expansion formula of the form*

$$g(\hat{t})h(\widehat{f(t)}) = \sum_{\lambda \geq \hat{0}} t^\lambda p_\lambda(\widehat{D})f(\hat{0}), \tag{5.71}$$

where $p_\lambda(\widehat{D})$ are called Sheffer-type r-variate differential operators of degree λ associated with $g(\hat{t})$ and $whf(t)$, and

$$p_\lambda(\widehat{D})h(\hat{0}) = p_\lambda\left(\frac{\partial}{\partial t_1}, \cdots, \frac{\partial}{\partial t_r}\right) h(t_1, \cdots, t_r)\bigg|_{\hat{t}=\hat{0}}.$$

Proof. Clearly, the conditions imposed on $\widehat{f(t)}$ ensure that the method of formal power series applies to the composite formal power series $g(\hat{t})h(\widehat{f(t)})$. Thus, using symbolic calculus with operators $\widehat{D}$ and E, we find via (5.59)

$$g(\hat{t})h(\widehat{f(t)}) = g(\hat{t})E^{\widehat{f(t)}}h(\hat{0}) = g(\hat{t})e^{\widehat{f(t)}\cdot\widehat{D}}h(\hat{0}) = \sum_{\lambda \geq \hat{0}} t^\lambda p_\lambda(\widehat{D})h(\hat{0}).$$

This is the desired expression given by (5.71).

■

Remark 5.1.49 *For dimension $r = 1$, $p_k(D)$ $(k = 0, 1, 2, \ldots)$ satisfy the recurrence relation*

$$(k+1)p_{k+1}(D) = \sum_{j=0}^{k} (\alpha_j + \beta_j D)p_{k-j}(D) \tag{5.72}$$

with $p_0(D) = I$, where α_j and β_j are given by

$$\alpha_j = (j+1)[t^{j+1}] \log g(t) \quad \textit{and} \quad \beta_j = (j+1)[t^{j+1}]f(t), \tag{5.73}$$

respectively. Accordingly we have

$$(k+1)p_{k+1}(x) = \sum_{j=0}^{k} \lambda_{j+1}(x)p_{k-j}(x), \tag{5.74}$$

where $\lambda_{j+1}(x)$ *are given by*

$$\lambda_{j+1}(x) = (j+1)[t^{j+1}] \log\left(g(t)e^{xf(t)}\right). \tag{5.75}$$

Thus, from (5.74) *and* (5.75) *we may infer that the differential operators* $p_k(D)$*'s satisfy the relations* (5.72) *with* α_j *and* β_j *being defined by* (5.73).

Remark 5.1.50 *If* $h(x) = \sum_{n=0}^{\infty} h_n x^n$ *be a formal power series and* $d_{n,k} = [t^n]g(t)(f(t))^k$*, then*

$$g(t)h(f(t)) = \sum_{k\geq 0} h_k g(t)(f(t))^k = \sum_{n\geq 0}\sum_{k\geq 0} h_k d_{n,k}$$

yields the generating function of the column of $(g, f)h$*. Thus,* $g(t)h(f(t))$ *is an infinite linear combination of column sums* gf^k *of the Riordan array* (g, f)*, i.e., the FTRA.*

Several immediate consequences of Theorem 2.1 may be stated as examples as follows.

Example 5.1.51 *For* $g(\hat{t}) = 1$ *and* $\widehat{f(t)} = \hat{t}$ *we see that* (5.71) *yields the Maclaurin expansion* $h(\hat{t}) = \sum_{\lambda\geq\hat{0}} t^\lambda D^\lambda h(\hat{0})/\lambda!$*, where* $\lambda! = (\lambda_1)!\cdots(\lambda_r)!$.

Example 5.1.52 *If* $g(\hat{z})$*,* $\widehat{f(z)}$*, and* $h(\hat{z})$ *are entire functions with* $g(\hat{0}) = 1$*,* $f_i(0) = 0$*,* $f_i'(0) \neq 0$ *for* $i = 1, 2, \cdots, r$*, then the equation* (5.71) *holds for the entire function* $g(\hat{z})h(\widehat{f(z)})$.

Example 5.1.53 *For* $j = 0, 1, \cdots, m$*, let* $f_j(t)$ *be a formal power series satisfying the conditions* $f_j(0) = 0$ *and* $f_j'(0) \neq 0$*. Denote* $(f_j \circ f_{j-1})(t) = f_j(f_{j-1}(t))$ $(j \geq 1)$*, then the power series expansion of* $(f_m \circ f_{m-1} \circ \cdots \circ f_0)(t)$ $(m \geq 2)$ *can be obtained recursively via the implicative relations*

$$\begin{aligned} p_{jk}(x) &= [t^k]e^{x\cdot(f_j\circ\cdots\circ f_0)(t)} \\ \Rightarrow &p_{jk}(D)f_{j+1}(0) = [t^k](f_{j+1}\circ f_j \circ\cdots\circ f_0)(t), \end{aligned} \tag{5.76}$$

where $1 \leq j \leq m-1$*, and* $p_{jk}(D)$ *are Sheffer-type operators.*

Denote the compositional inverse of f_i *by* $\bar{f}_i$*. Suppose that* $f_1, \cdots, f_m$ *are given and that* $h_m(t) = (f_m \circ f_{m-1} \circ \cdots \circ f_0)(t)$ *is a known series, but* $f_0(t)$ *is unknown to be determined. Certainly one may get* $f_0(t)$ *via computing*

$$f_0(t) = (\bar{f}_1 \circ \bar{f}_2 \circ \cdots \circ \bar{f}_m \circ h_m)(t). \tag{5.77}$$

Here it may be worth mentioning that the processes (5.76) *and* (5.77) *suggest a kind of compositional power-series-techniques that could be used to devise a certain procedure for the modification of a sequence represented by the coefficient sequence of* $f_0(t)$.

Example 5.1.54 *For the case of* $r = 1$, *let* $B_n(x)$, $\widehat{C}_n^{(\alpha)}(x)$, *and* $T_n^{(p)}(x)$ *be Bernoulli, Charlier, and Touchard polynomials, respectively. Then, for any given formal power series* $h(t)$ *over* $\mathbb{C}$ *we have three weighted expansion formulas as follows.*

$$\frac{t}{e^t - 1}h(t) = \sum_{n=0}^{\infty} \frac{t^n}{n!} B_n(D)h(0),$$

$$e^{-\alpha t}h(\log(1+t)) = \sum_{n=0}^{\infty} t^n \widehat{C}_n^{(\alpha)}(D)h(0) \quad (\alpha \neq 0),$$

$$(1-t)^p h(e^t - 1) = \sum_{n=0}^{\infty} t^n T_n^{(p)}(D)h(0) \quad (p > 0),$$

where $B_n(D)$, $\widehat{C}_n^{(\alpha)}(D)$, *and* $T_n^{(p)}(D)$ *may be called Bernoulli, Charlier, and Touchard differential operators, respectively. The above formulae are just three instances drawn from the table of special Sheffer-type polynomials as shown in Subsection 5.1.2.*

More expansion examples have been shown in examples 5.1.44-5.1.46.

As an application of Theorem 5.1.40, we now turn to the problem for finding an expansion of a multivariate analytic function f in terms of a sequence of higher Sheffer-type polynomials (p_n).

Theorem 5.1.55 *(Second Expansion Theorem) The Sheffe-type operator* $p_{\hat{n}}(\widehat{D})$ *defined in* (5.71) *has an expression of the form*

$$p_{\hat{n}}(\widehat{D}) = \frac{1}{\hat{n}!} \sum_{\hat{n} \geq \hat{k} \geq \hat{0}} \sigma(\hat{n}, \hat{k}) \widehat{D}^{\hat{k}}, \tag{5.78}$$

where $\sigma(\hat{n}, \hat{k})$ *is defined by* (5.62) *in Definition 5.1.38.*

Proof. Using the multivariate Taylor's formula and (5.62) we have formally

$$g(\hat{t})h(\widehat{f(t)}) = g(\hat{t}) \sum_{\hat{k} \geq \hat{0}} \frac{1}{\hat{k}!} \Pi_{i=1}^r f_i^{k_i}(t_i) \widehat{D}^{\hat{k}} h(\hat{0}) = \sum_{\hat{k} \geq \hat{0}} \left(\sum_{\hat{n} \geq \hat{k}} \sigma(\hat{n}, \hat{k}) \frac{t^{\hat{n}}}{\hat{n}!} \right) \widehat{D}^{\hat{k}} h(\hat{0})$$

$$= \sum_{\hat{n} \geq \hat{0}} \left(\sum_{\hat{n} \geq \hat{k} \geq \hat{0}} \sigma(\hat{n}, \hat{k}) \widehat{D}^{\hat{k}} h(\hat{0}) \right) \frac{t^{\hat{n}}}{\hat{n}!}.$$

Thus the rightmost expression, comparing with (5.71), leads us to (5.78).

We now establish the following theorem.

Theorem 5.1.56 *Let $h(\hat{z})$ be a multivariate entire function defined on $\mathbb{C}$, then we have the formal expansion of h in terms of a sequence of multivariate Sheffer-type polynomials $(p_{\hat{k}})$ as*

$$h(\hat{z}) = \sum_{k\geq 0} \alpha_{\hat{k}} p_{\hat{k}}(\hat{z}), \tag{5.79}$$

where

$$\alpha_{\hat{k}} = \sum_{\hat{n}\geq\hat{k}} \frac{\hat{k}!}{\hat{n}!}\sigma^*(\hat{n},\hat{k})\widehat{D}^{\hat{n}}h(\hat{0}) \tag{5.80}$$

or

$$\alpha_{\hat{k}} = \Lambda_{\hat{k}}(\widehat{D})h(\hat{0}) \tag{5.81}$$

with

$$\Lambda_{\hat{k}}(\widehat{D}) = \sum_{\hat{n}\geq\hat{k}} \frac{\hat{k}!}{\hat{n}!}\sigma^*(\hat{n},\hat{k})\widehat{D}^{\hat{n}}. \tag{5.82}$$

Proof. Let $h(\hat{z})$ be a multivariate analytic function defined on $\mathbb{C}^r$. Then, we can write its Taylor's series expansion as

$$\begin{aligned} h(\hat{z}) &= \sum_{\hat{n}\geq\hat{0}} \frac{\widehat{D}^{\hat{n}}h(\hat{0})}{\hat{n}!}\hat{z}^{\hat{n}} = \sum_{\hat{n}\geq\hat{0}} \frac{\widehat{D}^{\hat{n}}h(\hat{0})}{\hat{n}!} \sum_{\hat{n}\geq\hat{k}\geq\hat{0}} \sigma^*(\hat{n},\hat{k})\hat{k}!p_{\hat{k}}(\hat{z}) \\ &= \sum_{\hat{k}\geq\hat{0}} p_{\hat{k}}(\hat{z})\hat{k}! \sum_{\hat{n}\geq\hat{k}} \frac{1}{\hat{n}!}\sigma^*(\hat{n},\hat{k})\widehat{D}^{\hat{n}}h(\hat{0}) = \sum_{\hat{k}\geq\hat{0}} \alpha_{\hat{k}} p_{\hat{k}}(\hat{z}), \end{aligned}$$

where $\alpha_{\hat{k}}$ can be written as the forms of (5.80) or (5.81)–(5.82), which completes the proof of the theorem.

■

Remark 5.1.57 *The univariate setting of Theorem 5.1.56 can help us to find an expansion of a univariate analytic function f in terms of a sequence of Sheffer-type polynomials $(p_n(x))_{n\in\mathbb{N}_0}$. From the expression of α_k in Theorem 5.1.56, it is easy to derive Boas-Buck formulas (7.3) and (7.4) (cf. [21]) of the coefficients of the series expansion of an entire function in terms of polynomial $p_k(z)$. Indeed, for the fixed k, using the expression of α_k in the univariate setting and* (5.63) *and Cauchy's residue theorem yields*

$$\begin{aligned} \alpha_k &= \sum_{n=k}^{\infty} \frac{k!}{n!}\sigma^*(n,k)h^{(n)}(0) = \sum_{n=k}^{\infty} [t^n]\left((g(\bar{f}(t)))^{-1}(\bar{f}(t))^k\right)h^{(n)}(0) \\ &= \frac{1}{2\pi i}\oint_\Gamma \sum_{n=k}^{\infty} \frac{(\bar{f}(t))^k}{g(\bar{f}(t))t^{n+1}}h^{(n)}(0)dt = \frac{1}{2\pi i}\oint_\Gamma \frac{(\bar{f}(t))^k}{g(\bar{f}(t))}\left(\sum_{n=k}^{\infty}\frac{h^{(n)}(0)}{t^{n+1}}\right)dt \end{aligned}$$

$$= \frac{1}{2\pi i}\oint_\Gamma \frac{(\bar{f}(t))^k}{g(\bar{f}(t))}\left(\sum_{n=k}^\infty \frac{n!h_n}{t^{n+1}}\right)dt = \frac{1}{2\pi i}\oint_\Gamma \frac{(\bar{f}(\zeta))^k}{g(\bar{f}(\zeta))}H(\zeta)d\zeta,$$

where $H(\zeta)$ is the Borel's transform of $(h_n = h^{(n)}(0)/n!)$.

We now give algorithms to derive the series expansion of $f(z)$ in terms of a Sheffer type polynomial set $(p_n(x))_{n\in\mathbb{N}_0}$.

Algorithm 3.1

Step 1 For given Sheffer type polynomial $(p_n(x))_{n\in\mathbb{N}_0}$, we determine its generating function pair $(g(t), f(t))$ and the compositional inverse $\bar{f}(t)$ of $f(t)$.

Step 2 Use (5.63) to evaluate set $(\sigma^*(n,k))_{n\geq k}$ and substitute it into the corresponding expression (5.80) to find α_k $(k \geq 0)$.

Algorithm 3.2

Step 1 For given Sheffer type polynomial $(p_n(x))_{n\in\mathbb{N}_0}$, apply the first equation in (5.62) to obtain set $(\sigma(n,k))_{n\geq k\geq 0}$.

Step 2 Use (5.63) to solve for set $(\sigma^*(n,k))_{n\geq k}$ and substitute it into the corresponding expression (5.80) to find α_k $(k \geq 0)$.

It is easy to see the equivalence of the two algorithms. However, the first algorithm is more readily applied than the second one.

Example 5.1.58 *If $p_n(x) = x^n/n!$, then the corresponding generating function pair is $(g(t), f(t)) = (1, t)$. Hence, noting $\bar{f}(t) = t$ and $g(\bar{f}(t))^{-1} = 1$, from Definition 5.1.38 we have*

$$\sigma^*(\hat{n},\hat{k}) = \delta_{n,k},$$

the Kronecker delta. Therefore,

$$\alpha_k = \sum_{n=k}^\infty \frac{k!}{n!}\sigma^*(n,k)h^{(n)}(0) = h^{(k)}(0)$$

and the expansion of f is its Maclaurin expansion.

Example 5.1.59 *We now use Algorithm* 3.1 *to find the expansion of an entire function in terms of Bernoulli polynomials.*

Since the generating function of the Bernoulli polynomials is $g(t)exp(xf(t))$ with $g(t) = t/(e^t - 1)$ and $f(t) = t$, we have the compositional inverse of $f(t)$ as $\bar{f}(t) = t$ and

$$g(\bar{f}(t))^{-1} = (e^t - 1)/t.$$

Hence, from Definition 5.1.38 we can present

$$\frac{1}{k!}g(\bar{f}(t))^{-1}(\bar{f}(t))^k = \frac{1}{k!}(e^t-1)t^{k-1} = \frac{1}{k!}\sum_{n=1}^\infty \frac{t^{n+k-1}}{n!}$$

$$= \sum_{n=k}^\infty \frac{1}{k!(n-k+1)!}t^n = \sum_{n=k}^\infty \frac{1}{n+1}\binom{n+1}{k}\frac{t^n}{n!}.$$

Hence, $\sigma^*(\hat{n},\hat{k}) = \binom{n+1}{k}/(n+1)$ *and*

$$\alpha_k = \sum_{n=k}^{\infty} \frac{k!}{n!}\sigma^*(\hat{n},\hat{k})h^{(n)}(0) = \sum_{n=k}^{\infty} \frac{k!}{(n+1)!}\binom{n+1}{k}h^{(n)}(0).$$

Noting $h(t) = \sum_{n=0}^{\infty} h^{(n)}(0)t^n/n!$ *formally, for* $k = 0$*, we can write* α_0 *as*

$$\alpha_0 = \sum_{n=0}^{\infty} \frac{1}{(n+1)!}h^{(n)}(0) = \int_0^1 h(t)dt$$

and for $k > 0$ *we have*

$$\alpha_k = \sum_{n=k}^{\infty} \frac{1}{(n-k+1)!}h^{(n)}(0) = \sum_{n=k-1}^{\infty} \frac{1}{(n-k+1)!}h^{(n)}(0) - h^{(k-1)}(0)$$

$$= \sum_{n=k-1}^{\infty} \frac{h^{(n)}(0)}{n!} D_x^{k-1}x^n\Big|_{x=1} - h^{(k-1)}(0) = h^{(k-1)}(1) - h^{(k-1)}(0),$$

which are exactly the expressions of the expansion coefficients obtain on Page 29 of [26], which were derived by using contour integrals.

Example 5.1.60 *Let* $p_n(x)$ *be the Laguerre polynomial with its generating function pair* $(g(t), f(t)) = ((1-t)^{-p}, t/(t-1))$, $p > 0$, *then* $\bar{f}(t) = t/(t-1)$ *and* $g(\bar{f}(t))^{-1} = (1-t)^{-p}$. *Thus, using a similar argument of Example 5.1.59, we obtain*

$$\sigma^*(n,k) = (-1)^k \frac{n!}{k!}\binom{n+p-1}{n-k}.$$

Hence, the coefficients of the corresponding expansion can be written as

$$\alpha_k^{(p)} = \sum_{n=k}^{\infty} \frac{k!}{n!}\sigma^*(n,k)h^{(n)}(0)$$

$$= \sum_{n=k}^{\infty} (-1)^k \binom{n+p-1}{n-k} h^{(n)}(0)$$

($k = 0, 1, \cdots$*)*

Example 5.1.61 *If expansion basis polynomials are Angelescu polynomial* $A_n(x)$ $(n \in \mathbb{N}_0)$*, then their generating function pair is* $(A(t), g(t)) = (1/(1+t), t/(t-1))$ *and the dual* $\sigma^*(n,k)$ *can be found as follows:*

$$\sigma^*(n,k) = (-1)^{k+1}n!\left[2\binom{n-1}{k} - \binom{n}{k}\right].$$

Substituting the above expression of $\sigma^*(n,k)$ *into the corresponding formula in Theorem 11 yields the coefficients* α_k *as*

$$\alpha_k = \sum_{n=k}^{\infty} (-1)^{k+1}k!\left[2\binom{n-1}{k} - \binom{n}{k}\right] h^{(n)}(0)$$

($k = 0, 1, \cdots$*).*

5.1.4 Dual sequences and pseudo-Riordan involutions

Krattenthaler defines a class of matrix inverses in [147], which has numerous famous special cases represented by Gould and Hsu [77], Carlitz [29], Bressoud [24], Chu and Hsu [43], Ma[155], etc. We have presented some Riordan matrix inverses in subsection 5.1.2 (cf. also [105]). Let infinite low triangle matrices $(d_{n,k})_{0\leq k\leq n}$ and $(\bar{d}_{n,k})_{0\leq k\leq n}$ be *matrix inverses*, i.e., $(d_{n,k})(\bar{d}_{n,k}) = I$. Then a *sequence inverse relationship* can be defined by

$$f_n = \sum_{k=0}^{n} d_{n,k} g_k \quad \textit{if and only if} \quad g_n = \sum_{k=0}^{n} \bar{d}_{n,k} f_k, \tag{5.83}$$

where the sequences $(f_n)_{n\geq 0}$ and $(g_n)_{n\geq 0}$ are called the *inverse sequences* with respect to the inverse matrices $(d_{n,k})_{0\leq k\leq n}$ and $(\bar{d}_{n,k})_{0\leq k\leq n}$. If $d_{n,k} = \bar{d}_{n,k}$, then we say $(d_{n,k})_{0\leq k\leq n}$ is *self-inverse*, and $(f_n)_{n\geq 0}$ and $(g_n)_{n\geq 0}$ are said to be a pair of *dual sequences* (or they are dual to each other) with respect to the self-inverse matrix $(d_{n,k})_{0\leq k\leq n}$. In other words, $\{f_n\}$ and $\{g_n\}$ satisfy

$$f_n = \sum_{k=0}^{n} d_{n,k} g_k \quad \textit{if and only if} \quad g_n = \sum_{k=0}^{n} d_{n,k} f_k. \tag{5.84}$$

If there exists a sequence $(f_n)_{n\geq 0}$ that is dual to itself with respect to a self-inverse matrix $(d_{n,k})_{0\leq k\leq n}$, i.e.,

$$f_n = \sum_{k=0}^{n} d_{n,k} f_k, \tag{5.85}$$

then $(f_n)_{n\geq 0}$ is called a *self-dual sequence* with respect to the matrix $(d_{n,k})_{0\leq k\leq n}$.

Let $d_{n,k} = \binom{n}{k}(-1)^k$. Then $(d_{n,k})_{0\leq k\leq n}$ is a self-inverse matrix because

$$\sum_{k=0}^{n}\sum_{j=0}^{k} \binom{n}{k}\binom{k}{j}(-1)^{k+j} = \delta_{n,j},$$

where $\delta_{n,j}$ is the Kronecker delta. Let $(a_n)_{n\geq 0}$ be a sequence of complex numbers. Then $(a_n^*)_{n\geq 0}$ defined by

$$a_n^* = \sum_{k=0}^{n} \binom{n}{k}(-1)^k a_k \tag{5.86}$$

is the dual sequence of $(a_n)_{n\geq 0}$ with respect to $(\binom{n}{k}(-1)^k)$ (cf. for example, Graham, Knuth, and Patashnik [79]). Hence,

$$a_n = \sum_{k=0}^{n} \binom{n}{k}(-1)^k a_k^*. \tag{5.87}$$

Note that $(a_n)_{n\geq 0}$ and $(a_n^*)_{n\geq 0}$ are a pair of inverse sequences with $(a^*)_n^* = a_n$. If $a_n^* = a_n$, then $(a_n)_{n\geq 0}$ is a self-dual sequence with respect to $((-1)^k\binom{n}{k})$ (cf. Sun [205]). For instance, the following number sequences are self-dual sequences with respect to the dual relationship (5.86) for $(d_{n,k})_{n,k\geq 0} = ((-1)^k\binom{n}{k})$ (cf. for example, [205]):

$$\left(\frac{1}{2^n}\right), \left(\frac{1}{\binom{n+2m-1}{m}}\right), ((-1)^n B_n), (L_n), (nF_{n-1}),$$

where (B_n), (L_n), and (F_n) are the Bernoulli number sequence, Lucas number sequence, and Fibonacci number sequence, respectively.

Bernoulli polynomials $B_n(x)$ and Euler polynomials $E_n(x)$ for $n = 0, 1, \ldots$ are defined by (cf. (4.24) and (4.243))

$$\frac{te^{xt}}{e^t - 1} = \sum_{n\geq 0} B_n(x)\frac{t^n}{n!} \text{ and } \frac{2e^{xt}}{e^t + 1} = \sum_{n\geq 0} E_n(x)\frac{t^n}{n!}, \tag{5.88}$$

respectively. Bernoulli numbers B_n and Euler numbers E_n for $n = 0, 1, \ldots$ are defined by

$$B_n = B_n(0) \quad \text{and} \quad E_n = 2^n E_n\left(\frac{1}{2}\right), \tag{5.89}$$

respectively. Numerous literature scatters widely in combinatorics, number theory, approximation theory, etc. on Bernoulli and Euler numbers and polynomials. The following binomial expressions connect Bernoulli polynomials and numbers and Euler polynomials and numbers, respectively:

$$B_n(x) = \sum_{k=0}^{n} \binom{n}{k} B_k x^{n-k}, \quad n \geq 0, \tag{5.90}$$

$$E_n(x) = \sum_{k=0}^{n} \binom{n}{k} \left(x - \frac{1}{2}\right)^{n-k} \frac{E_k}{2^k}, \quad n \geq 0, \tag{5.91}$$

where $E_k = 2^k E_k(1/2)$. There are numerous results of the identities of Bernoulli and Euler numbers and polynomials. Many well-known results can be found in Buijs, Carrasquel-Vera, and Murillo [27], Dilcher [57], Gessel [66], Pan and Sun [173], Miki [163], Sun [205], Sun [206], Sun and Pan [207], Zheng and the author [114], etc. Some applications of the dual Bernoulli and Euler numbers and polynomials shown below are inspired by those results, particularly, [114, 173, 205, 206].

This subsection will study the duals of Bernoulli number sequence and Euler number sequence with respect to $(\binom{n}{k}(-1)^k)$ and other self-inverse matrices by using Riordan arrays. The dual sequence of $(B_n)_{n\geq 0}$ with respect to $(\binom{n}{k}(-1)^k)$ is denoted by $(B_n^*)_{n\geq 0}$, and the corresponding *dual Bernoulli polynomials* denoted by $(B_n^*(x))_{n\geq 0}$ is defined similarly to (5.90) by using

$$B_n^*(x) = \sum_{k=0}^{n} \binom{n}{k} B_k^* x^{n-k}, \quad n \geq 0, \tag{5.92}$$

where B_n^* are the *dual Bernoulli numbers* $(B_n)_{n\geq 0}$, i.e.,

$$B_n^* = \sum_{k=0}^{n} \binom{n}{k}(-1)^k B_k, \quad n \geq 0. \tag{5.93}$$

Hence, there are three questions raised: (1) What is the self-inverse matrix with respect to which the Bernoulli number sequence (B_n) is self-dual? (2) What is the relationship between the self-inverse matrix found in (1) and the self-inverse matrix, $(\binom{n}{k}(-1)^k)$, for $((-1)^n B_n)_{n\geq 0}$? And (3) does there exist a unified approach to construct self-inverse matrices and self-duals? We will answer those questions by using the Riordan array theory.

In this subsection, we shall present a unified approach to construct a class of self-inverse matrices and their applications to the construction of dual number sequences and dual polynomial sequences. Then we shall discuss the algebraic structure of dual number sequences with respect to the established dual relationships. Some identities of self-dual number sequences will be found accordingly. Finally, advanced applications of the dual relationships in the construction of identities for dual number sequences will be given.

We recall that a Riordan array $(g(t), f(t))$ and its inverse

$$(g(t), f(t))^{-1} = \left(\frac{1}{g(\bar{f}(t))}, \bar{f}(t)\right)$$

is a pair of inverse matrices. If $(g(t), f(t))^{-1} = (g(t), f(t))$, i.e., $(g(t), f(t))$ is an *involution*, or equivalently, it has an order of 2, then $(g(t), f(t))$ is a self-inverse matrix. Here, if g is an element of a group, then the smallest positive integer n such that g^n equals to the identity e of the group, if it exists, is called the order of g. If no such integer exists, then g is said to have infinite order. It is well-known (see Shapiro [194]) that if we restrict all entries of a Riordan array to be integers, then any element of finite order in the Riordan group must have order 1 or 2, and each element of order 2 generates a subgroup of order 2. Sprugnoli and the author find the sequence characterization of Riordan arrays of order 2 in [113].

The Riordan arrays related to combinatorics often have non-negative integer entries, and hence they can not have order 2. Therefore, we consider the possibility for an element $R \in \mathcal{R}$ possessing a pseudo-order 2. Here, we say R has a pseudo-order 2 if RM has order 2 with $M = (1, -t)$. Those R are called *pseudo-Riordan involutions* or briefly *pseudo-involutions* (see Cameron and Nkwanta [28] and [194]). We now present some sufficient and necessary conditions to identify pseudo-involutions.

Theorem 5.1.62 *Let $(g(t), f(t))$ be a pseudo-involution, and let $A(t)$ and $Z(t)$ be the generating functions of the A-sequence and Z-sequence of $(g(t), f(t))$, where $A(t)$ and $Z(t)$ are called the A-function and Z-function, respectively. Then the following statements are equivalent to the statement that the Riordan array $(g(t), f(t))$ is a pseudo-involution:*

(i) $\pm(g(t), f(t))(1, -t) = (\pm g(t), -f(t))$ *are involutions.*
(ii) $(1, -t)(g(t), f(t))(1, -t) = (g(t), f(t))^{-1}$, *the inverse of* $(g(t), f(t))$.
(iii) $\pm(1, -t)(g(t), f(t)) = (\pm g(-t), f(-t))$ *are involutions.*
(iv) $A(t) = \frac{-t}{f(-t)}$ *and* $Z(t) = \frac{g(-t)-1}{f(-t)}$.

Proof. Let $(g(t), f(t))$ be a pseudo-involution. Then $(g(t), f(t))(1, -t) = (g(t), -f(t))$ is an involution, i.e.,

$$(g(t), -f(t))(g(t), -f(t)) = (g(t)g(-f(t)), -f(-f(t)) = (1, t).$$

Additionally, $-(g(t), f(t))(1, -t) = (-g(t), -f(t))$ satisfies

$$(-g(t), -f(t))(-g(t), -f(t)) = (g(t), -f(t))(g(t), -f(t)) = (1, t),$$

which implies $-(g(t), f(t))(1, -t) = (-g(t), -f(t))$ is also an involution. Hence, we have proven that $(g(t), f(t))$ is a pseudo-involution implies (i). Equation (ii) follows from (i) because

$$\begin{aligned} I &= ((g(t), f(t))(1, -t))\,((g(t), f(t))(1, -t)) \\ &= (g(t), f(t))\,((1, -t)(g(t), f(t))(1, -t)). \end{aligned}$$

From the above equations, we have

$$\begin{aligned} &(\pm(1, -t)(g(t), f(t)))\,(\pm(1, -t)(g(t), f(t))) \\ =\ &((1, -t)(g(t), f(t))(1, -t))\,(g(t), f(t)) = (1, t) = I. \end{aligned}$$

Hence, we obtain (iii) from (ii). Similarly, (i) follows from (iii).

To find the A-function and Z-function of a pseudo-involution $(g(t), f(t))$, we recall (see Theorem 3.3 of [113]) the A-sequence of

$$(d_3(t), h_3(t)) = (d_1(t), h_1(t))(d_2(t), h_2(t))$$

has the A-function

$$A_3(t) = A_2(t)A_1\left(\frac{t}{A_2(t)}\right),$$

where $A_1(t)$ and $A_2(t)$ are the A-functions of $(d_1(t), h_1(t))$ and $(d_2(t), h_2(t))$, respectively. Since the A-functions of $(g(t), f(t))$ and $(1, -t)$ are respectively $A(t)$ and -1, the A-function of $(g(t), f(t))(1, -t)$ is

$$A_3(t) = -A\,(-t)\,. \tag{5.94}$$

On the other hand, from Theorem 4.3 of [113] we know the A-function of the involution $(g(t), f(t))(1, -t) = (g(t), -f(t))$ is

$$A_3(t) = \frac{t}{-f(t)}. \tag{5.95}$$

Comparing (5.94) and (5.95) we have

$$A(t) = \frac{-t}{f(-t)}.$$

From Theorem 3.4 of [113], the Z-function of

$$(d_3(t), h_3(t)) = (g(t), f(t))(1, -t) = (g(t), -f(t))$$

is $Z_3(t) = Z(-t)$ because the Z-functions of $(g(t), f(t))$ and $(1, -t)$ are $Z(t)$ and 0, respectively. Since $(d_3(t), h_3(t))$ is an involution, from Theorem 4.3 of [113] we also have

$$Z_3(t) = \frac{1 - g(t)}{-f(t)}.$$

Comparing the last two equations, we immediately know

$$Z(t) = \frac{g(-t) - 1}{f(-t)}.$$

From the definitions of A-sequence and Z-sequence, a Riordan array $(g(t), f(t))$ that possesses the above A-function and Z-function is a pseudo-involution.

■

Corollary 5.1.63 *Let nonzero $g(t) \in \mathcal{F}_0$ and $f(t) \in \mathcal{F}_1$. Then the infinite lower triangle matrix $(g(t), f(t))$ is a pseudo-involution if and only if*

$$\bar{f}(t) = -f(-t) \quad and \quad g(t) = \frac{f(-t)}{f(-t) - tg(-t) + t},$$

where the denominator is assumed not to be zero.

Proof. Equation $\bar{f}(t) = -f(-t)$ holds if and only if

$$\frac{t}{\bar{f}(t)} = \frac{-t}{f(-t)},$$

or equivalently, there exists the function $A(t)$ such that

$$A(t) = \frac{t}{\bar{f}(t)} \quad \text{and} \quad A(t) = \frac{-t}{f(-t)},$$

which implies that $(g(t), f(t))$ has an A-function (cf. for example, [113]) satisfying (iv) of Theorem 5.1.62.

Note that $g(t) = f(-t)/(f(-t) - tg(-t) + t)$ if and only if

$$\frac{g(-t) - 1}{f(-t)} = \frac{g(t) - 1}{tg(t)},$$

or equivalently, there exists the function $Z(t)$ such that

$$Z(t) = \frac{g(t)-1}{tg(t)} \quad \text{and} \quad Z(t) = \frac{g(-t)-1}{f(-t)},$$

which implies that $(g(t), f(t))$ has a Z-function (cf. for example, [113]) satisfying (iv) of Theorem 5.1.62. Combining all above statements together we know that $(g(t), f(t))$ is a pseudo-involution, which completes the proof of Corollary 5.1.63.

■

Corollary 5.1.63 also provides an algorithm for finding pseudo-involutions. Generally, one can accomplish this by carrying through the procedure demonstrated by the following example. For instance, it is clear that $f(t) = t/(1-t)$ (or $t/(1+t)$) has the compositional inverse $\bar{f}(t) = t/(1+t)$ (or $t/(1-t)$) and satisfies $\bar{f}(t) = -f(-t)$. From Corollary 5.1.63, we may also find that $g(t) = 1/(1-t)$ (or $1/(1+t)$) satisfies $g(t) = f(-t)/(f(-t) - tg(-t) + t)$. Therefore $(1/(1-t), t/(1-t))$ (or $(1/(1+t), t/(1+t))$) is a pseudo-involution.

Remark 5.1.64 *It can be seen that $(g(t), f(t))$ is a pseudo-involution if and only if there exist A-function and Z-function satisfying four relationships*

$$A(t) = \frac{t}{\bar{f}(t)}, \ A(t) = \frac{-t}{f(-t)}, \ Z(t) = \frac{g(t)-1}{tg(t)}, \ \textit{and } Z(t) = \frac{g(-t)-1}{f(-t)},$$

which give not only the formulas as shown in Corollary 5.1.63 but also the following useful formulas in the construction of pseudo-involutions:

$$f(t) = \frac{tZ(t)}{Z(-t)(1-tZ(t))} \quad \textit{and} \quad g(t) = \frac{f(t)Z(-t)}{tZ(t)}.$$

The proof is straightforward from the above four relationships.

From (i) and (iii) of Theorem 5.1.62 we obtain the following results.

Corollary 5.1.65 *Two Riordan arrays $(1/(1-t), t/(1-t)) = \left(\binom{n}{k}\right)_{n,k\geq 0}$ and $(1/(1+t), t/(1+t)) = \left((-1)^{n-k}\binom{n}{k}\right)_{n,k\geq 0}$ are pseudo-involutions. Hence, from (i) and (iii) of Theorem 5.1.62, we generate the following four Riordan involutions, denoted by R_1, R_2, R_3, and R_4, respectively:*

$$\begin{aligned} R_1 &= \left(\frac{1}{1-t}, \frac{t}{1-t}\right)(1,-t) = (1,-t)\left(\frac{1}{1+t}, \frac{t}{1+t}\right) \\ &= \left(\frac{1}{1-t}, \frac{-t}{1-t}\right) = \left((-1)^k\binom{n}{k}\right)_{n,k\geq 0}, \\ R_2 &= -\left(\frac{1}{1-t}, \frac{t}{1-t}\right)(1,-t) = -(1,-t)\left(\frac{1}{1+t}, \frac{t}{1+t}\right) \\ &= \left(\frac{1}{t-1}, \frac{t}{t-1}\right) = \left((-1)^{k+1}\binom{n}{k}\right)_{n,k\geq 0}, \end{aligned}$$

$$\begin{aligned}
R_3 &= (1,-t)\left(\frac{1}{1-t},\frac{t}{1-t}\right)=\left(\frac{1}{1+t},\frac{t}{1+t}\right)(1,-t)\\
&= \left(\frac{1}{1+t},\frac{-t}{1+t}\right)=\left((-1)^n\binom{n}{k}\right)_{n,k\geq 0},\\
R_4 &= -(1,-t)\left(\frac{1}{1-t},\frac{t}{1-t}\right)=-\left(\frac{1}{1+t},\frac{t}{1+t}\right)(1,-t)\\
&= \left(-\frac{1}{1+t},\frac{-t}{1+t}\right)=\left((-1)^{n+1}\binom{n}{k}\right)_{n,k\geq 0}.
\end{aligned} \tag{5.96}$$

Theorem 5.1.66 *Let R_i $(i=1,2,3,4)$ be the Riordan arrays as shown in Corollary 5.1.65. Then there hold the following four dual relationships, denoted by D_1, D_2, D_3, and D_4, respectively:*

$$D_1: a_n=\sum_{k=0}^n(-1)^k\binom{n}{k}a_k,\quad D_2: a_n=\sum_{k=0}^n(-1)^{k+1}\binom{n}{k}a_k,$$
$$D_3: a_n=\sum_{k=0}^n(-1)^n\binom{n}{k}a_k,\quad D_4: a_n=\sum_{k=0}^n(-1)^{n+1}\binom{n}{k}a_k.$$

Furthermore, $((-1)^nB_n)_{n\geq 0}$ and $(B_n)_{n\geq 0}$ are self-dual sequences with respect to D_1 and D_3, respectively, while $\left(E_n\left(\frac{1}{2}\right)-\frac{1}{2^n}\right)$ and $\left((-1)^n\left(E_n\left(\frac{1}{2}\right)-\frac{1}{2^n}\right)\right)$ are self-dual sequences with respect to D_2 and D_4, respectively.

Proof. Since (cf. for example, Apostol [10] and Milton and Stegun [165])

$$\sum_{k=0}^n\binom{n}{k}B_k=B_n(1)=(-1)^nB_n, \tag{5.97}$$

we can conclude that $((-1)^nB_n)_{n\geq 0}$ and $(B_n)_{n\geq 0}$ are self-dual sequences with respect to D_1 and D_3, respectively. Similarly, from (5.91) and $(-1)^nE_n(-x)=-E_n(x)+2x^n$ (cf. for example [10, 165]), we get

$$\begin{aligned}
&\sum_{k=0}^n(-1)^{k+1}\binom{n}{k}\left(E_k\left(\frac{1}{2}\right)-\frac{1}{2^k}\right)\\
&= -(-1)^n\sum_{k=0}^n\binom{n}{k}\left(-\frac{1}{2}-\frac{1}{2}\right)^{n-k}E_k\left(\frac{1}{2}\right)+\sum_{k=0}^n(-1)^k\binom{n}{k}\frac{1}{2^k}\\
&= -(-1)^nE_n\left(-\frac{1}{2}\right)+\frac{1}{2^n}\\
&= E_n\left(\frac{1}{2}\right)-2\frac{1}{2^n}+\frac{1}{2^n}=E_n\left(\frac{1}{2}\right)-\frac{1}{2^n}.
\end{aligned}$$

Similarly, we have

$$\sum_{k=0}^n(-1)^{n+1}(-1)^k\binom{n}{k}\left(E_k\left(\frac{1}{2}\right)-\frac{1}{2^k}\right)$$

$$= -\sum_{k=0}^{n}(-1)^{n-k}\binom{n}{k}E_k\left(\frac{1}{2}\right)+(-1)^n\sum_{k=0}^{n}(-1)^k\binom{n}{k}\frac{1}{2^k}$$

$$= -E_n\left(-\frac{1}{2}\right)+(-1)^n\frac{1}{2^n}=(-1)^n\left(E_n\left(\frac{1}{2}\right)-\frac{1}{2^n}\right),$$

completing the proof of the theorem.

■

We now consider the *duals of Bernoulli and Euler numbers* and the corresponding *duals of Bernoulli and Euler polynomials* with respect to different dual relationships as shown in Theorem 5.1.66.

Theorem 5.1.67 *Let B_n^* be the duals of Bernoulli numbers B_n with respect to R_1 defined by (5.93), and let $B_n^*(x)$ be the corresponding dual Bernoulli polynomials defined by (5.92). Then, there hold*

$$B_n^*(x)=(-1)^nB_n(-x-1) \tag{5.98}$$

and

$$B_n^*=(-1)^nB_n+n \tag{5.99}$$

for all $n\geq 0$.

Proof. From (5.90) and (5.92), we have

$$B_n^*(x)=\sum_{k=0}^{n}\binom{n}{k}(x)^{n-k}B_k^*=\sum_{k=0}^{n}\binom{n}{k}(x)^{n-k}\sum_{j=0}^{k}\binom{k}{j}(-1)^jB_j$$

$$= \sum_{j=0}^{n}\binom{n}{j}(-1)^jB_j\sum_{k=j}^{n}\binom{n-j}{k-j}(x)^{n-k}$$

$$= \sum_{j=0}^{n}\binom{n}{j}(-1)^jB_j(x+1)^{n-j}=(-1)^nB_n(-x-1).$$

Hence,

$$B_n^*=B_n^*(0)=(-1)^nB_n(-1)$$
$$= B_n(1)+n=(-1)^nB_n(0)+n,$$

which implies (5.99).

■

Corollary 5.1.68 *Let $(B_n^*(x))_{n\in\mathbb{N}_0}$ be the sequence of the duals of Bernoulli polynomial sequence. Then its generating function is*

$$\sum_{n\geq 0}B_n^*(x)\frac{t^n}{n!}=\sum_{n\geq 0}(-1)B_n(-x-1)\frac{t^n}{n!}$$

$$= \frac{-te^{(x+1)t}}{e^{-t}-1}=\frac{e^{(x+1)t}}{1+\frac{1-t-e^{-t}}{t}}. \tag{5.100}$$

Theorem 5.1.69 *With respect to the dual relationship D_3, the duals of numbers $(-1)^n B_n$, denoted by $((-1)^n B_n)^*$, can be written as*

$$((-1)^n B_n)^* = \sum_{k=0}^{n} (-1)^n (-1)^k \binom{n}{k} B_k = B_n + (-1)^n n, \tag{5.101}$$

and the corresponding dual Bernoulli polynomials can be expressed as

$$\sum_{k=0}^{n} \binom{n}{k} \left((-1)^k B_k\right)^* x^{n-k} = B_n(x) - n(x-1)^{n-1}. \tag{5.102}$$

With respect to the dual relationship D_4, the duals of numbers $E_n(1/2) - (1/2)^n$, denoted by $(E_n(1/2) - (1/2)^n)^*$, have the expression

$$\begin{aligned}\left(E_n\left(\frac{1}{2}\right) - \left(\frac{1}{2}\right)^n\right)^* &= \sum_{k=0}^{n} (-1)^{n+1} \binom{n}{k} \left(E_k\left(\frac{1}{2}\right) - \left(\frac{1}{2}\right)^k\right) \\ &= (-1)^n \left(E_n\left(\frac{1}{2}\right) + \frac{3^n - 2}{2^n}\right).\end{aligned} \tag{5.103}$$

The corresponding dual Euler polynomials can be written as

$$\begin{aligned}&\sum_{k=0}^{n} \binom{n}{k} \left(x - \frac{1}{2}\right)^{n-k} \left(E_k\left(\frac{1}{2}\right) - \left(\frac{1}{2}\right)^k\right)^* \\ =&(-1)^n E_n(-x+1) - 2(x-1)^n + (x-2)^n.\end{aligned} \tag{5.104}$$

With respect to the dual relationship D_2, the duals of numbers $(-1)^n (E_n(1/2) - (1/2)^n)$, denoted by $((-1)^n (E_n(1/2) - (1/2)^n))^*$, have the expression

$$\begin{aligned}\left((-1)^n \left(E_n\left(\frac{1}{2}\right) - \frac{1}{2^n}\right)\right)^* &= \sum_{k=0}^{n} (-1)^{k+1} \binom{n}{k} (-1)^k \left(E_k\left(\frac{1}{2}\right) - \frac{1}{2^k}\right) \\ &= E_n\left(\frac{1}{2}\right) + \frac{3^n - 2}{2^n},\end{aligned} \tag{5.105}$$

and the corresponding dual Euler polynomials can be represented as

$$\begin{aligned}&\sum_{k=0}^{n} \binom{n}{k} \left(x - \frac{1}{2}\right)^{n-k} \left((-1)^k \left(E_k\left(\frac{1}{2}\right) - \frac{1}{2^k}\right)\right)^* \\ =&E_n(x) + (x+1)^n - 2x^n.\end{aligned} \tag{5.106}$$

Proof. The proof is directly from the definitions and is omitted. ■

We now give some structures of self-dual number sequences, which include their characterizations, relationships, generating functions, and some properties. The first half of the following theorem is represented in Prodinger [179].

Theorem 5.1.70 *Let $a(x) = \sum_{n\geq 0} a_n x^n$. Then its coefficient sequence is a self-dual sequence with respect to D_1 (D_2), i.e., it satisfies*

$$\sum_{k=0}^{n} (-1)^k \binom{n}{k} a_k = a_n \quad (-a_n)$$

for $n \in \mathbb{N}_0$ if and only if $a(x)$ satisfies the equation

$$a\left(\frac{x}{x-1}\right) = (1-x)a(x) \quad (-(1-x)a(x)).$$

The coefficient sequence of $a(x)$ is a self-dual sequence with respect to D_3 (D_4), i.e., it satisfies

$$\sum_{k=0}^{n} (-1)^n \binom{n}{k} a_k = a_n \quad (-a_n)$$

for $n \in \mathbb{N}_0$ if and only if $a(x)$ satisfies the equation

$$a\left(-\frac{x}{1+x}\right) = (1+x)a(x) \quad (-(1+x)a(x)).$$

Proof. We leave the proof of the first half of the theorem as Exercise 5.5 and prove its second half as follows:

$$\begin{aligned}
\pm \quad \frac{1}{1+x} a\left(-\frac{x}{1+x}\right) &= \pm \sum_{k=0}^{\infty} a_k (-1)^k x^k (1+x)^{-k-1} \\
&= \pm \sum_{k=0}^{\infty} a_k (-1)^k x^k \sum_{j=0}^{\infty} \binom{-k-1}{j} x^j \\
&= \pm \sum_{k=0}^{\infty} \sum_{j=0}^{\infty} a_k (-1)^{k+j} \binom{k+j}{j} x^{k+j} \\
&= \pm \sum_{n=0}^{\infty} \left((-1)^n \sum_{j=0}^{n} \binom{n}{j} a_{n-j} \right) x^n = \pm \sum_{n=0}^{\infty} a_n x^n = \pm a(x).
\end{aligned}$$

The proof is complete. ■

Corollary 5.1.71 *Let $(a_n)_{n\geq 0}$ be a given sequence with ordinary generating function $a(x) = \sum_{n\geq 0} a_n x^n$. Then*

(i) $(a_n)_{n\geq 0}$ is a self-dual sequence with respect to D_1 if and only if $(2a_{n+1} - a_n)_{n\geq 0}$ is a self-dual sequence with respect to D_2.

(ii) $(a_n)_{n\geq 0}$ is a self-dual sequence with respect to D_3 if and only if $(2a_{n+1} + a_n)_{n\geq 0}$ is a self-dual sequence with respect to D_4.

Proof. We leave the proof of (i) as Exercise 5.6 and give a proof of (ii) as follows: Let

$$b_n = 2a_{n+1} + a_n$$

with its ordinary generating function $b(x) = \sum_{n\geq 0} b_n x^n$. Thus,

$$b(x) = \frac{2(a(x) - a_0)}{x} + a(x) = \frac{x+2}{x}a(x) - \frac{2}{x}a_0,$$

which implies

$$b\left(-\frac{x}{1+x}\right) = -\frac{x+2}{x}a\left(-\frac{x}{1+x}\right) + \frac{2}{x}(1+x)a_0$$

$$= -(1+x)\left(\frac{x+2}{x}\frac{1}{1+x}a\left(-\frac{x}{1+x}\right) - \frac{2}{x}a_0\right).$$

Therefore,

$$a\left(-\frac{x}{1+x}\right) = (1+x)a(x)$$

if and only if

$$b\left(-\frac{x}{1+x}\right) = -(1+x)\left(\frac{x+2}{x}a(x) - \frac{2}{x}a_0\right) = -(1+x)b(x).$$

Based on Theorem 5.1.70 we have finished the proof.

■

Theorem 5.1.72 *Let $(a_n)_{n\geq 0}$ be a given sequence with exponential generating function $a^*(x) = \sum_{n\geq 0} a_n x^n/n!$. Then*

(i) $(a_n)_{n\geq 0}$ is a self-dual sequence with respect to D_1 if and only if $a^*(x)e^{-x/2}$ is an even function.

(ii) $(a_n)_{n\geq 0}$ is a self-dual sequence with respect to D_2 if and only if $a^*(x)e^{-x/2}$ is an odd function.

(iii) $(a_n)_{n\geq 0}$ is a self-dual sequence with respect to D_3 if and only if $a^*(x)e^{x/2}$ is an even function.

(iv) $(a_n)_{n\geq 0}$ is a self-dual sequence with respect to D_4 if and only if $a^*(x)e^{x/2}$ is an odd function.

Proof. The proof of (i) and (ii) are left as Exercise 5.7. Here, we only prove (iii) and (iv).

$$a^*(-x)e^{-x} = \sum_{k\geq 0}(-1)^k a_k \frac{x^k}{k!}\sum_{j\geq 0}(-1)^j\frac{x^j}{j!}$$

$$= \sum_{k\geq 0}\sum_{j\geq 0}(-1)^{k+j}a_k\frac{x^{k+j}}{k!j!} = \sum_{n\geq 0}\left(\sum_{k=0}^{n}(-1)^n\binom{n}{k}a_k\right)\frac{x^n}{n!}.$$

Hence, if

$$\sum_{k=0}^{n}(-1)^n\binom{n}{k}a_k = a_n \quad \text{and} \quad -a_n,$$

then

$$a^*(-x)e^{-x} = a(x) \quad \text{and} \quad -a(x),$$

respectively, which can be written briefly as

$$a^*(-x)e^{-x/2} = \pm a(x)e^{x/2}.$$

If the case of the positive sign on the right-hand side holds, i.e., (a_n) is a self-dual sequence with respect to D_3, then $a^*(x)e^{x/2}$ is an even function; while the negative sign holds, or equivalently, (a_n) is a self-dual sequence with respect to D_4, then the function $a^*(x)e^{x/2}$ is odd. It is easy to see the sufficiencies of (iii) and (iv) are also true. This concludes the proof of the theorem.

■

Sun [205] uses Theorem 5.1.72 to derive numerous identities including (i) in the following theorem. Wang [211] uses umbral calculus to extend (i) to (ii). We now survey their results and extend them to (iii) and (iv) for other self-dual sequences.

Theorem 5.1.73 *For any function f, we have*

(i) $\sum_{k=0}^{n}\binom{n}{k}\left(f(k)-(-1)^{n-k}\sum_{j=0}^{k}\binom{k}{j}f(j)\right)a_{n-k} = 0$ *for* $n \in \mathbb{N}_0$ *if* $(a_n)_{n\geq 0}$ *is a self-dual sequence with respect to* D_1.

(ii) $\sum_{k=0}^{n}\binom{n}{k}\left(f(k)+(-1)^{n-k}\sum_{j=0}^{k}\binom{k}{j}f(j)\right)a_{n-k} = 0$ *for* $n \in \mathbb{N}_0$ *if* $(a_n)_{n\geq 0}$ *is a self-dual sequence with respect to* D_2.

(iii) $\sum_{k=0}^{n}\binom{n}{k}\left(f(k)-\sum_{j=0}^{k}(-1)^{n-j}\binom{k}{j}f(j)\right)a_{n-k} = 0$ *for* $n \in \mathbb{N}_0$ *if* $(a_n)_{n\geq 0}$ *is a self-dual sequence with respect to* D_3.

(iv) $\sum_{k=0}^{n}\binom{n}{k}\left(f(k)+\sum_{j=0}^{k}(-1)^{n-j}\binom{k}{j}f(j)\right)a_{n-k} = 0$ *for* $n \in \mathbb{N}_0$ *if* $(a_n)_{n\geq 0}$ *is a self-dual sequence with respect to* D_4.

Proof. The proofs of (i) and (ii) can be found from [205] and [211], respectively. The proofs of (iii) and (iv) are similar as the proofs of (i) and (ii) by using either Theorem 5.1.72 or umbral calculus. We leave them as Exercise 5.8.

■

From Theorem 5.1.66, we know that $((-1)^n B_n)_{n\geq 0}$ and $(B_n)_{n\geq 0}$ are self-dual sequences with respect to D_1 and D_3, respectively. In addition, $\left(E_n\left(\frac{1}{2}\right)-\frac{1}{2^n}\right)$ and $\left((-1)^n\left(E_n\left(\frac{1}{2}\right)-\frac{1}{2^n}\right)\right)$ are self-dual sequences with respect to D_2 and D_4, respectively. Therefore, we may use Theorem 5.1.73 to obtain the following identities.

Theorem 5.1.74 *Let B_n and E_n be Bernoulli numbers and Euler numbers, respectively. For any function f, we have the following identities:*

$$\sum_{k=0}^{n}\binom{n}{k}\left((-1)^{n-k}f(k)-\sum_{j=0}^{k}\binom{k}{j}f(j)\right)B_{n-k}=0,$$

$$\sum_{k=0}^{n}\binom{n}{k}\left(f(k)+(-1)^{n-k}\sum_{j=0}^{k}\binom{k}{j}f(j)\right)\left(E_{n-k}\left(\frac{1}{2}\right)-\frac{1}{2^{n-k}}\right)=0,$$

$$\sum_{k=0}^{n}\binom{n}{k}\left(f(k)-\sum_{j=0}^{k}(-1)^{n-j}\binom{k}{j}f(j)\right)B_{n-k}=0,$$

$$\sum_{k=0}^{n}\binom{n}{k}\left((-1)^{n-k}f(k)+\sum_{j=0}^{k}(-1)^{k-j}\binom{k}{j}f(j)\right)\left(E_{n-k}\left(\frac{1}{2}\right)-\frac{1}{2^{n-k}}\right)$$
$$=0. \tag{5.107}$$

The first identity of (5.107) is given in [205], and the others are shown in [114], which proofs are left as Exercise 5.9.

5.2 On an Extension of Riordan Array and Its Application in the Construction of Convolution-type and Abel-type Identities

Using the basic fact that any formal power series over the real or complex number field can be expressed in terms of a given polynomial sequence $(p_n(t))_{n\geq 0}$, where $p_n(t)$ is of degree n, we extend the ordinary Riordan array (resp. Riordan group) to a generalized Riordan array (resp. generalized Riordan group) associated with $(p_n(t))$. A rather general Vandermonde-type convolution formula and certain of its particular forms are represented in this section. The construction of the Abel type identities using the generalized Riordan arrays will also be discussed.

5.2.1 Generalized Riordan arrays with respect to basic sequences of polynomials

In [84], the author defined a *generalized Sheffer-type polynomial sequences* as follows.

Definition 5.2.1 *Let $g(t)$, $U(t)$, and $f(t)$ be any formal power series over the real number field $\mathbb{R}$ or complex number field $\mathbb{C}$ with $g(0) = 1$, $U(0) = 1$, $f(0) = 0$, and $f'(0) \neq 0$. Then the polynomials $u_n(x)$ $(n = 0, 1, 2, \cdots)$ defined by the generating function*

$$g(t)U(xf(t)) = \sum_{n\geq 0} u_n(x)t^n \tag{5.108}$$

are called the generalized Sheffer-type polynomials associated with $(g(t), f(t))_{U(t)}$. Accordingly, $u_n(D)$ with $D \equiv d/dt$ is called Sheffer-type differential operator of degree n associated with $(g(t), f(t))_{U(t)}$. Particularly, $u_0(D) \equiv I$ is the identity operator due to $u_0(x) = 1$.

The author [85] shows that for every $U(t)$ there exists a one-to-one correspondence between $(g(t), f(t))$ and $(u_n(x))_{n\geq 0}$, and the collection P_U of all polynomial sequences $(u_n(x))_{n\geq 0}$ with respect to $V(t) = \sum_{n\geq 0} a_n t^n$, defined by (5.108), forms a group $(P_U, \tilde{\#})$ under the operation $\tilde{\#}$, defined by

$$\{p_n(x)\}\tilde{\#}\{q_n(x)\} = \{r_n(x) = \sum_{k=0}^{n} r_{n,k}x^k : r_{n,k} = \sum_{\ell=k}^{n} p_{n,\ell}q_{\ell,k}/a_\ell, n \geq k\},$$

which is isomorphic to the Riordan group. Hence, for different power series $U(t)$ and $V(t)$, groups $(P_U, \tilde{\#})$ and $(P_V, \tilde{\#})$, defined by (5.108) associated with $U(t)$ and $V(t)$, respectively, are isomorphic.

Let $c = (c_0, c_1, c_2, \ldots)$ be a sequence satisfying $c_0 = 1$ and $c_k > 0$ for all $k = 1, 2, \ldots$. We call the element $A \in \mathcal{F}$ with the form $A(x) = \sum_{k\geq 0} \frac{x^k}{c_k}$ a *generalized power series* associated with $(c_n)_{n\geq 0}$ or, simply, a (c)-GPS, where $\mathcal{F}$ is the GPS set associated with $(c_n)_{n\geq 0}$. In [76], a (c)-Riordan array generated by $g(t) \in \mathcal{F}_0$ and $f(t) \in \mathcal{F}_1$ with respect to $A(x)$ and $(c_n)_{n\geq 0}$ is an infinite complex matrix $(d_{n,k})_{0\leq k\leq n}$, whose bivariate generating function has the form

$$\sum_{n,k\geq 0} d_{n,k}\frac{t^n}{c_n}x^k = g(t)g(xf(t)). \tag{5.109}$$

Hence, we denote $(d_{n,k}) = (g(t), f(t))$. Particularly, if $f'(0) \neq 0$, the corresponding Riordan array is called a proper Riordan array. Otherwise, it is called an improper Riordan array.

Furthermore, if $c_k = 1$ $(k = 0, 1, 2, \ldots)$, then the corresponding series $g(t)$ and $f(t)$ are ordinary power series. Hence, expression (5.109) is written as

$$\sum_{n,k\geq 0} d_{n,k}t^n x^k = \frac{g(t)}{1 - xf(t)}, \tag{5.110}$$

which defines the classical Riordan array, called (1)-Riordan array. If $c_k = k!$ $(k = 0, 1, 2, \ldots)$, i.e., the corresponding series $g(t)$ and $f(t)$ are exponential power series, then expression (5.109) is written as

$$\sum_{n,k\geq 0} d_{n,k}\frac{t^n}{n!}x^k = g(t)e^{xf(t)}, \tag{5.111}$$

which defines the Sheffer-type (or exponential) Riordan array. If $c_0 = 1$ and $c_k = k$ $(k = 1, 2, \ldots)$, then the corresponding series $g(t)$ and $f(t)$ are Dirichlet series. Hence, expression (5.109) is written as

$$d_{0,0} + \sum_{1\leq k\leq n} d_{n,k}\frac{t^n}{n}x^k = g(t)\left(1 - \ln(1 - xf(t))\right), \tag{5.112}$$

which is called the Dirichlet Riordan series. The improper Riordan arrays derived from the bivariate generating function $g(t)\ln(1/(1 - xf(t))$ is worth being investigated.

In the above definitions, the (n, k) entry of (c)-Riordan array $(d_{n,k})$ is

$$d_{n,k} = \left[\frac{t^n}{c_n}\right] g(t)\frac{f(t)^k}{c_k} = [p_n(t)]g(t)p_k(f(t)), \quad p_j(t) = \frac{t^j}{c_j}, \tag{5.113}$$

for all $0 \leq k \leq n$, and $d_{n,k} = 0$ otherwise. Obviously, we have $d_{n,k} = [t^n]g(t)(f(t))^k$ and $d_{n,k} = [t^n]g(t)(f(t))^k/n!$, $0 \leq k \leq n$, for the classical (1)–Riordan arrays and the Sheffer-type Riordan arrays, respectively, and $d_{n,k} = [t^n/n]g(t)(f(t))^k/n$ $(1 \leq k \leq n)$ for the Dirichlet Riordan arrays. Notation $[t^n/c_n]f(t)$ was introduced by Knuth [145] in 1993.

Gould and the author [76] considered the characterization of (c)-Riordan arrays by means of the A- and Z-sequences. They also showed a one-to-one correspondence between *Gegenbauer-Humbert-type polynomial sequences* and the set of (c)-Riordan arrays, which generates the sequence characterization of Gegenbauer-Humbert-type polynomial sequences.

In this subsection, we will consider the (c)- extension of Riordan arrays, the change of the basic sequence, and the algebraic structure of the Riordan group. Here a basic sequence is an extension of polynomial sequence $(p_n(t) = t^n/c_n)_{n\geq 0}$ defined below. As an application, we will give a new method to construct convolution-type identities using the extended Riordan arrays.

Given a polynomial $p(t)$ in t of degree n, we may denote $deg\, p(t) = n$. If $f(t)$ is a formal power series in $\mathcal{F}_m$, then its lowest order (i.e., exponent) is m, which is denoted by $ord\, f(t) = m$. Particularly, the case $ord\, f(t) = 1$ means that $f(t) \in \mathcal{F}_1$, a delta series. We need the following definitions (cf.[21]).

Definition 5.2.2 *A sequence of polynomials* $(p_n(t))_{n\geq 0}$ *is called a normal sequence, if* $p_0(t) \equiv 0$ *and* $deg\, p_n(t) = n$ $(n \geq 1)$.

Definition 5.2.3 *$(p_n(t))_{n\geq 0}$ is said to be a basic sequence of polynomials if it is a normal sequences and every formal power series $f(t)$ can be written uniquely as*

$$f(t)=\sum_{n=0}^{\infty} a_n p_n(t),$$

in terms of $\{p_n(t)\}$ and real or complex coefficients $\{a_n\}_{n\geq 0}$, in the sense that $f(t)=0$ implies all coefficients $a_n=0$, $n=0,1,\ldots$.

Note that a normal sequences may not be a basic sequence. The simple example is given by

$$q_0(t)=1, q_1(t)=-t, q_n(t)=t^{n-1}/(n-1)!-t^n/n!\ (n\geq 2).$$

Clearly, $\{q_n(t)\}$ is a normal sequence, but it is not a basic sequence since $f(t)=1$ has two representations:

$$f(t)=q_0(t) \text{ and } f(t)=q_0(t)+\sum_{n=1}^{\infty} q_n(t).$$

Also, it is known that $\{t^n\}$ and $\{(t)_n\}$ are the simplest basic sequences of polynomials, where $(t)_n$ are the falling factorial polynomials:

$$(t)_0=1, (t)_n=t(t-1)\cdots(t-n+1) \text{ for } n\geq 1.$$

Similarly, we have basic sequences $\{t^n/c_n\}$ and $\{(t)_n/c_n\}$, where $c_n\neq 0$. Particularly, if $c_n=1$, $n!$, and n, the corresponding basic sequences are called the classical, Sheffer-type, and Dirichlet-type basic sequences. For example, the $\{p_n(t)=t^n/c_n\}$ defined by (5.110)–(5.112) with $x=1$ are classical, Sheffer-type, and Dirichlet-type basic sequences, respectively.

We shall show that a basic sequence is an infinite linearly independent normal sequence based on the following definition of the infinite linearly independent sequence.

Definition 5.2.4 *Let $(f_n(t))_{n\geq 0}$ be a function sequence defined on a region Ω in $\mathbb{R}$ or $\mathbb{C}$ with its every finite subsequence being linearly independent on Ω, or equivalently, for every finite subset $N\subseteq \mathbb{N}_0$, $(f_n(t))_{n\in N}$ is linearly independent on Ω. We say $(f_n(t))_{n\geq 0}$ is infinite linearly independent on Ω, if every series $\sum_{n\geq 0}\gamma_n f_n(t)$ vanishing on Ω has zero partial sum sequence, or equivalently, $\sum_{n\geq 0}\gamma_n f_n(t)=0$ implies its partial sum sequence $(s_n(t)=\sum_{k=0}^{n}\gamma_k f_k(t))_{n\geq 0}$ is a zero sequence at every point $t\in\Omega$.*

Roughly speaking, an infinite function sequence is said to be linearly independent if its zero linear combination implies its every partial sum is identically zero. Or equivalently, an infinite function sequence is linearly dependent if there exists a vanishing linear combination of the sequence that has a non-vanishing partial sum. Thus, the sequence $\{1, t, t^2-t, t^3-t^2, \ldots\}$ is linearly dependent, while both sequences $\{1, t/c_1, t^2/c_2, t^3/c_3, \ldots\}$, $c_n = 1$, $n!$ and n, respectively, are linearly independent. If $c_n = 1$, then the corresponding basic sequence is called the standard basic sequence, which is also a standard normal polynomial sequence.

Proposition 5.2.5 *A normal polynomial sequence defined in Definition 5.2.2 is a basic sequence if and only if it is linearly independent.*

Proof. Let $(p_n(t))_{n\geq 0}$ be a normal polynomial sequence. Then any formal power series can be written as a linear combination of $(p_n(t) = \sum_{j=1}^{n} \alpha_{n,j} t^j)_{n\geq 0}$, where $\alpha_{n,n} \neq 0$. More precisely, in a formal power series $f(t) = \sum_{n\geq 0} f_n t^n$, we may take a transformation of the standard normal polynomial sequence $(t^n)_{n\geq 0}$ to an arbitrary normal polynomial sequence $(p_n(t))_{n\geq 0}$. First, we have $t = p_1(t)/\alpha_{1,1}$. Assume that $t^i = \sum_{j=1}^{i} \beta_{i,j} p_j(t)$ for all $1 \leq i \leq k-1$. Then,

$$\begin{aligned} t^k &= \frac{1}{\alpha_{k,k}} p_k(t) - \frac{1}{\alpha_{k,k}} \sum_{i=1}^{k-1} \alpha_{k,i} t^i \\ &= \frac{1}{\alpha_{k,k}} p_k(t) - \frac{1}{\alpha_{k,k}} \sum_{i=1}^{k-1} \alpha_{k,i} \sum_{j=1}^{i} \beta_{i,j} p_j(t) \\ &= \frac{1}{\alpha_{k,k}} p_k(t) - \frac{1}{\alpha_{k,k}} \sum_{j=1}^{k-1} \left(\sum_{i=j}^{k-1} \alpha_{k,i} \beta_{i,j} \right) p_j(t). \end{aligned}$$

Hence we have proved that the transform $t^n \mapsto \{p_k(t)\}_{1\leq k\leq n}$ holds by using the mathematical induction. Therefore, any formal power series can be written as a linear expression in terms of $\{p_n(t) = \sum_{j=1}^{n} \alpha_{n,j} t^j\}$. We should note that the expression may not be unique.

If $(p_n(t))_{n\geq 0}$ is not a basic sequence, i.e., there exists a formal power series $f(t)$ defined on $\Omega \subset \mathbb{R}$ (or $\mathbb{C}$) has two different linear sums in terms of $(p_n(t))_{n\geq 0}$, say

$$f(t) = \sum_{n\geq 0} a_n p_n(t) \quad f(t) = \sum_{n\geq 0} b_n p_n(t),$$

and assume n_0 is the first subindex such that $a_{n_0} \neq b_{n_0}$, then the vanishing series $\sum_{n\geq 0}(a_n - b_n)p_n(t)$ has a partial sum $s_{n_0} = \sum_{k=0}^{n_0}(a_k - b_k)p_k(t)$ not being identically zero on Ω because $a_{n_0} \neq b_{n_0}$. Thus, $(p_n(t))_{n\geq 0}$ is not linearly independent from the definition.

Conversely, if $(p_n(t))_{n\geq 0}$ defined on $\Omega \subset \mathbb{R}$ is not an infinite linearly independent sequence, then there exists a vanishing series in terms of $(p_n(t))_{n\geq 0}$, denoted by $\sum_{n\geq 0} a_n p_n(t) = 0$, such that it has a partial sum $s_{n_0} = \sum_{k=0}^{n_0} a_k p_k(t)$ not being zero at some point $t_0 \in \Omega$. Hence, there exists at least one non-zero coefficient a_{n_0} in the partial sum s_{n_0}. If a formal power series

$f(t)$ is a linear sum in terms of $(p_n(t))_{n\geq 0}$ shown as $f(t) = \sum_{n\geq 0} b_n p_n(t)$, then $\sum_{n\geq 0}(a_n + b_n)p_n(t)$ is also a linear expression of $f(t)$ in terms of $(p_n(t))_{n\geq 0}$. Therefore, we have

$$f(t) = \sum_{n\geq 0} b_n p_n(t) = \sum_{n\geq 0}(a_n + b_n)p_n(t),$$

which implies that $f(t)$ has two different linear expressions in terms of $(p_n(t))_{n\geq 0}$ because of $b_{n_0} \neq a_{n_0} + b_{n_0}$.

■

From Proposition 5.2.5, we know that $\{1, t, t^2 - t, t^3 - t^2, \ldots\}$ is not a basic sequence while $\{1, t/c_1, t^2/c_2, t^3/c_3, \ldots\}$ with $c_n = n!$ and n are basic sequences.

Remark 5.2.6 *Definition 5.2.3 states that a normal polynomial sequence is a basic sequence if every formal power series $f(t) = \sum_{n\geq 0} f_n t^n$ can be written as a unique linear sum in terms of $(p_n(t))_{n\geq 0}$:*

$$f(t) = \sum_{n\geq 0} a_n p_n(t), \tag{5.114}$$

or equivalently, the above coefficient set $\{a_n\}$ is the unique solution set of the system of the linear equations

$$a_0 = f_0, \quad \sum_{i\geq n} a_i \alpha_{i,n} = f_n, \quad n \geq 1, \tag{5.115}$$

which is called the characterization system for $f(t)$ associated with $(p_n(t))_{n\geq 0}$. In the proof of Proposition 5.2.5, we have shown that every formal power series can be written as a linear sum as shown in (5.114). Thus, system (5.115) is always solvable provided that $(p_n(t))_{n\geq 0}$ is a normal polynomial sequence. The following example shows the solution may not be unique: For $(p_n(t) = t^n - t^{n-1})_{n\geq 1}$ and $p_0(t) = 1$, system (5.115) becomes

$$a_0 = f_0, \quad a_n - a_{n+1} = f_n, \quad n \geq 1,$$

which has infinitely many solutions. Thus a normal polynomial sequence is a basic sequence can be characterized as follows.

Proposition 5.2.7 *A normal polynomial sequence $(p_n(t))_{n\geq 0}$ is a basic sequence if for every formal power series, the characterization system associated with $(p_n(t))_{n\geq 0}$ has a unique solution set.*

Next are three useful type of basic sequences of polynomials frequently used in the book.

(i) $\{p_n(t) = t^n/c_n : c_0 = 1, c_n \neq 0, n \geq 1\}$ (cf. also [85]).

(ii) $\{p_n(t) = (t)_n/c_n : c_0 = 1, c_n \neq 0, n \geq 1\}$ including a special case $\{p_n(t) = (t)_n/n! = \binom{t}{n}\}_{n\geq 0}$.

(iii) Note that both $t^n/n!$ and $\binom{t}{n}$ are the simplest Sheffer-type polynomials. Certainly every special kind of Sheffer-type polynomial sequence $(p_n(t))_{n\geq 0}$ could be used as a basic sequence.

In subsection 5.2.3, we will present a method (and an algorithm) to find more basic sequences based on Proposition 5.2.7.

We now define the generalized Riordan arrays with respect to basic sequences and prove the set of those arrays with a basic sequence forms a group, called the Riordan group with respect to the basic sequence. We will also show some subgroups of the Riordan group. Different Riordan groups with respect to several different basic sequences are given. And the isomorphism between the Riordan group and the Sheffer group with the same basic sequence and the isomorphism between two Riordan groups with different basic sequences are also shown. In next subsection, we shall construct a general class of convolution-type identities using the formal expressions of entire functions in terms of the Sheffer-type polynomials and the generalized Riordan arrays (cf. [103, 105]),. Three different classes identities with respect to three different type basic sequences as well as the corresponding algorithms are given. A general method to identify basic sequences will also be given in next subsection. Finally, we present Abel type identities using the generalized Riordan arrays in Subsection 5.2.3.

Given a normal basic sequence of polynomials $(p_n(t))_{n\geq 0}$ with $p_0(t) = 1$ and $deg\, p(t) = n$ $(n \geq 1)$. Let $g(t) \in \mathcal{F}_0$ and $f(t) \in \mathcal{F}_1$. Consequently, $g(t)$ has reciprocal $g(t)^{-1}$ and $f(t)$ has compositional inverse $\bar{f}(t)$. Especially noteworthy is that $g(t)p_k(f(t))$ is a formal power series so that it can be expressed uniquely as linear sums in terms of $p_n(t)'s$ with coefficients $d_{n,k}$, viz.,

$$g(t)p_k(f(t)) = \sum_{n=0}^{\infty} d_{n,k}p_n(t).$$

Thus, based on this relation we may write, upon using of the extracting-coefficient operator $[p_n(t)]$, by

$$d_{n,k} = [p_n(t)]g(t)p_k(f(t)), \tag{5.116}$$

where $n, k \in \mathbb{N}_0$, the set of non-negative integers. A combination of these facts allows us to introduce

Definition 5.2.8 *The matrix $(d_{n,k})$ defined in* (5.116) *is said to be a generalized Riordan array with respect to the basic sequence $(p_n(t))_{n\geq 0}$. Following Shapiro–Getu–Woan–Woodson [196], we write it by*

$$(d_{n,k}) = (g(t), f(t)).$$

Although we use the same notation for the generalized Riordan arrays $(d_{n,k})$ and the classical Riordan arrays $(\bar{d}_{n,k})$, the entries are usually different, where $d_{n,k}$ is defined by (5.116), while $\bar{d}_{n,k} = [t^n]g(t)(f(t))^k$. Let $(p_n(t))_{n\geq 0}$ be a basic sequence, where $p_0(t) = 1$ and $p_k(t) = \sum_{j=1}^k \alpha_{k,j}t^j$ ($\alpha_{k,j} \neq 0$, $k \geq 1$). From (5.116) we obtain a relationship between $d_{n,k}$ and $\bar{d}_{n,k}$:

$$d_{n,k} = \sum_{\ell=1}^{n} \frac{1}{\alpha_{n,\ell}}[t^\ell]g(t)\sum_{j=1}^{k}\alpha_{k,j}(f(t))^j = \sum_{\ell=1}^{n}\sum_{j=1}^{k}\frac{\alpha_{k,j}}{\alpha_{n,\ell}}\bar{d}_{\ell,j}. \tag{5.117}$$

Remark 5.2.9 *Some coefficients $\alpha_{k,j}$, $1 \leq j \leq k-1$, may be vanishing in $p_k(t) = \sum_{j=1}^k \alpha_{k,j}t^j$. If it happens, then the corresponding terms in the expression (5.117) are also missing. However, we emphasize that $\alpha_{k,k}$ $(k \geq 0)$ never be zero.*

As may be expected, two arbitrary Riordan matrices, $(d_{n,k}) = (g(t), f(t))$ and $(c_{n,k}) = (f(t), g(t))$, $g(t), f(t) \in \mathcal{F}_0$ and $f(t), g(t) \in \mathcal{F}_1$, of such sort can also carry out the usual matrix multiplications. To make this effect, let us denote

$$\xi_{n,k} = \sum_{\lambda=0}^{\infty} d_{n,\lambda}c_{\lambda,k}, \quad (n,k) \in \mathbb{N}\times\mathbb{N},$$

where $\xi_{n,k}$ may be real or complex numbers. Particularly, we denote $\xi_{n,k} = \infty$ in case $\left|\sum_{\lambda=0}^{\infty} d_{n,\lambda}c_{\lambda,k}\right|$ diverges to $+\infty$.

Theorem 5.2.10 *Let $(d_{n,k}) = (d(t), h(t))$ and $(c_{n,k}) = (g(t), f(t))$ be two generalized Riordan arrays with respect to the basic sequence $(p_n(t))_{n\geq 0}$, where $d(t), g(t) \in \mathcal{F}_0$ and $h(t), f(t) \in \mathcal{F}_1$. There holds the matrix multiplication*

$$(d_{n,k})(c_{n,k}) =: (\xi_{n,k}) = (d(t)g(h(t)), f(h(t))), \tag{5.118}$$

or equivalently,

$$(d(t), h(t))(g(t), f(t)) = (\xi_{n,k}) = (d(t)g(h(t)), f(h(t))). \tag{5.119}$$

Proof. The formal proof is entirely similar to that for the ordinary Riordan matrices. In order to justify the relation (5.118) formally, we make the replacement $t \to h(t)$ in $c_{n,k} = [p_n(t)]g(t)p_k(f(t))$ so that we have

$$g(h(t))p_k(f(h(t)) = \sum_{\lambda=0}^{\infty} c_{\lambda,k}p_\lambda(h(t)). \tag{5.120}$$

It is worthy of note that both the left-hand side of (5.120) and $p_\lambda(h(t))$ are formal power series, and $d(t)g(h(t))|_{t=0} = d(0)g(0) = 1, f(h(t))|_{t=0} = f(0) = 0$ and $\left(f(h(t))\right)'|_{t=0} = \left(\frac{d}{dt}\right)f(h(t))|_{t=0} = f'(0)h'(0) = 1$. Thus, in view of (5.120) we may compute

$$[p_n(t)]d(t)g(h(t))p_k(f(h(t))) = [p_n(t)]d(t)\sum_{\lambda=0}^{\infty} c_{\lambda,k}p_\lambda(h(t))$$

$$= [p_n(t)]\sum_{\lambda=0}^{\infty} c_{\lambda,k}\sum_{m=0}^{\infty} d_{m,\lambda}p_m(t) = [p_n(t)]\sum_{m=0}^{\infty} p_m(t)\left\{\sum_{\lambda=0}^{\infty} d_{m,\lambda}c_{\lambda,k}\right\}$$

$$= \sum_{\lambda=0}^{\infty} d_{n,\lambda}c_{\lambda,k} = \xi_{n,k}.$$

This shows that

$$(\xi_{n,k}) = (d(t)g(h(t)), f(h(t))).$$

Hence the theorem is verified.

■

As usual, we have the inverse for the matrix $(g(t), f(t))$:

$$(g(t), f(t))^{-1} = \left(\frac{1}{g(\bar{f}(t))}, \bar{f}(t)\right). \tag{5.121}$$

Indeed

$$(g(t), f(t))\left(\frac{1}{g(\bar{f}(t))}, \bar{f}(t)\right) = \left(\frac{g(t)}{g(f(\bar{f}(t)))}, f(\bar{f}(t))\right) = (1, t).$$

Here $(1, t)$ denotes the unit matrix $(\delta_{n,k}) = \text{diag}(1, 1, 1, \cdots)$, since for the case $g(t) = 1, f(t) = t$, we have

$$d_{n,k} = [p_n(t)]p_k(t) = \delta_{n,k}, \quad (d_{n,k}) = (\delta_{n,k}).$$

Moreover, the associative law for the matrix multiplication rule as given by (5.119) can also be confirmed without any difficulty. Thus for given basic sequence $(p_n(t))_{n\geq 0}$ of polynomials, all matrices $(g(t), f(t))$ formed by formal power series $g(t)$ and $f(t)$ satisfying those conditions mentioned previously just yields a group with the multiplication as shown in (5.119), in which the inverse and the unit are given by (5.121) and $(1, t)$, respectively.

All these can be summarized by

Proposition 5.2.11 *All the matrices $(d_{n,k}) = (g(t), f(t))$ associated with a given basic sequence $(p_n(t))_{n\geq 0}$ and with $d_{n,k}$ being defined by (5.116) form a generalized Riordan group, denoted by $\mathcal{R}$, associated with $(p_n(t))_{n\geq 0}$ with respect to the multiplication as given by (5.119), and with the inverse element $(d_{n,k})^{-1}$ and the unit element being given by (5.121) and $(1, t) = (\delta_{n,k})$, respectively.*

Particular two subgroups of $\mathcal{R}$ are important and have been considered in the this section:

- the set $\mathcal{A}$ of *Appell arrays*, that is the (c)-Riordan arrays $R = (g(t), f(t))$ for which $f(t) = t$; it is an invariant subgroup and is isomorphic to the group $\mathcal{F}_0$ with the usual product as the group operation;

- the set $\mathcal{L}$ of *Lagrange arrays*, that is the (c)-Riordan arrays $R = (g(t),\ f(t))$ for which $g(t) = 1$; it is also called the *associated subgroup*; it is isomorphic with the group $\mathcal{F}_1$ with composition as the group operation;

Next are three useful cases of Riordan groups associated with three type of basic sequences of polynomials given before.

(i) If the basic sequence $(p_n(t))_{n\geq 0}$ is taken to be $\{t^n/c_n\}$, where $c_0 = 1$ and $c_n \neq 0$ for all $n \geq 1$, then we get the (c)- Riordan matrices and generalized Riordan group (cf. [85]).

(ii) If the basic sequence $(p_n(t))_{n\geq 0}$ is taken to be $\{(t)_n/c_n\}$, where $c_0 = 1$ and $c_n \neq 0$ for all $n \geq 1$, which includes the special case $\{(t)_n/n! = \binom{t}{n}\}$, then we get a new type of Riordan group, which will be discussed later.

(iii) Note that both $t^n/n!$ and $\binom{t}{n}$ are the simplest Sheffer-type polynomials. Certainly, every special kind of Sheffer-type polynomial sequence $(p_n(t))_{n\geq 0}$ can be used as a basic sequence, thereby producing a kind of Riordan groups related to Sheffer-type polynomial sequences.

Remark 5.2.12 *Analog to the isomorphic property of the generalized Sheffer-type polynomial groups (cf. [85]), we have an isomorphic property for our generalized Riordan groups. Let $P = (p_n(t))_{n\geq 0}$ and $Q = (q_n(t))_{n\geq 0}$ be two distinct basic sequences of polynomials, and let G_P and G_Q be the generalized Riordan groups associated with P and Q, respectively. Then it can be shown that G_P and G_Q are isomorphic in the sense that a one-to-one correspondence can be established between the elements of G_P and G_Q.*

More precisely, let $(d_{n,k})_P = (d(t), h(t))_P$ and $(c_{n,k})_P = (g(t), f(t))_P$ be two generalized Riordan arrays with respect to the basic sequence $P = (p_n(t))_{n\geq 0}$, where $d(t), g(t) \in \mathcal{F}_0$ and $h(t), f(t) \in \mathcal{F}_1$. From Theorem 5.2.10, we have their product

$$(\xi_{n,k})_P = (d(t), h(t))_P \cdot (g(t), f(t))_P = (d(t)g(h(t)), f(h(t)))_P.$$

Similarly, $(d(t), h(t))_Q$, $(g(t), f(t))_Q$, and $(d(t)g(h(t)), f(h(t)))_Q$ denote the corresponding elements of G_Q associated with $Q = (q_n)_{n\geq 0}$, where $(d(t), h(t))_Q = (d'_{n,k})_Q$ with $d'_{n,k} = [q_n(t)]d(t)q_k(h(t))$, etc. Then G_P and G_Q are isomorphic under the one-to-one correspondence relations $(d(t), h(t))_P \leftrightarrow (d(t), h(t))_Q$, $(g(t), f(t))_P \leftrightarrow (g(t), f(t))_Q$, which imply

$$(d(t), h(t))_P(g(t), f(t))_P \leftrightarrow (d(t), h(t))_Q(g(t), f(t))_Q$$

and

$$(g(t), f(t))_P^{-1} = \left(\frac{1}{g(\bar{f}(t))}, \bar{f}(t)\right)_P \leftrightarrow (g(t), f(t))_Q^{-1} = \left(\frac{1}{g(\bar{f}(t))}, \bar{f}(t)\right)_Q.$$

Note the unit element $(1, t)$ *is common to both* G_P *and* G_Q. *Consequently, all the generalized Riordan groups are isomorphic to the ordinary Riordan group with* $\{t^n/n!\}$. *However, concrete structures and analytic and combinatorial implications between different groups are different. More details and applications demonstrating the differences are worthy of further study.*

5.2.2 A general class of convolution-type identities

In order to secure convergence in power series expansions, throughout this subsection, $\alpha(t), \beta(t), \phi(t)$ and $\psi(t)$ might be assumed to be real entire functions having power series expansions in $\mathbb{R}$. We may also assume $\phi(t), \psi(t) \in \mathcal{F}_1$ so that both $\phi(t)$ and $\psi(t)$ are compositionally invertible functions. Hence $\psi(\phi(t))$ is also an entire function in $\mathcal{F}_1$. In addition, some weak conditions for the convergence can be found by using the similar arguments as shown in previous contents (cf. also [102, 104] and [107]).

Let $(p_n(t))_{n\geq 0}$ be a basic sequence of polynomials with real coefficients, and let $\alpha(0) = \beta(0) = 1$. Then parallel to (5.116), we define

$$\begin{aligned} d_{n,k} &= [p_n(t)]\alpha(t)p_k(\phi(t)) \\ c_{n,k} &= [p_n(t)]\beta(t)p_k(\psi(t)) \end{aligned}$$

By Theorem 5.2.10 we have

$$(\alpha(t), \phi(t))(\beta(t), \psi(t)) = (\alpha(t)\beta(\phi(t)), \psi(\phi(t))). \tag{5.122}$$

This implies

Theorem 5.2.13 (Vandermonde-type convolution formula) *With all the same assumptions as above, we have*

$$\begin{aligned} &\sum_{\lambda\geq 0} ([p_n(t)]\alpha(t)p_\lambda(\phi(t))) ([p_\lambda(t)]\beta(t)p_k(\psi(t))) \\ &\qquad = [p_n(t)]\alpha(t)\beta(\phi(t))p_k(\psi(\phi(t))). \end{aligned} \tag{5.123}$$

Remark 5.2.14 *Let* $(\alpha(t), \phi(t))$ *and* $(\beta(t), \psi(t))$ *be the generalized Riordan arrays with respect to a basic sequence* $(p_n(t))_{n\geq 0}$. *In (5.122), if* $\psi(\phi(t)) = \phi(\psi(t)) = t$ *and* $\alpha(t) = 1/\beta(\phi(t))$, *then a pair of inverse matrices,* $(\alpha(t), \phi(t))$ *and* $(\beta(t), \psi(t))$, *with respect to the basic sequence* $(p_n(t))_{n\geq 0}$ *is defined. Following Definition 5.2.8, we denote* $(\alpha(t), \phi(t))$ *and its inverse by* $(d_{n,k})$ *and* $(e_{n,k})$, *respectively. Then a pair of inverse matrices* $(d_{n,k})$ *and* $(e_{n,k})$ *can be used to generalize the combinatorial inversion*

$$f_n = \sum_{k=0}^{n} d_{n,k}g_k \Leftrightarrow g_n = \sum_{k=0}^{n} e_{n,k}f_k, \tag{5.124}$$

or equivalently,

$$F(t) = \alpha(t)G(\phi(t)) \Leftrightarrow G(t) = \beta(t)F(\psi(t)),$$

where $F(t)$ and $G(t)$ are the generating functions of the sequences (f_n) and (g_n), respectively. The inverting combinatorial sums by means of Riordan arrays was introduced and discussed in [105] by He, Hsu, and Shiue. This problem was also studied by Merlini, Sprugnoli, and Verri in [162]. It is easy to see

$$(d_{n,k})^{-1} = (g(t), f(t))^{-1} = (\tilde{g}(t), \bar{f}(t)) = (e_{n,k}), \tag{5.125}$$

where

$$\tilde{g}(t) = \frac{1}{g(\bar{f}(t))}. \tag{5.126}$$

As an example, for the Pascal triangle $(g(t), f(t)) = (1/(1-t), t/(1-t)) = (\binom{n}{k})$ from (5.126) there holds

$$(\tilde{g}(t), \bar{f}(t)) = \left(\frac{1}{1+t}, \frac{t}{1+t}\right) = \left((-1)^{n-k}\binom{n}{k}\right),$$

which yields the well known binomial inversion (cf. Subsection 5.1.4)

$$f_n = \sum_{k=0}^{n} \binom{n}{k} g_k \Leftrightarrow g_n = \sum_{k=0}^{n} (-1)^{n-k} \binom{n}{k} f_k.$$

Another example can be found in Catalan matrix $(C(t), tC(t)) = (d^c_{n,k})$, where

$$C(t) = \frac{1-\sqrt{1-4t}}{2t}$$

and the entries of the Catalan matrix are (cf., for example, [42, 86, 151, 193])

$$d^c_{n,k} = [t^n]t^k C(t)^{k+1} = \frac{k+1}{n+1}\binom{2n-k}{n}, \quad 0 \le k \le n.$$

It is easy to find

$$(C(t), tC(t))^{-1} = (1-t, t(1-t)) = (e^c_{n,k}),$$

where the matrix entries are

$$e^c_{n,k} = [t^n]t^k(1-t)^{k+1} = (-1)^{n-k}\binom{k+1}{n-k}, \quad 0 \le k \le n.$$

Thus we have the sum inversion

$$f_n = \sum_{k=0}^{n} \frac{k+1}{n+1}\binom{2n-k}{n} g_k \Leftrightarrow g_n = \sum_{k=0}^{n} (-1)^{n-k} \binom{k+1}{n-k} f_k.$$

Remark 5.2.15 *It may be shown that (5.123) is a valid finite convolution-type identity for the case where $\alpha(t)$, $\beta(t)$, $\phi(t)$, and $\psi(t)$ are arbitrary polynomials. This is not a particular case of Theorem 5.2.13, but can be proved* *similarly. Let $\alpha(t)$ and $\phi(t)$ be arbitrary polynomials. Especially noteworthy is that $\alpha(t)p_k(\phi(t))$ is a polynomial so that it can be expressed uniquely as linear sums in terms of $p_n(t)'s$ with coefficients $d_{n,k}$, viz.,*

$$\alpha(t)p_k(\phi(t)) = \sum_{n=0}^{\infty} d_{n,k}p_n(t).$$

Thus, based on this relation we may write, upon using of the extracting-coefficient operator $[p_n(t)]$, by

$$d_{n,k} = [p_n(t)]\alpha(t)p_k(\phi(t)),$$

where $n, k \in J$, $J \subset \mathbb{N}_0$. We denote the matrix $(d_{n,k})$ by $(\alpha(t), \phi(t))$. For arbitrary polynomials $\beta(t)$ and $\psi(t)$, we may define

$$c_{n,k} = [p_n(t)]\beta(t)p_k(\psi(t))$$

and denote $(c_{n,k}) = (\beta(t), \psi(t))$, where $n, k \in J$. Thus there holds matrix multiplication

$$(\alpha(t), \phi(t))(\beta(t), \psi(t)) = (\alpha(t)\beta(\phi(t)), \psi(\phi(t)),$$

which implies a finite convolution-type identity (5.123), in which there is no infinite summation involved.

Before proceeding to applications of (5.123), we need Theorem 3.7 of Hsu, Shiue and the author [103] represented as follows.

Lemma 5.2.16 *Let $f(t)$ be an entire function. Then we have a formal expansion of $f(t)$ in terms of a sequence of Sheffer-type polynomial sequence $(p_n(t))_{n\geq 0}$, namely*

$$f(t) = \sum_{k\geq 0} \alpha_k \, p_k(t), \tag{5.127}$$

where

$$\begin{aligned} \alpha_k &= \Lambda_k(D)f(0), D = \frac{d}{dt}, &(5.128)\\ \Lambda_k(D) &= \sum_{n\geq k} \frac{k!}{n!}\sigma^*(n,k)D^n, &(5.129)\\ \sigma^*(n,k) &= \left[\frac{t^n}{n!}\right]\frac{1}{k!}\frac{(h^*(t))^k}{g(h^*(t))}, &(5.130)\end{aligned}$$

and

$$p_k(t) = [\tau^k]g(\tau)\exp(tf(\tau)) \tag{5.131}$$

is derived from the definition $g(\tau)\exp(tf(\tau)) = \sum_{k\geq 0} p_k(t)\tau^k$.

Applying Lemma 5.2.16 to (5.123) leads us to the following corollary.

Corollary 5.2.17 *Let the differential operator $\Lambda_k(D)$ be defined by (5.129), (5.130), and (5.131), where $A(t)$ and $g(t)$ satisfies the conditions $A(0) = 1, ord\, g(t) = 1$. Then there holds the general convolution formula of the form*

$$\begin{aligned}&\sum_{\lambda\geq 0}\Lambda_n(D)\big(\alpha(t)p_\lambda(\phi(t))\big)_{t=0}\Lambda_\lambda(D)\big(\beta(t)p_k(\psi(t))\big)_{t=0}\\ &= \Lambda_n(D)\big(\alpha(t)\beta(\phi(t))p_k(\psi(\phi(t))\big)_{t=0}.\end{aligned} \tag{5.132}$$

In fact, all the functions appearing in (5.132) are entire functions so that differential operators can apply. However, a practical application of formula (5.132) would seem quite complicated.

The key to construct convolution formulas using (5.132) is to find the basic sequences with respect to corresponding operators Λ_n similar to those defined by Lemma 5.2.16. More precisely, we call $\Lambda = (\Lambda_n)_{n\geq 0}$ a sequence of the *function-to-sequence operators* if there exist $\Lambda_n : f \mapsto a_n$ $(n \geq 0)$ mapping every real entire function $f(t)$ to a real number sequence $(a_n)_{n\geq 0}$ with the property that the particular case $f(t) = 0$ leads to the zero-sequence $(a_n = 0)_{n\geq 0}$. Given a simple normal polynomial sequence $(p_n(t))_{n\geq 0}$, suppose that there is a function-to-sequence operator sequence $(\Lambda_n)_{n\geq 0}$ that makes every real entire function $f(t)$ expressible formally or analytically in the form

$$f(t) = \sum_{n\geq 0} a_n p_n(t),$$

where the sequence $(a_n)_{n\geq 0}$ is determined from $f(t)$ via Λ_n. Then from Proposition 5.2.7 $(p_n(t))_{n\geq 0}$ is a basic sequence, called a basic sequence for entire functions with respect to the operator sequence $\Lambda = (\Lambda_n)_{n\geq 0}$.

Here are some examples of basic sequences with respect to some function-to-sequence operators.

Example 5.2.18 *Generally, every Sheffer-type polynomial sequence $(p_n(t))_{n\geq 0}$ is a basic sequence with respect to $\Lambda(D) = (\Lambda_n(D))_{n\geq 0}$ defined by (5.129). Let the Sheffer-type sequence $(p_n(t))_{n\geq 0}$ be given by $(g(z), f(z))$ as*

$$g(z)exp(tf(z)) = \sum_{n\geq 0} p_n(t)z^n.$$

Let $g(z)$ and $f(z)$ have convergence radii ρ_1 and ρ_2, respectively, and let $\rho = \min\{\rho_1, \rho_2\}$. Then the absolute convergence of $\sum_{n\geq 0} p_n(t)z^n$ with $|z| < \rho$ implies the convergence of $\sum_{n\geq 0} a_n p_n(t)$ with $\overline{\lim}_{n\to\infty}|a_n|^{1/n} < \rho$. Accordingly, for the basic sequence $(p_n(t))_{n\geq 0}$ with respect to the operator $\Lambda(D)$ we see that the expression

$$h(t) = \sum_{n\geq 0}(\Lambda_n(D)h(0))p_n(t)$$

is an absolutely convergent expansion for every entire function $h(t)$ satisfying the condition

$$\overline{\lim}_{n\to\infty} |\Lambda_n(D)h(0)|^{1/n} < \rho.$$

Example 5.2.19 *Bernoulli polynomial sequence $(B_n(t)/n!)_{n\geq 0}$ and Euler polynomial sequence $(E_n/n!)_{n\geq 0}$ are basic sequences for entire functions with respect to the operators $\Lambda(D) = (\Delta D^{n-1})_{n\geq 0}$ and $\Lambda(D) = (MD^n)_{n\geq 0}$ acting to $h(t)$ at $t = 0$, respectively so that there hold formal expressions*

$$h(t) = \sum_{n\geq 0} \left(\Delta D^{n-1}h(0)\right) \frac{B_n(t)}{n!} \text{ and } h(t) = \sum_{n\geq 0} \left(MD^n h(0)\right) \frac{E_n(t)}{n!},$$

where $Mh(t) = (h(t) + h(t+1))/2$. Particularly, the above expressions are convergent analytic expressions in case $h(t)$ are taken to be entire functions satisfying the following conditions, respectively:

$$\overline{\lim}_{n\to\infty} \left|\Delta D^{n-1}h(0)\right|^{1/n} < 2\pi, \quad \overline{\lim}_{n\to\infty} |MD^n h(0)|^{1/n} < \pi.$$

These follow from the following generating functions, respectively,

$$\left(\frac{z}{e^z - 1}\right) e^{tz} = \sum_{n\geq 0} \frac{B_n(t)}{n!} z^n \ (|z| < 2\pi),$$

$$\left(\frac{2}{e^z + 1}\right) e^{tz} = \sum_{n\geq 0} \frac{E_n(t)}{n!} z^n \ (|z| < \pi).$$

In the rest of this subsection, we consider in detail the convolution formulas constructed by using the following basic sequences.

(i) $p_n(t) = t^n/n!$;

(ii) $p_n(t) = (t)_n/n!$;

(iii) $p_n(t) = t^{[n]}/n!$, where $t^{[n]} = t\left(t + \frac{n}{2} - 1\right)_{n-1}$;

(iv) $p_n(t) = B_n(t)/n!$, where $B_n(t)$ is the nth order Bernoulli polynomial of the first kind;

(v) $p_n(t) = E_n(t)$, where $E_n(t)$ is the nth order Euler polynomial;

(vi) $p_n(t) = \xi_n(t)$, where $\xi_n(t)$ is the nth order Boole's polynomial;

(vii) $p_n(t) \equiv p_n(t,x) := t(t-nx)^{n-1}/n!$ $(n \geq 1)$ and $p_0(t) \equiv p_0(t,x) = 1$ for fixed x, where $p_n(t)$ is the nth order Appel polynomial.

For (i), the Maclaurin expansion

$$h(t) = \sum_{n\geq 0} h^{(n)}(0)\frac{t^n}{n!} = \sum_{n\geq 0} D^n h(0) p_n(t)$$

shows that $\Lambda_n(D) = D^n$. Consequently, we get the convolution formula

$$\sum_{\lambda\geq 0} D^n\left(\alpha(t)\frac{\phi^\lambda(t)}{\lambda!}\right)_0 D^\lambda\left(\beta(t)\frac{\psi^k(t)}{k!}\right)_0$$
$$=D^n\left(\alpha(t)\beta(\phi(t))\frac{\psi(\phi(t))^k}{k!}\right)_0, \tag{5.133}$$

where the index 0 in the above equation means the corresponding functions take their values at $t = 0$.

For Lagrange arrays $(\alpha(t), \phi(t))$ and $(\beta(t), \psi(t))$, i.e., $\alpha(t) = \beta(t) = 1$, we have the following lemma to evaluate the high order derivatives in (5.133).

Proposition 5.2.20 *Let $f(t) \in \mathcal{F}_1$ with compositional inverse $\bar{f}(t)$. Then*

$$D^n\left(f^k(t)\right)\big|_{t=0} = (n-1)!k[u^{n-k}]\left(\frac{u}{\bar{f}(u)}\right)^n \tag{5.134}$$

for all $n \geq k \geq 0$.

Proof. First, for $p_n(t) = t^n/n!$ and $h(t) \in \mathcal{F}$ we observe the nth coefficient of Maclaurin expansion of

$$h(t) = \sum_{k\geq 0} D^k h(0) p_k(t)$$

can be written as

$$D^n h(0) = [p_n(t)]h(t).$$

Let $h(u) = u^k$ and $u = f(t)$. Then

$$D^n h(f(t))|_{t=0} = [p_n(t)]h(f(t)) = n![t^n]h(f(t)).$$

Using the Lagrange inversion formula (4.179) (cf. also Theorem 5.1 in [216]) and noting the compositional inverse of $u = f(t)$ satisfies $u = ut/\bar{f}(u)$ with $u/\bar{f}(u) \in \mathcal{F}_0$, we can write the last equation as

$$\begin{aligned} D^n h(f(t))|_{t=0} &= D^n f^k(t)\big|_{t=0} = n![t^n]h(f(t)) \\ &= n!\frac{1}{n}[u^{n-1}]Dh(u)\left(\frac{u}{\bar{f}(u)}\right)^n \\ &= (n-1)!k[u^{n-1}]u^{k-1}\left(\frac{u}{\bar{f}(u)}\right)^n, \end{aligned}$$

which implies (5.134). ∎

Corollary 5.2.21 *Let $f(t) \in \mathcal{F}_1$, and let $g(t) \in \mathcal{F}_0$ be the generating function of sequence $\{g_n\}_{n\geq 0}$. Then*

$$D^n\left(g(t)f^k(t)\right)\big|_{t=0} = kn! \sum_{\ell=\max\{k,1\}}^{n} \frac{g_{n-\ell}}{\ell}[u^{\ell-k}]\left(\frac{u}{\bar{f}(u)}\right)^{\ell} \tag{5.135}$$

for all $n \geq k \geq 0$.

Proof. Noting $f(t) \in \mathcal{F}_1$, from the Leibniz formula

$$D^n(\alpha(t)\beta(t)) = \sum_{\ell=0}^{n}\binom{n}{\ell}D^{\ell}\alpha(t)D^{n-\ell}\beta(t)$$

and formula (5.134) we have

$$D^n\left(g(t)f^k(t)\right)\big|_{t=0} = \sum_{\ell=\max\{k,1\}}^{n}\binom{n}{\ell}D^{n-\ell}g(t)\big|_{t=0}D^{\ell}f^k(t)\big|_{t=0}$$

$$= \sum_{\ell=\max\{k,1\}}^{n}\binom{n}{\ell}(n-\ell)!g_{n-\ell}(\ell-1)!k[u^{\ell-k}]\left(\frac{u}{\bar{f}(u)}\right)^{\ell},$$

which implies (5.135). ■

Denote the compositional inverses of $\phi(t)$, $\psi(t)$, and $(\psi\circ\phi)(t)$ by $\overline{\phi}(t)$, $\overline{\psi}(t)$, and $\overline{(\psi\circ\phi)}(t)$, respectively. Using Proposition 5.2.20, we can immediately write Identity (5.133) as

$$\sum_{\lambda=k}^{n}\left([u^{n-\lambda}]\left(\frac{u}{\overline{\phi}(u)}\right)^{n}\right)\left([u^{\lambda-k}]\left(\frac{u}{\overline{\psi}(u)}\right)^{\lambda}\right) = [u^{n-k}]\left(\frac{u}{\overline{(\psi\circ\phi)}(u)}\right)^{n} \tag{5.136}$$

for $\alpha(t) = \beta(t) = 1$.

Let $\alpha(t), \beta(t) \in \mathcal{F}_0$ with $\alpha(t)\beta(t) \neq 1$ and $\phi(t), \psi(t) \in \mathcal{F}_1$. Using Corollary 5.2.21, we can modify Identity (5.133) as

$$\begin{aligned}&\sum_{\lambda=0}^{n}\lambda\left(\sum_{\ell=\max\{\lambda,1\}}^{n}\frac{1}{\ell}\left([t^{n-\ell}]\alpha(t)\right)\left([u^{\ell-\lambda}]\left(\frac{u}{\overline{\phi}(u)}\right)^{\ell}\right)\right)\\ &\times\left(\sum_{j=\max\{k,1\}}^{\lambda}\frac{1}{j}\left([t^{\lambda-j}]\beta(t)\right)\left([u^{j-k}]\left(\frac{u}{\overline{\psi}(u)}\right)^{j}\right)\right)\\ =&\sum_{\ell=\max\{k,1\}}^{n}\frac{1}{\ell}\left([t^{n-\ell}]\alpha(t)\beta(\phi(t))\right)\left([u^{\ell-k}]\left(\frac{u}{\overline{(\psi\circ\phi)}(u)}\right)^{\ell}\right)\end{aligned} \tag{5.137}$$

for all $n \geq k \geq 0$.

As an application of identity (5.137), if $\alpha(t) = \beta(t) = 1/(1-t)$, and $\phi(t) = \psi(t) = t/(1-t)$, then $\alpha(t)\beta(\phi(t)) = 1/(1-2t)$. Similarly, there holds identity

$$\begin{aligned}&\sum_{\lambda=k}^{n}\lambda\left(\sum_{\ell=\max\{\lambda,1\}}^{n}\frac{1}{\ell}\binom{\ell}{\ell-\lambda}\right)\left(\sum_{j=\max\{k,1\}}^{\lambda}\frac{1}{j}\binom{j}{j-k}\right)\\ =\ & 2^{n-k}\sum_{\ell=\max\{k,1\}}^{n}\frac{1}{\ell}\binom{\ell}{\ell-k}. \qquad (5.138)\end{aligned}$$

For the basic sequence (ii), the Newton series

$$f(t) = \sum_{n\geq 0}\Delta^n f(0)\frac{(t)_n}{n!} = \sum_{n\geq 0}\Delta^n f(0)p_n(t)$$

implies that $\Lambda_n(D) = \Delta^n$. Consequently, we obtain the convolution formula

$$\begin{aligned}&\sum_{\lambda\geq 0}\Delta^n\left(\alpha(t)\binom{\phi(t)}{\lambda}\right)_0\Delta^\lambda\left(\beta(t)\binom{\psi(t)}{k}\right)_0\\ =\ & \Delta^n\left(\alpha(t)\beta(\phi(t))\binom{\psi(\phi(t))}{k}\right)_0, \qquad (5.139)\end{aligned}$$

among which the particular case with $\alpha(t) = \beta(t) = 1$ seems more interesting, since

$$\sum_{\lambda\geq 0}\Delta^n\binom{\phi(t)}{\lambda}_0\Delta^\lambda\binom{\psi(t)}{k}_0 = \Delta^n\binom{\psi(\phi(t))}{k}_0. \qquad (5.140)$$

This is actually a "deep generalization" of the Vandermonde convolution formula. Indeed, let $k > n$ and take $\phi(t) = t + a$ and $\psi(t) = t + b$. It follows from Remark 5.2.15 that

$$\begin{aligned}\sum_{\lambda\geq 0}\Delta^n\binom{t+a}{\lambda}_0\Delta^\lambda\binom{t+b}{k}_0 &= \sum_{\lambda\geq 0}\binom{a}{\lambda-n}\binom{b}{k-\lambda}\\ &= \Delta^n\binom{t+a+b}{k}_0. \qquad (5.141)\end{aligned}$$

The last equality gives

$$\sum_{\lambda\geq 0}\binom{a}{\lambda-n}\binom{b}{k-\lambda} = \binom{a+b}{k-n}. \qquad (5.142)$$

For the basic sequence (iii), the Newton series in terms of central difference

$$f(t) = \sum_{n\geq 0}\delta^{(n)}f(0)\frac{t^{[n]}}{n!} = \sum_{n\geq 0}\delta^n f(0)p_n(t),$$

where $\delta f(t) = f(t+(1/2)) - f(t-(1/2))$, implies that $\Lambda_n(D) = \delta^n$. Consequently, noting

$$p_n(t) = t^{[n]}/n! = \frac{t}{t-\frac{n}{2}}\binom{t+\frac{n}{2}-1}{n},$$

we obtain the convolution formula

$$\sum_{\lambda\geq 0}\delta^n\left(\frac{\alpha(t)\phi(t)}{\phi(t)-\frac{\lambda}{2}}\binom{\phi(t)+\frac{\lambda}{2}-1}{\lambda}\right)_0\delta^\lambda\left(\frac{\beta(t)\psi(t)}{\psi(t)-\frac{k}{2}}\binom{\psi(t)+\frac{k}{2}-1}{k}\right)_0$$
$$= \delta^n\left(\frac{\alpha(t)\beta(\phi(t))\psi(\phi(t))}{\psi(\phi(t))-\frac{k}{2}}\binom{\psi(\phi(t))+\frac{k}{2}-1}{k}\right)_0, \tag{5.143}$$

where $2\phi(0), 2\psi(0), 2\psi(\phi(0)) \notin \mathbb{N}_0$, among which the particular case with $\alpha(t) = \beta(t) = 1$ yields

$$\sum_{\lambda\geq 0}\delta^n\left(\frac{\phi(t)}{\phi(t)-\frac{\lambda}{2}}\binom{\phi(t)+\frac{\lambda}{2}-1}{\lambda}\right)_0\delta^\lambda\left(\frac{\psi(t)}{\psi(t)-\frac{k}{2}}\binom{\psi(t)+\frac{k}{2}-1}{k}\right)_0$$
$$= \delta^n\left(\frac{\psi(\phi(t))}{\psi(\phi(t)-\frac{k}{2}}\binom{\psi(\phi(t))+\frac{k}{2}-1}{k}\right)_0. \tag{5.144}$$

Taking $\phi(t) = t+a$ and $\psi(t) = t+b$, $2a, 2b, 2(a+b) \notin \mathbb{N}_0$, we obtain the convolution formula

$$\sum_{\lambda\geq 0}\delta^n\left(\frac{t+a}{t+a-\frac{\lambda}{2}}\binom{t+a+\frac{\lambda}{2}-1}{\lambda}\right)_0\delta^\lambda\left(\frac{t+b}{t+b-\frac{k}{2}}\binom{t+b+\frac{k}{2}-1}{k}\right)_0$$
$$= \delta^n\left(\frac{t+a+b}{t+a+b-\frac{k}{2}}\binom{t+a+b+\frac{k}{2}-1}{k}\right)_0 \tag{5.145}$$

for $k > n$, or equivalently,

$$\sum_{\lambda\geq 0}\frac{ab}{(a-\frac{\lambda-n}{2})(b-\frac{k-\lambda}{2})}\binom{a+\frac{\lambda-n}{2}-1}{\lambda-n}\binom{b+\frac{k-\lambda}{2}-1}{k-\lambda}$$
$$= \frac{a+b}{a+b-\frac{k-n}{2}}\binom{a+b+\frac{k-n}{2}-1}{k-n} \tag{5.146}$$

for all $k \geq n$ due to $\delta^k p_n(t) = p_{n-k}(t)$.

For the basic sequence (iv), using the property of Bernoulli polynomials (cf. [44]) we have the following series expansion

$$f(t) = \sum_{n\geq 0}(f^{(n-1)}(1) - f^{(n-1)}(0))\frac{B_n(t)}{n!} = \sum_{n\geq 0}[D^{n-1}f(t)]_0^1\frac{B_n(t)}{n!}$$

where

$$[D^{-1}f(t)]_0^1 = \int_0^1 f(t)dt.$$

Consequently, we obtain the convolution formula

$$\sum_{\lambda\geq 0}\left[D^{n-1}\frac{B_\lambda(\phi(t))}{\lambda!}\right]_0^1\left[D^{\lambda-1}\frac{B_k(\phi(t))}{k!}\right]_0^1=\left[D^{n-1}\frac{B_k(\psi(\phi(t)))}{k!}\right]_0^1, \quad (5.147)$$

where we may assume $n \geq 2$.

For the basic sequence (v), we apply the property of Euler polynomials (cf. [44]) to have the following series expansion

$$f(t)=\sum_{n\geq 0}\frac{1}{2}(f^{(n)}(1)+f^{(n)}(0))E_n(t)=\sum_{n\geq 0}[MD^n f(0)]E_n(t),$$

where $Mf(t) = (f(t) + f(t+1))/2$. Consequently, we obtain the convolution formula

$$\sum_{\lambda\geq 0}\left[MD^n E_\lambda(\phi(t))\right]_0\left[MD^\lambda E_k(\phi(t))\right]_0=\left[MD^n E_k(\psi(\phi(t)))\right]_0, \quad (5.148)$$

where we assume $n \geq 2$.

For the basic sequence (vi), similar to (iv) and (v), from the property of Boole's polynomials (cf. [44]) we have the series expansion

$$f(t)=\sum_{n\geq 0}\frac{1}{2}(\Delta^n f(1)+\Delta^n f(0))\xi_n(t)=\sum_{n\geq 0}[M\Delta^n f(0)]\xi_n(t).$$

Thus there holds the convolution formula

$$\sum_{\lambda\geq 0}\left[M\Delta^n \xi_\lambda(\phi(t))\right]_0\left[M\Delta^\lambda \xi_k(\phi(t))\right]_0=\left[M\Delta^n \xi_k(\psi(\phi(t)))\right]_0, \quad (5.149)$$

where we assume $n \geq 2$.

Finally, for the basic sequence (vii), from Theorem C on page 130 of [44] there holds

$$f(t)=\sum_{n\geq 0}f^{(n)}(nx)p_n(t,x)$$

for all $f \in \mathcal{F}$, where $f^{(k)}$ is the kth derivative of f and x is fixed. If $x = 0$, then the above expansion is the ordinary (formal) Taylor formula. Noting

$$[p_n(t)]\, f(t) = [p_n(t,x)]\, f(t) = f^{(n)}(nx),$$

from Vandermonde-type convolution formula (5.123) in Theorem 5.2.13, we find out

$$\sum_{\lambda\geq 0}D_t^n\left(\alpha(t)p_\lambda(\phi(t))\right)\big|_{t=nx}\, D_t^\lambda\left(\beta(t)p_k(\psi(t))\right)\big|_{t=nx}$$
$$=D_t^n\left(\alpha(t)\beta(\phi(t))p_k(\psi(\phi(t)))\right)\big|_{t=nx}. \quad (5.150)$$

Remark 5.2.22 *Clearly, (5.122) subject to (5.128) reduces to finite summations whenever $\psi(t)$ is polynomial. The final result is:*

$$\sum_{\lambda\geq 0}\Delta^n\binom{\phi(t)}{\lambda}_0\Delta^\lambda\binom{\psi(t)}{k}_0=\Delta^n\binom{\psi(\phi(t))}{k}_0. \tag{5.151}$$

Remark 5.2.23 *The Lagrange inversion formula* (4.175) *(cf. also Theorem A of [44]) guarantees the relation*

$$\left[\frac{t^n}{n!}\right]\frac{\bar{f}(t)^k}{k!}=D_0^n\frac{\bar{f}(t)^k}{k!}=\binom{n-1}{k-1}D_0^{n-k}\left(\frac{t}{f(t)}\right)^k,$$

suggesting that for the case $g(t)=1$, (5.130) may be reformulated in the form

$$\begin{aligned}\sigma^*(n,k) &= \left[\frac{t^n}{n!}\right]\frac{\bar{f}(t)^k}{k!}\\ &= \binom{n-1}{k-1}\left[\frac{t^{n-k}}{(n-k)!}\right]\left(\frac{t}{f(t)}\right)^n\\ &= \binom{n-1}{k-1}D^{n-k}\left(\frac{t}{f(t)}\right)^n_{t=0}.\end{aligned} \tag{5.152}$$

This applies to the case when $\bar{f}(t)$, the compositional inverse of $f(t)$, is not easily computed. For instance, [86] consider $(g(t),f(t))=\left(\frac{1-et}{1+(c-e)t},\frac{t(1-et)}{1+(c-e)t}\right)$, which has a complicated inverse $(d_{c,e}(t),td_{c,e}(t))$, where

$$d_{c,e}(t)=\frac{1-(c-e)t-\sqrt{1-2(c+e)t+(c-e)^2t^2}}{2et},\quad e\neq 0.$$

5.2.3 A general class of Abel identities

Inspired by previous subsection, we will use the fundamental property of the generalized Riordan arrays and its alternating form with respect to basic sequences to give two ways for the construction of a general class of Abel identities. First, we present the fundamental property of the generalized Riordan arrays with respect to a basic sequence using a similar argument of [193]. Let $(g(t),f(t))$, $g(t)\in\mathcal{F}_0$ and $f(t)\in\mathcal{F}_1$, be a generalized Riordan array with respect to the basic sequence $(p_n(t))_{n\geq 0}$, and let $h(t)$ be the generating function of any sequence $\{h_n\}_{n\geq 0}$ in terms of $(p_n(t))_{n\geq 0}$, i.e., $h(t)=\sum_{n\geq 0}h_np_n(t)$. We have

$$\sum_{k\geq 0}d_{n,k}h_k=[p_n(t)]g(t)h(f(t)), \tag{5.153}$$

which comes from the following observation

$$[p_n(t)]g(t)h(f(t)) = [p_n(t)]g(t)\sum_{k\geq 0} h_k p_k(f(t))$$

$$= \sum_{k\geq 0} h_k[p_n(t)]g(t)p_k(f(t)) = \sum_{k\geq 0} d_{n,k}h_k. \tag{5.154}$$

Proposition 5.2.24 *Let $(g(t),\ f(t)) = (d_{n,k})_{n\geq k\geq 0}$ be a generalized Riordan array with respect to the basic sequence $(p_n(t))_{n\geq 0}$ with $p_0(t) = 1$ and $p_n(t) = \sum_{j=1}^n \alpha_{n,j}t^j$ $(n \geq 1,\ \alpha_{n,j} \neq 0)$, where $g(t) \in \mathcal{F}_0$ and $f(t) \in \mathcal{F}_1$, and let $h(t) = \sum_{n\geq 0} h_n p_n(t)$. Then there holds*

$$\sum_{k=1}^n d_{n,k}h_k = \sum_{j=1}^n \sum_{\ell=1}^j \frac{1}{\alpha_{n,j}}[t^\ell]h(f(t))[t^{j-\ell}]g(t). \tag{5.155}$$

Furthermore, if $\bar{f}(t)$ is the compositional inverse of $f(t)$, then (5.155) can be written as

$$\sum_{k=1}^n d_{n,k}h_k = \sum_{j=1}^n \sum_{\ell=1}^j \frac{1}{\ell\alpha_{n,j}}[t^{\ell-1}]h'(t)\left(\frac{t}{\bar{f}(t)}\right)^\ell [t^{j-\ell}]g(t). \tag{5.156}$$

Particularly, if $p_n(t) = t^n/c_n$, $c_0 = 1$, $c_n \neq 0$ for all $n \geq 1$, then (5.156) is specified to

$$\sum_{k=1}^n d_{n,k}h_k = c_n \sum_{\ell=1}^n \frac{1}{\ell}[t^{\ell-1}]h'(t)\left(\frac{t}{\bar{f}(t)}\right)^\ell [t^{n-\ell}]g(t). \tag{5.157}$$

Proof. For $n \geq 1$, substituting $p_n(t) = \sum_{j=1}^n \alpha_{n,j}t^j$ ($\alpha_{n,n} \neq 0$ or see the following Remark 5.115) into (5.153), we can write it as

$$\sum_{k=0}^n d_{n,k}h_k = \sum_{j=1}^n \frac{1}{\alpha_{n,j}}[t^j]g(t)h(f(t))$$

$$= \sum_{j=1}^n \sum_{\ell=0}^j \frac{1}{\alpha_{n,j}}[t^\ell]h(f(t))[t^{j-\ell}]g(t)$$

$$= \sum_{j=1}^n \frac{1}{\alpha_{n,j}}[1]h(f(t))[t^j]g(t) + \sum_{\ell=1}^n \sum_{j=\ell}^n \frac{1}{\alpha_{n,j}}[t^\ell]h(f(t))[t^{j-\ell}]g(t)$$

$$= h_0[p_n(t)]g(t) + \sum_{\ell=1}^n \sum_{j=\ell}^n \frac{1}{\alpha_{n,j}}[t^\ell]h(f(t))[t^{j-\ell}]g(t),$$

which implies (5.155). We may use the Lagrange inversion formula (4.179) to write

$$[t^\ell]h(f(t)) = \frac{1}{\ell}[t^{\ell-1}]h'(t)\left(\frac{t}{\bar{f}(t)}\right)^\ell, \quad \ell \geq 1.$$

Thus (5.156) is derived from (5.155).

■

Remark 5.2.25 *From the proof of (5.155), it is easy to see that the initial terms of both sides of (5.155) can be extended to $k = 0$ and $\ell = 0$, respectively, namely,*

$$\sum_{k=0}^{n} d_{n,k}h_k = \sum_{j=1}^{n}\sum_{\ell=0}^{j} \frac{1}{\alpha_{n,j}}[t^\ell]h(f(t))[t^{j-\ell}]g(t).$$

The fundamental property (5.153) of the generalized Riordan arrays has an alternating form shown in the following theorem.

Theorem 5.2.26 *Let $(g(t), f(t))$, $g(t) \in \mathcal{F}_0$, $f(t) \in \mathcal{F}_1$, be a generalized Riordan array with respect to $(p_n(t))_{n\geq 0}$, where $p_0(t) = 1$ and $p_n(t) = \sum_{k=1}^{n} \alpha_{n,j}t^j$, $\alpha_{n,n} \neq 0$, for $n \geq 1$, and let $h(t)$ be a formal power series. Denote $h(f(t))$ by $v(t)$, and $[p_n(t)]v(\bar{f}(t))$ by v_n. Then there holds*

$$\sum_{k\geq 0} d_{n,k}v_k = [p_n(t)]g(t)v(t) \tag{5.158}$$

where $v_0 = h_0$ and

$$v_k = [p_k(t)]v(\bar{f}(t)) = \sum_{j=1}^{k} \frac{1}{j\alpha_{k,j}}[t^{j-1}]v'(t)\left(\frac{t}{f(t)}\right)^j \tag{5.159}$$

for all $k \geq 1$. Particularly, if $p_n(t) = \alpha_{n,n}t^n$, $\alpha_{n,n} \neq 0$, then

$$v_k = \frac{1}{k\alpha_{k,k}}[t^{k-1}]v'(t)\left(\frac{t}{f(t)}\right)^k \tag{5.160}$$

for all $k \geq 1$.

Proof. Since $v(t) = h(f(t))$, then $v(\bar{f}(t)) = h(t)$. Thus, (5.154) becomes

$$[p_n(t)]g(t)v(t) = \sum_{k\geq 0} d_{n,k}h_k,$$

where $h_k = [p_k(t)]h(t) = [p_k(t)]v(\bar{f}(t)) = v_k$. Using the Lagrange inversion formula (4.179), in which $\bar{h}(t) = tA(\bar{h}(t))$ or $A(t) = t/h(t)$, we may further obtain

$$\begin{aligned} v_k &= [p_k(t)]v(\bar{f}(t)) = \sum_{j=1}^{k} \frac{1}{\alpha_{k,j}}[t^j]v(\bar{f}(t)) \\ &= \sum_{j=1}^{k} \frac{1}{j\alpha_{k,j}}[t^{j-1}]v'(t)\left(\frac{t}{f(t)}\right)^j \end{aligned}$$

for all $k \geq 1$. As for $k = 0$, $v_0 = [1]v(\bar{f}(t)) = v(\bar{f}(0)) = v(0) = h(0) = h_0$, completing the proof.

■

Substituting (5.158) into (5.159), we obtain

Corollary 5.2.27 *Let $(g(t), f(t))$, $g(t) \in \mathcal{F}_0$, $f(t) \in \mathcal{F}_1$, be a generalized Riordan array with respect to $(p_n(t))_{n\geq 0}$, where $p_0(t) = 1$ and $p_n(t) = \sum_{k=1}^{n} \alpha_{n,j} t^j$, $\alpha_{n,n} \neq 0$, for $n \geq 1$, and let $h(t)$ be a formal power series. Denote $h(f(t))$ by $v(t)$, and $[p_n(t)]v(\bar{f}(t))$ by v_n. Then, we have the following identity:*

$$\sum_{k\geq 0} d_{n,k} v_k = d_{n,0} v_0 + \sum_{k=1}^{n} \sum_{j=1}^{k} \frac{d_{n,k}}{j\alpha_{k,j}} [t^{j-1}] v'(t) \left(\frac{t}{f(t)}\right)^j = [p_n(t)] g(t) v(t). \tag{5.161}$$

Particularly, if $p_n(t) = t^n/c_n$, then (5.161) becomes

$$\sum_{k\geq 0} d_{n,k} v_k = d_{n,0} v_0 + \sum_{k=1}^{n} d_{n,k} \frac{c_k}{k} [t^{k-1}] v'(t) \left(\frac{t}{f(t)}\right)^k = [p_n(t)] g(t) v(t). \tag{5.162}$$

Example 5.2.28 *As an example of (5.162) with $p_n(t) = t^n/c_n$, we now consider $(g(t), f(t)) = (e^{\alpha t}, te^{\beta t})$ and $v(t) = e^{\gamma t}$. Thus,*

$$\begin{aligned}
d_{n,k} &= \left[\frac{t^n}{c_n}\right] e^{\alpha t} \frac{(te^{\beta t})^k}{c_k} = \frac{c_n}{c_k} \frac{(\alpha + \beta k)^{n-k}}{(n-k)!}, \\
v_k &= \left[\frac{t^k}{c_k}\right] v(\bar{f}(t)) = \frac{c_k}{k} [t^{k-1}] v'(t) \left(\frac{t}{f(t)}\right)^k = \frac{c_k \gamma}{k!} (\gamma - \beta k)^{k-1}, \\
[p_n(t)]\, g(t) v(t) &= c_n [t^n] e^{(\alpha+\gamma)t} = \frac{c_n}{n!} (\alpha + \gamma)^n,
\end{aligned}$$

which implies the following well-known Abel identity:

$$\frac{c_n}{c_0} \frac{(\alpha)^n}{n!} c_0 \gamma (\gamma)^{-1} + \sum_{k=1}^{n} \frac{c_n}{c_k} \frac{(\alpha + \beta k)^{n-k}}{(n-k)!} \frac{c_k \gamma}{k!} (\gamma - \beta k)^{k-1} = \frac{c_n}{n!} (\alpha + \gamma)^n,$$

or equivalently,

$$\sum_{k=0}^{n} \gamma \binom{n}{k} (\gamma - \beta k)^{k-1} (\alpha + \beta k)^{n-k} = (\alpha + \gamma)^n.$$

From (5.158), we can find another type identities. Let $(g(t), f(t)) = (d_{n,k})_{n\geq k\geq 0}$ be a (c)-generalized Riordan array with respect to a (c)-sequence $\{c_k\}_{k\geq 0}$ with $c_0 = 1$ and $c_k \neq 0$ for all $k > 0$, where $g(t) \in \mathcal{F}_0$ and $f(t) \in \mathcal{F}_1$, and let $h(t) = \sum_{n\geq 0} h_n t^n / c_n$. Then, there holds identity

$$\sum_{k=1}^{n} d_{n,k} \frac{h_k}{c_k} = \sum_{j=1}^{n} [t^j] h(f(t)) [t^{n-j}] g(t). \tag{5.163}$$

Furthermore, denote by $\bar{f}(t)$ is the compositional inverse of $f(t)$, then (5.163) can be written as

$$\sum_{k=1}^{n} d_{n,k}\frac{h_k}{c_k} = \sum_{j=1}^{n} \frac{1}{j}[t^{j-1}]h'(t)\left(\frac{t}{\bar{f}(t)}\right)^j [t^{n-j}]g(t). \tag{5.164}$$

Indeed, substituting $v_k = h_k$ and $v(t) = h(f(t))$ into (5.158), we can write it as

$$\begin{aligned}\sum_{k=0}^{n} d_{n,k}\frac{h_k}{c_k} &= [t^n]g(t)h(f(t))\\ &= \sum_{j=0}^{n}[t^j]h(f(t))[t^{n-j}]g(t)\\ &= f_0 d_{n,0} + \sum_{j=1}^{n}[t^j]h(f(t))[t^{n-j}]g(t),\end{aligned}$$

which implies (5.163). By using the Lagrange inversion formula (4.179) we have

$$[t^j]h(f(t)) = \frac{1}{j}[t^{j-1}]h'(u)\left(\frac{t}{\bar{f}(t)}\right)^j, \quad j \geq 1.$$

Thus (5.164) is derived from (5.163).

From the proof of (5.163), it is easy to see that the initial terms of both sides of (5.163) can be extended to $k = 0$ and $\ell = 0$, respectively, namely,

$$\sum_{k=0}^{n} d_{n,k}\frac{h_k}{c_k} = \sum_{j=0}^{n}[t^j]h(f(t))[t^{n-j}]g(t).$$

Example 5.2.29 *As an example, we consider $(g(t),\ f(t)) = (1/(1-t), t/(1-t))$. From (5.164). Noting formula (5.117) for evaluating $d_{n,k}$, from (5.157) we obtain the identity*

$$\sum_{k=1}^{n}\binom{n}{k}\left[t^k\right]h(t) = \sum_{j=1}^{n}\frac{1}{j}[t^{j-1}]h'(u)\,(1+t)^j\,, \tag{5.165}$$

which yields the identity

$$\sum_{k=1}^{n}\frac{1}{k!}\binom{n}{k} = \sum_{j=1}^{n}\sum_{\ell=0}^{j-1}\frac{1}{j\ell!}\binom{j}{j-1-\ell}$$

when $h(t) = e^t$. For $h(t) = (a+t)^m$, formula (5.165) yields the identity

$$\sum_{k=0}^{n}\binom{n}{k}\binom{m}{k}a^{m-k} = \sum_{l=0}^{m}\binom{m}{l}\binom{m+n-l}{m-l}(a-1)^l.$$

In fact, the right-hand side of (5.165) *becomes*

$$\begin{aligned}
&\sum_{j=1}^{n}\frac{m}{j}[t^{j-1}](a+t)^{m-1}(1+t)^{j}\\
=\ &\sum_{j=1}^{n}\frac{m}{j}[t^{j-1}]\sum_{l=0}^{m-1}\binom{m-1}{l}(a-1)^{l}(1+t)^{m-l+j-1}\\
=\ &m\sum_{l=0}^{m-1}\binom{m-1}{l}(a-1)^{l}\sum_{j=1}^{n}\frac{1}{j}\binom{m-l+j-1}{j-1}\\
=\ &\sum_{l=0}^{m-1}\frac{m}{m-\ell}\binom{m-1}{\ell}(a-1)^{\ell}\sum_{j=1}^{n}\binom{m-\ell+j-1}{m-\ell-1}\\
=\ &\sum_{l=0}^{m-1}\binom{m}{\ell}(a-1)^{\ell}\left(\sum_{j=0}^{n}\binom{m-\ell+j-1}{m-\ell-1}-1\right).
\end{aligned}$$

Applying these to make further simplification, in which we have used the summation formula

$$\sum_{k=0}^{n}\binom{m+k}{m}=\binom{n+m+1}{m+1}.$$

Thus, the right-hand side of (5.165) *can be written as*

$$\begin{aligned}
&\sum_{l=0}^{m-1}\binom{m}{\ell}(a-1)^{\ell}\left(\binom{m+n-\ell}{m-\ell}-1\right)\\
=\ &\sum_{l=0}^{m-1}\binom{m}{\ell}(a-1)^{\ell}\binom{m+n-\ell}{m-\ell}-\sum_{l=0}^{m-1}\binom{m}{\ell}(a-1)^{\ell}\\
=\ &\sum_{l=0}^{m}\binom{m}{\ell}(a-1)^{\ell}\binom{m+n-\ell}{m-\ell}-a^{m},
\end{aligned}$$

which yields the desired identity.

For $h(t)=\frac{1}{\sqrt{1+t}}$, *formula* (5.165) *yields the identity (cf. (3.85) in [73])*

$$\sum_{k=0}^{n}\binom{n}{k}\binom{2k}{k}\frac{(-1)^{k}}{4^{k}}=\sum_{j=0}^{n}\frac{(-3/2+j)_{j}}{j!}=\binom{-1/2+n}{n}$$

because that

$$h(t)=\sum_{k=0}^{\infty}\binom{2k}{k}\frac{(-t)^{k}}{4^{k}}$$

and

$$-\frac{1}{2}[t^{j-1}](1+t)^{-3/2+j}=\frac{(-3/2+j)_{j}}{j!}=\binom{-3/2+j}{j}.$$

In general, for $\alpha \in \mathbb{C}$, let $h(t) = (1+t)^\alpha$, we might deduce the following identity similarly:

$$\sum_{k=0}^{n} \binom{n}{k}\binom{\alpha}{k} = \sum_{j=0}^{n} \frac{(\alpha+j-1)_j}{j!} = \binom{\alpha+n}{n}.$$

In the following example, we consider parameter $0<q<1$ and denote

$$c_n = n!_q := \frac{(1-q)(1-q^2)\cdots(1-q^n)}{(1-q)^n}. \tag{5.166}$$

Roman [185] defines

$$\epsilon_a(t) = \sum_{k\geq 0} \frac{a^k}{c_k} t^k$$

with a non-zero constant a, and evaluates

$$\frac{1}{\epsilon_a(t)} = \sum_{k\geq 0} q^{\binom{k}{2}} \frac{(-at)^k}{c_k}.$$

Hence, we have the q-analog (c)-Riordan array

$$\left(\frac{1}{\epsilon_a(t)}, \frac{t}{1-t}\right) = \left(\sum_{\ell=k}^{n} \binom{\ell-1}{k-1} \frac{a^{n-\ell}}{c_{n-\ell}}\right)_{n\geq k\geq 0},$$

where we use the formula

$$[t^n]\alpha(t)\beta(t) = \sum_{\ell=0}^{n} [t^\ell]\alpha(t)[t^{n-\ell}]\beta(t)$$

to evaluate

$$d_{n,k} = [t^n]\epsilon_a(t)\left(\frac{t}{1-t}\right)^k.$$

Therefore we obtain $q-$identities

$$\sum_{k=1}^{n} \frac{1}{k!} \sum_{\ell=k}^{n} \binom{\ell-1}{k-1} \frac{a^{n-\ell}}{c_{n-\ell}} = \sum_{j=1}^{n} \frac{a^{n-j}}{c_{n-j}} \sum_{\ell=0}^{j-1} \frac{1}{j\ell!} \binom{j}{j-1-\ell}$$

and

$$\sum_{k=1}^{n} \sum_{\ell=k}^{n} \binom{\ell-1}{k-1} \frac{a^{n-\ell}}{c_{n-\ell}} = \sum_{j=1}^{n} \frac{a^{n-j}}{c_{n-j}} \sum_{\ell=0}^{j-1} \frac{\ell+1}{j} \binom{j}{j-1-\ell}$$

from (5.164) with $h(t) = e^t$ and $h(t) = 1/(1-t)$, respectively, where c_k are defined by (5.166) and note $\binom{0}{k} = \delta_{k,0}$.

For $(g(t), f(t)) = (\epsilon_a(t), \frac{t}{1-t})$, *the* q*–exponential function* $\epsilon_a(t) = \sum_{n\geq 0} \frac{(at)^k}{c_k}, c_k = n!_q$. *Assume further* $h(t) = 1/(1+t)$. *Then it is clear that*

$$d_{n,k} = [t^n]\epsilon_a(t)\left(\frac{t}{1-t}\right)^k = \sum_{l=k}^{n} \binom{l-1}{k-1} \frac{a^{n-l}}{c_{n-l}}.$$

All these leads us to the following q*–identity*

$$\sum_{k=1}^{n} (-1)^k \sum_{l=k}^{n} \binom{l-1}{k-1} \frac{a^{n-l}}{c_{n-l}} = -\frac{a^{n-1}}{c_{n-1}}$$

since

$$\sum_{j=1}^{n} \frac{-1}{j}[t^{j-1}](1+t)^{j-2}[t^{n-j}]\epsilon_a(t)$$
$$= -[t^{n-1}]\epsilon_a(t) = -\frac{(at)^{n-1}}{c_{n-1}}.$$

We now discuss more Abel-Gould identities using Classical Riordan arrays. At the end of Subsection 4.3.7, we used the Riordan arrays related Fuss-Catalan numbers to prove *Gould identity*:

$$\sum_{i=0}^{n} \frac{r}{r-qi}\binom{r-qi}{i}\binom{p+qi}{n-i} = \binom{r+p}{n}, \tag{5.167}$$

which can also be proved by using the following process as shown in [202].

Theorem 5.2.30 *Let* (g, f) *be a Riordan array, and let* $h = \sum_{n\geq 0} h_n t^n$ *be a formal power series. Denote the compositional inverse of* f *by* $\bar{f}$. *Then*

$$\sum_{k\geq 0} [t^n]g(t)f(t)^k[t^k]h(\bar{f})(t) = [t^n]g(t)h(t). \tag{5.168}$$

Proof. The left-hand side of (5.168) is $(g, f)h(\bar{f}) = gh(\bar{f} \circ f) = gh$, which gives the desired formula (5.168).

■

For the Riordan array $((1+\alpha t)^p, t(1+\alpha t)^q)$, $p, q \in \mathbb{R}$, its (n, k) entry is

$$d_{n,k} = [t^n](1+\alpha t)^p(t(1+\alpha t)^q)^k = [t^{n-k}](1+\alpha t)^{p+qk} = \binom{p+qk}{n-k}\alpha^{n-k}.$$

Let $h(t) = t^s(1+\alpha t)^r$. Then for $k \neq 0$ and $k \geq s$, from the Lagrange inversion formula (4.179), we have

$$\begin{aligned}
[t^k]h(\bar{f}) =& \frac{1}{k}[t^{k-1}]h'(t)\left(\frac{t}{f(t)}\right)^k = \frac{1}{k}[t^{k-1}]h'(t)(t/t(1+\alpha t)^q)^k \\
=& \frac{1}{k}[t^{k-1}](st^{s-1}(1+\alpha t)^r + r\alpha t^s(1+\alpha t)^{r-1})(1+\alpha t)^{-qk} \\
=& \frac{s}{k}[t^{k-s}](1+\alpha t)^{r-qk} + \frac{\alpha r}{k}[t^{k-s-1}](1+\alpha t)^{r-1-qk} \\
=& \frac{s}{k}\alpha^{k-s}\binom{r-qk}{k-s} + \frac{\alpha r}{k}\alpha^{k-s-1}\binom{r-qk-1}{k-s-1} \\
=& \alpha^{k-s}\binom{r-qk}{k-s}\left(\frac{s}{k} + \frac{r(k-s)}{k(r-qk)}\right) \\
=& \frac{r-qs}{r-qk}\alpha^{k-s}\binom{r-qk}{k-s}.
\end{aligned}$$

Meanwhile, we have

$$[t^n]g(t)h(t) = [t^n](1+\alpha t)^p t^s(1+\alpha t)^r = [t^{n-s}](1+\alpha t)^{p+r} = \alpha^{n-s}\binom{p+r}{n-s}.$$

Consequently, using (5.168) in Theorem 5.2.30 yields

$$\sum_{k=0}^{n} \frac{r-qs}{r-qk}\binom{r-qk}{k-s}\binom{p+qk}{n-k} = \binom{p+r}{n-s}, \tag{5.169}$$

which gives Gould identity (5.167) when we set $s = 0$.

As one of central identities, Gould identity has many applications. For an example related to the previous materials, we use it to prove an alternative form of Girard-Waring identity as shown in (4.266),

$$\frac{x^{n+1} - y^{n+1}}{x-y} = \sum_{0\leq k\leq n/2} (-1)^k \binom{n-k}{k} (xy)^k (x+y)^{n-2k}, \tag{5.170}$$

where $xy \neq 0$ and $x \neq y$. Identity (5.170) was proved in Subsection 4.3.5 using the Riordan arrays $(1/(1+at^2), bt/(1+at^2))$ with $b^2 \geq 4a > 0$. The alternative form of formula (5.170) that we are going to prove is

$$x^n + y^n = \sum_{0\leq k\leq [n/2]} (-1)^k \frac{n}{n-k}\binom{n-k}{k}(x+y)^{n-2k}(xy)^k, \tag{5.171}$$

First, we replace $n \to j$, $i \to k$, and $r \to n$ in (5.167) to obtain

$$\sum_{k=0}^{j} \frac{n}{n-k}\binom{n-k}{k}\binom{n-j-1+k}{j-k} = \binom{2n-j-1}{j}. \tag{5.172}$$

Then, from (5.172) we have

$$\sum_{0\le k\le [n/2]}\sum_{0\le i\le [n/2]}(-1)^{k+i}\frac{n}{n-k}\binom{n-k}{k}\binom{n-i-1}{i}$$
$$(x+y)^{2n-2(k+i)-1}(xy)^{k+i}$$
$$=\sum_{0\le k\le [n/2]}\sum_{0\le j\le n}(-1)^{j}\frac{n}{n-k}\binom{n-k}{k}\binom{n+k-j-1}{j-k}$$
$$(x+y)^{2n-2j-1}(xy)^{j}$$

$$=\sum_{0\le j\le n}(-1)^{j}\left(\sum_{0\le k\le [n/2]}\frac{n}{n-k}\binom{n-k}{k}\binom{n+k-j-1}{j-k}\right)$$
$$(x+y)^{2n-2j-1}(xy)^{j}$$
$$=\sum_{0\le k\le n}(-1)^{k}\binom{2n-k-1}{k}(x+y)^{2n-2k-1}(xy)^{k}.$$

Hence,

$$\sum_{0\le k\le [n/2]}(-1)^{k}\frac{n}{n-k}\binom{n-k}{k}(x+y)^{n-2k}(xy)^{k}$$
$$=\sum_{0\le k\le n}(-1)^{k}\binom{2n-k-1}{k}(x+y)^{2n-2k-1}(xy)^{k}/$$
$$\sum_{0\le k\le [n/2]}(-1)^{k}\binom{n-k-1}{k}(x+y)^{n-2k-1}(xy)^{k}$$
$$=\frac{x^{2n}-y^{2n}}{x-y}/\frac{x^{n}-y^{n}}{x-y}=x^{n}+y^{n},$$

which implies (5.171).

Similarly, we can use Theorem 5.2.30 to prove *Abel identity*

$$\sum_{k=0}^{n}a(a+k)^{k-1}(b+n-k)^{n-k}\binom{n}{k}=(a+b+n)^{n}. \tag{5.173}$$

The Riordan array we now considering is $(g,f)=(e^{pt},te^{qt})$ with $p,q\in\mathbb{R}$, the formal power series we are using is $h(t)=t^{s}e^{rt}$. For $k\neq 0$ and $k\geq s$, we use the Lagrange inversion formula (4.179) to obtain

$$[t^{k}]h(\bar{f})=\frac{1}{k}[t^{k-1}]h'(t)\left(\frac{t}{f(t)}\right)^{k}=\frac{1}{k}[t^{k-1}](st^{s-1}e^{rt}+rt^{s}e^{rt})e^{-qk}$$
$$=\frac{s}{k}\frac{(r-qk)^{k-s}}{(k-s)!}+\frac{r}{k}\frac{(r-qk)^{k-s-1}}{(k-s-1)!}$$

$$=\frac{s}{k}\frac{(r-qk)^{k-s}}{(k-s)!}+\frac{r(k-s)}{k(r-qk)}\frac{(r-qk)^{k-s}}{(k-s)!}$$
$$=\frac{r-qs}{r-qk}\frac{(r-qk)^{k-s}}{(k-s)!}.$$

Since the (n,k) entry of (e^{pt}, te^{qt}) is

$$d_{n,k}=[t^n]e^{pt}(te^{qt})^k=[t^{n-k}]e^{(p+qk)t}=\frac{(p+qk)^{n-k}}{(n-k)!}$$

and

$$[t^n]gh=[t^n]e^{pt}t^se^{rt}=[t^{n-s}]e^{(p+r)t}=\frac{(p+r)^{n-s}}{(n-s)!},$$

the formula (5.168) in Theorem 5.2.30 yields

$$\sum_{k=0}^{\infty}\frac{(p+qk)^{n-k}}{(n-k)!}\frac{r-qs}{r-qk}\frac{(r-qk)^{k-s}}{(k-s)!}=\frac{(p+r)^{n-s}}{(n-s)!}.$$

The above expression can be written as

$$\sum_{k=s}^{\infty}(r-qs)(r-qk)^{k-s-1}(p+qk)^{n-k}\binom{n-s}{n-k}=(p+r)^{n-s}. \tag{5.174}$$

By setting $s=0$, $r=a$, $q=-1$, and $p=b+n$ into (5.174), we obtain Abel identity (5.173).

If we start from (5.174) and specialize the values of p, q, r, and s, numerous familiar identities can be obtained. For example, by substituting $p=0$, $q=-1$, $r=x$, and $s=0$, we have

$$\sum_{k=0}^{\infty}(-1)^{n+k}k^{n-k}x(x+k)^{k-1}=x^n,$$

which is given in [183, (5)].

5.3 Various Methods for constructing Identities Related to Bernoulli and Euler Polynomials

Some powerful methods of evaluating in closed form the sums involving binomial coefficients or factorials such as *W-Z method*, *snack oil method*, *hypergeometric function method*, haven't introduced till now. It is partially because that they have been well represented in [9, 177, 189, 216], partially because that this book is focused on symbolic methods and Riordan array approach, and partially because that the summations dealing with involve Stirling numbers, Bernoulli numbers and polynomials, Euler numbers and polynomials, etc. for which other methods either hardly to be applied or offer no hope. Hence an extension of those methods are needed. This section only intends an initial and intuitive discussion of the recent related work. We just use examples to demonstrate snack oil method and hypergeometric series method and leave W-Z method and its extension in the second subsection. Finally, W-Z method, snack oil method, hypergeometric function method are foundation of computer proof of combinatorial identities, particularly, the identities with binomial coefficients. However, the computer proof for the identities involving Stirling numbers, Bernoulli numbers, Euler numbers, etc. is still on the way for developing.

The snack method might be called two variable generating function method, which is an extension method from known identities to develop more wide classes of identities. The book [216] phased it as following steps: (1) Name the sum in terms of n that we are working on, say $f(n)$. (2) Denote $F(x)=\sum_n f(n)x^n$, i.e., the ordinary power series generating function of $f(n)$, that is a double sum. (3) Interchange the order of two summation and perform the inner one in a closed form by using some known series sums such as

$$\sum_{r\geq 0}\binom{r}{k}x^r=\frac{x^k}{(1-x)^{k+1}}\ (k\geq 0),\quad \sum_{r\geq 0}\binom{n}{r}x^r=(1+x)^n,$$
$$\sum_{n\geq 0}\frac{k}{mn+k}\binom{mn+k}{n}=F_m^k\ (k\geq 1),\ \text{where } F_2=C(t)=\frac{1-\sqrt{1-4t}}{2t}, \tag{5.175}$$

etc. (4) Try to identify the coefficients of the generating function. The method procedure and the demonstrating example shown below are given in [216].

Let $f(n)=\sum_{k\geq 0}\binom{k}{n-k}$ for $n=0,1,2,\ldots$, and let $F(x)=\sum_n f(n)x^n$. Then, taking $r=n-k$ yields

$$F(x)=\sum_{k\geq 0}\sum_n\binom{k}{n-k}x^n=\sum_{k\geq 0}x^k\sum_{r\geq 0}\binom{k}{r}x^r=\sum_{k\geq 0}x^k(1+x)^k=\frac{1}{1-x-x^2},$$

which implies $f(n)=\sum_{k\geq 0}\binom{k}{n-k}=F_{n+1}$ for $n=0,1,2,\ldots$.

The process proving the following identity (cf. [212, 116]) may demonstrate the purely algorithmic method by using hypergeometric series.

$$\sum_k\binom{m}{2k}\binom{k}{n}=2^{m-2n-1}\left[\binom{m-n}{n}+\binom{m-n-1}{n-1}\right]. \tag{5.176}$$

Since any term in the sum on the left-hand side is non-zero, we require $m\geq 2k$ and $k\geq n$, so $m\geq 2n$ and the sum can be written as

$$S=\sum_{k\geq n}\binom{m}{2k}\binom{k}{n}=\sum_{k\geq 0}\binom{m}{2n+2k}\binom{n+k}{n}$$

$$=\sum_{k\geq 0}\binom{m}{2n}\frac{(2n)!(m-2n)!}{(2n+2k)!(m-2n-2k)!}\frac{(n+k)!}{n!k!}$$
$$=\binom{m}{2n}\sum_{k\geq 0}\frac{(m-2n-2k+1)_{2k}}{(2n+1)_{2k}}\frac{(n+1)_k}{k!},$$

where we use the raising factorial $(a)_n = a(a+1)\cdots(a+n-1)$. Furthermore, noting

$$(2n+1)_{2k} = 2^{2k}\left(n+\frac{1}{2}\right)_k (n+1)_k,$$
$$(m-2n-2k+1)_{2k} = 2^{2k}\left(\frac{-m+2n}{2}\right)_k\left(\frac{-m+2n+1}{2}\right)_k,$$

S can ba written as

$$S = \binom{m}{2n}\sum_{k\geq 0}\frac{\left(\frac{-m+2n}{2}\right)_k\left(\frac{-m+2n+1}{2}\right)_k}{k!\left(n+\frac{1}{2}\right)_k}. \tag{5.177}$$

Now define

$${}_2F_1\left(\begin{matrix} a, & b \\ & c \end{matrix};x\right) = \sum_{k\geq 0}\frac{(a)_k(b)_k}{k!(c)_k}x^k.$$

Then, from (5.177) we may write S in terms of hypergeometric series as

$$S = \binom{m}{2n}{}_2F_1\left(\begin{matrix}\frac{-m+2n}{2}, & \frac{-m+2n+1}{2} \\ & n+\frac{1}{2}\end{matrix};1\right). \tag{5.178}$$

The Chu-Vandermonde theorem (cf. [9, Corollary 2.2.3]) states

$${}_2F_1\left(\begin{matrix}-n, & a \\ & c\end{matrix};1\right) = \frac{(c-a)_n}{(c)_n}.$$

Thus,

$${}_2F_1\left(\begin{matrix}-\frac{N}{2}-\frac{1}{2}, & -\frac{N}{2} \\ & n+\frac{1}{2}\end{matrix};1\right) = 2^N\frac{(2n)!(n+N)!}{n!(2n+N)!}. \tag{5.179}$$

Applying (5.224) to the right-hand side of (5.178), we obtain (5.176).

In the first subsection, we shall present two approaches in the construction of identities related to Bernoulli numbers and polynomials and Euler numbers and polynomials. The first approach is using dual sequences and auxiliary formal power series. The second one is a computer algebraic approach, which provides identities related to Bernoulli polynomials and Euler polynomials by using the extended Zeilberger's algorithm (cf. [226, 35]). The key idea of this approach is to use the contour integral definitions of the Bernoulli and Euler numbers to establish recurrence relations on the integrands. Such recurrence relations have certain parameter free properties which lead to the required identities without computing the integrals. Furthermore two new identities related to Bernoulli numbers derived in [34] will be included.

5.3.1 Applications of dual sequences to Bernoulli and Euler polynomials

The first subsection diverts to the identities of Bernoulli numbers and polynomials and their conjugates by using dual sequences and an auxiliary formal power series method. Dual sequences related to Riordan pseudo-involutions have been introduced in Section 5.1.

Recall the Bernoulli polynomials, $B_n(x)$, and Euler polynomials, $E_n(x)$, defined by (1.67) and (1.68), respectively; i.e.,

$$\frac{t}{e^t-1}e^{xt}=\sum_{n\geq 0}\frac{B_n(x)}{n!}t^n \tag{5.180}$$

(5.181)

$$\frac{2}{e^t+1}e^{xt}=\sum_{n\geq 0}\frac{E_n(x)}{n!}t^n, \tag{5.182}$$

respectively. In addition, $B_n = B_n(0)$ and $E_n = 2^n E_n(1/2)$ are called the Bernoulli numbers and the Euler numbers, respectively. Consequently, from the definitions, we can easily deduce the following well known formulas:

$$B_n(1-x)=(-1)^nB_n(x),\quad B_n(x+1)-B_n(x)=nx^{n-1}, \tag{5.183}$$

$$E_n(1-x)=(-1)^nE_n(x),\quad E_n(x+1)+E_n(x)=2x^n. \tag{5.184}$$

In 1995 M. Kaneko [142] found that B_{2n} can be computed in terms of those B_k with $n\leq k\leq 2n$, namely, he proved the recurrence formula

$$\sum_{k=0}^{n}\binom{n+1}{k}(n+k+1)B_{n+k}=0 \tag{5.185}$$

for $n=1,2,3,\ldots$. In 2001 Momiyama [167] extended the above result as follows by using Volkenborn integral (cf. [190]): If $m,n\in\mathbb{N}_0$ and $m+n>0$, then

$$(-1)^m\sum_{k=0}^{m}\binom{m+1}{k}(n+k+1)B_{n+k}+(-1)^n\sum_{j=0}^{n}\binom{n+1}{j}(m+j+1)B_{m+j}=0. \tag{5.186}$$

Later, some extensions by using an auxiliary formal series method were given in [219] and [206]. We first introduce their work. Then, we present some further extensions, which result derive Momiyama identity as a congruence. At the end of Subsubsection 5.3.2.4, extended Zeilberger's creative telescoping algorithm will be used to prove again Momiyama's identity.

Theorem 5.3.1 *Let $(f_k(x))_{k\geq 0}$ be a sequence of polynomials defined by*

$$\sum_{k=0}^{\infty}f_k(x)\frac{z^k}{k!}=e^{(x-1/2)z}F(z), \tag{5.187}$$

where $F(z)$ is a formal power series. Let $m, n \in \mathbb{N}_0$. If F is even, i.e., $F(-z) = F(z)$, then

$$(-1)^m \sum_{k=0}^{m} \binom{m}{k} f_{n+k}(x) = (-1)^n \sum_{j=0}^{n} \binom{n}{j} f_{m+j}(-x); \tag{5.188}$$

if F is odd, i.e., $F(-z) = -F(z)$, then

$$(-1)^m \sum_{k=0}^{m} \binom{m}{k} f_{n+k}(x) = -(-1)^n \sum_{j=0}^{n} \binom{n}{j} f_{m+j}(-x). \tag{5.189}$$

Proof. Suppose that $F(-z) = \epsilon F(z)$ for all z where $\epsilon \in \{1, -1\}$. Consider the generating function

$$G(x, y, z) = \sum_{n=0}^{\infty} \sum_{m=0}^{\infty} \left((-1)^m \sum_{k=0}^{m} \binom{m}{k} f_{n+k}(x) \right) \frac{y^m}{m!} \frac{z^n}{n!}.$$

Thus, we have to show that $G(x, y, z) = \epsilon G(-x, z, y)$. By changing the order of summation in the above equation, we obtain

$$\begin{aligned}
G(x, y, z) =& \sum_{n=0}^{\infty} \sum_{k=0}^{\infty} \sum_{m=k}^{\infty} (-1)^m \binom{m}{k} f_{n+k}(x) \frac{y^m}{m!} \frac{z^n}{n!} \\
=& \sum_{n=0}^{\infty} \sum_{k=0}^{\infty} f_{n+k}(x) \frac{z^n}{n!} \sum_{m=k}^{\infty} (-1)^m \binom{m}{k} \frac{y^m}{m!} \\
=& \sum_{n=0}^{\infty} \sum_{k=0}^{\infty} f_{n+k}(x) \frac{z^n}{n!} \frac{(-y)^k}{k!} e^{-y} \\
=& e^{-y} \sum_{j=0}^{\infty} \sum_{k=0}^{j} \frac{f_j(x)}{j!} \binom{j}{k} z^{j-k} (-y)^k \\
=& e^{-y} \sum_{j=0}^{\infty} f_j(x) \frac{(z-y)^j}{j!} = e^{-y} e^{(x-1/2)(z-y)} F(z-y) \\
=& e^{x(z-y)-(y+z)/2} F(z-y),
\end{aligned}$$

which implies

$$\begin{aligned}
G(-x, z, y) &= e^{-x(y-z)-(z+y)/2} F(y-z) = e^{x(z-y)-(y+z)/2} \epsilon F(z-y) \\
&= \epsilon G(x, y, z),
\end{aligned}$$

as desired. ■

Here is a consequence of Theorem 5.3.1.

Corollary 5.3.2 *Let $F(z)$ be an even or odd formal power series, and let $f_k(x)$ $(k \in \mathbb{N}_0)$ be given by (5.187). Let $m, n \in \mathbb{N}_0$ and $\epsilon = 1$ or -1 depend on whether $F(z)$ is even or odd. Then*

$$\begin{aligned}
&(-1)^m \sum_{k=0}^{m+1} \binom{m+1}{k} (n+k+1) f_{n+k}(x) \\
&= -\epsilon(-1)^n \sum_{j=0}^{n+1} \binom{n+1}{j} (m+j+1) f_{m+j}(-x). \qquad (5.190)
\end{aligned}$$

Proof. Clearly, $-zF(-z) = -\epsilon z F(z)$ and

$$e^{(x-1/2)z} z F(z) = z \sum_{k=0}^{\infty} f_k(x) \frac{z^k}{k!} = \sum_{k=1}^{\infty} f_k^*(x) \frac{z^k}{k!},$$

where $f_k^*(x) = k f_{k-1}(x)$. In view of Theorem 5.3.1, we have

$$(-1)^{m+1} \sum_{k=0}^{m+1} \binom{m+1}{k} f_{n+1+k}^*(x) = -\epsilon(-1)^{n+1} \sum_{j=0}^{n+1} \binom{n+1}{j} f_{m+1+j}^*(-x),$$

which implies (5.190). ■

Observe that

$$\sum_{n=0}^{\infty} B_n(x) \frac{z^n}{n!} = e^{(x-1/2)z} \frac{z}{e^{z/2} - e^{-z/2}} \text{ and}$$
$$\sum_{n=0}^{\infty} E_n(x) \frac{z^n}{n!} = e^{(x-1/2)z} \frac{2}{e^{z/2} + e^{-z/2}}.$$

Also,

$$\begin{aligned}
B_{m+n+1}(x) + (-1)^{m+n} B_{m+n+1}(-x) &= B_{m+n+1}(x) - B_{m+n+1}(1+x) \\
&= -(m+n+1)x^{m+n}
\end{aligned}$$

and

$$\begin{aligned}
E_{m+n+1}(x) + (-1)^{m+n} E_{m+n+1}(-x) &= E_{m+n+1}(x) - E_{m+n+1}(1+x) \\
&= 2E_{m+n+1}(x) - 2x^{m+n+1}.
\end{aligned}$$

So Theorem 5.3.1 and Corollary 5.3.2 imply the following result.

Theorem 5.3.3 *Let $m, n \in \mathbb{N}_0$. Then we have*

$$(-1)^m \sum_{k=0}^{m} \binom{m}{k} B_{n+k}(x) = (-1)^n \sum_{j=0}^{n} \binom{n}{j} B_{m+j}(-x), \qquad (5.191)$$

$$(-1)^m \sum_{k=0}^{m} \binom{m}{k} E_{n+k}(x) = (-1)^n \sum_{j=0}^{n} \binom{n}{j} E_{m+j}(-x); \tag{5.192}$$

and

$$\begin{aligned}
&(-1)^m \sum_{k=0}^{m} \binom{m+1}{k} (n+k+1) B_{n+k}(x) \\
&+(-1)^n \sum_{j=0}^{n} \binom{n+1}{j} (m+j+1) B_{m+j}(-x) \\
&=(-1)^m (m+n+2)(m+n+1) x^{m+n},
\end{aligned} \tag{5.193}$$

$$\begin{aligned}
&(-1)^m \sum_{k=0}^{m} \binom{m+1}{k} (n+k+1) E_{n+k}(x) \\
&+(-1)^n \sum_{j=0}^{n} \binom{n+1}{j} (m+j+1) E_{m+j}(-x) \\
&=(-1)^m 2(m+n+2)(x^{m+n+1} - E_{m+n+1}(x)).
\end{aligned} \tag{5.194}$$

Clearly, (5.191) and (5.193) in the case of $x = 0$ yield

$$(-1)^m \sum_{k=0}^{m} \binom{m}{k} B_{n+k} = (-1)^n \sum_{j=0}^{n} \binom{n}{j} B_{m+j}$$

and Momiyama's formula (5.186), respectively, and the latter in the case of $m = n$ implies Kaneko's recurrent formula (5.185). Similarly, (5.192) and (5.194) in the case of $x = 1/2$ provides recurrent formulas for Euler polynomials, for instance,

$$(-1)^m \sum_{k=0}^{m} \binom{m}{k} \frac{E_{n+k}}{2^{n+k}} = (-1)^n \sum_{j=0}^{n} \binom{n}{j} E_{m+j}\left(-\frac{1}{2}\right).$$

We now extend the above identities of Bernoulli numbers to other dual and self-dual sequences with respect to $((-1)^k \binom{n}{k})$. Recall that $(a_n^*)_{n\geq 0}$ defined by (5.86) as $a_n^* = \sum_{k=0}^{n} \binom{n}{k}(-1)^k a_k$ is the dual sequence of $(a_n)_{n\geq 0}$ with respect to $(\binom{n}{k}(-1)^k)$ (cf. for example, Graham, Knuth, and Patashnik [79]). Hence, $a_n = \sum_{k=0}^{n} \binom{n}{k}(-1)^k a_k^*$ and $a_n^{**} = a_n$. If $a_n^* = a_n$, then $(a_n)_{n\geq 0}$ is a self-dual sequence with respect to $((-1)^k \binom{n}{k})$. For instance, $((-1)^n B_n)_{n\geq 0}$ is a self-dual sequence (cf. Subsection 5.1.4). Like the definition of Bernoulli polynomials, [206] introduce

$$A_n(x) = \sum_{k=0}^{n} \binom{n}{k} (-1)^k a_i x^{n-i} \quad \text{and} \quad A_n^*(x) = \sum_{k=0}^{n} \binom{n}{k} (-1)^k a_i^* x^{n-i}, \tag{5.195}$$

where $(a_n)_{n\geq 0}$ and $(a_n^*)_{n\geq 0}$ are dual sequences. Clearly, $A_n(0) = (-1)^n a_n$, $A_n(1) = \sum_{k=0}^{n} \binom{n}{k}(-1)_k^a = a_n^*$, and

$$A'_{n+1}(x) = \sum_{k=0}^{n} \binom{n+1}{k}(-1)^k(n+1-k)a_k x^{n-k} = (n+1)A_n(x).$$

Motivated by the results shown in Theorem 5.3.3, Sun [206] established the following extension.

Theorem 5.3.4 *Let A_n and A_n^* be the formal power series defined in* (5.195), *and let $k, \ell \in \mathbb{N}_0$ and $x+y+z=1$. Then*

$$(-1)^k \sum_{j=0}^{k} \binom{k}{j} x^{k-j} \frac{A_{\ell+j+1}(y)}{\ell+j+1} + (-1)^\ell \sum_{j=0}^{\ell} \binom{\ell}{j} x^{\ell-j} \frac{A^*_{k+j+1}(z)}{k+j+1} = \frac{a_0(-x)^{k+\ell+1}}{(k+\ell+1)\binom{k+\ell}{k}}. \tag{5.196}$$

In addition, we have

$$(-1)^k \sum_{j=0}^{k} \binom{k}{j} x^{k-j} A_{\ell+j}(y) = (-1)^\ell \sum_{j=0}^{\ell} \binom{\ell}{j} x^{\ell-j} A^*_{k+j}(z) \tag{5.197}$$

and

$$\begin{aligned}&(-1)^k \sum_{j=0}^{k} \binom{k+1}{j}(\ell+j+1)x^{k-j+1} A_{\ell+j}(y)\\ &+(-1)^\ell \sum_{j=0}^{\ell} \binom{\ell+1}{j}(k+j+1)x^{\ell-j+1} A^*_{k+j}(z)\\ =&(k+\ell+2)\left((-1)^{k+1}A_{k+\ell+1}(y) + (-1)^{\ell+1}A^*_{k+\ell+1}(z)\right).\end{aligned} \tag{5.198}$$

Here is an application of Theorem 5.3.4 given in [206].

Theorem 5.3.5 *Let k and ℓ be non-negative integer.*

(i) If $x+y+z=0$, then

$$(-1)^k \sum_{j=0}^{k} \binom{k}{j} x^{k-j} \frac{y^{\ell+j+1}}{\ell+j+1} + (-1)^\ell \sum_{j=0}^{\ell} \binom{\ell}{j} x^{\ell-j} \frac{z^{k+j+1}}{k+j+1} = \frac{(-x)^{k+\ell+1}}{(k+\ell+1)\binom{k+\ell}{k}}. \tag{5.199}$$

In particular,

$$\sum_{j=0}^{k} \binom{k}{j} \frac{(-1)^j}{(k+j+1)2^j} = \frac{2^k}{(2k+1)\binom{2k}{k}} \tag{5.200}$$

(ii) Let $B_n(x)$ denote the Bernoulli polynomial of degree n ($n \in \mathbb{N}_0$), and let $x+y+z=1$. Then

$$(-1)^k \sum_{j=0}^{k} \binom{k}{j} x^{k-j} \frac{B_{\ell+j+1}(y)}{\ell+j+1} + (-1)^\ell \sum_{j=0}^{\ell} \binom{\ell}{j} x^{\ell-j} \frac{B_{k+j+1}(z)}{k+j+1} = \frac{(-x)^{k+\ell+1}}{(k+\ell+1)\binom{k+\ell}{k}}. \tag{5.201}$$

In addition, we have

$$(-1)^k \sum_{j=0}^{k} \binom{k}{j} x^{k-j} B_{\ell+j}(y) = (-1)^\ell \sum_{j=0}^{\ell} \binom{\ell}{j} x^{\ell-j} B_{k+j}(z) \quad \text{and} \tag{5.202}$$

$$\begin{aligned}&(-1)^k \sum_{j=0}^{k} \binom{k+1}{j} (\ell+j+1) x^{k-j+1} B_{\ell+j}(y)\\ &+(-1)^\ell \sum_{j=0}^{\ell} \binom{\ell+1}{j} (k+j+1) x^{\ell-j+1} B_{k+j}(z)\\ &=(-1)^k (k+\ell+2)(B_{k+\ell+1}(x+y) - B_{k+\ell+1}(y)).\end{aligned} \tag{5.203}$$

(iii) The part (ii) remains valid if we replace all the Bernoulli polynomials in (5.201)–(5.203) by corresponding Euler polynomials defined in (5.88):

$$\frac{2e^{xt}}{e^t+1} = \sum_{n\geq 0} E_n(x) \frac{t^n}{n!}.$$

Proof. (i) Let $a_0 = 1$ and $a_k = 0$ for $k = 1, 2, \ldots$. For any $n \in \mathbb{N}_0$ it is clear that $a_n^* = 1$, $A_n(t) = t^n$, and $A_n^*(t) = (t-1)^n$. If $x + y + z = 0$, then $x + y + (1 + z) = 1$, and $A_n^*(1+z) = z^n$. Consequently, (5.199) follows from (5.196). When $\ell = k$, $x = -1$, and $y = z = 1/2$, (5.199) yields (5.200).

(ii) Let $a_n = (-1)^n B_n$ for $n \in \mathbb{N}_0$. Then $A_n(x) = A_n^*(x) = B_n(x)$, and Theorem 5.3.4 yields the identities (5.201)–(5.203), where the identity $B_n(1-x) = (-1)^n B_n(x)$ is applied in the rightmost term of (5.198) to find

$$\begin{aligned}&(-1)^{\ell+1} A^*_{k+\ell+1}(z) = (-1)^{\ell+1} B_{k+\ell+1}(z)\\ =&(-1)^{\ell+1} B_{k+\ell+1}(1-(x+y)) = (-1)^k B_{k+\ell+1}(x+y).\end{aligned}$$

(iii) By the definition of the Euler polynomials, we have $E_n(1-x) = (-1)^n E_n(x)$ and $E_n(x+y) = \sum_{k=0}^{n} E_k(x) y^{n-k}$ for all $n \in \mathbb{N}_0$. Let $a_n = (-1)^n E_n(0)$ for $n = 0, 1, 2, \ldots$. Then

$$a_n^* = \sum_{k=0}^{n} \binom{n}{k} E_k(0) = E_n(1) = a_n \quad \text{and} \quad A_n(x) = A_n^*(x) = E_n(x).$$

So we obtain (iii) from Theorem 5.3.4.

■

We now extend Theorems 5.3.4 and 5.3.5 to the dual relationship D_3 : $a_n = \sum_{k=0}^{n} (-1)^n \binom{n}{k} a_k$, which is shown in Theorem 5.1.66. We first define an analog pair of $(A_n(x), A_n^*(x))$ as follows.

Let $\alpha \in \mathbb{R}$, and let

$$C_{n,\alpha}(x) = (-1)^n \sum_{k=0}^{n} \binom{n}{k} a_k (x-\alpha)^{n-k} \quad \text{and}$$
$$C^*_{n,\alpha}(x) = (-1)^n \sum_{k=0}^{n} \binom{n}{k} a^*_k (x-\alpha)^{n-k}. \tag{5.204}$$

Then we have the following identities about $C_{n,\alpha}(x)$ and $C^*_{n,\alpha}(x)$, which are extensions of the main results shown in [206].

Theorem 5.3.6 *Let $C_{n,\alpha}(x)$ and $C^*_{n,\alpha}(x)$ be defined as (5.204), and let $k, \ell \in \mathbb{N}$ and $x+y+z = 1+2\alpha$. Then, there exists the following identity:*

$$\sum_{j=0}^{k}(-1)^j x^{k-j}\binom{k}{j}\frac{C_{\ell+j+1,\alpha}(y)}{\ell+j+1} + \sum_{j=0}^{\ell}(-1)^j x^{\ell-j}\binom{\ell}{j}\frac{C^*_{k+j+1,\alpha}(z)}{k+j+1}$$
$$= \frac{a_0 x^{k+\ell+1}}{(k+\ell+1)\binom{k+\ell}{k}}. \tag{5.205}$$

In addition, we have

$$\sum_{j=0}^{k}(-1)^j x^{k-j}\binom{k}{j} C_{\ell+j,\alpha}(y) = \sum_{j=0}^{\ell}(-1)^j x^{\ell-j}\binom{\ell}{j} C^*_{k+j,\alpha}(z) \tag{5.206}$$

and

$$\sum_{j=0}^{k}(-1)^j (\ell+j+1) x^{k-j+1}\binom{k+1}{j} C_{\ell+j,\alpha}(y)$$
$$+\sum_{j=0}^{\ell}(-1)^j (k+j+1) x^{\ell-j+1}\binom{\ell+1}{j} C^*_{k+j,\alpha}(z)$$
$$= (k+\ell+2)\left((-1)^k C_{k+\ell+1,\alpha}(y) + (-1)^\ell C^*_{k+\ell+1,\alpha}(z)\right). \tag{5.207}$$

The proof of Theorem 5.3.6 is left as Exercise 5.12.

The following corollary of Theorem 5.3.6 are analogs of the results as shown in Theorems 5.3.4 and 5.3.5.

Corollary 5.3.7 *Let $C_{n,0}(x)$ and $C^*_{n,0}(x)$ be defined as (5.204), and let $k, \ell \in \mathbb{N}$ and $x+y+z=1$. Then there holds*

$$\sum_{j=0}^{k}(-1)^j x^{k-j}\binom{k}{j}\frac{C_{\ell+j+1,0}(y)}{\ell+j+1} + \sum_{j=0}^{\ell}(-1)^j x^{\ell-j}\binom{\ell}{j}\frac{C^*_{k+j+1,0}(z)}{k+j+1} = \frac{a_0 x^{k+\ell+1}}{(k+\ell+1)\binom{k+\ell}{k}}. \tag{5.208}$$

In addition, we have

$$\sum_{j=0}^{k}(-1)^j x^{k-j}\binom{k}{j} C_{\ell+j,0}(y) = \sum_{j=0}^{\ell}(-1)^j x^{\ell-j}\binom{\ell}{j} C^*_{k+j,0}(z) \tag{5.209}$$

and

$$\begin{aligned}
&\sum_{j=0}^{k}(-1)^j(\ell+j+1)x^{k-j+1}\binom{k+1}{j}C_{\ell+j,0}(y)\\
&+\sum_{j=0}^{\ell}(-1)^j(k+j+1)x^{\ell-j+1}\binom{\ell+1}{j}C^*_{k+j,0}(z)\\
=\;& (k+\ell+2)\left((-1)^k C_{k+\ell+1,0}(y)+(-1)^\ell C^*_{k+\ell+1,0}(z)\right).
\end{aligned}$$

If $a_k = B_k$ *or* $(-1)^k B_k$*, then we have*

$$\begin{aligned}
&(-1)^{\ell+1}\sum_{j=0}^{k}x^{k-j}\binom{k}{j}\frac{B_{\ell+j+1}(y)}{\ell+j+1}+(-1)^{k+1}\sum_{j=0}^{\ell}x^{\ell-j}\binom{\ell}{j}\frac{B_{k+j+1}(z)}{k+j+1}\\
=\;& \frac{x^{k+\ell+1}}{(k+\ell+1)\binom{k+\ell}{k}}. \qquad (5.210)
\end{aligned}$$

In addition, we have

$$(-1)^\ell\sum_{j=0}^{k}x^{k-j}\binom{k}{j}B_{\ell+j}(y)=(-1)^k\sum_{j=0}^{\ell}x^{\ell-j}\binom{\ell}{j}B_{k+j}(z) \qquad (5.211)$$

and

$$\begin{aligned}
&(-1)^\ell\sum_{j=0}^{k}(\ell+j+1)x^{k-j}\binom{k+1}{j}B_{\ell+j}(y)\\
&+(-1)^k\sum_{j=0}^{\ell}(k+j+1)x^{\ell-j}\binom{\ell+1}{j}B_{k+j}(z)\\
=\;& (k+\ell+2)\left((-1)^k B_{k+\ell+1}(y)+(-1)^\ell B_{k+\ell+1}(z)\right).
\end{aligned}$$

The *conjugate Bernoulli polynomials* $\tilde{B}_n(x)$ are introduced in [92] by using their generating function shown below:

$$\frac{e^{xt}}{1+\frac{1+t-e^t}{t}}=\sum_{n=0}^{\infty}\tilde{B}_n(x)\frac{t^n}{n!}, \qquad (5.212)$$

where the first few terms of the conjugate polynomial sequence $(\tilde{B}_n)_{n\geq 0}$ are

$$\begin{aligned}
&\tilde{B}_0(x)=1,\\
&\tilde{B}_1(x)=x+\frac{1}{2},\\
&\tilde{B}_2(x)=x^2+x+\frac{5}{6},
\end{aligned}$$

$$\tilde{B}_3(x) = x^3 + \frac{3}{2}x^2 + \frac{5}{2}x + 2,$$
$$\tilde{B}_4(x) = x^4 + 2x^3 + 5x^2 + 8x + \frac{191}{30},$$
$$\tilde{B}_5(x) = x^5 + \frac{5}{2}x^4 + \frac{25}{3}x^3 + 20x^2 + \frac{191}{6}x + \frac{76}{3}, \textit{etc.}$$

Let $(\tilde{B}_n(x))_{n\geq 0}$ be the conjugate Bernoulli polynomial sequence defined by (5.212), and let the conjugate Bernoulli numbers $(\tilde{B}_n)_{n\geq 0}$ be defined by $\tilde{B}_n = \tilde{B}_n(0)$. Then we may find that

$$\tilde{B}_n(x) = \sum_{k=0}^{n} \binom{n}{k} x^{n-k} \tilde{B}_k \tag{5.213}$$

for all $n \geq 0$.

We define the dual sequence of the conjugate Bernoulli number sequence with respect to inverse matrix $R_1 = \left(\frac{1}{1-t}, \frac{t}{1-t}\right)(1, -t) = \left((-1)^k \binom{n}{k}\right)_{n,k\geq 0}$ as shown in Corollary 5.1.65. Then the corresponding dual polynomial sequence is

$$\tilde{B}_n^*(x) := \sum_{k=0}^{n} \binom{n}{k} x^{n-k} \tilde{B}_k^* \tag{5.214}$$

for all $n \geq 0$. From Theorem 5.3.6, we obtain the following results.

Corollary 5.3.8 *Let $k, \ell \in \mathbb{N}$ and $x + y + z = 1$. Then there holds*

$$\begin{aligned}
&(-1)^k \sum_{j=0}^{k} \binom{k}{j} x^{k-j} (-1)^{j+1} \frac{\tilde{B}_{\ell+j+1}(-y)}{\ell+j+1} + \sum_{j=0}^{\ell} \binom{\ell}{j} x^{\ell-j} \frac{\tilde{B}^*_{k+j+1}(-z)}{k+j+1} \\
= \;& \frac{(-1)^{k+1} x^{k+\ell+1}}{(k+\ell+1)\binom{k+\ell}{k}},
\end{aligned} \tag{5.215}$$

where $\tilde{B}_n(x)$ are defined by (5.212), and $\tilde{B}_n^*(x)$ are defined by (5.214). In addition,

$$(-1)^k \sum_{j=0}^{k} \binom{k}{j} (-x)^{k-j} \tilde{B}_{\ell+j}(-y) = (-1)^\ell \sum_{j=0}^{\ell} \binom{\ell}{j} (-x)^{\ell-j} \tilde{B}^*_{k+j}(-z) \tag{5.216}$$

and

$$\begin{aligned}
&(-1)^\ell \sum_{j=0}^{k} \binom{k+1}{j} (-x)^{k-j+1} (\ell+j+1) \tilde{B}_{\ell+j}(-y) \\
&+ (-1)^k \sum_{j=0}^{\ell} \binom{\ell+1}{j} (-x)^{\ell-j+1} (k+j+1) \tilde{B}^*_{k+j}(-z) \\
= \;& (k+\ell+2)\left((-1)^{\ell+1} \tilde{B}_{k+\ell+1}(-y) + (-1)^{k+1} \tilde{B}^*_{k+\ell+1}(-z)\right).
\end{aligned} \tag{5.217}$$

In (5.204), we substitute $\alpha = 1/2$ and get

$$a_k = E_k\left(\frac{1}{2}\right) - \frac{1}{2^k}, \quad \text{and} \quad a_k^* = (-1)^k\left(E_k\left(\frac{1}{2}\right) + \frac{3^k-2}{2^k}\right), \tag{5.218}$$

where the duals a_k^* are derived by using (5.103) of Theorem 5.1.69. Hence, from (5.195), the corresponding dual polynomials $A_n(x)$ and $A_n^*(x)$ are

$$\begin{aligned} A_n(x) = (-1)^n E_n(x) - (-1)^n x^n \quad \text{and} \quad & A_n^*(x) \\ = E_n(-x+1) + (2-x)^n - 2(1-x)^n, \end{aligned}$$

respectively.

Corollary 5.3.9 *Let $C_{n,\alpha}(x)$ and $C_{n,\alpha}^*(x)$ be defined as (5.204) with $\alpha = 1/2$, and let $k, \ell \in \mathbb{N}$ and $x + y + z = 2$. For a_k and a_k^* as shown in (5.218), there holds*

$$\begin{aligned} &\sum_{j=0}^{k}(-1)^{\ell+1}x^{k-j}\binom{k}{j}\frac{E_{\ell+j+1}(y)}{\ell+j+1} + \sum_{j=0}^{\ell}(-1)^j x^{\ell-j}\binom{\ell}{j}\frac{E_{k+j+1}(-z+1)}{k+j+1} \\ = \;& (-1)^{\ell+1}\int_0^y t^\ell(t+x)^k dt - \int_0^{x+y} t^k(x-t)^\ell dt + 2\int_0^{x+y-1} t^k(x-t)^\ell dt. \end{aligned} \tag{5.219}$$

In addition, we have

$$\begin{aligned} &(-1)^\ell\sum_{j=0}^{k}x^{k-j}\binom{k}{j}E_{\ell+j}(y) - (-1)^\ell y^\ell(x+y)^k \\ = \;& \sum_{j=0}^{\ell}(-1)^j x^{\ell-j}\binom{\ell}{j}E_{k+j}(1-z) + (x+y)^k(-y)^\ell - 2(x+y-1)^k(1-y)^\ell. \end{aligned} \tag{5.220}$$

The above identities of polynomial sequences can be used to establish identities of number sequences. For instance, if $x = 1$, $y = 0$, and $z = 1$, then (5.219) yields the number sequence identity

$$\begin{aligned} &\sum_{j=0}^{k}(-1)^{\ell+1}x^{k-j}\binom{k}{j}\frac{E_{\ell+j+1}(0)}{\ell+j+1} \\ &+ \sum_{j=0}^{\ell}(-1)^j x^{\ell-j}\binom{\ell}{j}\frac{E_{k+j+1}(0)}{k+j+1} = -\frac{1}{(k+\ell+1)\binom{k+\ell}{k}}. \end{aligned}$$

If $x = 1$ and $y = z = 1/2$, from (5.219) we have

$$\sum_{j=0}^{k}(-1)^{\ell+1}x^{k-j}\binom{k}{j}\frac{E_{\ell+j+1}(1/2)}{\ell+j+1}+\sum_{j=0}^{\ell}(-1)^{j}x^{\ell-j}\binom{\ell}{j}\frac{E_{k+j+1}(1/2)}{k+j+1}$$
$$= -\frac{1}{(k+\ell+1)\binom{k+\ell}{k}}+2B\left(\frac{1}{2},k+1,\ell+1\right)-2B\left(\frac{3}{2},k+1,\ell+1\right),$$

where $B(\alpha,a,b)=\int_0^\alpha t^{a-1}(1-t)^{b-1}dt$ is an incomplete beta function. From the last two identities, we have

$$\sum_{j=0}^{k}(-1)^{\ell+1}x^{k-j}\binom{k}{j}\frac{E_{\ell+j+1}(1/2)-E_{\ell+j+1}(0)}{\ell+j+1}$$
$$+\sum_{j=0}^{\ell}(-1)^{j}x^{\ell-j}\binom{\ell}{j}\frac{E_{k+j+1}(1/2)-E_{k+j+1}(0)}{k+j+1}$$
$$=2\left(B\left(\frac{1}{2},k+1,\ell+1\right)-B\left(\frac{3}{2},k+1,\ell+1\right)\right).$$

Let A and B be any $m\times m$ and $n\times n$ square matrices, respectively. Cheon and Kim [39] introduce the notation $\oplus$ for the direct sum of matrices A and B:

$$A\oplus B=\begin{bmatrix} A & 0\\ 0 & B\end{bmatrix}. \tag{5.221}$$

By using the notation, we may obtain the following results on the conjugate Bernoulli polynomials $\tilde{B}(x)$ defined by (5.214), which includes the matrix form of Lemma 3.2 of Pan and Sun [173] as a special case.

Theorem 5.3.10 *Let n be a positive integer, and let $\tilde{B}(t)$ and $P[t]$ be defined as*

$$\tilde{B}(t):=(\tilde{B}_0(t),\tilde{B}_1(t),\cdots)\quad \textit{and}$$
$$P[t]:=\left(\binom{n}{k}t^{n-k}\right)_{n,k\geq 0},$$

respectively. Let $(x)=(1,x,x^2,\ldots)^T$, $D=diag(0,1,1/2,\ldots)$, and $[X]=[0]\oplus\left[\frac{x^{n-k}}{k}\right]_{1\leq k\leq n}$, i.e.,

$$[X]=\begin{bmatrix} 0 & 0 & 0 & \cdots & 0\\ 0 & 1 & 0 & \cdots & 0\\ 0 & x & \frac{1}{2} & \cdots & 0\\ \vdots & \vdots & \vdots & \ddots & \vdots\\ 0 & x^{n-1} & \frac{x^{n-2}}{2} & \cdots & \frac{1}{n}\\ \vdots & \vdots & \vdots & \ddots & \vdots\end{bmatrix}.$$

Then we have

$$[X]\tilde{B}(x+y) = P[x]D\tilde{B}(y) + [X](x), \tag{5.222}$$

which implies

$$\sum_{k=1}^{n} \frac{\tilde{B}_k(x+y)}{k} x^{n-k} = \sum_{l=1}^{n} \binom{n}{l} \frac{\tilde{B}_l(y)}{l} x^{n-l} + H_n x^n \tag{5.223}$$

for all $n \geq 0$ *as shown in [173].*

Proof. Noting that $\tilde{B}_0(y) = 1$, the left-hand side of (5.222) can be written as

$$LHS\ of\ (5.222) = [X]P[x]\tilde{B}(y)$$
$$= [X]\begin{bmatrix} 1 \\ x + \binom{1}{1}\tilde{B}_1(y) \\ x^2 + \binom{2}{1}x\tilde{B}_1(y) + \binom{2}{2}\tilde{B}_2(y) \\ \vdots \\ x^n + \binom{n}{1}x^{n-1}\tilde{B}_1(y) + \cdots + \binom{n}{n}\tilde{B}_n(y) \\ \vdots \end{bmatrix}$$

$$= [X]\left(\begin{bmatrix} 1 \\ x \\ x^2 \\ \vdots \\ x^n \\ \vdots \end{bmatrix} + \begin{bmatrix} 0 \\ \binom{1}{1}\tilde{B}_1(y) \\ \binom{2}{1}x\tilde{B}_1(y) + \binom{2}{2}\tilde{B}_2(y) \\ \vdots \\ \binom{n}{1}x^{n-1}\tilde{B}_1(y) + \cdots + \binom{n}{n}\tilde{B}_n(y) \\ \vdots \end{bmatrix}\right)$$

$$= \begin{bmatrix} 0 \\ x \\ (1+\frac{1}{2})x^2 \\ \vdots \\ (1+\frac{1}{2}+\cdots+\frac{1}{n})x^n \\ \vdots \end{bmatrix} + [X]\begin{bmatrix} 0 \\ \binom{1}{1}\tilde{B}_1(y) \\ \binom{2}{1}x\tilde{B}_1(y) + \binom{2}{2}\tilde{B}_2(y) \\ \vdots \\ \binom{n}{1}x^{n-1}\tilde{B}_1(y) + \cdots + \binom{n}{n}\tilde{B}_n(y) \\ \vdots \end{bmatrix},$$

where the second term can be written as

$$\begin{bmatrix} 0 \\ \left(\binom{1}{1} + \frac{1}{2}\binom{2}{1}\right)\tilde{B}_1(y)x + \frac{1}{2}\binom{2}{2}\tilde{B}_2(y) \\ \vdots \\ \left(\binom{1}{1} + \cdots + \frac{1}{n}\binom{n}{1}\right)\tilde{B}_1(y)x^{n-1} + \left(\frac{1}{2}\binom{2}{2} + \cdots + \frac{1}{n}\binom{n}{2}\right)\tilde{B}_2(y)x^{n-2} + \cdots \\ \quad + \frac{1}{n}\binom{n}{n}\tilde{B}_n(y) \\ \vdots \end{bmatrix}$$

$$= \begin{bmatrix} 0 \\ \binom{2}{1}\tilde{B}_1(y)x + \frac{1}{2}\binom{2}{2}\tilde{B}_2(y) \\ \vdots \\ \binom{n}{1}\tilde{B}_1(y)x^{n-1} + \frac{1}{2}\binom{n}{2}\tilde{B}_2(y)x^{n-2} + \cdots \frac{1}{n}\binom{n}{n}\tilde{B}_n(y) \\ \vdots \end{bmatrix}.$$

In the last step we use the identities $\frac{1}{n}\binom{n}{k} = \frac{1}{k}\binom{n-1}{k-1}$ and

$$\binom{n}{k} = \binom{n-1}{k-1} + \binom{n-1}{k} = \binom{n-1}{k-1} + \binom{n-2}{k-1} + \binom{n-3}{k} = \cdots$$
$$= \binom{n-1}{k-1} + \binom{n-2}{k-1} + \cdots + \binom{k}{k-1} + \binom{k}{k}$$

for $k, n \in \mathbb{N}$ with $1 \leq k \leq n$. Since the last matrix can be written as $P[x]DB(y)$, we obtain (5.222). By accomplishing the multiplications of the matrices in (5.222) and using them in (5.223), we find that the left-hand side of (5.223) can be written as

$$LHS\ of\ (5.223) = H_n x^n + \sum_{l=1}^{n} \binom{n}{l} \frac{\tilde{B}_l(y)}{l} x^{n-l} = \sum_{l=1}^{n} \binom{n}{l} \frac{\tilde{B}_l(y)}{l} x^{n-l} + H_n x^n,$$

which is the right-hand side of (5.223).

■

5.3.2 Extended Zeilberger's algorithm for constructing identities related to Bernoulli and Euler polynomials

Gosper's algorithm (cf. [69]) is one of the landmarks in the history of computerization of the problem of closed form summation and also vital in the operation of the Zeilberger's *creative telescoping algorithm* and the W-Z algorithm given in [217, 218]. Hence, we first briefly introduce this algorithm based on the description as shown in [69, 177]. Then we describe W-Z algorithm, and Zeilberger's creative telescoping algorithm. Finally, we present the extended Zeilberger's algorithm given by Chen et al. in [35].

5.3.2.1 Gosper's algorithm

We are now looking for a closed form of

$$s_n = \sum_{k=0}^{n-1} t_k, \tag{5.224}$$

where t_k is a *hypergeometric term* (i.e., t_{k+1}/t_k is a rational function of k) that is independent of n. This problem is also an indefinite sum problem.

As an analog to antiderivative, we may define *antidifference* (also known as *indefinite sum*) to a function $f(x)$, denoted by $\Delta^{-1}f(x)$ or $\sum_x f(x)$, as

$$\Delta \sum_x f(x) = f(x). \tag{5.225}$$

More explicitly, if $\sum_x f(x) = F(x) + C$, then $F(x+1) - F(x) = f(x)$, which is similar to indefinite integration for a continuous function, $\int_a^x f(x)dx = F(x) - F(a)$, when $F'(x) = f(x)$. For instance, $\Delta^{-1}n = n(n-1)/2 + C$ because $\Delta(n(n-1)/2 + C) = n(n+1)/2 - n(n-1)/2 = n$. Hence, we may change our problem to be the following one: given a hypergeometric term t_n, find a hypergeometric indefinite sum z_n such that

$$\Delta z_n = z_{n+1} - z_n = t_n, \tag{5.226}$$

i.e., $\sum_{k=0}^{n-1} t_k$ becomes a telescoping series. Thus, the relation between z_n and s_n is

$$z_n = z_{n-1} + t_{n-1} = z_{n-2} + t_{n-2} + t_{n-1} = \cdots = z_0 + \sum_{k=0}^{n-1} t_k = s_n + c,$$

where $s_n = \sum_{k=0}^{n-1} t_k$ and $c = z_0$ is a constant. Since

$$\frac{z_n}{t_n} = \frac{z_n}{z_{n+1} - z_n} = \frac{1}{\frac{z_{n+1}}{z_n} - 1} \tag{5.227}$$

is a rational function of n denoted by $y(n)$, we have $z_n = y(n)t_n$. Substituting $y(n)t_n$ for z_n into (5.226) yields

$$r(n)y(n+1) - y(n) = 1, \quad \text{where} \quad r(n) = \frac{t_{n+1}}{t_n}. \tag{5.228}$$

Gosper then reduced the problem further to that of finding polynomial solutions of yet another first-order recurrence.

Express the ratio $r(n) = t_{n+1}/t_n$ as

$$r(n) = \frac{a(n)}{b(n)} \frac{c(n+1)}{c(n)}, \tag{5.229}$$

where $a(n)$, $b(n)$, and $c(n)$ are polynomials in n with $gcd(a(n), b(n+\ell)) = 1$ for all $\ell \in \mathbb{N}_0$. It is always possible to put a rational function in this form, for if $gcd(a(n), b(n+\ell)) = d(n)$ for all $\ell \in \mathbb{N}_0$, then the common factor can be eliminated with the change of variables: $a'_n = a_n/d(n)$, $b'_n = b(n)/d(n-\ell)$, and $c'(n) = c(n)d(n)d(n-1)\ldots d(n-\ell+1)$, which leaves the ratio unchanged. The values of ℓ for which such $d(n-i)$ $(0 \le i \le \ell - 1)$ exist can be detected as the non-negative integer roots of the resultant of $a(n)$ and $b(n+\ell)$ with respect to n.

If we write

$$y(n) = \frac{b(n-1)x(n)}{c(n)}, \tag{5.230}$$

where $x(n)$ is an unknown rational function of n. Substituting (5.229) and (5.230) into the first equation of (5.228) shows that $x(n)$ satisfies

$$a(n)x(n+1) - b(n-1)x(n) = c(n). \tag{5.231}$$

Then [69] proved that under the above conditions of $a(n)$, $b(n)$, and $c(n)$, the rational function $x(n)$ satisfying (5.231) is a polynomial in n. Hence, finding hypergeometric solutions of (5.226) is equivalently to find polynomial solutions of (5.231). This is because if $x(n)$ is a nonzero polynomial solution of (5.231), then

$$z_n = \frac{b(n-1)x(n)}{c(n)} t_n$$

is a hypergeometric solution of (5.226), and vice verse. A general way to find polynomial solutions of (5.231), if they exist, or to prove that there does not exist a solution, is given in Section 5.4 of [177]. In the following example, we can see $x(n)$ can be found by using test and try method. After finding z_n, one may have $s_n = z_n - z_0$ or $s_n = z_n - z_{k_0}$ if the sum starts from k_0.

As an example that is shown in [177], let us consider

$$S_n = \sum_{k=0}^{n} (4k+1)\frac{k!}{(2k+1)!},$$

and let $s_n = S_{n-1}$. Following Gosper's algorithm and noting $t_n = n!(4n+1)/((2n+1)!)$ we have

$$r(n) = \frac{t_{n+1}}{t_n} = \frac{4n+5}{2(4n+1)(2n+3)},$$

which is a rational function in n. We may choose

$$a(n) = 1, \quad b(n) = 2(2n+3), \quad \text{and} \quad c(n) = 4n+1,$$

because they satisfy equation (5.229) and the condition $gcd(a(n), b(n+\ell)) = 1$ for all $\ell \in \mathbb{N}_0$. Consequently, equation (5.231) becomes

$$x(n+1) - 2(2n+1)x(n) = 4n+1.$$

We might assume $x(n)$ is a constant, a linear function in n, etc. to test and try a polynomial solution. Here, $x(n) = -1$ works, and

$$z_n = \frac{b(n-1)x(n)}{c(n)} t_n = -\frac{2(2n+1)}{4n+1}(4n+1)\frac{n!}{(2n+1)!} = -2\frac{n!}{(2n)!},$$

which satisfies $z_{n+1} - z_n = t_n$. Hence,

$$s_n = z_n - z_0 = 2 - 2\frac{n!}{(2n)!} \quad \text{and} \quad S_n = s_{n+1} = 2 - \frac{n!}{(2n+1)!}.$$

5.3.2.2 W-Z algorithm

Because of (5.226), Gosper's algorithm is an extension of telescoping method. We now consider general sum

$$f(n) = \sum_k F(n,k), \tag{5.232}$$

where $F(n,k)$ is a hypergeometric term in both arguments, i.e., both $F(n+1,k)/F(n,k)$ and $F(n,k+1)/F(n,k)$ are rational functions of n and k, and the range of the summation index k may be the set of all integers. If the sum is telescoping with respect to k, say $F(n,k) = G(n,k+1) - G(n,k)$ for some nice G, then the sum is easily to be found. However, we cannot expect, in general, to have it. For instance, $\sum_k \binom{n}{k}$ is not Gosper-summable. We saw in [216, 217, 218] another type extended telescoping method, called W-Z method, is given for constant sums including (5.232), because after dividing $f(n)$ on both sides we get the new summation equaling to one. To apply this method for the constant sum $\sum_k F(n,k)$, we need the summands of $\sum_k F(n,k)$ satisfy

$$F(n+1,k) - F(n,k) = G(n,k+1) - G(n,k) \tag{5.233}$$

for some nice functions $G(n,k)$. Then, under the condition

$$\lim_{k\to\pm\infty} G(n,k) = 0 \tag{5.234}$$

for $n = 0, 1, 2, \ldots$, we obtain a telescoping summation in terms of G as

$$\begin{aligned}&\sum_{k=-L}^{k=K} (F(n+1,k) - F(n,k)) = \sum_{k=-L}^{k=K} (G(n,k+1) - G(n,k))\\ &= G(n,K+1) - G(n,-L) \to 0 \end{aligned}\tag{5.235}$$

as $K, L \to \infty$. Hence, $\sum_k F(n,k) = \textit{constant}$. The summation in (5.235) is definite summation (for discrete variables). Therefore, W-Z method and Zeilberger's algorithm shown later are based on definite summation, while Gosper algorithm is based on indefinite summation. Hence, the definite summation of a discrete function (or sequence) a_n is defined by

$$\sum_{k=-L}^{K} a(k) = \left[\Delta^{-1} a(k)\right]_{-L}^{K+1},$$

where Δ^{-1} is the indefinite summation operator, and Δ is the forward difference operator. As what we know, for some continuous functions, although their indefinite integrals have no a closed form, their definite integrals is calculable. For instance, indefinite integral $\int e^{-t^2} dt$ has no a closed form, but $\int_{-\infty}^{\infty} e^{-t^2} dt = \sqrt{\pi}$. Similarly, there exist non-Gosper-summable series that are W-Z or Zeilberger summable.

Now we go back to the non-Gosper-summable series $\sum_k \binom{n}{k}$ and change it to the series $\sum_k \binom{n}{k}/2^n$. Let $F(n,k) = \binom{n}{k}/2^n$ $(n \geq 0)$, and let $G(n,k) = -\binom{n}{k-1}/2^{n+1}$. $G(n,k)$ satisfies (5.234). Then from the relation

$$F(n+1,k) - F(n,k) = \frac{1}{2^{n+1}}\binom{n+1}{k} - \frac{1}{2^n}\binom{n}{k}$$

$$= -\frac{1}{2^{n+1}}\binom{n}{k} + \frac{1}{2^{n+1}}\binom{n}{k-1} = G(n,k+1) - G(n,k),$$

we may write the original summation as a telescoping series in terms of G and obtain

$$\sum_k \left(\frac{1}{2^{n+1}}\binom{n+1}{k} - \frac{1}{2^n}\binom{n}{k}\right) = 0$$

for $n = 0, 1, 2, \ldots$. Thus, $\sum_k F(n,k) = \sum_k F(0,k) = 1$, i.e., $\sum_k \binom{n}{k} = 2^n$.

We call (F, G) W-Z pair if the conditions (5.233) and (5.234) hold. W-Z method provide a way to find $G(n,k)$, or equivalently, $R(n,k)$ such that $G(n,k) = R(n,k)F(n,k-1)$, when $F(n,k)$ (resp. $\sum_k F(n,k)$) is a hypergeometric function (resp. *hypergeometric sum*) if both arguments, $F(n+1,k)/F(n,k)$ and $F(n,k+1)/F(n,k)$, are rational functions of n and k (cf. [216, 217, 218]).

For instance, to prove $\sum_k (-1)^k \binom{n}{k}\binom{2k}{k}4^{n-k} = \binom{2n}{n}$, we consider $F(n,k) = (-1)^k \binom{n}{k}\binom{2k}{k}4^{n-k}/\binom{2n}{n}$. Then $R(n,k) = (2k-1)/(2n+1)$. Thus, $G(n,k) = R(n,k)F(n,k-1)$ satisfies (5.233) and (5.234), which implies the desired result.

5.3.2.3 Zeilberger's creative telescoping algorithm

In general, we cannot expect (5.233) always happen. Hence, we need to take a somewhat more general difference operator in n on the left-hand side of (5.233), i.e., a general telescoping technique, called the method of *creative telescoping*, given by Zeilberger [225, 226].

More precisely, let N and K be the shift operators in n and k, respectively, i.e., $N(A(n,k)) = A(n+1,k)$ and $K(A(n,k)) = A(n,k+1)$ for any $A(n,k)$. Hence, $N^i K^j(A(n,k)) = A(n+i, k+j)$. Thus, the telescoping used in Gosper's algorithm can be written as $N^0 F(n,k) = (K-1)G(n,k)$, and the telescoping used in W-Z algorithm is $(N-1)F(n,k) = (K-1)G(n,k)$. The Zeilberger's creative telescoping is

$$p(n,N)F(n,k) = (K-1)G(n,k), \tag{5.236}$$

where $p(n,N) = \sum_{j=0}^J a_j(n)N^j$, in which the coefficients $\{a_j(n)\}_0^J$ are polynomials in n. Articles [225, 226] proves that if the general term $F(n,k)$ is a proper hypergeometric term, where a proper hypergeometric function is defined in Definition 4.4. of [218, 177], then F satisfies a nontrivial recurrence of the form (5.236), in which $G(n,k)/F(n,k)$ is a rational function of n and k, i.e., such that

$$\sum_{j=0}^{J} a_j(n)F(n+j,k) = G(n,k+1) - G(n,k). \tag{5.237}$$

Since the coefficients on the left-hand side of (5.237) are independent of k, we can take summation on both sides of the equation over all integer values of k and obtain

$$\sum_{j=0}^{J} a_j(n)f(n+j) = G(n,\infty) - G(n,-\infty) = 0, \tag{5.238}$$

if (5.234) holds or $G(n,k)$ has compact support in k for each n. Now we may solve the last recurrence equation for $f(n)$ in terms of n.

For instance, for $f(n) = \sum_{0\le k\le[n/3]} 2^k n\binom{n-k}{2k}/(n-k)$, [177] shows it satisfies (5.236) with $p(n,N) = (N-2)(N^2+1)$ and $G(n,k) = 2^k n\binom{n-k}{2k-2}/(n-3k+3)$. Thus, the sum $f(n)$ satisfies recurrence relation $(N-2)(N^2+1)f(n) = 0$, which gives solution $f(n) = c_1 2^n + c_2 i^n + c_3(-i)^n$. Substituting the values $f(1) = 1$, $f(2) = 1$, and $f(3) = 4$, we solve $c_1 = c_2 = c_3 = 1/2$. Therefore, $f(n) = 2^{n-1} + (i^n + (-i)^n)/2 = 2^{n-1} + \cos(n\pi/2)$.

Denote the left-hand side of (5.237) by t_k. Section 6.3 of [177] gives a procedure similar to Gosper's procedure to write

$$\frac{t_{k+1}}{t_k} = \frac{p_0(k+1)}{p_0(k)}\frac{p_1(k+1)p_2(k)}{p_1(k)p_3(k)}$$

where coefficients a_j only appear in the first polynomial ratio on the right-hand side of the above equation, the numerator and denominator in the second polynomial ratio on the right-hand side are coprime and $gcd(p_2(k), p_3(k+j)) = 1$ for all non-negative integer j. The expressions of those polynomials are shown in [177, 226]. If t_{k+1}/t_k is written as

$$\frac{t_{k+1}}{t_k} = \frac{p(k+1)}{p(k)}\frac{p_2(k)}{p_3(k)},$$

where $p(k) = p_0(k)p_1(k)$, then $G(n,k) = (b(k)p_3(k-1)/p(k))t_k$, where $b(k)$ is a polynomial solution of

$$p_2(k)b(k+1) - p_3(k-1)b(k) = p(k).$$

Furthermore, it was also proved that the last recurrence equation has a polynomial solution $b(k)$ if and only if t_k is an indefinitely summable hypergeometric term.

Remark 5.3.11 *The purpose of Zeilberger's algorithm is to establish* (5.238), *i.e.,* $f(n)$ *is holonomic in* n *by using creative telescoping* (5.237). *There is no reason to expect* (5.236) *holds for a given* $f(n)$, *and when it does, there is no guarantee that such a miraculous* $G(n,k)$ *that satisfies* (5.237) *exists. However, it was proved in Zeilberger [225] that if* $F(n,k)$ *is holonomic, (in particular if both* $F(n,\cdot)$ *and* $F(\cdot,k)$ *satisfy linear recurrences with polynomial coefficients that are "independent" in a certain technical sense), then it is guaranteed that* $f(n)$ *is holonomic in* n*, i.e., satisfies a homogeneous linear recurrence equation of the form* (5.238), *with polynomial coefficients. Zeilberger's general "slow" algorithm in [225] produces a recursion, and in principle it also enables us to find* $G(n,k)$.

In analysis, a holonomic function is a smooth function of several variables that is a solution of a system of linear homogeneous differential equations with polynomial coefficients and satisfies a suitable dimension condition in terms of D-modules theory. More precisely, a holonomic function is an element of a holonomic module of smooth functions. Holonomic functions can also be described as differentiably finite functions, also known as D-finite functions. When a power series in the variables is the Taylor expansion of a holonomic function, the sequence of its coefficients, in one or several indices, is also called holonomic. Holonomic sequences are also called P-recursive sequences: they are defined recursively by multivariate recurrences satisfied by the whole sequence and by its suitable specializations. The situation simplifies in the univariate case: Any univariate sequence that satisfies a linear homogeneous recurrence relation with polynomial coefficients, or equivalently a linear homogeneous difference equation with polynomial coefficients, is holonomic. Joseph N. Bernstein's theory of holonomic systems (cf. Bernstein, [19] and Björk [20]) forms a natural framework for proving a very large class of special function identities.

5.3.2.4 Extended Zeilberger's algorithm

We are now ready to introduce the work by Chen et al. in [46] on the extended Zeilberger's algorithm used to prove and establish identities related to Bernoulli and Euler polynomials and numbers. The Bernoulli polynomials and Euler polynomials can be defined by the generating functions as (5.181) and (5.182), respectively. Noting $B_n = B_n(0)$ and $E_n = 2^n E_n(1/2)$ are referred to as the Bernoulli numbers and the Euler numbers, respectively, then Bernoulli numbers and Euler numbers can be defined by the generating functions

$$\sum_{n=0}^{\infty} B_n \frac{z^n}{n!} = \frac{z}{e^z - 1} \quad \text{and} \quad \sum_{n=0}^{\infty} E_n \frac{z^n}{n!} = \frac{2e^z}{e^{2z}+1},$$

respectively. The contour integral definitions of the Bernoulli numbers and the Euler numbers can be represented with the Cauchy integrals as

$$B_n = \frac{n!}{2\pi i} \oint \frac{z}{e^z - 1} \frac{dz}{z^{n+1}} \quad \text{and} \quad E_n = \frac{n!}{2\pi i} \oint \frac{2e^z}{e^{2z}+1} \frac{dz}{z^{n+1}},$$

respectively, where the contour encloses the origin, has radius less than 2π (to avoid the poles at $\pm 2\pi i$), and is traversed in a counterclockwise direction. We will see there will be no need to compute the contour integrals, and one can formally treat the contour integrals as linear operators. The integral representation plays a crucial role in connecting the Bernoulli numbers and Euler numbers to hypergeometric terms. The Bernoulli numbers are also given by the recursion

$$\sum_{k=0}^{n} \binom{n+1}{k} B_k = 0, \quad n > 0, \quad \text{and} \quad B_0 = 1. \tag{5.239}$$

It is well known that $B_{2n+1} = 0$ for $n \geq 1$. The first few values of the Bernoulli numbers are $B_0 = 1$, $B_1 = -1/2$, $B_2 = 1/6$, $B_4 = -1/30$, $B_6 = 1/42$, etc. For the Euler numbers, we have $E_{2n+1} = 0$ for $n \geq 0$.

The Bernoulli polynomials and Euler polynomials obey the relations (5.90) and (5.91), which we repeat here for easy reference:

$$B_n(x) = \sum_{k=0}^{n} \binom{n}{k} B_k x^{n-k}, \quad n \geq 0, \tag{5.240}$$

$$E_n(x) = \sum_{k=0}^{n} \binom{n}{k} \left(x - \frac{1}{2}\right)^{n-k} \frac{E_k}{2^k}, \quad n \geq 0. \tag{5.241}$$

We need the properties of $B_n(x)$ and $E_n(x)$ given in (5.183) and (5.184), i.e.,

$$B_n(1-x) = (-1)^n B_n(x), \quad B_n(x+1) - B_n(x) = nx^{n-1}, \tag{5.242}$$

$$E_n(1-x) = (-1)^n E_n(x), \quad E_n(x+1) + E_n(x) = 2x^n. \tag{5.243}$$

We also need the following well known binomial expansions for Bernoulli and Euler polynomials to calculate initial values for the recurrence relations derived by the algorithm given in [35, 46].

$$B_n(x+y) = \sum_{k=0}^{n} \binom{n}{k} B_k(x) y^{n-k} \quad \text{and} \quad E_n(x+y) = \sum_{k=0}^{n} \binom{n}{k} E_k(x) y^{n-k}. \tag{5.244}$$

The original Zeilberger's algorithm is devised to find recurrence relations of the summation $\sum_k F(n,k)$ by solving the equation (5.236), i.e.,

$$a_0(n)F(n,k) + a_1(n)F(n+1,k) + + a_J(n)F(n+J,k) = G(n,k+1) - G(n,k),$$

where $F(n,k)$ is a hypergeometric term in n and k, $a_i(n)$ are polynomials in n and are k-free, $G(n,k)/F(n,k)$ is a rational function in n and k. It is known that Zeilberger's algorithm can be applied to summands with parameters in order to establish multiple index recurrence relations, for example, see Section 4.3.1 of [9] and [35]. The algorithm as shown in the latter will become apparent when it is being used to prove some identities related to Bernoulli and Euler polynomials. Let us consider Gessel (cf. [66, Lemma 7.2]) identity as an example

$$\sum_{k=0}^{m}\binom{m}{k}B_{n+k}=(-1)^{m+n}\sum_{k=0}^{n}\binom{n}{k}B_{m+k}, \tag{5.245}$$

where m and n are non-negative integers.

To justify the above identity, we aim to find recurrence relations for both sides. If they agree with each other, then the equality is established by considering the initial values. There are four steps to compute the recurrence relations for the above summations. We will give detailed steps for the left-hand side of (5.245).

Denote the left-hand side of (5.245) by $L(n,m)$. The first step is to extract the hypergeometric sum from the Cauchy integral formula of Bernoulli numbers. From

$$L(n,m)=\frac{1}{2\pi i}\oint\frac{1}{e^z-1}\sum_{k=0}^{m}\binom{m}{k}\frac{(n+k)!}{z^{n+k}}dz,$$

we take a summation $S(n,m)=\sum_{k=0}^{m}C(n,m,k)$, where

$$C(n,m,k)=\binom{m}{k}\frac{(n+k)!}{z^{n+k}}.$$

In the second step, we construct an extended telescoping equation with a shift on the parameter m of the summand $C(n,m,k)$, and solve this equation by the extended Zeilberger's algorithm. More precisely, we set the hypergeometric term

$$\begin{aligned}F(n,m,k)&=b_0C(n,m,k)+b_1C(n,m+1,k)\\&+b_2C(n+1,m,k)+b_3C(n+1,m+1,k),\end{aligned} \tag{5.246}$$

where b_i's are k-free rational functions of n and m, namely, k does not appear in b_i's. Moreover, we require that the rational functions b_i's are independent of the variable z. By Gosper's algorithm, it is easy to check that $C(n,m,k)=z_{k+1}-z_k$ has no hypergeometric solution for z_k. Moreover, since the Bernoulli numbers are not P-recursive, Zeilberger's algorithm does not work in this case. Then, we will try to solve the equation

$$F(n,m,k)=G(n,m,k+1)-G(n,m,k), \tag{5.247}$$

where $G(n,m,k)$ is a hypergeometric term. By Gosper's algorithm, we have

$$r(k)=\frac{F(n,m,k+1)}{F(n,m,k)}=\frac{a(k)}{b(k)}\frac{c(k+1)}{c(k)}, \tag{5.248}$$

where $a(k)=(m-k+1)(n+k+1)$, $b(k)=z(k+1)$, $c(k)=b_2k^2+(b_2n-b_2m+b_0z-b_3m-b_3)k-b_0z(m+1)-b_1z(m+1)-b_2(n+1)(m+1)-b_3(n+1)(m+1)$.

Assume that $G(n,m,k) = y(k)F(n,m,k)$, where $y(k)$ is an unknown rational function of k. Substituting $y(k)F(n,m,k)$ for $G(n,m,k)$ in (5.247) reveals that $y(k)$ satisfies

$$r(k)y(k+1) - y(k) = 1.$$

From Zeilberger's algorithm, by substituting the factorization (5.248) into the above equation and setting

$$x(k) = \frac{y(k)c(k)}{b(k-1)},$$

we reduce the problem to be the problem of finding polynomial solutions of the following equation (cf. Theorem 5.2.1 of [177] or Subsubsection 5.3.2.3)

$$a(k)x(k+1) - b(k-1)x(k) = c(k). \tag{5.249}$$

Notice that the coefficients $a(k)$ and $b(k)$ are independent of the unknowns b_i's, and $c(k)$ is a linear combination of b_i's. One can estimate the degree of the polynomial $x(k)$, as in Gosper's algorithm. In this case, $x(k)$ is of degree 0. Assume that $x(k) = a_0$. Then Equation (5.249) becomes

$$\begin{aligned}(-a_0 - b_2)k^2 + (mb_2 - nb_2 + mb_3 - b_0 z - (n-m)a_0 - a_0 z + b_3)k \\ +((n+1)(m+1)a_0 + b_0 z + nb_3 + b_3 nm + b_0 zm + b_2 nm + nb_2 \\ +b_1 zm + b_1 z + b_2 + mb_2 + mb_3 + b_3) = 0\end{aligned}$$

By setting the coefficient of each power of k to zero, we get a system of linear equations in a_0 and b_i's. Note that in the solution of this system, a_0 and b_i's may contain the variable z. To prevent z from appearing in b_i's, we should go one step further to impose that the coefficient of any positive power of z in b_i's is zero. This may also lead to additional equations. Combining all these equations, if we can find a nonzero solution, then take this solution to the next step. Otherwise, we may try recurrences of higher order. In this case, we get a nonzero solution $a_0 = -1$, $b_0 = 1$, $b_1 = -1$, $b_2 = 1$, $b_3 = 0$. Note that in general the b_i's are polynomials in n and m.

Thirdly, we find the recurrence relation for $L(n,m)$. Since the solution of $a_0, b_0, \ldots, b_3$ leads to the following telescoping equation

$$C(n,m,k) - C(n,m+1,k) + C(n+1,m,k) = G(n,m,k+1) - G(n,m,k), \tag{5.250}$$

where

$$G(n,m,k) = \frac{m!(n+k)!}{(k-1)!(m-k+1)!z^{n+k}}. \tag{5.251}$$

Summing the above recurrence over k from 0 to $m+1$, we obtain $S(n,m) - S(n,m+1) + S(n+1,m) = 0$. Substituting this recurrence relation to the contour integral definition of B_n, we find that $L(n,m)$ satisfies $L(n,m) - L(n,m+1) + L(n+1,m) = 0$.

By the same procedure, we see that the right-hand side of (5.245), denoted by $R(n,m)$, satisfies the same recurrence relation as $L(n,m)$, namely, $R(n,m) - R(n,m+1) + R(n+1,m) = 0$.

Finally, we verify initial values. By considering the parity of B_m, we see that $(-1)^m B_m = B_m$ unless $m = 1$. Hence, $L(0,1) = R(0,1) = 1/2$ and $L(0,m) = R(0,m) = B_m$ for $m \neq 1$. The proof is complete.

We now consider Momiyama identity

$$(-1)^m \sum_{k=0}^{m} \binom{m+1}{k} (n+k+1) B_{n+k} + (-1)^n \sum_{j=0}^{n} \binom{n+1}{j} (m+j+1) B_{m+j} = 0, \tag{5.252}$$

where $m + n > 0$. The identity was proved by using Volkenborn integral in [167] and, later, by using the auxiliary formal series method in [219]. We now show that it can also be proved by using extended Zeilberger's algorithm after noting that the identity contains parameters m and n.

Denote the left-hand side and the right-hand side of (5.252) by $L(n,m)$ and $R(n,m)$, respectively. By the contour integral definition of the Bernoulli numbers, we have

$$L(n,m) = \frac{1}{2\pi i} \oint \frac{1}{e^z - 1} \left(\sum_{k=0}^{m} (-1)^m \binom{m+1}{k} (n+k+1) \frac{(n+k)!}{z^{n+k}} \right) dz.$$

Denote the summand in the above summation by $F(n,m,k)$, that is,

$$F(n,m,k) = (-1)^m \binom{m+1}{k} (n+k+1) \frac{(n+k)!}{z^{n+k}}.$$

Applying the extended Zeilberger's algorithm to $F(n,m,k)$ and assuming that the output is independent of z, we obtain

$$F(n,m,k) + F(n,m+1,k) + F(n+1,m,k) = G(n,m,k+1) - G(n,m,k), \tag{5.253}$$

where

$$G(n,m,k) = \frac{(-1)^m (m+1)!(n+k+1)!}{(k-1)!(m+2-k)! z^{n+k}}.$$

Summing the telescoping equation (5.253) over k from 0 to m, we are led to the following recurrence relation for $L(n,m)$

$$L(n,m) + L(n,m+1) + L(n+1,m) = -(-1)^m (n+m+2) B_{n+m+1}.$$

Similarly, we find that $R(n,m)$ also satisfies

$$R(n,m) + R(n,m+1) + R(n+1,m) = (-1)^n (n+m+2) B_{n+m+1}.$$

Considering the parity of B_n. It is easy to see that

$$((-1)^m + (-1)^n)(n+m+2) B_{n+m+1} = 0.$$

Therefore, both sides of Momiyama's identity (5.252) satisfy the same recurrence relation.

To compute the initial values, setting $m = 0$, we get $L(n, 0) = (n + 1)B_n$. It follows from the recurrence relation (5.239) that

$$0 = \sum_{k=0}^{n} \binom{n+1}{k} B_k = \sum_{k=0}^{n} \binom{n}{k} B_k + \sum_{k=0}^{n} \binom{n}{k-1} B_k = B_n + \sum_{k=0}^{n} \binom{n}{k-1} B_k,$$

which implies $\sum_{k=0}^{n} \binom{n}{k-1} B_k = -B_n$. On the other hand, for $n \neq 1$, we have

$$\begin{aligned} R(n,0) &= -(-1)^n \sum_{k=0}^{n} \binom{n+1}{k} (k+1) B_k \\ &= (-1)^{n+1} \left(\sum_{k=0}^{n} \binom{n+1}{k} k B_k + \sum_{k=0}^{n} \binom{n+1}{k} B_k \right) \\ &= (-1)^{n+1} (n+1) \sum_{k=0}^{n} \binom{n}{k-1} B_k = (-1)^n (n+1) B_n = (n+1) B_n. \end{aligned}$$

It is easily checked that $L(1, 0) = R(1, 0) = -1$. So we deduce that $L(n, 0) = R(n, 0)$ for all $n \geq 0$. This completes entire proof.

Exercises

5.1 Consider a special case of (5.22) for $k=0$: Let $D=(g,f)$ be a Riordan array, and let $\bar{f}=\sum_{k\geq 0}\bar{f}_k t^k$ be the compositional inverse of f. Then

$$\sum_{k=0}^{\infty} d_{n,k}\bar{f}_k = [t^{n-1}]g(t) = g_{n-1}$$

for $n\geq 1$. Use this formula to prove the following identities.

(i) $\sum_{k=1}^{n}(-1)^{k-1}\binom{n}{k}=1$.
Hint: consider $D=(1/(1-t),t/(1-t))$.
(ii) $\sum_{k=1}^{n}(-1)^{k-1}\left[{n \atop k}\right]=n!\delta_{n,1}, \quad n\geq 1$.
Hint: Consider $D=(1,-\ln(1-t))$.
(iii) $\sum_{k=0}^{n}(-1)^{k-1}\binom{m+k}{k}\left[{n \atop m+k}\right]=n\left[{n-1 \atop m}\right]. \quad n\geq 1$.
Hint: Consider $D=((-\ln(1-t))^m,-\ln(1-t))$.
(iv) $\sum_{k=1}^{n}(-1)^{k-1}(k-1)!\left\{{n \atop k}\right\}=n!\delta_{n,1}, \quad n\geq 1$.
Hint: Consider $D=(1,e^t-1)$.
(v) $\sum_{k=1}^{n}\frac{(-1)^{k-1}(k+p)!}{k}\left\{{n \atop k+p}\right\}=np!\left\{{n-1 \atop p}\right\}, \quad n\geq 1$.
Hint: Consider $D=((e^t-1)^p,e^t-1)$.

5.2 Let the generalized Stirling numbers $S(n,k;\alpha,\beta,\gamma)$ be defined by

$$\frac{1}{k!}(1+\alpha t)^{\gamma/\alpha}\left(\frac{(1+\alpha t)^{\beta/\alpha}-1}{\beta}\right)^k=\sum_{n\geq 0}S(n,k;\alpha,\beta,\gamma)\frac{t^n}{n!} \tag{5.254}$$

for $(\alpha,\beta,\gamma)\in\mathbb{R}^3$ and $\beta\neq 0$, where $\lim_{\beta\to 0}S(n,k;1,\beta,0)=s(n,k)$, the Stirling numbers of the first kind, and $\lim_{\alpha\to 0}S(n,k;\alpha,1,0)=S(n,k)$, the Stirling numbers of the second kind.

(i) Use Theorem 5.1.13 to find the recurrence relation of Stirling numbers of the second kind.

(ii) Use Theorem 5.1.13 to find the recurrence relation of Stirling numbers $S(n,k;\alpha,\alpha,\gamma)$.

(ii) Prove the numbers $S(n,k;\alpha,\beta,\gamma)$ satisfy the following recursive relation.

$$S(n+1,k;\alpha,\beta,\gamma)=S(n,k-1;\alpha,\beta,\gamma)+(\gamma+k\beta-n\alpha)S(n,k;\alpha,\beta,\gamma) \tag{5.255}$$

for $n\geq 1$.

Hint: (i) Since the exponential Riordan array $(S(n,k))_{n,k\geq 0}$ is $(1,e^t-1)$, the generating functions of the corresponding c-sequence and r-sequence can be found by using formulae in (5.27) as $c(t)=0$ and $r(t)=1+t$. Then the recurrence relation of Stirling numbers of the second kind can be obtained from (5.30):

$$S(0,0)=1,\quad S(n,0)=0\ (n\geq 1),\quad S(n,n)=1\ (n\geq 1),$$
$$S(n+1,k)=S(n,k-1)+kS(n,k).$$

(ii) Noting $(S(n,k;\alpha,\beta,\gamma) = (d(t),h(t))$, where

$$d(t) = (1+\alpha t)^{\gamma/\alpha} \quad \text{and} \quad h(t) = \frac{(1+\alpha t)^{\beta/\alpha}-1}{\beta},$$

from (5.27) we get

$$c(t) = \gamma(1+\beta t)^{-\alpha/\beta} \quad \text{and} \quad r(t) = (1+\beta t)^{1-\alpha/\beta}.$$

If $\alpha = \beta$, then $c(t) = \gamma/(1+\alpha t)$ and $r(t) = 1$. Hence, by using (5.30) we obtain the following recurrence relation of Stirling numbers of the second kind $S(n,k;\alpha,\alpha,\gamma)$.

$$\begin{aligned}
&S(n+1,0;\alpha,\alpha,\gamma) = \gamma\sum_{i=0}^{n} i!(-\alpha)^i S(n,i;\alpha,\alpha,\gamma),\\
&S(n,n;\alpha,\alpha,\gamma) = 1\ (n\geq 0),\\
&S(n+1,k;\alpha,\alpha,\gamma) = S(n,k-1;\alpha,\alpha,\gamma)\\
&+\frac{\gamma}{k!}\sum_{i=k}^{n} i!(-\alpha)^{i-k}S(n,i;\alpha,\alpha,\gamma)\ (n\geq k\geq 0).
\end{aligned}$$

(iii) Denote the LHS of (5.254) by $\phi_k(t)$. Evidently, $\phi_k(0) = 0$ for $k \geq 1$, and $\phi_0(t) = (1+\alpha t)^{\gamma/\alpha}$. Moreover, using elementary differentiation and algebraic computations, one can verify that

$$(1+\alpha t)\frac{d}{dt}\phi_k(t) - (\gamma+k\beta)\phi_k(t) = \phi_{k-1}(t) \quad (k\geq 1). \tag{5.256}$$

Since the above differential equation has a unique solution satisfying the initial condition $\phi_k(0) = 0$, the RHS of (5.254), $\sum_{n\geq 0} S(n,k;\alpha,\beta,\gamma)t^n/n!$, also satisfies the above equation. Consequently,

$$\begin{aligned}
&(1+\alpha t)\sum_{n\geq 0} S(n,k;\alpha,\beta,\gamma)\frac{nt^{n-1}}{n!} - (\gamma+k\beta)\sum_{n\geq 0} S(n,k;\alpha,\beta,\gamma)\\
&=\sum_{n\geq 0} S(n,k-1;\alpha,\beta,\gamma).
\end{aligned}$$

The LHS of the last equation can be written as

$$LHS = \sum_{n\geq 1} S(n+1,k;\alpha,\beta,\gamma)\frac{t^n}{n!} - \sum_{n\geq 0}(\gamma+k\beta-n\alpha)S(n,k;\alpha,\beta,\gamma)\frac{t^n}{n!}.$$

Comparing the coefficients of the nth term on the both sides, we obtain (5.255). For $\alpha = 1$ and $\beta = 0$, we have the recursive relation for the Stirling numbers of the first kind, while the recursive relation of the Stirling numbers of the second kind $S(n,k) = S(n,k;0,1,0)$ as shown in (i) can also be derived from (5.255).

5.3 Denote $A(s) = \sum_{n\geq 0} a_n s^n$. Show

$$A(s) = \frac{a_0 + (a_1 - pa_0)s}{1 - ps - qs^2}. \tag{5.257}$$

Furthermore, prove the Taylor expansion A165998 [197] of $A(s)$ is

$$\begin{aligned} &A(s) \\ = \;& a_0 + \sum_{n\geq 1}\left(a_1 p^{n-1} + \sum_{j=1}^{[n/2]} \frac{1}{j}\binom{n-j-1}{j-1} p^{n-2j-1} q^j \left(jpa_0 + (n-2j)a_1\right)\right) s^n. \end{aligned} \tag{5.258}$$

Hint: See Proposition 1 in [112].

5.4 Prove (5.170),

$$\frac{x^{n+1} - y^{n+1}}{x - y} = \sum_{0\leq k\leq n/2} (-1)^k \binom{n-k}{k} (xy)^k (x+y)^{n-2k},$$

where $xy \neq 0$ and $x \neq y$, using the difference of a polynomial.

Hint: See Lemma 3.1 of [112].

5.5 Prove the first half of Theorem 5.1.70.

Hint: Use a similar argument in the proof of the second half of the theorem or see Prodinger [179].

5.6 Prove (i) of Corollary 5.1.71.

Hint: Use a similar argument in the proof of (ii) of the corollary or see Sun [205].

5.7 Prove (i) and (ii) of Theorem 5.1.72.

Hint: Use an argument similar to the proof of (iii) and (iv) of the theorem or see [205].

5.8 Prove Theorem 5.1.73.

Hint: See the statement in the book.

5.9 Prove Theorem 5.1.74.

Hint: See [205] and [114].

5.10 Let $d(t) \in \mathcal{F}_0$, $h(t) \in \mathcal{F}_1$, and $U = (\delta_{i+1,j})_{i,j\geq 0}$. Then a lower triangle matrix $R = (g(t), f(t)) = (d_{n,k})_{n,k\geq 0}$ is a Riordan array if and only if

$$UR = RP, \tag{5.259}$$

or equivalently,

$$P = R^{-1}UR,$$

where P, called the production matrix of R, has the expression

$$P = \begin{bmatrix} z_0 & a_0 & 0 & 0 & 0 & \cdots \\ z_1 & a_1 & a_0 & 0 & 0 & \cdots \\ z_2 & a_2 & a_1 & a_0 & 0 & \cdots \\ z_3 & a_3 & a_2 & a_1 & a_0 & \cdots \\ \vdots & \vdots & \vdots & \vdots & \vdots & \ddots \end{bmatrix} = \left(Z(t), A(t), tA(t), t^2A(t), \ldots\right), \tag{5.260}$$

where the rightmost expression is the presentation of P by using its column generating functions.

Hint: Let Riordan array $R = (g(t), f(t)) = (d_{n,k})_{n,k\geq 0}$ have the generating functions $A(t)$ and $Z(t)$ of its A-sequence and Z-sequence, respectively. Then,

$$g(t) = \frac{d_{0,0}}{1 - tZ(f(t))} \quad \text{and} \quad f(t) = tA(f(t)). \tag{5.261}$$

We may write (5.261) as

$$\frac{g(t) - d_{0,0}}{t} = g(t)Z(f(t)) \quad \text{and} \quad \frac{g(t)f^n(t)}{t} = g(t)f^{n-1}(t)A(f(t)). \tag{5.262}$$

Noting

$$U = (\delta_{i+1,j})_{i,j\geq 0} = \begin{bmatrix} 0 & 1 & 0 & 0 & 0 & \cdots \\ 0 & 0 & 1 & 0 & 0 & \cdots \\ 0 & 0 & 0 & 1 & 0 & \cdots \\ 0 & 0 & 0 & 0 & 1 & \cdots \\ \vdots & \vdots & \vdots & \vdots & \vdots & \ddots \end{bmatrix},$$

we may re-write (5.262) by using upper shift matrix U as

$$U(g(t), f(t)) = (g(t), f(t))P.$$

5.11 Let $D = (g(t), f(t))$ be a generalized Riordan array with respect to $(k!)_{k\geq 0}$. Then the (n,k) entry of its production matrix $(p_{n,k})_{n,k\geq 0}$ satisfies

$$p_{n,k} = \frac{n!}{k!}\left(c_{n-k} + kr_{n-k+1}\right), \tag{5.263}$$

where the sequences $(r_k)_{k\in\mathbb{N}}$ and $(c_k)_{k\in\mathbb{N}}$ are defined by the ordinary generating functions

$$r(t) = f'(\bar{f}(t)) = \sum_{k=0}^{\infty} r_k t^k \quad \text{and} \quad c(t) = \frac{g'(\bar{f}(t))}{g(\bar{f}(t))} = \sum_{k=0}^{\infty} r_k t^k, \tag{5.264}$$

respectively. Namely,

$$P = \begin{bmatrix} c_0 & r_0 & 0 & 0 & 0 & \cdots \\ 1!c_1 & \frac{1!}{1!}(c_0 + r_1) & r_0 & 0 & 0 & \cdots \\ 2!c_2 & \frac{2!}{1!}(c_1 + r_2) & \frac{2!}{2!}(c_0 + 2r_1) & r_0 & 0 & \cdots \\ 3!c_3 & \frac{3!}{1!}(c_2 + r_3) & \frac{3!}{2!}(c_1 + 2r_2) & \frac{3!}{3!}(c_0 + 3r_1) & r_0 & \cdots \\ \vdots & \vdots & \vdots & \vdots & \vdots & \ddots \end{bmatrix}.$$

Then, use (5.263) to reprove Theorem 5.1.13.

Hint: From Exercise 5.10, the production matrix of R is given by

$$P = D^{-1}UD\,.$$

Then, the exponential generating function of the k-th column of UD is

$$\frac{\mathrm{d}}{\mathrm{d}t}g(t)\frac{(f(t))^k}{k!} = g'(t)\frac{(f(t))^k}{k!} + kg(t)\frac{(f(t))^{k-1}}{k!}f'(t)\,.$$

Hence, the exponential generating function of the k-th column of the production matrix P is

$$\begin{aligned} p_k(t) &= \left[\frac{1}{g(\bar{f}(t))}, \bar{f}(t)\right]\frac{\mathrm{d}}{\mathrm{d}t}g(t)\frac{(f(t))^k}{k!} = \frac{g'(\bar{f}(t))}{g(\bar{f}(t))}\frac{t^k}{k!} + k\frac{t^{k-1}}{k!}f'(\bar{f}(t)) \\ &= \frac{1}{k!}(c(t)t^k + kt^{k-1}r(t))\,. \end{aligned}$$

The (n,k)-th entry (5.263) of production matrix P is determined accordingly.

From Exercise 5.10, we have $DP = UD$, which implies

$$d_{n+1,k} = \sum_{i=k-1}^{n}\frac{i!}{k!}(c_{i-k} + kr_{i-k+1})d_{n,i}\,,$$

for all $k \geq 0$. Consequently, we have

$$\begin{aligned} d_{n+1,0} &= \sum_{i=0}^{n} i!c_i d_{n,i}\,, \\ d_{n+1,k} &= r_0 d_{n,k-1} + \sum_{i=k}^{n}\frac{i!}{k!}(c_{i-k} + kr_{i-k+1})d_{n,i} \quad \text{for } k \geq 1\,. \end{aligned}$$

5.12 Prove Theorem 5.3.6.

Hint: See Theorem 9 of [114].

Bibliography

[1] M. Abramowitz and I. A. Stegun, (Eds.). *Handbook of Mathematical Functions with Formulas, Graphs, and Mathematical Tables*, 9th printing. New York: Dover, pp. 16 and 806 (Euler-Mac.), 1972.

[2] M. Aigner, *Combinatorial Theory*, New York: Springer-Verlag, 1979.

[3] M. Aigner, Catalan-like numbers and determinants, *J. Combin. Theory Ser. A*, 87 (1999), 33–51.

[4] M. Aigner, Catalan and other numbers: a recurrent theme, in: *Algebraic Combinatorics and Computer Science*, Springer Italia, Milan, 2001, 347–390.

[5] M. Aigner, *A Course in Enumeration*, Graduate Texts in Mathematics, Vol. 238, Springer, Berlin, 2007.

[6] M. Aigner, Enumeration via ballot numbers, *Discrete Math.* 308 (2008), 2544–2563.

[7] G. E. Andrews, Some formulae for the Fibonacci sequence with generalizations, *Fibonacci Quart.* 7 (1969), 113–130.

[8] G. E. Andrews, A polynomial identity which implies the Rogers-Ramanujan identities, *Scripta Math.* 28 (1970), 297–305.

[9] G. E. Andrews, R. Askey, and R. Roy, *Special Functions*, Encyclopedia of Mathematics and its Applications Vol. 71, Cambridge University Press, 1999.

[10] T. M. Apostol, *Introduction to Analytic Number Theory*, Undergraduate Texts in Mathematics, New York-Heidelberg: Springer-Verlag, 1976.

[11] T. M. Apostol, An elementary view of Euler's summation formula, *Amer. Math. Monthly*, 106 (1999), 409–418.

[12] L. F. A. Arbogast, *Du calcul des derivations [On the calculus of derivatives]* (in French), Strasbourg: Levrault, pp. xxiii+404, 1800.

[13] G. Arfken, Bernoulli Numbers, Euler-Maclaurin formula, in *Mathematical Methods for Physicists*, 3rd ed. Orlando, FL: Academic Press, 327-338, 1985.

[14] P. Barry, On the halves of a Riordan array and their antecedents, *Linear Algebra Appl.* 582 (2019), 114–137.

[15] P. Barry, On the r-shifted central triangles of a Riordan array, https://arxiv.org/abs/1906.01328.

[16] P. Barry, On the central coefficients of Riordan matrices, *J. Integer Seq.* 16 (2013), 13.5.1.

[17] P. Barry, *Riordan Arrays: A Primer*, LOGIC Press, Kilcock, 2016.

[18] F. R. Bernhart, Catalan, Motzkin, and Riordan numbers, *Discrete Math.* 204 (1999), 73–112.

[19] I. N. Bernstein, Modules over a ring of differential operators, study of the fundamental solutions of equations with constant coefficients, *Functional Anal. Appl.* 5 (2) (1971) 1–16 (in Russian); 89–101 (English translation).

[20] J.-E. Björk, *Rings of Differential Operators*, North-Holland Publishing Com. Amsterdam, 1979.

[21] R. P. Boas and R. C. Buck, *Polynomial Expansions of Analytic Functions* Springer, New York, 1964.

[22] J. M. Borwein, P. B. Borwein, and K. Dilcher, Pi, Euler numbers, and asymptotic expansions, *Amer. Math. Monthly*, 96 (1989), 681–687.

[23] N. Bourbaki, *Algèbre*, Hermann, 1959.

[24] D. M. Bressoud, A matrix inverse, *Proc. Amer. Math. Soc.* 88 (1983), 446–448.

[25] E. H. M. Brietzke, An identity of Andrews and a new method for the Riordan array proof of combinatorial identities, *Discrete Math.* 308 (2008), 4246–4262.

[26] A. Z. Broder, The r-Stirling numbers, *Discrete Math.* 49(1984), 241–259.

[27] U. Buijs, J. G. Carrasquel-Vera, and A. Murillo, The gauge action, DG Lie Algebra and identities for Bernoulli numbers, arXiv:1403.3630 [math.NT].

[28] N. T. Cameron and A. Nkwanta, On some (pseudo) involutions in the Riordan group, *J. Integer Seq.* 8 (2005), no. 3, Article 05.3.7, 16 pp.

[29] L. Carlitz, Some inverse relations, *Duke Math. J.* 40 (1972), 893–901.

[30] L. Carlitz, Note on some convolved power Sums, *SIAM J. Math. Anal.*, 8(1977), no. 4, 701–709.

[31] L. Carlitz, Degenerate Stirling, Bernoulli and Eulerian numbers, *Utilitas Mathematica*, 15 (1979), 51–88.

[32] L. Carlitz, Weighted Stirling numbers, I, II, *Fibonacci Quart.* 18 (1980), 147–162, 242–257.

[33] Ch. A. Charalambides and M. Koutras, On the differences of the generalized factorials at an arbitrary point and their combinatorial applications, *Discrete Math.* 47 (1983), no. 2-3, 183–201.

[34] W. Y. C. Chen and L. H. Sun, Extended Zeilberger's algorithm for identities on Bernoulli and Euler polynomials, *J. Number Theory*, 129 (2009), 2111–2132.

[35] W. Y. C. Chen, Q.-H. Hou, Y.-P. Mu, Extended Zeilberger's algorithm with parameters, *J. Symbolic Comput.* 47 (2012), no. 6, 643–654.

[36] X. Chen, H. Liang and Y. Wang, Total positivity of Riordan arrays, *European J. Combin.* 46 (2015) 68–74.

[37] X. Chen and Y. Wang, Notes on the total positivity of Riordan arrays, *Linear Algebra Appl.* 569 (2019) 156–161.

[38] G.-S. Cheon and S.-T. Jin, Structural properties of Riordan matrices and extending the matrices, *Linear Algebra Appl.*, 435 (2011), 2019–2032.

[39] G.-S. Cheon and J.-S. Kim, Stirling matrix via Pascal matrix, *Linear Algebra Appl.*, 329 (2001), no. 1-3, 49-59.

[40] G.-S. Cheon, S.-T. Jin, H. Kim, L. W. Shapiro, Riordan group involutions and the Δ-sequence, *Discrete Appl. Math.* 157 (2009), no. 8, 1696–1701.

[41] G.-S. Cheon, H. Kim, L. W. Shapiro, An algebraic structure for Faber polynomials, *Linear Algebra Appl.*, 433 (2010), 1170–1179.

[42] G.-S. Cheon, H. Kim, L. W. Shapiro, Combinatorics of Riordan arrays with identical A and Z sequences, *Disc. Math.*, 312 (2012), 2040–2049.

[43] W. C. Chu and L. C. Hsu, Some new applications of Gould-Hsu inversions, *J. Combin. Inform. System Sci.* 14 (1989), 1–4.

[44] L. Comtet, *Advanced Combinatorics*, Dordrecht: Reidel, 1974.

[45] J. H. Conway and R. Guy, *The Book of Numbers*, Springer, 1996.

[46] G. M. Constantine and T. H. Savits, A multivariate Faà di Bruno formula with applications, *Trans. Amer. Math. Soc.* 348 (1996), no. 2, 503–520.

[47] C. Corsani, D. Merlini, R. Sprugnoli, Left-inversion of combinatorial sums, *Discrete Math.* 180 (1998), 10–122.

[48] M. Dancs and T.-X. He, Q-analogues of symbolic operators, *J. Discrete Math.* 2013 (2013), DOI:10.1155/2013/487546, 6 pages.

[49] D. E. Davenport, L. W. Shapiro, and L. C. Woodson, The double Riordan group, *Electronic J. combin.* 18 (2012), no. 2, 33 pages.

[50] F. N. David and D. E. Barton, *Combinatorial Chance*, Hafner Publishing Co., New York, 1962.

[51] H. T. Davis, *The Summation of Series*, Principia Press of Trinity University, San Antonio, 1962.

[52] P. J. Davis, *Interpolation and Approximation*, Blaisdell Publishing Company, London, 1963.

[53] E. Deutsch, Dyck path enumeration, *Discrete Math.* 204 (1999), 167–202.

[54] E. Deutsch, L. Ferrari, and S. Rinaldi, Production matrices, *Adv. Appl. Math.*, 34 (2005), no. 1, 101–122.

[55] E. Deutsch, L. Ferrari, and S. Rinaldi, Production matrices and Riordan arrays, *Ann. Comb.*, 13 (2009), 65–85.

[56] E. Deutsch, L. Shapiro, A survey of fine numbers, *Discrete Math.* 241 (2001), 241–265.

[57] K. Dilcher, Sums of products of Bernoulli numbers, *J. Number Theory*, 60 (1996), 23–41.

[58] G. P. Egorychev, *Integral Representation and the Computation of Combinatorial Sums*, Translation of Math. Monographs, Vol. 59, AMS, 1984.

[59] A. Erdelyi, *Higher Transcendental Functions*, New York, McGraw-Hill, 1953.

[60] L. Euler, Methodus universalis serierum convergentium summas quam proxime inveniendi, *Commentarii academie scientiarum Petropolitanae*, Vol. 8(1736), 3-9; Opera Omnia, Vol. XIV. 101-107.

[61] L. Euler, Methodus universalis series summandi ulterius promota, *Commentarii academie scientiarum Petropolitanae*, Vol. 8(1736), 147-158; Opera Omnia, Vol. XIV. 124-137.

[62] L. Euler, *Comm. Acad. Sci. Imp. Petrop.* 6, 68, 1738.

[63] Q. Fang, M. Xu, and T. Wang, A symbolic operator approach to Newton series, *J. Math. Research & Exposition*, 31 (2011), no. 1, 67–72.

[64] P. J. Forrester and D.-Z. Liu, Raney distributions and random matrix theory, *J. Stat. Phys.* 158 (2015), 1051–1082.

[65] S. Fu and Y. Wang, Bijective proofs of recurrences involving two Schröder triangles, *European J. Combin.* 86 (2020), 103077, 18 pp.

[66] I. Gessel, On Miki's Identity for Bernoulli numbers, *J. Number Theory,* 110 (2005), no. 1, 75–82.

[67] J. W. L. Glaisher, On a class of relations connecting any consecutive Bernoullian functions, *Quart. J. Pure and Appl. Math.*, 42(1911), 86–157.

[68] J. W. L. Glaisher, $1^n(x-1)^m + 2^n(x-2)^m + \cdots + (x-1)^n 1^m$ and other similar series, *Quart. J. Pure and Appl. Math*, 43(1912), 101–122.

[69] R. W. Gosper, Jr. Decision procedure of indefinite summation, *Proc. Natl. Acad. Sci USA*, 75 (1978), 40–42.

[70] H. W. Gould, A series transformation for finding convolution identities, *Duke Math. J.* 28 (1961), 193–202.

[71] H. W. Gould, A new series transform with applications to Bessel, Legendre, and Tchebycheff polynomials, *Duke Math. J.* 31 (1964), 325–334.

[72] H. W. Gould, Inverse series relations and other expansions involving Humbert polynomials, *Duke Math. J.* 32 1965, 697–711.

[73] H. W. Gould, Combinatorial Identities, Revised Edition, Morgantown, W. Va., 1972.

[74] H. W. Gould, Evaluation of sums of convolved powers using Stirling and Eulerian numbers, *Fibonacci Quart.* 16 (1978), no. 6, 488–497.

[75] H. W. Gould, The Girard-Waring power sum formulas for symmetric functions and Fibonacci sequences, *Fibonacci Quart.* 37 (2) (1999), 135–140.

[76] H. W. Gould and T.-X. He, Characterization of (c)-Riordan arrays, Gegenbauer-Humbert type polynomial sequences, and (c)-Bell polynomials, *J. Math. Res. Appl.* 33 (2013), no. 5, 505–527.

[77] H. W. Gould and L. C. Hsu, Some new inverse series relations, *Duke Math. J.* 40 (1973), 885–891.

[78] H. W. Gould and J. Wetweerapong, Evaluation of some classes of binomial identities and two new sets of polynomials, *Indian J. Math.* 41(1999), no. 2, 159–190.

[79] R. L. Graham, D. E. Knuth, and O. Partashnik, *Concrete Mathematics,* Addison-Wesley Publishing Company, Reading, MA, 1994.

[80] D. H. Greene and D. E. Knuth, *Mathematics for the Analysis of Algorithms*, Basel: Birkhauser, 1981, Boston, Birkhauser, 1982.

[81] H. Gupta, The Andrews formula for Fibonacci numbers, *Fibonacci Quart.* 16 (1978), 552–555.

[82] J. Havil, Euler-Maclaurin Summation, in *Gamma: Exploring Euler's Constant*, Princeton, NJ: Princeton University Press, pp. 85-86, 2003.

[83] T.-X. He, *Dimensionality Reducing Expansion of Multivariate Integration*, Birkhäuser, Boston, 2001.

[84] T.-X. He, A Symbolic operator approach to power series transformation -expansion formulas, *J. Integer Seq.* 11(2008), no. 2, Article 08.2.7, 19 pages.

[85] T.-X. He, Riordan arrays associated with Laurent series and generalized Sheffer-type groups, *Linear Algebra Appl.* 435 (2011) 1241–1256.

[86] T.-X. He, Parametric Catalan numbers and Catalan triangles, *Linear Algebra Appl.* 438 (2013), no. 3, 1467–1484.

[87] T.-X. He, A unified approach to generalized Stirling functions, *J. Math. Res. Appl.* 32 (2012), no. 6, 631–646.

[88] T.-X. He, Expression and computation of generalized Stirling numbers, *J. Combin. Math. Combin. Comput.* 86 (2013), 239–268.

[89] T.-X. He, Matrix Characterizations of Riordan arrays, *Linear Algebra Appl.* 465 (2015), no. 1, 15–42.

[90] T.-X. He, Shift operators defined in the Riordan group and their applications, *Linear Algebra Appl.* 496 (2016), 331–350.

[91] T.-X. He, Row sums and alternating sums of Riordan arrays, *Linear Algebra Appl.* 507 (2016), 77–95.

[92] T.-X. He, Applications of Riordan Matrix Functions to Bernoulli and Euler Polynomials, *Linear Algebra Appl.* 507 (2016), 208–228.

[93] T.-X. He, Sequence characterizations of double Riordan arrays and their compressions, *Linear Algebra Appl.* 549 (2018), 176–202.

[94] T.-X. He, A-sequences, Z-sequence, and B-sequences of Riordan matrices. Discrete Math. 343 (2020), no. 3, 111718, 18 pp.

[95] T.-X. He, Half Riordan array sequences, *Linear Algebra Appl.* 604 (2020), 236–264.

[96] T.-X. He, One-pth Riordan arrays in the construction of identities, *J. Math. Res. Appl.* 41 (2013),

[97] T.-X. He, L. C. Hsu, and X. R. Ma, An extension of Riordan array and its application in the construction of convolution-type and Abel-type identities, *European J. Combin.* 42 (2014), 112–134.

[98] T.-X. He, L. C. Hsu, and P. J.-S. Shiue, On Abel-Gontscharoff-Gould's polynomials, *Anal. Theory Appl.* 19 (2003), no. 2, 166–184.

[99] T.-X. He, L. C. Hsu, and P. J.-S. Shiue, and D. C. Torney, A symbolic operator approach to several summation formulas for power series, *J. Comput. Appl. Math.* 177 (2005), no. 1, 17–33.

[100] T.-X. He, L. C. Hsu, and P. J.-S. Shiue, On an extension of Abel-Gontscharoff's expansion formula, *Anal. Theory Appl.* 21 (2005), no. 4, 359–369.

[101] T.-X. He, L. C. Hsu, and P. J.-S. Shiue, On generalized Möbius inversion formulas, *Bull. Austral. Math. Soc.* 73 (2006), no. 1, 79–88.

[102] T.-X. He, L. C. Hsu, and P. J.-S. Shiue, Convergence of the summation formulas constructed by using a symbolic operator approach, *Comput. Math. Appl.* 51 (2006), no. 3-4, 441–450.

[103] T.-X. He, L. C. Hsu, and P. J.-S. Shiue, Multivariate expansions associated with Sheffer-type polynomials and operators, *Bull. Inst. Math. Acad. Sin. (N.S.)* 1 (2006), no. 4, 451–473.

[104] T.-X. He, L. C. Hsu, and P. J.-S. Shiue, Symbolization of generating functions; an application of the Mullin-Rota theory of binomial enumeration, *Comput. Math. Appl.* 54 (2007), no. 5, 664–678.

[105] T.-X. He, L. C. Hsu, and P. J.-S. Shiue, The Sheffer group and the Riordan group, *Discrete Appl. Math.* 155 (2007), no. 15, 1895–1909.

[106] T.-X. He, L. C. Hsu, and P. J.-S. Shiue, A symbolic operator approach to several summation formulas for power series, II, *Discrete Math.* 308 (2008), no. 16, 3427–3440.

[107] T.-X. He, L. C. Hsu, and D. Yin, A pair of operator summation formulas and their applications, *Comput. Math. Appl.* 58 (2009), no. 7, 1340–1348.

[108] T.-X. He and L. W. Shapiro, Row sums and alternating sums of Riordan arrays, *Linear Algebra Appl.* 507 (2016), 77–95.

[109] T.-X. He and L. W. Shapiro, Fuss-Catalan matrices, their weighted sums, and stabilizer subgroups of the Riordan group, *Linear Algebra Appl.* 532 (2017), 25–42.

[110] T.-X. He and L. W. Shapiro, Palindromes and pseudo-involution multiplication, *Linear Algebra Appl.* 593 (2020), 1–17.

[111] T.-X. He, P.J.-S. Shiue, On sequences of numbers and polynomials defined by linear recurrence relations of order 2, *Int. J. Math. Math. Sci.* 2009 (2009), 709386.

[112] T. X. He, P. J.-S. Shiue, S. Nie, and M. Chen, Recursive sequences and Girard-Waring identities with applications in sequence transformation, *Electron. Res. Arch.* 28 (2020), no. 2, 1049–1062.

[113] T.-X. He and R. Sprugnoli, Sequence characterization of Riordan arrays, *Discrete Math.* 309 (2009), no. 12, 3962–3974.

[114] T.-X. He and J. Zheng, Duals of the Bernoulli numbers and polynomials and the Euler numbers and polynomials, *Integers*, 17 (2017), Paper No. A1, 26 pp.

[115] M. D. Hirschhorn, The Andrews formula for Fibonacci numbers, *Fibonacci Quart.* 19 (1981), 373–375.

[116] M. D. Hirschhorn, A notes posted online, https://web.maths.unsw.edu.au/ ~ mikeh/ webpapers/ paper87.pdf.

[117] M. D. Hirschhorn, Solution to problem 1621, *Math. Mag.* 75 (2002), 149–150.

[118] J. Hirshfeld, Nonstandard combinatorics, *Studia Logica*, 47 (1988), no. 3, 221–232.

[119] F. T. Howard, Degenerate weighted Stirling numbers, *Discrete Math.* 57(1985), no. 1, 45–58.

[120] L. C. Hsu, Some combinatorial formulas with applications to probable values of a polynomial-product and to differences of zero, *Ann. Math. Statistics*, 15 (1944), 399–413.

[121] L. C. Hsu, Application of a symbolic operator to the evaluation of certain sums, *Sci. Rep. Nat. Tsing Hua Univ. Ser. A.*, 5 (1948), 139–149.

[122] L. C. Hsu, Some remarks on a generalized Newton interpolation formula, *Math. Student*, 19 (1951), 25–29.

[123] L. C. Hsu, *Selected Topics of Methods and Examples of Mathematical Analysis*, Business Publishers, Beijing, 1955.

[124] L. C. Hsu. Generalized Möbius-Rota inversion theory associated with nonstandard analysis, *Sci. Exploration*, 3 (1983), no.1, 1–8.

[125] L. C. Hsu, Generalized Stirling number pairs associated with inverse relations, *Fibonacci Quart.* 25 (1987), 346–351.

[126] L. C. Hsu, Theory and application of general Stirling number pairs, *J. Math. Res. & exposition*, 9 (1989), no. 2, 211–220.

[127] L. C. Hsu, A summation rule using Stirling numbers of the second kind, *Fibonacci Quart.* 31 (1993), no. 3, 256–262.

[128] L. C. Hsu, Concerning two formulaic classes in computational combinatorics, *J. Math. Res. Appl.* 32 (2012), no. 2, 127–142.

[129] L. C. Hsu, On a pair of operator series expansions implying a variety of summation formulas, *Anal. Theory Appl.* 31 (2015), no. 3, 260–282.

[130] L. C. Hsu, Concerning a general source formula and its applications, *J. Math. Res. Appl.* 36 (2016), no. 5, 505–514.

[131] L. C. Hsu, and X. Ma, Some combinatorial series and reciprocal relations involving multifold convolutions, *Integers*, 14 (2014), 1–20.

[132] L.C. Hsu and P. J.-S. Shiue, A unified approach to generalized Stirling numbers, *Adv. Appl. Math.* 20 (1998), 366–384.

[133] L. C. Hsu and P. J.-S. Shiue, On certain summation problems and generalizations of Eulerian polynomials and numbers, *Discrete Mathematics*, 204 (1999), 237–247.

[134] L. C. Hsu and P. J.-S. Shiue, Cycle indicators and special functions, *Annals of Combinatorics*, 5 (2001), 179–196.

[135] L. C. Hsu and X. H. Wang, *Selected Topics of Methods and Examples of Mathematical Analysis*, 2nd Edition, Higher Education Publishers, Beijing, 1983.

[136] I.-C. Huang, Residue methods in combinatorial analysis, in *Local Cohomology and Its Applications (Guanajuato, 1999)*, Vol. 226 of Lecture Notes in Pure and Appl. Math., Marcel Dekker, New York, NY, USA, 2002, 255–342.

[137] A. E. Hurd, *Nonstandard methods in combinatorics and graph-theory*, Abstracts, *Notices Amer. Math. Soc.* 1975, 22: A20.

[138] C. Jean-Louis and A. Nkwanta, Some algebraic structure of the Riordan group, *Linear Algebra Appl.*, 438 (2013), no. 5, 2018–2035.

[139] L. B. W. Jolley, *Summation of series*, 2d Revised Edition, Dover Publications, New York, 1961.

[140] C. H. Jones, Generalized hockey stick identities and N-dimensional blockwalking, *Fibonacci Quart.* 34 (1996), no. 3, 280–288.

[141] Ch. Jordan, *Calculus of Finite Differences*, Chelsea, 1965.

[142] M. Kaneko, A recurrence formula for the Bernoulli numbers, *Proc. Japan Acad. Ser. A. Math. Sci.* 71 (1995), 192–193.

[143] K. Knopp, *Theory and Application of Infinite Series*, New York: Dover, 1990.

[144] D. E. Knuth, Convolution Polynomials, *Math. J.* 24 (1992), 67–68.

[145] D. E. Knuth, Bracket notation for the 'coefficient of' operator, ArXiv Math., arXiv:math/9402216, Correct copy, 19 August 1993.

[146] W. Koepf, *Hypergeometric Summation: An Algorithmic Approach to Summation and Special Function Identities*, Vieweg-Verlag, Braunschweig, Germany, 1998.

[147] C. Krattenthaler, Operator Methods and Lagrange Inversions: A Unified Approach to Lagrange Formulas, *Trans. Amer. Math. Soc.* 305 (1988), 431–465.

[148] G. Liu, Higher-order multivariable Euler's polynomial and higher-order multivariable Bernoulli's polynomial, *Appl. Math. Mech. (English Ed.)* 19 (1998), no. 9, 895–906; translated from *Appl. Math. Mech.* 19 (1998), no. 9, 827–836 (Chinese).

[149] J.-G. Liu and R. Pego, On generating functions of Hausdorff moment sequences, *Trans. Am. Math. Soc.*, 368 (2016), no. 12, 8499–8518.

[150] D. E. Loeb and G.-C. Rota, Recent advances in the calculus of finite differences, in: Geometry and Complex Variables, S. Coen, ed., *Lecture Notes in Pure and Applied Mathematics*, Vol. 132, Marcel Dekker, New York, 1991, 239–276.

[151] A. Luzón, D. Merlini, M. A. Morón, and R. Sprugnoli, Identities induced by Riordan arrays, *Linear Algebra Appl.*, 436 (2012), no. 3, 631–647.

[152] A. Luzón, D. Merlini, M. A. Morón, and R. Sprugnoli, Complementary Riordan arrays, *Discrete Appl. Math.* 172 (2014), 75–87.

[153] A. Luzón, M. A. Morón, and L. F. Prieto-Martinez, The group generated by Riordan involutions, arXiv:1803.06872.

[154] X. Ma and L. C. Hsu, A further investigation of a pair of series transformation formulas with applications, *J. Difference Equ. Appl.* 17 2011, no.10, 1519–1535.

[155] X. Ma, Two matrix inversions associated with the Hagen-Rothe formula, their q-analogues and applications, *J. Combin. Theory Ser. A*, 118 (2011), no. 4, 1475–1493.

[156] C. Maclaurin, *Treatise of Fluxions*, Edinburgh, p. 672, 1742.

[157] W. Magnus, F. Oberhettinger, R. P. Soni, *Formulas and Theorems for the Special Functions of Mathematical Physics,* Springer, New York, 1966.

[158] D. Merlini, D. G. Rogers, R. Sprugnoli, and M. C. Verri, On some alternative characterizations of Riordan arrays, *Canad. J. Math.* 49 (1997), no. 2, 301–320.

[159] D. Merlini, R. Sprugnoli, and M. C. Verri, An algebra for proper generating trees. Mathematics and computer science (Versailles, 2000), 127–139.

[160] D. Merlini, R. Sprugnoli, and M. C. Verri, Lagrange inversion: When and how, *Acta Appl. Math.* 94 (2006), 233–249.

[161] D. Merlini, R. Sprugnoli, and M. C. Verri, The method of coefficients, *Amer. Math. Monthly*, 114 (2007), no. 1, 40–57.

[162] D. Merlini, R. Sprugnoli, and M. C. Verri, Combinatorial sums and implicit Riordan arrays, *Discrete Math.* 309 (2009), no. 2, 475–486.

[163] H. Miki, A relation between Bernoulli numbers, *J. Number Theory*, 10 (1978), 297–302.

[164] L. M. Milne-Thomson, *The Calculus of Finite Differences*, London, 1933.

[165] A. Milton and I. A. Stegun Eds., *Handbook of Mathematical Functions with Formulas*, Graphs, and Mathematical Tables, Dover, New York, 1972.

[166] W. Mlotkowski, Fuss-Catalan numbers in noncommutative probability, *Documenta Mathematica*, 15 (2010), 939–955.

[167] H. Momiyama, A new recurrence formula for Bernoulli numbers, *Fibonacci Quart.* 39 (2001), 285–288.

[168] R. Mullin and G.-C. Rota, On the foundations of combinatorial theory: III. Theory of binomial enumeration, in: Graph Theory and its Applications, B. Harris (ed.), Academic Press, New York and London, 1970, 167–213.

[169] C. P. Neuman, D. I. Schonbach, Evaluation of sums of convolved powers using Bernoulli numbers, *SIAM Rev.*, 19 (1977), no. 1, 90–99.

[170] A. Nkwanta and N. Knox, A note on Riordan matrices, *African Americans in mathematics, II* (Houston, TX, 1998), 99–107, *Contemp. Math.* 252, Amer. Math. Soc., Providence, RI, 1999.

[171] A. Nkwanta, A Riordan matrix approach to unifying a selected class of combinatorial arrays, Proceedings of the Thirty-Fourth Southeastern International Conference on Combinatorics, *Graph Theory and Computing, Congressus Numerantium*, 160 (2003), 33–45.

[172] N. Obreschkoff, *Neue Quadraturformeln,* Abhandl. d. preuss. Akad. d. Wiss., Math. Natur. wiss. K1., 4 (1940), 1–20.

[173] H. Pan and Z.-W. Sun, New identities involving Bernoulli and Euler polynomials, *J. Combin. Theory Ser. A*, 113 (2006), no. 1, 156–175.

[174] P. Peart, and W.-J. Woan, A divisibility property for a subgroup of Riordan matrices, *Discrete Appl. Math.*, 98 (2000), 255–263.

[175] P. Peart and W.-J. Woan, Generating functions via Hankel and Stieltjes matrices, *J. Integer Seq.* 3 (2000), no. 2, Article 00.2.1, 1 HTML document (electronic).

[176] K. A. Penson and K. Życzkowski, Product of Ginibre matrices: Fuss-Catalan and Raney distributions, *Phys. Rev. E*, 83 (2011) 061118, 9 pp.

[177] M. Petkovšek, H. S. Wilf, and D. Zeilberger, *A=B*, AK Peters, Ltd. Wellesley, Massachusetts, 1996.

[178] D. Phulara, L. Shapiro, Constructing pseudo-involutions in the Riordan group, *J. Integer Seq.* 20 (2017), 17.4.7.

[179] H. Prodinger, Some information about the binomial transform, *Fibonacci Quart.* 32 (1994), 412–415.

[180] J. Quaintance and H. W. Gould, *Combinatorial Identities for Stirling Numbers*, World Scientific Publ. Co., Pte. Ltd. Singapore, London, Hong Kong, Tokyo, 2016.

[181] J. B. Remmel and M. L. Wachs, Rook theory, generalized Stirling numbers and (p, q)-analogues, *Electron. J. Combin.* 11 (2004), no. 1, Research Paper 84, 48 pp.

[182] J. Riordan, *Introduction to Combinatorial Analysis*, Published by the John Wiley & Sons, Inc., New York, 1958.

[183] J. Riordan, *Combinatorial Identities*, R. E. Krieger Pub. Co., Huntington, New York , 1968.

[184] D. G. Rogers, Pascal triangles, Catalan numbers and renewal arrays, *Discrete Math.*, 22 (1978), 301–310.

[185] S. Roman, The theory of the umbral calculus, *I. J. Math. Anal. Appl.* 87 (1982), no. 1, 58 –115.

[186] S. M. Roman, *The Umbral Calculus*, New York, Acad. Press, 1984.

[187] S. Roman and G.-C. Rota, The Umbral Calculus, *Adv. in Math.* 27 (1978), no. 2, 95–188.

[188] G.-C. Rota, *Finite Operator Calculus*, Academic Press, New York, 1975.

[189] R. Roy, Binomial identities and hypergeometric series, *American Math. Monthly*, 94 (1987), 36–46.

[190] W. H. Schikhof, *Ultrametric Calculus* Cambridge: Cambridge University Press, 1984.

[191] I. J. Schwatt, *An Introduction to the Operations with Series*, Chelsea Publishing Company, New York, 1962.

[192] I. Schur, On Faber polynomials, *Amer. J. Math.* 67 (1945), 33–41.

[193] L. W. Shapiro, A survey of the Riordan group, Talk at a meeting of the American Mathematical Society, Richmond, Virginia, 1994.

[194] L. W. Shapiro, Some open questions about random walks, involutions, limiting distributions and generating functions, *Adv. in Appl. Math.*, 27 (2001), 585–596.

[195] L. Shapiro, Bijections and the Riordan group, *Theoretical Computer Science*, 307 (2003), 403–413.

[196] L. W. Shapiro, S. Getu, W. J. Woan and L. Woodson, The Riordan group, *Discrete Appl. Math.* 34 (1991), 229–239.

[197] N. J. A. Sloane, The On-Line Encyclopedia of Integer Sequences, https://oeis.org/, founded in 1964.

[198] R. Sowik, Sequence characterization of 3-dimensional Riordan arrays and some application, *Results Math.* 74 (2019), no. 4, Paper No. 169, 7 pp.

[199] R. Sowik, Some (counter)examples on totally positive Riordan arrays, *Linear Algebra Appl.* 594 (2020), 117–123.

[200] A. Sofo, *Computational Techniques for the Summation of Series*, Kluwer Acad. New York, 2003.

[201] R. Sprugnoli, Riordan arrays and combinatorial sums, *Discrete Math.* 132 (1994), no. 1-3, 267–290.

[202] R. Sprugnoli, Riordan arrays and the Abel-Gould identity, *Discrete Math.* 142 (1995), no. 1-3, 213–233.

[203] R. P. Stanley, *Enumerative Combinatorics*, Vol. 2, Cambridge University Press, 1999.

[204] R. P. Stanley, *Catalan Numbers*, Cambridge University Press, New York, 2015.

[205] Z.-H. Sun, Invariant sequences under binomial transformations, *Fibonacci Quart.* 39 (2001), 324–333.

[206] Z.-W. Sun, Combinatorial identities in dual sequences, *European J. Combin.* 24 (2003), no. 6, 709–718.

[207] Z.-W. Sun, and H. Pan, Identities concerning Bernoulli and Euler polynomials, *Acta Arith.* 125 (2006), no. 1, 21-39.

[208] I. Vardi, The Euler-Maclaurin Formula, in *Computational Recreations in Mathematica Reading*, MA: Addison-Wesley, 1991, 159–163.

[209] W. Wang and T. Wang, Generalized Riordan arrays, *DiscreteMath.* 308 (2008), no. 24, 6466–6500.

[210] X.-H. Wang and L. C. Hsu, A summation formula for power series using Eulerian fractions, *Fibonacci Quart.* 41 (2003), no. 1, 23–30.

[211] Y. Wang, Self-inverse sequences related to a binomial inverse pair, *Fibonacci Quart.* 43 (2005), 46–52.

[212] Z. X. Wang and D. R. Guo, *Special Functions*, World Scientific, Singapore, 1989.

[213] W. Eric Weisstein, Euler-Maclaurin Integration Formulas, From *MathWorld–A Wolfram Web Resource*, http://mathworld.wolfram.com/ Euler-Maclaurin IntegrationFormulas.html

[214] E. T. Whittaker and G. Robinson, The Euler-Maclaurin Formula, in *The Calculus of Observations: A Treatise on Numerical Mathematics*, 4th ed. New York: Dover, 1967, 134–136.

[215] E. T. Whittaker and G. N. Watson, The Euler-Maclaurin Expansion, in *A Course in Modern Analysis*, 4th ed. Cambridge, England: Cambridge University Press, 1990, 127–128.

[216] H. S. Wilf, *Generatingfunctionology*, Acad. Press, New York, 1990.

[217] H. S. Wilf and D. Zeilberger, Rational functions certify combinatorial identities, *J. Amer. Math. Soc.* 3 (1990), 147–158.

[218] H. S. Wilf and D. Zeilberger, An algorithmic proof theory for hypergeometric (ordinary and "q") multisum/integral identities, *Invent. Math.* 108 (1992), 575–633.

[219] K.J. Wu, Z.W. Sun, and H. Pan, Some identities for Bernoulli and Euler polynomials, *Fibonacci Quart.* 42 (2004), no. 4, 295–299.

[220] S.-L. Yang, Y.-N. Dong, L. Yang, and J. Yin, Half of a Riordan array and restricted lattice paths, *Linear Algebra Appl.* 537 (2018), 1–11.

[221] S.-L. Yang, Y.-X. Xu, and X. Gao, On the half of a Riordan array, *Ars Combin.* 133 (2017), 407–422.

[222] S.-L. Yang, Y.-X. Xu, and T.-X. He, (m, r)-central Riordan arrays and their applications, *Czechoslovak Math. J.* 67(142) (2017), no. 4, 919–936.

[223] S.-L. Yang, S.-N. Zheng, S.-P. Yuan, and T.-X. He, Schröder matrix as inverse of Delannoy matrix, *Linear Algebra Appl.* 439 (2013), no. 11, 3605–3614.

[224] D. A. Zave, A series expansion involving the harmonic numbers, *Inform. Process. Lett.* 5 (1976), 75–77.

[225] D. Zeilberger, A holonomic systems approach to special functions identities, *J. Comput. Appl. Math.* 32 (1990), no. 3, 321–368.

[226] D. Zeilberger, The method of creative telescoping, *J. Symbolic Comput.* 11 (1991), 195–204.

[227] X. Zhao and T. Wang, Some identities related to reciprocal functions, *Discrete Math.* 265 (2003), 323–335.

Index

For Product Safety Concerns and Information please contact our EU representative GPSR@taylorandfrancis.com
Taylor & Francis Verlag GmbH, Kaufingerstraße 24, 80331 München, Germany

www.ingramcontent.com/pod-product-compliance
Lightning Source LLC
Chambersburg PA
CBHW060744220326
41598CB00022B/2319

* 9 7 8 1 0 3 2 1 9 5 0 0 1 *